Mountains: Physical, Human-Environmental, and Sociocultural Dynamics

Mountains have captured the interests and passions of people for thousands of years. Today, millions of people live within mountain regions, and mountain regions are often areas of accelerated environmental change. This edited volume highlights new understanding of mountain environments and mountain peoples around the world. The understanding of mountain environments and peoples has been a focus of individual researchers for centuries; more recently the interest in mountain regions among researchers has been growing rapidly. The chapters contained herein are from a wide spectrum of researchers from different parts of the world who address physical, political, theoretical, social, empirical, environmental, methodological, and economic issues focused on the geography of mountains and their inhabitants. The chapters in this volume are organized into three themed sections with very loose boundaries between themes: (1) physical dynamics of mountain environments; (2) coupled human–physical dynamics; and (3) socio-cultural dynamics in mountain regions.

This book was first published as a special issue of the *Annals of the American Association of Geographers*.

Mark A. Fonstad is an Associate Professor of Geography at the University of Oregon, USA. He specializes in studies of riverine and mountain environments. His areas of research are the physical geography of rivers and mountains, the fusion of physical geography with geographic information science, geomorphology, hydrology, and remote sensing.

Mountains: Physical, Human-Environmental, and Sociocultural Dynamics

Edited by
Mark A. Fonstad

Routledge
Taylor & Francis Group

LONDON AND NEW YORK

First published 2018 by Routledge

2 Park Square, Milton Park, Abingdon, Oxfordshire OX14 4RN
52 Vanderbilt Avenue, New York, NY 10017

Routledge is an imprint of the Taylor & Francis Group, an informa business

First issued in paperback 2019

Introduction, Chapters 1–12 & Chapters 14–27 © 2018 Taylor & Francis
Chapter 13 © 2018 Sven Fuchs, Veronika Röthlisberger, Thomas Thaler,
Andreas Zischg, and Margreth Keiler. Originally published as Open Access.
Chapter 15 © 2018 Marcus Nüsser and Susanne Schmidt. Originally published
as Open Access.

Notice:
Product or corporate names may be trademarks or registered trademarks,
and are used only for identification and explanation without
intent to infringe.

British Library Cataloguing in Publication Data
A catalogue record for this book is available from the British Library

ISBN 13: 978-1-138-06697-7 (hbk)
ISBN 13: 978-0-367-89050-6 (pbk)

Typeset in Goudy Old Style
by RefineCatch Limited, Bungay, Suffolk

Publisher's Note
The publisher accepts responsibility for any inconsistencies that may have
arisen during the conversion of this book from journal articles to book chapters,
namely the possible inclusion of journal terminology.

Disclaimer
Every effort has been made to contact copyright holders for their permission to
reprint material in this book. The publishers would be grateful to hear from any
copyright holder who is not here acknowledged and will undertake to rectify
any errors or omissions in future editions of this book.

Contents

CONTENTS

CONTENTS

Citation Information

The chapters in this book were originally published in the *Annals of the American Association of Geographers*, volume 107, issue 2 (March 2017). When citing this material, please use the original page numbering for each article, as follows:

Chapter 27

Khumbi yullha *and the Beyul: Sacred Space and the Cultural Politics of Religion in Khumbu, Nepal*

Lindsay A. Skog

Annals of the American Association of Geographers, volume 107, issue 2 (March 2017), pp. 546–554

For any permission-related enquiries please visit:
http://www.tandfonline.com/page/help/permissions

Notes on Contributors

Kevin J. Anchukaitis is an Associate Professor in the School of Geography and Development at the University of Arizona, USA.

Marcos F. Andrade-Flores is Director of the Laboratory for Atmospheric Physics at University Mayor de San Andres, La Paz, Bolivia.

Sandro Arias is a Hydrometeorological Researcher in the National Meteorology and Hydrology Service of Peru, Cusco, Peru.

Andrew J. Bach is an Associate Professor in the Department of Environmental Studies at Western Washington University, USA.

Antonia Barreau is an Associate Researcher in the Center for Local Development, Education and Interculturality, Villarrica Campus, Pontifical Catholic University of Chile.

Karl W. Birkeland is the Director of the U.S. Department of Agriculture Forest Service National Avalanche Center and an Adjunct Professor in the Earth Sciences Department at Montana State University, USA.

Marti Bonshoms is a Meteorologist at the National Meteorology and Hydrology Service of Peru and co-investigator in the CRYOPERU.

Sophia L. Borgias is a doctoral student in the School of Geography and Development at the University of Arizona, USA.

Leah L. Bremer is a Conservation Scientist at The Natural Capital Project, Stanford University, USA.

Christy E. Briles is an Assistant Professor at the University of Colorado Denver, USA.

Eric J. Burton is an undergraduate student in the Department of Geography and Planning at Appalachian State University, USA.

Mark Carey is an Associate Professor of History in the Robert D. Clark Honors College and an Associate Professor of Environmental Studies, both at the University of Oregon, USA.

Kristy C. Carter is a co-major PhD student in Meteorology and Wind Energy Science, Engineering, and Policy at Iowa State University, USA.

Edwin J. Castellanos is a Professor and Director of the Center for Environmental Studies and Biodiversity at Universidad del Valle de Guatemala.

Ashwini Chhatre is a Visiting Professor and Senior Research Fellow at the Indian School of Business, Hyderabad, India.

Nathan Clay is a PhD student in the Department of Geography at Pennsylvania State University, University Park, USA.

Courtney M. Cooper is a PhD student and NSFIGERT Fellow in the Water Resource Program at the University of Idaho, USA.

Hildegardo Córdova-Aguilar is Executive Director of the Research Center for Applied Geography in the Institute for Research in Nature Sciences, Territory, and Renewable Energies and is Professor at the Pontificia Universidad Católica del Perú, Lima, Peru.

Jason Michael Covert is a PhD student at the University of Albany, SUNY, Atmospheric Science Research Center, USA.

Arica Crootof is a doctoral student in the School of Geography and Development at the University of Arizona, USA.

Stephen F. Cunha is a Professor in the Department of Geography at Humboldt State University, USA.

Rafael de Grenade is a Postdoctoral Research Associate at the Udall Center for Studies in Public Policy, and Research Associate in the Southwest Center at the University of Arizona, USA.

Mabel Denzin Gergan is a Postdoctoral Research Associate in the Department of Geography at the University of North Carolina at Chapel Hill, USA.

Ronald I. Dorn is a Professor of Geography in the School of Geographical Sciences and Urban Planning at Arizona State University, USA.

John Douglass is a faculty member in Geography at Paradise Valley Community College, USA.

Jason L. Endries is a graduate student and Researcher in the Department of Geography and Planning at Appalachian State University, USA.

Kathleen A. Farley is an Associate Professor in the Department of Geography at San Diego State University, USA.

Alfonso Fernández is an Assistant Professor in the Department of Geography at the Universidad de Concepción, and Research Associate at The Ohio State University, USA.

Harry W. Fischer is an Associate Lecturer in the Department of Social Enquiry at La Trobe University and a New Generation Network (NGN) Fellow at the Australia India Institute, University of Melbourne, Australia.

Mark A. Fonstad is an Associate Professor of Geography at the University of Oregon, USA.

Andrew G. Fountain is a Professor in the Departments of Geography and Geology at Portland State University, USA.

Larry M. Frolich is a Professor in the Department of Natural Sciences at Miami Dade College, Wolfson Campus, USA.

Sven Fuchs is a Senior Scientist in the Institute of Mountain Risk Engineering at the University of Natural Resources and Life Sciences, Vienna, Austria.

Ricardo Jesús Gomez is Director of Huascaran National Park and World Biosphere Reserve (the most glacierized tropical mountain range on earth), Peru.

Juan A. González is a Full Professor in the Institute of Ecology, Miguel Lillo Foundation, Argentina.

Daniel Griffin is an Assistant Professor in the Department of Geography, Environment and Society at the University of Minnesota, USA.

Jon Harbor is a Professor in the Department of Earth, Atmospheric, and Planetary Sciences at Purdue University, USA.

Robert Åke Hellström is a Professor in the Geography Department at Bridgewater State University, USA.

Edward C. Holland is an Assistant Professor in the Department of Geosciences at the University of Arkansas, USA.

J. Tomás Ibarra is an Assistant Professor in the Centre for Local Development, Education and Interculturality, Villarrica Campus, and in the Fauna Australis Wildlife Laboratory, School of Agriculture and Forest Sciences, Pontifical Catholic University of Chile.

M Jackson is a PhD student in Geography at the University of Oregon, USA.

Esther Jacobson-Tepfer is Maude Kerns Professor Emeritus in the Department of History of Art and Architecture at the University of Oregon, USA.

Margreth Keiler is Associate Professor in the Institute of Geography at the University of Bern, Switzerland.

Gregory Knapp is an Associate Professor in the Department of Geography and the Environment at the University of Texas at Austin, USA.

Shikha Lakhanpal is a graduate Research Scholar at the Department of Geography at the University of Illinois, USA.

Phillip H. Larson is an Assistant Professor and Director of Earth Science Programs in the Department of Geography at Minnesota State University, USA.

Yanan Li is an Assistant Professor in the Department of Geography at South Dakota State University, USA.

Yingkui Li is an Associate Professor in the Department of Geography at the University of Tennessee, USA.

Andrew M. Linke is an Assistant Professor in the Department of Geography at the University of Utah, USA.

Jennifer K. Lipton is an Associate Professor in the Department of Geography at the Central Washington University, USA.

Xiaoyu Lu is a PhD student in the Department of Geography at the University of Tennessee, USA.

Guido Mamani is the Information Technology Manager at the National Meteorology and Hydrology Service of Peru, Cusco, Peru.

Bryan Greenwood Mark is a Professor in the Geography Department and Research Scientist at the Byrd Polar and Climate Research Center at Ohio State University, USA.

James E. Meacham is a Senior Research Associate in the Geography Department at the University of Oregon, USA.

Norman Meek is Chair and Professor in the Department of Geography and Environmental Studies at California State University, USA.

Megan Mills-Novoa is a doctoral student in the School of Geography and Development at the University of Arizona, USA.

Cary J. Mock is a Professor in the Department of Geography at the University of South Carolina, USA.

Olivia C. Molden is a doctoral student in Geography at the University of Oregon, USA.

Nilton Montoya is an Adjunct Professor in the Department of Agriculture at the Universidad Nacional de San Antonio Abád del Cusco, Cusco, Peru.

Galen Murton is a PhD student in the Department of Geography at the University of Colorado Boulder, USA.

Anne W. Nolin is a Professor in the College of Earth, Ocean and Atmospheric Sciences at Oregon State University, USA.

Marcus Nüsser is a Professor in the Department of Geography, South Asia Institute, Heidelberg University, Heidelberg, Germany.

Yolanda Jiménez Olivencia is a Professor in the Department of Regional Geographic Analysis and Physical Geography and Director of the Institute of Regional Planning at the Universidad de Granada, Spain.

Rafael Mata Olmo is Professor and Director of the Department of Geography at the Universidad Autónoma de Madrid, Spain.

John O'Loughlin is a College Professor of Distinction, Professor of Geography, and Faculty Research Associate (Program on International Development) in the Institute of Behavioral Science at the University of Colorado at Boulder, USA.

L. Baker Perry is Graduate Program Director and Associate Professor in the Department of Geography and Planning at Appalachian State University, USA.

J. Cristóbal Pizarro is a Postdoctoral Researcher at the Austral Center of Scientific Research in Ushuaia, Tierra del Fuego, Argentina, and is Associate Researcher in the Laboratory of Ecology at the Universidad de Los Lagos, Chile.

Molly H. Polk is Associate Director of Sustainability Studies in the Department of Geography and the Environment at the University of Texas, USA.

Alexandra G. Ponette-González is an Assistant Professor in the Department of Geography and the Environment at the University of North Texas, USA.

Diego Pons is a PhD student in the Department of Geography and the Environment at the University of Denver, USA.

Satya Prasanna is an independent scholar and activist working with researchers and civil society groups on the politics of access to natural resources.

Nelson Quispe is Director of the Center for Weather Prediction Branch of the National Meteorology and Hydrology Service of Peru.

Maxwell Rado is a Professor in the Department of Geography and member of the climate change scientific team at the Universidad Nacional de San Antonio Abád del Cusco, Cusco, Peru.

Alejo Cochachín Rapre is Director of the Division of Glaciers and Water Resources (UGRH), Peruvian National Water Authority (ANA), Peru.

Mattias Borg Rasmussen is an Assistant Professor in the Department of Food and Resource Economics at the University of Copenhagen, Denmark.

Veronika Röthlisberger is a PhD student in the Institute of Geography at the University of Bern, Switzerland.

Ricardo Rozzi is a Full Professor in the Department of Philosophy and Religion at the University of North Texas, USA.

Fausto O. Sarmiento is a Full Professor in the Department of Geography and Director of the Neotropical Montology Collaboratory at the University of Georgia, USA.

Susanne Schmidt is a Senior Lecturer in the Department of Geography, South Asia Institute, Heidelberg University, Heidelberg, Germany.

Christopher A. Scott is a Professor in the School of Geography and Development and Research Professor at the Udall Center for Studies in Public Policy at the University of Arizona, USA.

Anton Seimon is a Research Assistant Professor in the Department of Geography and Planning at Appalachian State University, USA.

Yeong Bae Seong is an Associate Professor of Geography in the Department of Geography Education at Korea University, Seoul, Republic of Korea.

Lindsay A. Skog is a Senior Instructor in the Department of Geography at Portland State University, USA.

Matthew J. Taylor is a Professor in the Department of Geography and the Environment at the University of Denver, USA.

Thomas Thaler is a Research Fellow in the Institute of Mountain Risk Engineering at the University of Natural Resources and Life Sciences, Austria.

Bhuwan Thapa is a doctoral student in the School of Geography and Development at the University of Arizona, USA.

Steven J. Vanek is a Research Associate in the GeoSyntheSES Lab and Visiting Scholar in the Department of Geography at Pennsylvania State University, USA.

Fernando Velarde is a Physical Meteorologist working as a Researcher in the Laboratory for Atmospheric Physics at the Universidad Mayor de San Andres, La Paz, Bolivia.

Kenneth R. Young is a Professor in the Department of Geography and the Environment at the University of Texas, USA.

Sandra E. Yuter is a Professor in the Department of Marine, Earth and Atmospheric Sciences at North Carolina State University, USA.

Paul Whelan is the Eastside GIS and Data Analyst with the Washington Department of Fish and Wildlife, Wildlife Science Division, USA.

I. Ronald Winkelmann is an Associate Researcher in the Atmospheric Physics Laboratory at the Universidad Mayor de San Andres, La Paz, Bolivia.

Frank D. W. Witmer is an Assistant Professor in the Department of Computer Science and Engineering at the University of Alaska Anchorage, USA.

Karl S. Zimmerer is a Professor in the Department of Geography and Director of the GeoSyntheSES Lab at Pennsylvania State University, University Park, USA.

Andreas Zischg a Scientist in the Institute of Geography at the University of Bern, Switzerland.

Mountains: A Special Issue

Mark A. Fonstad

The special issues of the *Annals* allow the editors to highlight themes of international significance that showcase the breadth and depth of geography in a format accessible to a broad array of readers. This ninth special issue of the *Annals of the AAG* focuses on mountains. The understanding of mountain environments and peoples has been a focus of individual geographers for centuries and for the organized discipline of geography for more than a century; more recently, the geographical interest in mountain regions among researchers has been growing rapidly. The articles contained within are from a wide spectrum of researchers from different parts of the world who address physical, political, theoretical, social, empirical, environmental, methodological, and economic issues focused on the geography of mountains and their inhabitants. The articles in this special issue are organized into three themed sections with very loose boundaries between themes: (1) physical dynamics of mountain environments, (2) coupled human–physical dynamics, and (3) sociocultural dynamics in mountain regions.

美国地理学家协会年鑑的特刊, 让编辑得以强调具有国际重要性的主题, 透过对更广泛的读者而言具可及性的形式, 展现地理学的深度与广度。本期美国地理学家协会年鑑的第九个特刊将聚焦山岳。对山岳环境及人们的理解, 是各别地理学者们数百年来, 以及地理学的组织化学门一百多年来的关注焦点; 近年来, 研究者对山岳地区的地理兴趣正迅速地成长中。本特刊的文章, 包含来自世界各地处理聚焦山岳地理及其居住者的物理、政治、理论、社会、经验、环境、方法论和经济议题的广泛研究者。本特刊中的文章, 以三大主题章节进行组织, 各主题之间具有鬆散的边界: (1) 山岳环境的物理动态, (2) 对偶的人类—物理动态, 以及 (3) 山岳区域的社会文化动态。关键词: 对偶的人类—环境系统, 环境变迁, 山岳, 区域, 系统。

Los números especiales de los *Annals* permiten a los editores destacar temas de valor internacional, lo cual resalta la amplitud y profundidad de la geografía en un formato accesible para una amplia variedad de lectores. Este noveno número especial de los *Annals de la* AAG está enfocado a las montañas. La comprensión de los ambientes y gente de montaña ha sido del interés especializado de geógrafos durante siglos y para la disciplina organizada de la geografía durante más de un siglo; más recientemente, el interés geográfico por las regiones montañosas entre investigadores se ha incrementado rápidamente. Los artículos que se incluyen en este número provienen de un amplio espectro de investigadores de diferentes partes del mundo que abocan temas físicos, políticos, teóricos, sociales, empíricos, ambientales, metodológicos y económicos, enfocados sobre la geografía de las montañas y sus habitantes. Los artículos de este número especial están organizados en tres secciones temáticas, con límites muy sueltos entre los temas: (1) dinámica física de los entornos montañosos, (2) dinámica humano-física acoplada, y (3) dinámica sociocultural en regiones montañosas.

The understanding of mountain environments and peoples has been a focus of individual geographers for centuries and for the organized discipline of geography for more than a century. Two hundred years ago, Alexander von Humboldt was undertaking some of the first large, systematic studies of mountain environments. One hundred years ago, William Morris Davis's theories on mountain landscape evolution became a centerpiece of physical geography teaching for decades, including an article on the Colorado Front Range in the very first issue of the *Annals* (Davis 1911). Since those times, geographers have diversified in their scholarship of mountains to include such wide-ranging and seemingly disparate topics as the relationships between land use and mountain geomorphology (Marston 2008), mountain climatology (Barry 2008), specialized geographic information science tools and approaches for mountains (Bishop and Shroder 2004), mountains as sacred spaces (Tuan 1974; Bernbaum 1997) and tourism centers (Price, Moss, and Williams 1997), the place of mountains in the geopolitical realm and the conflict over borders (Libiszewski and Bachler 1997), the response of mountains to a changing global environment (Orlove, Weigandt, and Luckman 2008), the education of geographers of the physical and human dimensions of mountains (Price et al. 2013), and many other areas. Over the past fifty years, mountain studies have moved from being the purview of a few individual scholars to a social process involving global agencies (Funnell and Price 2003). With such a growing, global interest in the mountains of the world, the time is right for an *Annals* special issue on this subject.

This ninth special issue of the *Annals of the AAG* focuses on mountains. The call for abstracts went out in May 2015, and the paper invitation, submission, and revision process lasted until October 2016. Papers were sought from a wide spectrum of researchers from different parts of the world who address physical, political, theoretical, social, empirical, environmental, methodological, and economic issues focused on the geography of mountains and their inhabitants. These articles include cutting-edge research with themes such as mountains as regions highly sensitive to climate change, as sites and corridors of cultural and environmental diversity and gradients, as sacred spaces, as sources of hazard and risk, as spaces of geopolitical conflict, as "water towers of the world," and many other major geographical themes. The goal of this special issue is to highlight the breadth of research in mountain geographies that spans systematic, regional, and synthetic approaches and to showcase major mountain issues around the globe to which geographical researchers contribute.

The articles in this special issue are organized into three themed sections with very loose boundaries between themes: (1) physical dynamics of mountain environments (nine articles), (2) coupled human–physical dynamics (nine articles), and (3) sociocultural dynamics in mountain regions (nine articles). These sections are artificial and do represent well the fuzzy boundaries between these themes. These artificial divisions also do not include major potential thematic areas such as division by regions or emphasize methods in understanding mountain areas geographically. As one example, multiple articles in this issue highlight the theme of sacred spaces in mountains, and these are often strongly linked to the physical landscape and landscape processes, but the themed sections in this issue do not perfectly reflect these interconnections. Luckily, many of the articles themselves work to highlight some of these crosscutting or poorly represented thematic areas. It is clear from the articles in this special issue that many of the authors have worked hard to connect disparate mountain processes and ways of knowing into their research.

Whether mountains are best viewed as particular geographic systems or as regions unto themselves is a question for which an answer is not generally agreed on. Many of the articles in this special issue, though, find commonalities among different mountain areas, such as their inhabitants being politically and economically marginalized from lowland populations, and

populations who are vulnerable to natural hazards that are often large and frequent. The physical environments of mountains are often highly sensitive to climate change and resource development but also contain a rich diversity of processes, forms, and organisms. Across these articles, it is clear that to better understand the connections between different mountain environments and the peoples that inhabit them will require a variety of approaches; "muddy boots" fieldwork, geographical modeling, social theory, and new tools and analytical techniques are and will be needed for the diversity of mountain questions we have today. The other clear theme among the articles in this special issue is the importance of place-based geographical research. Finding systematic commonalities between the mountains of the world has been and should continue to be an important geographic endeavor. Understanding the uniqueness of mountain places and how and why this uniqueness arises is of similar importance. In many respects, the genesis for this special issue began with Richard Marston's Presidential Address in the *Annals*, "Land, Life, and Environmental Change in Mountains" (Marston 2008). In the conclusion, Marston made an impassioned case to geographers: "Let us endeavor to use geographic theory, knowledge, and techniques to improve the human condition in the mountains and for all who live downstream" (517). That is a fitting endpoint for this special issue as well.

Acknowledgments

I wish to express my sincere thanks to several individuals who helped make this special issue possible. Dick Marston and Don Friend provided early support and excitement for the special issue's topical focus. The experiences and advice from my fellow *Annals* editors were very useful in helping me refine the special issue topic, develop its process, and make suggestions for how to avoid difficulties. The authors of this special issue were given a very tight schedule and deadline, and they have my gratitude for not only meeting these deadlines but also for producing high-quality articles that now make up this special issue. In addition, the special issue required the services of a large number of reviewers and editorial board members who contributed important and helpful reviews that improved the many articles herein. Alec Murphy helped me with the structural organization of the issue, and the other members of the Department of Geography at the University of Oregon provided positive emotional support throughout the development of this issue. Most important, the Managing Editors of the *Annals*, Jennifer Cassidento and Jenny Lunn, and our Production Manager at Taylor and Francis, Lea Cutler, have been extraordinarily important in getting this issue from the beginning to the end. To these people and a great many others, I am extremely grateful.

References

Barry, R. G. 2008. *Mountain weather and climate*. 3rd ed. Cambridge, UK: Cambridge University Press.

Bernbaum, E. 1997. *Sacred mountains of the world*. Berkeley: University of California Press.

Bishop, M. P., and J. F. Shroder, Jr., eds. 2004. *Geographic information science and mountain geomorphology*. Berlin: Springer-Praxis.

Davis, W. M. 1911. The Colorado Front Range: A study in physiographic presentation. *Annals of the Association of American Geographers* 1:21–83.

Funnell, D. C., and M. F. Price. 2003. Mountain geography: A review. *The Geographical Journal* 169 (3): 183–90.

Libiszewski, S., and G. Bachler. 1997. Conflicts in mountain areas: A predicament for sustainable development. In *Mountains of the world: A global priority*, ed. B. Messerli and J. D. Ives, 103–30. Carnforth, UK: Parthenon.

Marston, R. A. 2008. Land, life, and environmental change in mountains. *Annals of the Association of American Geographers* 98 (3): 507–20.

Orlove, B., E. Weigandt, and B. H. Luckman. 2008. *Darkening peaks: Glacier retreat, science, and society*. Berkeley: University of California Press.

Price, M. F., A. C. Byers, D. A. Friend, T. Kohler, and L. W. Price, eds. 2013. *Mountain geography: Physical and human dimensions*. Berkeley: University of California Press.

Price, M. F., L. A. G. Moss, and P. W. Williams. 1997. Tourism and amenity migration. In *Mountains of the world: A global priority*, ed. B. Messerli and J. D. Ives, 249–80. Carnforth, UK: Parthenon.

Tuan, Y. 1974. *Topophilia: A study in environmental perception, attitudes, and values*. Englewood Cliffs, NJ: Prentice-Hall.

Controls on Mountain Plant Diversity in Northern California: A 14,000-Year Overview

Christy E. Briles

A network of eight Holocene paleoenvironmental records from lakes in the Klamath Mountains of Northern California provides insights on how diverse coniferous forests are maintained in the face of climate change. Pollen data suggest that in most cases plants kept pace with climate change. The steep costal-to-inland precipitation gradient resulted in asynchronous responses to climate change with coastal forests responding before inland sites. This was likely due to the proximity to oceans, warm valleys, and the differential responses to changes in ocean upwelling. Plants growing on soils with heavy metals showed little response to Holocene climate variability, suggesting that they experienced stability during the Holocene, which helps explain the localized plant diversity on the harsh soils. Plant communities on soils without heavy metals adjusted their ranges along elevational gradients in response to climate change, however. Fires were a common occurrence at all sites and tracked climate; however, sites that were more coastal experienced fewer fires than inland sites. Fire severity remained similar through the Holocene at individual sites; however, it was low to moderate at southern locations and higher at more northern locations. The article highlights historical factors that help explain the diversity of plant species in the forests of Northern California and provides insights for managing these complex ecosystems.

来自北加州克拉马斯山区的八个全新世古环境记录网络, 提供了不同的针叶林在气候变迁中如何续存的洞见。花粉数据指出, 在多数案例中, 植物与气候变迁维持同步。从海岸到内陆的陡峭降雨梯度, 导致对气候变迁的非同期回应, 其中海岸森林较内陆地点更早做出回应, 而这有可能是由邻近海洋、暖谷和对海洋涌升流改变的差异回应所导致。生长于含有重金属的土壤的植物, 对全新世的气候变异鲜少产生回应, 显示它们在全新世中经历了稳定性, 并有助于解释硬土上的在地化植物多样性。但在未含有重金属的土壤生长的植物群落, 则依照海拔梯度调适其范围, 以此回应气候变迁。火灾是所有地点和追溯的气候中经常出现的事件; 但更接近海岸的地点, 则较内陆地点经历较少的火灾。在全新世中, 各个地点火灾的严重性皆相近, 但在南部地点的严重性是从低度到适度, 而更为北部的地点则较为严重。本文强调有助于解释北加州森林中的植物种类多样性的历史因素, 并为管理这些复杂的生态系统提供洞见。 关键词: *生物多样性, 火灾, 山区, 古生态学, 植物。*

Una cadena de ocho registros paleoambientales del Holoceno en lagos de las Montañas Klamath de la parte norte de California nos dan claves sobe la manera como se mantienen diferentes bosques de coníferas ante el cambio climático. Los datos de polen sugieren que en la mayoría de los casos las plantas siguen el paso al cambio de clima. El fuerte gradiente de precipitación de la costa hacia el interior generó respuestas asincrónicas al cambio climático, en las que se notó que los bosques litorales respondían antes de que lo hicieran los sitios del interior. Tal circunstancia probablemente se debió a la proximidad de los océanos, valles cálidos y a las diferentes respuestas a cambios en la surgencia del mar. Las plantas que crecen en suelos de metales pesados mostraron poca respuesta a la variabilidad climática del Holoceno, sugiriendo que ellos experimentaron estabilidad durante este período, lo cual ayuda a explicar la diversidad de plantas localizadas en suelos ásperos. Sin embargo, las comunidades de plantas desarrolladas en suelos carentes de metales pesados ajustaron su ámbito junto con los gradientes de altitud en respuesta al cambio climático. Los incendios forestales fueron una ocurrencia común en todos los sitios y reflejaban el clima; pero los sitios más costaneros experimentaban menos incendios que los sitios del interior. La severidad del incendio se mantuvo similar a través del Holoceno en sitios individuales; sin embargo, estaba entre baja y moderada en las localidades meridionales, y más alta en localidades situadas más al norte. El artículo destaca los factores históricos que ayudan a explicar la diversidad de especies vegetales en los bosques del norte de California y brinda buenas estrategias para el manejo de estos ecosistemas complejos.

Biodiversity is not evenly distributed across the globe and the factors involved in maintaining it are not well understood and debated (e.g., Warren et al. 2014). The Klamath Mountains of Northern California and southern Oregon harbor some of the most diverse coniferous forests in North America. They have been referred to as the "Galapagos of North America" (DellaSala 2003), yet are located thousands of kilometers north of the islands and the tropics. The factors that have been hypothesized to explain the diversity include (1) a complex geology and unusual soil makeup (Whittaker 1960), (2) a refuge during the cooling and drying of the Cenozoic (e.g., Douglas fir; Stebbins and Major 1965; Wolfe 1994; Gugger, Sugita, and Cavender-Bares 2010), (3) a steep temperature and precipitation gradient due to the spatial variability of climate (G. Taylor and

Hannan 1999), and (4) a spatially heterogeneous disturbance regime (A. H. Taylor and Skinner 1998, 2003; Skinner, Taylor, and Agee 2006). How these factors play out and are intertwined in the past is critically important for understanding the maintenance of biological diversity in the region today but also into the future under a changing climate. For example, understanding how species responded through migration, or in refugia, helps devise forest management plans that acknowledge the immense spatial variability of the region. Environmental archives derived from pollen and charcoal preserved in lake sediments can help discern the relative role of different mechanisms that maintain the floristic diversity.

A large network of vegetation and fire reconstructions from lake sediments exists in the Klamath Mountains (see Figure 1; Mohr, Whitlock, and

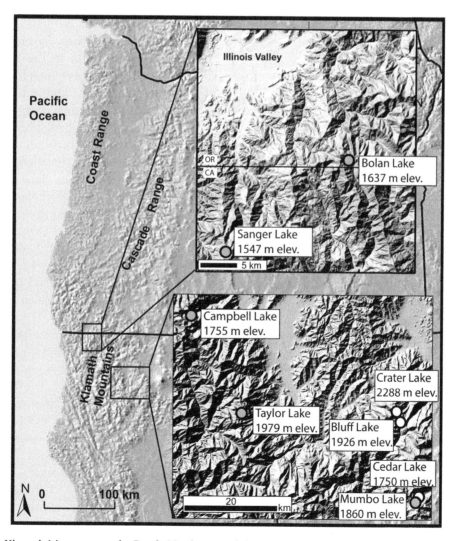

Figure 1. Map of the Klamath Mountains in the Pacific Northwest and the sites discussed in the text. Sites with white circles are lakes on ultramafic substrates and those with gray circles are lakes on nonultramafic substrates.

Skinner 2000; Briles, Whitlock, and Bartlein 2005; Daniels, Anderson, and Whitlock 2005; Briles et al. 2008; Colombaroli and Gavin 2010; Briles et al. 2011; Crawford et al. 2015; A. White, Briles, and Whitlock 2015). These sites provide a wide-angle snapshot into the past and a chance to understand environmental variability in a complex mountainous environment. Long-term climate records based on different proxies including ocean sediment–derived alkenones (Barron et al. 2003), speleothem oxygen isotope (Vacco et al. 2005), and tree ring and lake sediment records (Graumlich 1993; Graham and Hughes 2007; Steinman et al. 2012) exist in the region and provide records of climate variability that can be used to interpret the vegetation and fire history reconstructions. The objective of this article is to review the current records that exist since the last glacial period (~14,000 years ago) and discuss the long-term mechanisms that have maintained the diverse Klamath Mountain ecosystem. Management implications of the research are also discussed.

Regional Environmental Characteristics and Explanations for Diversity

The Klamath Mountains are located on the California–Oregon border west of Interstate 5 (I-5; see Figure 1). The forests harbor the tallest and largest trees in the world, *Sequoia sempervirens* (coast redwood), an impressive thirty-nine conifer species and subspecies, including 3,500 plant species (~220 only found there), and a host of other organisms including 115 butterfly, 235 mollusk, and 33 fish species (DellaSala et al. 1999).

Spatial variability of climate is a notable environmental characteristic of the Klamath Mountains. There is a steep ocean-to-inland climate gradient that results from close proximity to the Pacific Ocean and rain shadows created by the mountainous coastal complex. Annual precipitation averages >2,500 mm along the coast, whereas east of I-5 only 300 to 800 mm falls on average. Coastal locations are mild through the year. Valleys are cool in winter but hot during the summer months. The mountains experience freezing temperatures during the winter, with significant snowfall, and warm dry summers.

The climate of the Klamath Mountains is strongly influenced by high- and low-pressure systems that develop in the northeast Pacific Ocean. During the winter, a strong Aleutian low entrains moisture from the subtropical and tropical Pacific and delivers it to the Klamath Mountains. In summer, the Eastern Pacific Subtropical High creates subsidence and much drier conditions. Surface heating during the summer causes steep environmental lapse rates and consequently atmospheric instability and convectional storms, which are associated with lightning. In summer, fog along the coast is associated with the southward-moving California current that results in offshore flowing winds and ocean upwelling.

The diverse geology is a result of accreted terrains and the collision of the North American and Pacific continental plates (Norris and Webb 1990; Harden 1997). The mountains are composed of mudstone, limestone, sandstone, chert, schist, serpentine, peridotite, several plutonic deposits (e.g., igneous intrusions of diorite and granodiorite rocks), and volcanic rocks (e.g., basalts). A complex of mountain ranges contributes to the topographic heterogeneity of the region, giving it the nickname the Klamath Knot (Wallace 1983). Elevations range from sea level to 2,800 m (on Mount Eddy) over short distances. Several exposed sheets of ultramafic bedrock occur in the Klamaths, including the Trinity Ultramafic Sheet and Josephine and Sexton ophiolites (Irwin 1981). The soils derived from these rocks are deficient in critical minerals for plant growth, including calcium, nitrogen, phosphorus, and potassium, and they contain high concentrations of heavy metals, including nickel, magnesium, chromium, and iron, that inhibit or restrict plant growth (Kruckeburg 2002; Alexander et al. 2007). Forests on ultramafic substrates are usually open and dry compared with those growing on other soil types (Kruckeburg 1985, 2002; Alexander et al. 2007). Plants have evolved traits (e.g., hairy or waxy leaves, shallow roots) that allow them to tolerate drought conditions and methods for limiting their uptake by accumulating the minerals in their tissues (Alexander et al. 2007). The ultramafic substrates have more than forty endemic plant species, making them diversity hotspots within the region.

Fire is a common natural disturbance in the Klamath Mountains. Recently, the region has received attention as large areas of forest and protected wilderness have experienced large fires (e.g., Biscuit and Silver fires) that in some cases burned at high severity, specifically if they had burned in previous decades (Thompson, Spies, and Ganio 2007). Most of our understanding about fire in the Klamath Mountains comes from tree-ring studies extending back ~500 years. Fire regimes in the region have historically ranged from low to moderate severity with occasional high-severity fires in different forest types

(A. H. Taylor and Skinner 1998, 2003; Skinner, Taylor, and Agee 2006). These studies suggest that fire activity is governed by the topographic and climatic heterogeneity of the region and also the spatial pattern of previous fires and their severity (A. H. Taylor and Skinner 1998, 2003; Skinner, Taylor, and Agee 2006; Thompson, Spies, and Ganio 2007). For example, fire severity tends to be highest on the upper third of dry south- and west-facing slopes and lowest on the lower third of slopes on north- and east-facing slopes. The size of fire is influenced by forest patch size and barriers to fire spread (e.g., ridgetops, aspect changes, riparian zones, substrate differences). Fire severity also varies with vegetation type and structure, with lower severity in forests with medium to large trees and higher severity in stands dominated by small trees (Miller et al. 2012). All of these factors combine to create a patchwork of stands of different age and composition, which helps maintain the biological diversity (Skinner, Taylor, and Agee 2006).

The Role of Historical Climate Variability on Floristic Diversity

The factors discussed earlier have contributed or helped maintain the extraordinary plant diversity in the Klamath Mountains; however, one of the least understood is the role of long-term climate change. Paleoenvironmental research during the past two decades has focused on how paleoclimate variability has influenced the mountainous plant communities in the Klamath Mountains. Mohr, Whitlock, and Skinner (2000) and Daniels, Anderson, and Whitlock (2005) laid the groundwork by reconstructing the vegetation and fire history from the southeastern Klamath Mountains using lake sediments from Bluff, Crater, and Mumbo lakes on or surrounded by ultramafic substrates. These studies established the general sequence of vegetation changes including subalpine parkland during the late glacial period (> 11,000 cal yr BP) when climate was cooler and wetter than today, a xerophytic mixed-conifer woodland in the early Holocene (11,000 to ~5,000 cal yr BP) when climate was warmer and drier than today, and a diploxylon *Pinus* spp. (yellow pine species) forest with some *Abies* spp. (fir species) in the late Holocene (< 5,000 cal yr BP) as modern conditions became established. Fires were prevalent over the length of the records and were more frequent during drier periods, such as the early Holocene and Medieval Climate Anomaly.

Briles, Whitlock, and Bartlein (2005) reconstructed the vegetation and fire history from Bolan Lake on diorite soils in the northern Klamath Mountains and showed that plant communities were more strongly affected by variations in climate occurring on centennial and millennial timescales than the sites in the southeastern Klamath Mountains. Bolan Lake also supported more mesophytic species, such as *Picea breweriana* (Brewer spruce), *Tsuga mertensiana* (mountain hemlock), *Abies* spp., and haploxylon *Pinus* spp. (white pine species), and less frequent fires during the late glacial period and late Holocene than the sites in the southeastern Klamath Mountains. Another important finding of the research was that individual plant species and taxon moved along elevational gradients to track changing climate. The differences in plant communities and contrasting responses through time in the northern and southern Klamath Mountains suggested that there was spatial variability and the temporal responses to climate change were not synchronous and need additional research.

An in-depth study was conducted to examine the spatial and temporal variability in the Klamath Mountains through the last glacial period. A network of mid- to high-elevation (> 1,500 m) sites was chosen with similar elevations and aspects from the coast to more inland locations. First, two nearby sites (Bolan and Sanger lakes) situated along a coastal-to-inland (wet to dry) moisture gradient were compared to determine whether the present differences in forest composition arose from distinct vegetation, fire, and climate histories or merely reflected local site differences. Second, three sites on ultramafic substrates (Crater, Cedar and Bluff lakes) and five sites on nonultramafic substrates (Taylor, Campbell, Mumbo, Bolan, and Sanger lakes) were compared to examine how Holocene climate variations influenced vegetation and fire on different substrates. Due to the striking differences in forest composition and structure today on the different substrate types, it was of interest to understand whether the forests remained distinctly different over the Holocene as a result of soil characteristics or if climate overrode the soil influences and resulted in similar forest communities across the region. The following review builds on these studies.

The Moisture Factor

Comparison of the vegetation and fire histories at Bolan and Sanger lakes along the coastal-to-inland moisture gradient suggested that the histories diverged as a result of the persistent differences in climate along the gradient (Figure 2; see Briles et al. 2008). At both sites, the late glacial period (> 11,500 cal yr BP) was characterized by subalpine parkland and infrequent fire (although less frequent at Sanger Lake). Subalpine parkland was replaced by closed forest of haploxylon *Pinus* (likely *P. monticola* [western white pine] and/or *P. lambertiana* [sugar pine]), Cupressaceae (likely *Calocedrus decurrens* [incense cedar]) or *Chamaecyparis lawsoniana* [Port Orford cedar]), *Abies* (likely *Abies concolor* [white fire] or *A. magnifica* [red fir]), and *Pseudotsuga menziesii* (Douglas fir) and more frequent fires at both sites. The shift occurred 1,000 years earlier at the wetter coastal site (Sanger Lake) than at the drier, more inland site (Bolan Lake). The earlier coastal response is attributed to the influence of reduced Pacific Ocean upwelling between 12,000 and 11,000 cal yr BP, which created warmer, drier conditions at the coast. The inland site, in contrast, was influenced more by locally retreating glaciers and cooler conditions. In the early Holocene (11,500 to ~5,000 cal yr BP), *Pinus* was less abundant and fire activity was less frequent than before at the coastal site, whereas *Pinus* and higher fire frequencies predominated at the inland site. In addition, *Quercus*

vaccinifolia (huckleberry oak) became abundant at the inland site ~2,000 years earlier than at the coastal site in the early Holocene. The coastal site was likely influenced by increased coastal upwelling and fog production at this time, whereas hot, dry conditions from valley heating likely affected the more inland site. In the late Holocene (5,000 cal yr BP to present), *Abies* and *Pseudotsuga* increased and *Pinus* and *Quercus* decreased in the forest at both sites, suggesting a widespread response to cooler and wetter conditions than before. At the coastal site, though, wetter conditions and fewer fires than before allowed for more Cupressaceae (likely *Chamaecyparis lawsoniana*) and *Picea* (likely *Picea breweriana*). The inland site supported more *Tsuga mertensiana* in the late Holocene, which was likely the result of higher snowpack and proximity of high-elevation peaks where the species is most abundant today.

The comparison of the two records implies that large-scale controls in climate over the Holocene resulted in major changes in vegetation and fire regimes at both sites but the mesoscale contrast between coastal and inland locations resulted in somewhat different timing of climate change and ecosystem response. Thus, variations in the timing of changes in effective moisture and temperature, ultimately resulting from the influence of ocean upwelling and inland heating, are important in creating different vegetation histories and consequently supported different plant communities since the last glacial period.

Across the Western United States, differences in the timing of responses of vegetation and fire at nearby sites has been attributed to a variety of controls. These controls include topography and aspect that create a diversity of microclimates, spatial variability of precipitation, and the autecology of individual species that governs their responses to climate change (Whitlock and Bartlein 1993; Clark and Gillespie 1997; Heinrichs et al. 2002; Brunelle et al. 2005; Gavin et al. 2006). Heinrichs et al. (2002) compared the vegetation response to Holocene climate change for four sites at similar elevations along a climate gradient east of Vancouver, British Columbia. They found in the late Holocene that cool, wet sites responded earlier to cooling and the development of more mesic conditions than did sites at warmer and drier locations because plant species reached environmental thresholds that allowed for an earlier response at the cool sites. On Vancouver Island, British Columbia, Brown et al. (2006) reconstructed variations in Holocene precipitation from a series of sites along an east–west

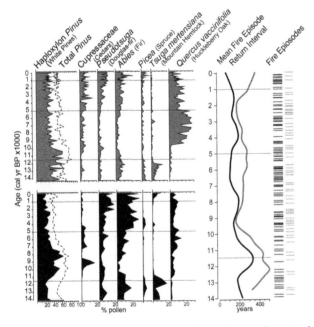

Figure 2. Pollen percentages, fire episodes, and mean fire episode return intervals for Sanger Lake (gray) and Bolan Lake (black) lake. The difference in timing in the pollen data and frequency of burning reflects meso- to microscale variability in climate within the region.

transect. Whereas the most western and eastern sites' vegetation reflected consistently wet or dry conditions, respectively, the central sites were drier during the early Holocene, suggesting a steep coastal-to-inland precipitation gradient. The central sites became wetter in the late Holocene, resulting in a more homogenous climate regime across the island. Both Heinrichs et al. (2002) and Brown et al. (2006) provided evidence for a coastal-to-inland moisture gradient during the Holocene that caused differential responses in vegetation and fire over short distances.

In summary, the results from the Klamath Mountains and British Colombia suggest that the coastal-to-inland moisture gradient can either enhance or subdue the response of vegetation and fire to larger scale climate variations during the Holocene by influencing both effective moisture and temperature. The significant difference between the environmental histories in the Klamaths and those in southern British Columbia is the influence of ocean upwelling and valley heating between Bolan and Sanger lakes that resulted in asynchronous responses in plants at the two sites.

The topographic diversity, ocean currents, fog production, and west–east moisture gradient create a diversity of climate conditions in the Klamaths that in turn support numerous plants within a fairly small area. When climate does change, some locations are more responsive than others, and some will be affected before others. Overall, the Klamath Mountains have a low climate-change velocity, or local displacement of climate conditions, due to the topographic heterogeneity and diversity of microclimates, which allows for the development and maintenance of endemism and biodiversity (Sandel et al. 2011). Both the historical and modern data, however, suggest that sensitivity to climate change is highly localized (Damschen, Harrison, and Grace 2010; Harrison et al. 2015). For forest practitioners, a local-scale analysis of climate velocities is needed through an examination of climate-sensitive and climate-tolerant species, along with landscape characteristics, to help identify potential climate-change refugia and strengthen forest management plans for the Klamath Mountains (Dobrowski 2011; Olson et al. 2012; Wilkin, Ackerly, and Stephens 2016).

The Soil Factor

Given the geologic diversity of the Klamaths, with ultramafic bedrock that harbors distinct plant life, it was of interest to understand the relative role of

Holocene climate change and substrates on forest development (Briles et al. 2011). Lakes were chosen in mixed-conifer forests at similar elevations (between 1,600 and 2,100 m) and with similar modern climates. In the southeastern Klamath Mountains, sites on ultramafic substrates (Crater, Cedar, and Bluff lakes) were located on an extensive peridotite outcrop, sites on nonultramafic substrates (Taylor and Campbell lakes) were located on granodiorite and sedimentary bedrock, and Mumbo Lake was located on diorite bedrock but surrounded by extensive peridotite outcrops. Sanger and Bolan lakes were also included in the comparison and located on diorite bedrock (see Briles et al. [2011] for site details; Figures 3 and 4).

During the late glacial period (> 11,500 cal yr BP), summer insolation increased to near-modern values, but conditions were cooler and wetter than today. Nonultramafic substrates supported an open forest of haploxylon *Pinus* (likely *P. monticola* or *P. lambertiana*), *Abies* (either *A. concolor* or *A. magnifica*), *Tsuga*, and *Pseudotsuga*, whereas ultramafics supported open forest of diploxylon *Pinus* (likely *P. jeffreyi* [Jeffrey pine] or *P. contorta* [lodgepole pine]) and *Abies*. This implies that ultramafics supported a more open forest of xerophytic taxa than other substrates (Figure 4). Charcoal data suggest that fire episodes (periods of fire) occurred infrequently and irregularly at all sites, indicating no widespread control on fire, and low charcoal accumulation rates indicate low- to moderate-severity surface fires (Figure 3).

Vegetation changed significantly on both substrates after 11,500 cal yr BP in response to increased summer insolation and warmer and drier conditions than before. On nonultramafics, closed forests with mesophytic species between 11,500 and 9,000 cal yr BP gave way to open forest of xerophytic taxa, such as *Pinus monticola/lambertiana*, Cupressaceae, and *Quercus*, between 9,000 and 7,000 cal yr BP. On ultramafics, open forests of Cupressaceae and *Quercus*, with less abundant *Pinus jeffreyi/contorta* and *Abies*, developed after 11,500 cal yr BP, and the composition has remained unchanged until present. This implies that climate had little influence on forest composition after 11,500 cal yr BP on ultramafic substrates. Fire episodes were more frequent than before on all substrates during the early Holocene, suggesting that warm, dry conditions had a widespread influence on fire regimes. At the sites influenced by ocean upwelling and valley heating (i.e., Bolan and Sanger lakes), charcoal accumulation rates (CHAR) and charcoal peaks (i.e., CHAR above BCHAR) increased, suggesting that fires

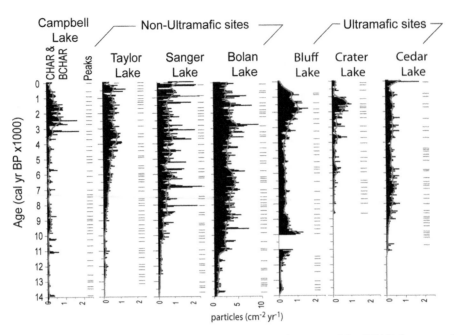

Figure 3. Charcoal accumulation rates (black bars), background CHAR (weighted average of the CHAR data, smooth gray line), and charcoal peaks (periods where CHAR exceeds BCHAR; dashes). CHAR levels above BCHAR represent biomass burned during each fire event and are an indicator of fire severity. Although there is no clear difference between fire regimes on the different soil types, southern sites (Campbell, Taylor, Bluff, Crater, and Cedar) all have smaller charcoal peaks than the sites further north (Bolan and Sanger). This suggests that fires burned at low to moderate severity in the southern Klamath Mountains and at a higher severity in the north during the Holocene. CHAR = charcoal accumulation rates; BHCAR = background charcoal accumulation rates.

burned more biomass and were likely more severe, whereas charcoal levels at all other sites remained low, indicating low- to moderate-severity burns.

After 7,000 cal yr BP, *Abies* increased and was abundant by 6,000 cal yr BP in forests on nonultramafics as summer insolation decreased and conditions became cooler and wetter than before, indicating that the forests were becoming more closed. Fire episodes were less frequent than before, between 7,000 and 5,000 cal yr BP, suggesting more mesic conditions were limiting fires. In contrast, fire activity was more variable in forests on ultramafics, between 7,000 and 4,000 cal yr BP, indicating that climate was not affecting fire regimes uniformly. *Pseudotsuga* returned to the forests on nonultramafics after 5,000 cal yr BP, during a period of frequent fire, and *Tsuga* was present after 2,000 cal yr BP, during a period of fewer fire episodes, indicating a transition from xerophytic fire-tolerant species to mesophytic fire-intolerant species in the forest. On ultramafics, forest composition remained the same as before; however, abundances of Cupressaceae and *Quercus* declined and *Pinus jeffreyi/contorta* dominated after 5,000 cal yr BP. Fire became less frequent on ultramafic substrates after 5,000 years likely due to the loss of the *Quercus* understory. Over the last 1,000 years, fire activity decreased to historical lows at

mid-elevation sites, whereas it increased at higher elevations in forests on both substrates, suggesting that humans might have been modifying fire activity at low elevations, especially during the fire suppression era (also see Colombaroli and Gavin 2010; Crawford et al. 2015; A. White, Briles, and Whitlock 2015). CHAR increased at most sites over the last 7,000 years, suggesting increased biomass burned on all substrate types, but charcoal peaks were small, indicating that many locations experienced low- to moderate-severity burns. The exception was in the northwestern Klamaths, where Bolan and Sanger lakes recorded larger charcoal peaks, suggesting higher severity fires than at the sites to the south.

The role of substrate in maintaining diverse plant communities and influencing fire regimes has been addressed in other paleostudies over the last three decades (Brubaker 1975; Whitlock and Bartlein 1993; Millspaugh, Whitlock, and Bartlein 2000; Oswald et al. 2003). Brubaker (1975) compared the vegetation history on glacial till and outwash in upper Michigan to determine how substrate differences (i.e., soil texture influencing soil moisture and nutrient availability) influenced past vegetation. Pollen records from lakes on the different substrates indicated that the vegetation history on dry outwash sites supported open

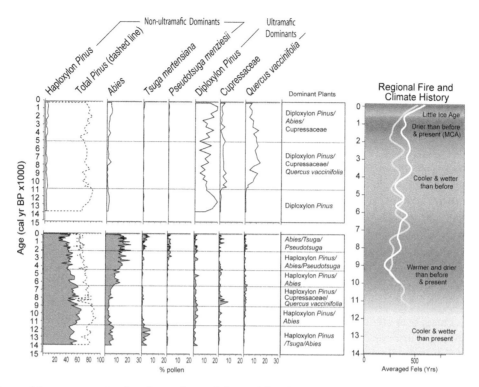

Figure 4. Vegetation and fire reconstruction based on pollen and charcoal data from the Klamath Mountains. The data highlighted in white are from lakes on ultramafic substrates and in gray lakes on nonultramafic substrates. Example pollen records from ultramafics (Bluff Lake, white) and nonultramafics (Taylor Lake, gray) are shown. Mean fire episode intervals for sites on ultramafics and nonultramafics are averages of the individual site fire episode records (see Figure 3). The climate reconstruction is based on July 45° N insolation anomaly, alkenone-derived sea-surface temperatures, speleothem δ18O-derived temperatures, and paleoclimate simulations (Bartlein et al. 1998; Barron et al. 2003; Vacco et al. 2005).

jack pine (*Pinus banksiana*) woodland for the last ~10,000 years. More mesic sites on till showed a more dynamic plant response and diverse vegetation history, including shifts from a jack pine forest following deglaciation to a white pine (*Pinus strobus*), deciduous tree forest after 8,000 radiocarbon years. The vegetation and fire history of Yellowstone and Grand Teton National Park also shows the long-term influence of substrate on vegetation dynamics (Whitlock and Bartlein 1993). Well-drained, infertile rhyolite soils in central Yellowstone supported *Pinus contorta* through the Holocene. In contrast, more fertile andesite and nonvolcanic substrates in Grand Teton maintained a diversity of species and communities including *Picea*, *Abies*, and *Pinus albicaulis* (whitebark pine) during the late glacial period; *Pinus contorta*, *Pseudotsuga*, and *Populus* in the early Holocene; and *Pinus albicaulis*, *Picea*, and *Abies* during the late Holocene. These vegetation changes on andesite soils correspond with long-term variations in climate. Another study from the central Arctic Foothills of northern Alaska showed that the edaphic diversity in the region supported different tundra communities through the Holocene

(Oswald et al. 2003). Flat surfaces with fine-textured soils were dominated by dwarf-shrub tundra throughout the Holocene, whereas coarse-textured soils supported a sparser, shorter canopy with nontussock sedges, prostrate shrubs, and non-Sphagnum mosses. Oswald et al. (2003) suggested that even in harsh tundra environments, substrate differences influenced the response of plants to long-term climate change.

The studies in upper Michigan, Yellowstone and Grand Teton National Park, and Alaska document differences in soil moisture and fertility to explain the long-term differences in vegetation history. In contrast, an analysis of available water-holding capacities in the Klamath Mountains revealed no significant differences between ultramafics and nonultramafics (Alexander et al. 2007). Studies have shown, however, that plants on ultramafics have developed traits (e.g., reduced root growth, sclerophyllous/hairy leaves) suggesting that the environment is moisture limiting (Poschenrieder and Barceló 2004; Alexander et al. 2007). Studies also show that plants on ultramafics are limited by low calcium and high magnesium and nickel concentrations contained in the parent rock

(Vlamis 1949; C. D. White 1971; Kruckeburg 1985; Proctor 1990). The calcium deficiency results in sparse shrub and herb communities with few species (many of which are endemic) that can tolerate the harsh environments. Surficial soils are drier due to the sparse plant cover and thinner soils due to increased runoff, which further limits plant growth. Therefore, mineral deficiencies (calcium) and the presence of heavy metals (magnesium and nickel) in ultramafics, as opposed to soil texture that results in leaching of nutrients mentioned in the other studies, hinder plant growth and create sparser communities than on nonultramafics. A recent study found that there is a resource colimitation on ultramafics between moisture and nutrients in the Klamaths (Eskelinen and Harrison 2015). When water or nutrients were added there was little response in biomass, but adding both resulted in a large biomass increase, a decrease in diversity, and almost complete species turnover. The corresponding effect of mineral deficiencies, heavy metals, and limited moisture on ultramafics likely created local conditions that exceeded more regional conditions imposed by Holocene climate variability and resulted in minor vegetation changes on the ultramafic soils.

Alternatively, forests on nonultramafics have shifted their elevational ranges up and downslope in response to climate change, but plant communities remained relatively intact as inferred from comparisons with modern pollen samples (Briles, Whitlock, and Bartlein 2005; Minckley, Whitlock, and Bartlein 2007). The vegetation reconstructions on nonultramafics have modern pollen analogues that can be found at different elevations today. For example, xerophytic species in the mixed-evergreen zone today were 500 m higher during the early Holocene when conditions were warmer than today. Species moved individually in response to climate change but likely found niches nearby, resulting in similar plant communities found today. The pollen records also suggest that plant species were able to keep pace with climate change. For example, during the last glacial–Holocene transition (~11,500 years ago), plants adjusted their ranges 800 to 1,000 m upslope in response to an abrupt ~5°C temperature increase based on a comparison of current phytogeography (Briles et al. 2011). These adjustments involved species that currently grow in the Klamath Mountains and there is no evidence of large-scale biogeographic range changes, such as those observed in the eastern United States (Webb 1988). In the Klamath Mountains, elevational adjustments on nonultramafic soils provide a localized way for plant species to avoid extirpation during periods of climate change. Ultramafic locations in the Klamath Mountains will likely experience the lowest climate-change velocity and need to be considered in planning future climate-change refugia (Dobrowski 2011; Sandel et al. 2011; Olsen et al. 2012; Wilkin, Ackerly, and Stephens 2016).

In summary, the different substrates in the Klamath Mountains have supported distinct plant communities over the Holocene, and these communities responded differently to climate variations. On ultramafics, plant species (at least those identified in the pollen record) were fairly unresponsive to changes in climate during the Holocene. Modern ecological studies in the region have documented similar findings (Harrison et al. 2015). In regard to fire activity, ultramafic plant communities are fuel limited and hence fire activity is driven by the levels of available fuels. On nonultramafics, in contrast, plant species were very responsive to climate change, resulting in a progression of different vegetation assemblages over time. Fire regimes were similar across sites and, due to the abundance of fuels through the Holocene, fire activity tracked large-scale climate changes resulting from variations in summer insolation. Interestingly, fire severity remained fairly consistent during the Holocene at the sites (see Figure 3). Bolan and Sanger lakes maintained moderate to high-severity burns, whereas the others were of low to moderate severity. The finding suggests that fire severity was not defined by substrate type but rather by terrain and microclimates and, to an extent, long-term change in climate (i.e., low-severity burns occurred at all sites prior to 11,500 cal year BP). Fire severity in mixed-conifer forests in Western North America is not well understood and debated (e.g., Hanson and Odion 2014, 2015; Odion et al. 2014; Safford, Miller, and Collins 2015). The long-term charcoal records from the Klamath Mountains demonstrate regional variability, yet local persistence in fire severity across the region.

Plant diversity in the Klamath Mountains has been, in part, maintained by substrate differences that influence forest structure and composition. Plants on ultramafics are able to withstand nutrient deficiencies, and this adaptation to edaphic conditions seems to limit their sensitivity to climate change, resulting in stable plant communities through time. Given the results of Eskelinen and Harrison (2015), it would be unwise and unnecessary to supplement these communities under future climate change to maintain diversity. Restricting mining and logging practices, invasive species, and habitat loss, however, will keep many of the small populations of endemics from extinction. Forest species on nonultramafics will require more monitoring under future climate change to make sure that they are able to move along elevational gradients unimpeded by human activities (i.e., logging, mining, homes).

Conclusions

Paleoenvironmental reconstructions using lake sediments from mid- to high-elevation sites in the Klamath Mountains help resolve some historical factors that explain how climate change has affected diverse mountain coniferous forests. Although the findings cannot be used as exact analogues for the future, they do provide evidence of the nature and extent of response to climate change and how diverse locations were maintained in the past. In the Klamath Mountains, high levels of plant diversity persisted despite significant fluctuations in climate during the last 14,000 years, including a transition from glacial to interglacial conditions ($\sim5°C$ increase) around 11,500 years ago. The close proximity of the oceans and upwelling provide a significant source of moisture and the mountains intercept it, creating meso- and microscale climates that supported different plant communities. This spatial variability in moisture due to the topographic diversity affects the timing and degree of response of plant communities and fire regimes to long-term climate change. Therefore, it is expected that changes in ocean circulation patterns and movement of hot, dry air resulting from valley heating will have localized impacts on vegetation into the future; these, along with determining climate velocities and identifying climate-change refugia, need to be considered in management plans. The range of edaphic conditions in the region has supported different plant communities over the Holocene. Communities on ultramafic substrates showed less sensitivity to Holocene climate changes, whereas communities on nonultramafics adjusted their ranges up and downslope. With current and future climate change, plants on nonultramafics will need to be closely monitored to assure that they are able to track climate, whereas for those on ultramafics management should focus on limiting the impacts of invasive species and natural resource extraction. Humans are another more recent control on forest composition and structure and although their role was likely fairly localized at lower elevations (Crawford et al. 2015) during the last

millennia, in the last century their role has significantly increased even at high elevations (Colombaroli and Gavin 2010). The human impact, compounded by climate change, is the greatest threat to plant diversity in the Klamath Mountains.

Acknowledgments

Thanks to Cathy Whitlock, Patrick Bartlein, and Carl Skinner for their contributions to the studies that resulted in the summary and conclusions in this article. I also thank the two anonymous reviewers who provided helpful suggestions on the article.

Funding

This research was supported by a U.S. Department of Agriculture Forest Service Challenge Cost-Share Agreement grant (10-JV-11272162-044).

References

Alexander, E. B., R. G. Coleman, T. Keeler-Wolfe, and S. P. Harrison. 2007. *Serpentine geoecology of western North America*. New York: Oxford University Press.

Barron, J. A., L. Heusser, T. Herbert, and M. Lyle. 2003. High-resolution climate evolution of coastal northern California during the past 16,000 years. *Paleoceanography* 18:1020–29.

Bartlein, P. J., K. H. Anderson, P. M. Anderson, M. E. Edwards, C. J. Mock, R. S. Thompson, R. S. Webb, T. Webb III, and C. Whitlock. 1998. Paleoclimate simulations for North America over the past 21,000 years: Features of the simulated climate and comparisons with paleoenvironmental data. *Quaternary Science Reviews* 17:549–85.

Briles, C., C. Whitlock, and P. J. Bartlein. 2005. Postglacial vegetation, fire and climate history of the Siskiyou Mountains, Oregon, USA. *Quaternary Research* 64:44–56.

Briles, C., C. Whitlock, P. J. Bartlein, and P. Higuera. 2008. Regional and local controls on postglacial vegetation and fire in the Siskiyou Mountains, northern California, USA. *Palaeogeography, Palaeoclimatology, Palaeoecology* 265 (1–2): 159–69.

Briles, C., C. Whitlock, C. N. Skinner, and J. Mohr. 2011. Holocene forest development and maintenance on different substrates in the Klamath Mountains, northern California, USA. *Ecology* 92 (3): 590–601.

Brown, K. J., R. J. Fitton, G. Schoups, G. B. Allan, K. A Wahl, and R. J. Hebda. 2006. Holocene precipitation in the coastal temperate rainforest complex of southern British Columbia, Canada. *Quaternary Science Reviews* 25:2762–79.

Brubaker, L. B. 1975. Postglacial forest patterns associated with till and outwash in northcentral upper Michigan. *Quaternary Research* 5:499–527.

Brunelle, A., C. Whitlock, P. J. Bartlein, and K. Kipfmuller. 2005. Postglacial fire, climate, and vegetation history along an environmental gradient in the northern Rocky Mountains. *Quaternary Science Reviews* 24:2281–2300.

Clark, D. H., and A. R. Gillespie. 1997. Timing and significance of late-glacial and Holocene cirque glaciation in the Sierra Nevada, California. *Quaternary International* 38/39:21–38.

Colombaroli, D., and D. G. Gavin. 2010. Highly episodic fire and erosion regime over the past 2,000 y in the Siskiyou Mountains, Oregon. *Proceedings of the National Academy of Sciences* 107:18909–14.

Crawford, J. N., S. Mensing, F. K. Lake, and S. R. Zimmerman. 2015. Late Holocene fire and vegetation reconstruction from the western Klamath Mountains, California, USA: A multi-disciplinary approach for examining potential human land-use impacts. *The Holocene* 25 (8): 1341–57.

Damschen, E. I., S. P. Harrison, and J. B. Grace. 2010. Climate change effects on an endemic-rich edaphic flora: Resurveying Robert H. Whittaker's Siskiyou sites (Oregon, USA). *Ecology* 91:3609–19.

Daniels, M. L., R. S. Anderson, and C. Whitlock. 2005. Vegetation and fire history since the Late Pleistocene from the Trinity Mountains, northwestern California, USA. *The Holocene* 15:1062–71.

DellaSala, D. A. 2003. Conservation planning for US National Forests: Conducting comprehensive biodiversity assessments. *BioScience* 53:1217–20.

DellaSala, D. A., S. R. Reid, T. J. Frest, J. R. Strittholt, and D. M. Olson. 1999. A global perspective on the biodiversity of the Klamath-Siskiyou ecoregion. *Natural Areas Journal* 19:300–19.

Dobrowski, S. Z. 2011. A climatic basis for microrefugia: The influence of terrain on climate. *Global Change Biology* 17:1022–35.

Eskelinen, A., and S. P. Harrison. 2015. Resource colimitation governs plant community responses to altered precipitation. *Proceedings of the National Academy of Sciences* 112 (42): 13009–14.

Gavin, D. G., F. S. Hu, K. Lertzman, and P. Corbett. 2006. Weak climatic control of stand-scale fire history during the late Holocene. *Ecology* 87:1722–32.

Graham, N. E., and M. K. Hughes. 2007. Tropical Pacific–mid-latitude teleconnections in medieval times. *Climate Change* 83:241–85.

Graumlich, L. J. 1993. A 1000-year record of temperature and precipitation in the Sierra Nevada. *Quaternary Research* 39:249–55.

Gugger, P. F., S. Sugita, and J. Cavender-Bares. 2010. Phylogeography of Douglas-fir based on mitochondrial and chloroplast DNA sequences: Testing hypotheses from the fossil record. *Molecular Ecology* 19:1877–97.

Hanson, C. T., and D. C. Odion. 2014. Is fire severity increasing in the Sierra Nevada, California, USA? *International Journal of Wildland Fire* 23:1–8.

———. 2015. Sierra Nevada fire severity conclusions are robust to further analysis: A reply to Safford et al. *International Journal of Wildland Fire* 24:294–95.

Harden, D. R. 1997. *California geology*. Upper Saddle River, NJ: Prentice Hall.

Harrison, S., E. Damschen, B. Fernandez-Going, A. Eskelinen, and S. Copeland. 2015. Plant communities on infertile soils are less sensitive to climate change. *Annals of Botany* 116:1017–22.

Heinrichs, M. L, R. J. Hebda, I. R. Walker, and S. L. Palmer. 2002. Postglacial paleoecology and inferred paleoclimate in the Engelmann spruce-subalpine fir forest of south-central British Columbia, Canada. *Palaeogeography, Palaeoclimateology, Palaeoecology* 184:347–69.

Irwin, W. P. 1981. Tectonic accretion of the Klamath Mountains. In *The geotectonic development of California*, ed. W. G. Ernst. Englewood Cliffs, NJ: Prentice-Hall.

Kruckeburg, A. R. 1985. *California serpentines: Flora, vegetation, geology, and management problems.* Oakland: University of California Press.

———. 2002. *Geology and plant life: The effects of landforms and rock types on plants.* Seattle: University of Washington Press.

Miller, J. D., C. N. Skinner, H. D. Safford, E. E. Knapp, & C. M. Ramirez. 2012. Trends and causes of severity, size, and number of fires in northwestern California, USA. *Ecological Applications* 22 (1): 184–203.

Millspaugh, S. H., C. Whitlock, and P. J. Bartlein. 2000. Variations in fire frequency and climate over the last 17,000 years in central Yellowstone National Park. *Geology* 28:211–14.

Minckley, T. A., C. Whitlock, and P. J. Bartlein. 2007. Vegetation, fire, and climate history of the northwestern Great Basin during the last 14,000 years. *Quaternary Science Reviews* 26 (17–18): 2167–84.

Mohr, J. A., C. Whitlock, and C. N. Skinner. 2000. Postglacial vegetation and fire history, Eastern Klamath Mountains, California. *The Holocene* 10:587–601.

Norris, R. M., and R. W. Webb. 1990. *Geology of California.* 2nd ed. New York: Wiley.

Odion, D. C., C. T. Hanson, A. Arsenault, W. L. Baker, D. A. DellaSala, R. L. Hutto, W. Klenner, et al. 2014. Examining historical and current mixed-severity fire regimes in ponderosa pine and mixed-conifer forests of western North America. *PLoS ONE* 9:e87852.

Olson, D., D. A. DellaSala, R. F. Noss, J. R. Strittholt, J. Kass, M. E. Koopman, and T. F. Allnutt. 2012. Climate change refugia for biodiversity in the Klamath-Siskiyou ecoregion. *Natural Areas Journal* 32:65–74.

Oswald, W. W., L. B. Brubaker, F. S. Hu, and G. W. Kling. 2003. Holocene pollen records from the central Arctic Foothills, northern Alaska: Testing the role of substrate in the response of tundra to climate change. *Journal of Ecology* 91:1034–48.

Poschenrieder, C. H., and J. Barceló. 2004. Water relations in heavy metal stressed plants. In *Heavy metal stress in plants*, ed. M. V. Prasad. 249–70. Berlin: Springer.

Proctor, J. 1990. Magnesium as a toxic element. *Nature* 227:742–43.

Safford, H. D., J. D. Miller, and B. M. Collins. 2015. Differences in land ownership, fire management objectives and source data matter: A reply to Hanson and Odion (2014). *International Journal of Wildland Fire* 24:286–93.

Sandel, B., L. Arge, B. Dalsgaard, R. G. Davies, K. J. Gaston, W. J. Sutherland, and J.-C. Svenning. 2011. The influence of late Quaternary climate-change velocity on species endemism. *Science* 334:660–64.

Skinner, C. N., A. H. Taylor, and J. K. Agee. 2006. Fire in the Klamath Mountains bioregion. In *Fire in California ecosystems*, ed. N. S. Sugihara, J. W. van Wagtendonk, J. Fites-Kaufmann, K. Shaffer, and A. Thode. 170–94. Berkeley: University of California Press.

Stebbins, G. L., and J. Major. 1965. Endemism and speciation in the California flora. *Ecological Monographs* 35:1–35.

Steinman, B. A., M. B. Abbott, M. E. Mann, N. D. Stansell, and B. P. Finney. 2012. 1,500 year quantitative reconstruction of winter precipitation in the Pacific Northwest. *Proceedings of the National Academy of Sciences* 109 (29): 11619–23.

Taylor, A. H., and C. N. Skinner. 1998. Fire history and landscape dynamics in a late-successional reserve, Klamath Mountains, California, USA. *Forest Ecology and Management* 111:285–301.

———. 2003. Spatial patterns and controls on historical fire regimes and forest structure in the Klamath Mountains. *Ecological Applications* 13:704–19.

Taylor, G., and C. Hannan. 1999. *The climate of Oregon: From rain forest to desert*. Corvallis: Oregon State University Press.

Thompson, J. R., T. A. Spies, and L. M. Ganio. 2007. Reburn severity in managed and unmanaged vegetation in a large wildfire. *Proceedings of the National Academy of Sciences* 104 (25): 10743–48.

Vacco, D. A., P. U. Clark, A. C. Mix, H. Cheng, and R. L. Edwards. 2005. A speleothem record of Younger Dryas cooling, Klamath Mountains, Oregon, USA. *Quaternary Research* 64:249–56.

Vlamis, J. 1949. Growth of lettuce and barley as influenced by degree of calcium saturation of soil. *Soil Science* 67:453–66.

Wallace, D. R. 1983. *The Klamath Knot: Explorations of myth and evolution*. Oakland: University of California Press.

Warren, D. L., M. Cardillo, D. F. Rosauer, and D. I. Bolnick. 2014. Mistaking geography for biology: Inferring processes from species distributions. *Trends in Ecology and Evolution* 29 (10): 572–80.

Webb, T., III. 1988. Eastern North America. In *Vegetation history*. Vol. 7, ed. B. Huntley and T. Webb, 385–414. Amsterdam: Kluwer Academic.

White, A., C. E. Briles, and C. Whitlock. 2015. Postglacial vegetation and fire history of the southern Cascade Range, Oregon. *Quaternary Research* 84:348–57.

White, C. D. 1971. *Vegetation—Soil chemistry correlations in serpentine ecosystems*. PhD dissertation, University of Oregon, Eugene.

Whitlock, C., and P. J. Bartlein. 1993. Spatial variations of Holocene climatic change in the Yellowstone region. *Quaternary Research* 39:231–38.

Whittaker, R. H. 1960. Vegetation of the Siskiyou Mountains, Oregon and California. *Ecological Monographs* 30:279–338.

Wilkin, K. M., D. D. Ackerly, and S. L. Stephens. 2016. Climate change refugia, fire ecology and management. *Forests* 7:77.

Wolfe, J. A. 1994. Tertiary climate changes at middle latitudes of western North America. *Palaeogeography, Palaeoclimatology, Palaeoecology* 108:195–205.

The Scientific Discovery of Glaciers in the American West

Andrew G. Fountain

The American West has been the proving ground for a number of earth sciences, including the study of glaciers. From their discovery by Western science in the late 1800s and continuing to the present day, studies of these glaciers have made important contributions to our understanding of glacial processes and to the recent assessments of global sea level rise. The growth of this science was founded on the interplay between trained scientists and dedicated nonprofessionals. This report summarizes the early history of glacier discovery and explorations in the West.

美国西部长久以来是若干地球科学的实证场域，包含冰川研究。从西方科学自 1800 年代晚期发现冰川至今，冰川研究对我们对于冰川过程与晚近对全球海平面上升的评估，做出了重要的贡献。此一科学进展，建立在受训练的科学家和勤奋的非专业者之间的互动之上。本报告摘要西部发现及探索冰川的早期历史。关键词：气候变迁, 环境历史, 冰川, 历史地理学, 美国西部。

El Oeste norteamericano ha sido campo de pruebas para un número de ciencias de la tierra, el estudio de los glaciares incluido. Desde su descubrimiento para la ciencia occidental a finales de los años 1800, hasta el presente, los estudios de estos glaciares han contribuido de modo importante a nuestro entendimiento de los procesos glaciales y a las recientes evaluaciones del ascenso del nivel del mar. El desarrollo de esta ciencia se fundamentó en la interacción entre científicos de formación y legos dedicados. Este informe resume la historia temprana del descubrimiento y exploración de los glaciares del Oeste.

The scientific history of the glaciers in the American West, defined by the Rocky Mountains from California north to Washington, east to Montana and south to Colorado, has not been well explored. Previous reports have either emphasized mountain explorations (e.g., Farquhar 1965; Becky 2003) or focused on the broader careers and context of the explorers (e.g., Goetzmann 1966; Wilson 2006). A notable exception is O'Connor's (2013) examination of the history of glacier observations on the Three Sisters volcano in Oregon. This report examines the "discovery" of glaciers in the American West, encountering the notion of discovery in a scientific sense, and summarizes the history of glacier observations from these early years to the rise of "modern" glaciology after World War II.

The discovery of glaciers in the American West is somewhat clouded and follows a common experience in the earth sciences—the locals knew what was there before science announced the discovery. A classic example is the discovery of Lascaux cave and its paintings by two local teenagers and the subsequent scientific investigation (Bahn 2007). For commercial discoveries like oil and minerals, their importance is self-evident by financial investment, exploitation, and profits obtained. For noncommercial, scientific discoveries, like a new species of butterfly, verification and importance are established by science via publication of a peer-reviewed journal article. From that article, credit of discovery is bestowed. Where the layperson might make a discovery, it is typically a scientist who understands its relevance to science. The discovery of glaciers in the American West follows this theme, in addition to a relatively unusual situation, in which an amateur challenged the established scientists.

The Native Americans clearly encountered glaciers prior to the arrival of Europeans (Cruikshank 2005). More than 10,000 years ago the Bering Land Bridge connected Russia to Alaska and migrating peoples taking the land route had to pass through the ice-free corridor between the Laurentide and Cordilleran ice sheets to make their way into the warmer climates to the south (Hopkins 1967). Alternatively, a coastal route passed by immense maritime glaciers in southern Alaska (Fladmark 1979). Along either route, glaciers were encountered, but no written or pictorial record of these experiences remains. Since that time, archeological and historical evidence testifies to the presence of Native Americans in alpine glacial environments (Coleman 1869; Kautz 1875; Lee 2012).

For Western science, the presence of glaciers in southeast Alaska was well known. These glaciers, large and terminating in the ocean, were reported by the first Russian fur traders and later mapped by Captain Vancouver (Vancouver 1798). The presence of glaciers in the American West was less clear. The settlement of the West, between the 1840s and the 1880s, occurred during the end of the Little Ice Age, that period of time from about 1450 to 1850 when global air temperatures were cool and cold, snowy winters were common across the Northern Hemisphere (Masson-Delmotte et al. 2013). In the high alpine landscape, winter snows typically lasted until late in the summer, blanketing the glaciers and hiding them from view, much like it does today after a snowy winter and cool summer. As we see later, this became the central question asked of amateur sightings of glaciers: Were they true glaciers or just accumulations of seasonal snow?

A few observations of glaciers were recorded as early as the 1840s, but they were buried in internal military reports or land survey reports, and only a few made it to newspaper accounts (Coleman 1877). In no case were the reports particularly detailed or widely distributed. Moreover, these claims, if known to science, were not followed up and critically examined. According to the informal rules of scientific "discovery," the person who first publishes the findings in a peer-reviewed journal receives the credit. Just announcing it in a newspaper is insufficient; the claim has to be critically examined by professionals. The discovery of glaciers in the American West is credited to Clarence King, who walked on a glacier in September 1870 and published his account in March 1871 (King 1871b; California Academy of Sciences 1872). The first credible published account predates King by two years, however, in the November 1869 issue of *Harper's Magazine* by E. T. Coleman, a landscape artist and enthusiastic climber living in Victoria, British Columbia, Canada (Coleman 1869; Stevens 1876). He described a mountaineering trip in 1868 to reach the top of Mount Baker, across the sound from Victoria in the Washington Territory of the United States (Figure 1). His narration describes a number of glaciers encountered, including observations of glaciers on nearby mountains. Coleman, whose climbing experience in Switzerland was known at the time (Stevens 1876), must have had experiences with glaciers in Europe and

was readily able to identify one. *Harper's Magazine* might not be a scientific journal but, unlike today, there were few science-dedicated journals, and it was common for alpine observations to appear in publications such as *Harper's*, *The Atlantic Monthly*, *Overland Monthly*, or the *Sierra Club Bulletin*. Coincidently, the same month that the Coleman article appeared, George Gibbs, a naturalist on the Northwest Boundary Survey (1857–1862) between Canada and the United States, reported on his landscape observations at a meeting of the American Geographical Society of New York. His report included observations of glaciers in the Cascade Range of Washington (not far from Mount Baker), and his written report, which was published four years later, included a sketch of one of the glaciers he observed (Gibbs 1873).

The Scientist

Clarence King was a graduate of Yale College in 1862, where he studied chemistry and the relatively new discipline of geology (Goetzmann 1966; Wilson 2006). After graduation, rather than enlist in the Army and serve in the Civil War, King traveled west, finding volunteer employment with another Yale man, Josiah Whitney, director of the California Geological Survey. The Survey spent the summers exploring and mapping the high Sierra Nevada including Yosemite Valley. Returning east in 1866, King conceived a plan to map the geology of the Western United States and lobbied Congress for the funding. By 1867, he was leading the United States Geological Exploration of the 40th Parallel, sponsored by the War Department (King and Gardiner 1878). The composition of his survey team departed from the practice of previous surveys by hiring specialists (e.g., geologists, botanists), rather than naturalists, helping to usher in a new era of scientific exploration (Goetzmann 1966). By late summer, King's survey had reached California and he journeyed to Mt. Shasta, thought to be the highest mountain in the continental United States (Wilson 2006). While exploring the mountain he encountered a glacier and was photographed standing it on 11 September 1870 (Figure 2). Like Coleman, he had also been to Switzerland and knew a glacier when he saw one. He named the glacier Whitney Glacier, after his friend, supervisor, and state geologist of California (Guyton 1998). This must have been a bit of fun on King's part at Whitney's expense because prior to the discovery Whitney wrote that no "living" glaciers

Figure 1. Image of Mount Baker from Coleman's (1869) article on climbing Mount Baker and observing glaciers. *Source*: Courtesy of *Harper's Magazine*.

Figure 2. Clarence King on the Whitney Glacier, 11 September 1870. *Source*: Photograph by C. E. Watkins, courtesy of the U.S. Geological Survey.

existed in California (Whitney 1869). In addition to exploring Mt. Shasta, King had sent team members north to Mt. Hood, outside of Portland, Oregon, and to Mt. Rainier, not far from Seattle, Washington, where they found more glaciers. He quickly announced his discovery at a February 1871 meeting of the Connecticut Academy of Sciences and the report was printed in newspapers across the nation (e.g., "Discovery of Glaciers" 1871a; "Discovery of Glaciers" 1871b; "The Glaciers of the Northwest" 1871).

The news of King's discovery in February 1871 must have come as some surprise to Coleman, who was living in Portland, Oregon, at the time. Coleman immediately wrote a letter to the editor of the *Morning Oregonian*, published two days after King's news, pointing out that he discovered a glacier on Mount Baker years earlier and published the account in *Harper's* (Coleman 1871). After doing their own research into the issue, a month later the *Morning Oregonian* published a short article about Coleman, Gibbs, and King, concluding that the proposed existence of glaciers as opposed to perennial snowfields would evoke much discussion among geologists ("Oregon Glaciers" 1871). Not letting the matter rest there, Coleman revisited the issue with an article in the *Alpine Journal* six years later summarizing the various reported sightings of glaciers that predated King's (Coleman 1877). Although there was some discussion in the scientific community about King's claims of discovery relative to earlier reports (California Academy of Sciences 1872), these discussions did not lead to any formal reexamination of King's claim. King himself largely ignored Coleman's claims in his official reports of his geological exploration of the 40th parallel (King and Gardiner 1878). He saved his criticism for John Muir.

Today, King is the acknowledged discoverer of the first glacier in the American West (Goetzmann 1966; Wilson 2006), whereas Coleman and Gibbs are overlooked. A number of factors might explain why. First, King's report is entitled, "On the Discovery of Actual Glaciers on the Mountains of the Pacific Slope" (King 1871b), and the content was focused on describing the glaciers and their environment. The title states his claim. In contrast, the reports of Coleman and Gibbs were titled with their respective journeys, "Mountaineering on the Pacific" (Coleman 1869), and "Physical Geography of the North-western Boundary of the United States" (Gibbs 1873). Their glacier observations were cursory, just another interesting feature found on this otherwise unknown landscape. They did not

highlight the uniqueness of their observation, nor did they include context for the importance of their observation. Perhaps they were unaware that the presence of glaciers was unknown and felt it only natural to find them in high alpine environments. The scientifically uninformed writer or reader would probably not recognize the significance either.

The second reason was that Clarence King published his findings prodigiously, making his claim well known. His observation was made in September 1870 and by March 1871 he had published in the *American Journal of Science and Arts* (King 1871b), perhaps the best scientific journal in the United States at that time ("Silliman's American Journal of Science and Arts" 1871). He went on that year to publish his discovery twice more in the *Atlantic Monthly*, a popular literary journal still in publication today. The first article appeared in March, summarizing his discovery, followed by a thrilling travelogue piece about climbing Shasta published in December (King 1871a, 1871c). Also, it was the practice for many newspapers to list the contents of the most recent issue of the *Atlantic Monthly*, which included King's article. Essentially, it was a media blitz and it was probably difficult for the reading public not to know that Clarence King discovered a glacier in California. Aside from the publications, another important factor favoring King's recognition was his strong social and scientific connections, as he was a member of many elite social and scientific societies in the Eastern United States (Goetzmann 1966; Wilson 2006).

The Poet

After King's published account, John Muir's (Figure 3) publication followed closely. He discovered the first glaciers in the Sierra Nevada in early autumn 1871 (Muir 1872), about the same time King was on Mt. Shasta. The amateur Muir recognized the presence of glaciers by asking whether the landforms fit the definition of a glacier—perennial snow or ice that moves (Cogley et al. 2011). Muir observed two indicators of movement, glacial "flour"—fine sediment in suspension that gives glacial streams a milky greenish color—and crevasses (Cuffey and Paterson 2010). The flour results from glacial sliding—rocks embedded in the ice abrade the bedrock floor. Crevasses result from differential movement: Some parts of the glacier move faster than other parts, causing tension that exceeds the ice strength. By peering into the crevasses, Muir observed ice beneath the seasonal snow, an indication

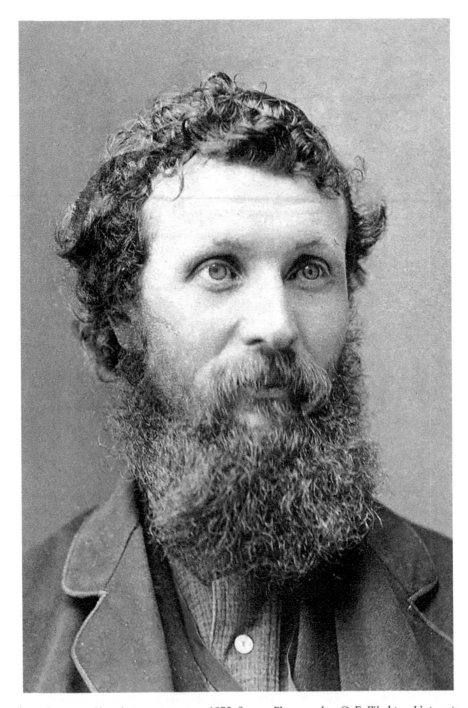

Figure 3. John Muir about the time of his glacier writings, ca. 1875. *Source*: Photographer C. E. Watkins, University of the Pacific Digital Collections. Licensed under Public Domain via Commons: https://commons.wikimedia.org/wiki/File:John_Muir_by_Carleton_Watkins, _c1875.jpg#/media/File:John_Muir_by_Carleton_Watkins,_c1875.jpg

of their perennial nature and clinching the notion that he indeed discovered a glacier. His friends, when told of the discovery, remained unconvinced (Muir 1874). Responding to their skepticism, Muir did not include his glacier observations in his December 1871 article for the *New York Tribune* describing his recent explorations of the Yosemite region (Muir 1871). To remove

any doubt, Muir returned to the glacier the following summer to *measure* its movement, the first scientific measures of a glacier in the American West. He set a line of alder poles into the glacier across its width. Returning forty-six days later, all of the stakes moved down slope and the one in the middle moved almost four feet (Muir, 1872, 1875). Muir had done all the things he needed to do to prove it was a glacier, it was perennial—evidenced by the observation of ice in the crevasse—and it moved, as indicated by the stakes, the crevasse, and the glacial flour.

During this second trip to the glacier, Muir encountered Professor Joseph LeConte, a geology professor from the University of California (LeConte 1873; Muir 1875). Muir guided him to the glacial features he had observed in the valleys of the Yosemite region and to the source of these features—the "living" glacier he found the year before (LeConte 1873). LeConte was surprised and almost convinced, saying,

> Here, then, on Mt. Lyell we have now existing, not a true glacier perhaps, certainly not a typical glacier ...; yet, nevertheless, in some sense a glacier, since there is true differential motion and a well-marked terminal moraine. It is in fact a glacier in its feeble old age. (332)

LeConte's concern was the lack of visible ice, particularly at the terminus of the glacier. This was the scientific concern of the discipline; the feature might be just an accumulation of seasonal snow that is slowly moving downslope. His conclusions about the glacier were first read before the California Academy of Sciences in late 1872, subsequently appearing in the *American Journal of Science and the Arts* (LeConte 1873), the same journal in which King's discovery paper was published. Muir later took the professor to task in his next article (Muir 1875), saying that LeConte had never seen a glacier before and did not look into crevasses for the ice below the snow. Muir went on saying that in August during LeConte's visit, the glaciers are commonly entirely covered in snow and had LeConte visited a few weeks later the seasonal snow would have been gone, revealing bare ice. In short, LeConte's conclusions were premature and based on little evidence. In any case, LeConte's support, if somewhat equivocal, must have been important to Muir, as he was an amateur in the new discipline of geology that was rapidly establishing professional standards.

When Muir's (1872) article appeared there must have been a small firestorm in geological circles. His observations challenged those of the state geologist of California, Whitney, and of the scientist in charge of the U.S. Geological Exploration of the 40th Parallel, King. Whitney's professional survey parties, including King at the time, extensively explored and mapped the Sierra and there was no way they could have missed any glaciers. Muir, on the other hand, was a complete amateur, as pointed out independently by both Whitney and King. To make matters worse, Muir was advocating the glacier origin of Yosemite rather than the faulting origin promoted by Whitney (Muir 1874). King eloquently expressed their views of Muir in his official and public expedition report describing their geological findings (King and Gardiner 1878):

> It is to be hoped that Mr. Muir's vagaries will not deceive geologists who are personally unacquainted with California, and that the ambitious amateur himself may divert his evident enthusiastic love of nature into a channel, if there is one, in which his attainments would save him from hopeless floundering. (478)

What separates Muir from other amateur discoveries and why he is recognized today were his prodigious writings in nationally recognized outlets and his connections to the science community (e.g., LeConte) and to the luminaries of the era. Yosemite, unlike Mt. Baker and the northwestern United States, was and is currently of popular national interest to the public and science community. The majesty of the new national park and the scientific discussion regarding its formation created a national focal point.

The irony in Whitney's and King's dismissal of Muir's study is that Muir did the science better than his scientifically trained critics. Whitney entirely overlooked the glaciers and King merely stated that the glaciers looked just like the ones he saw in Switzerland. Muir identified the perennial nature and its movement unambiguously. King's criticism ignored Muir's findings and focused on the issue of *neve*, or what is now called *firn*. Firn (neve) is snow that survived the summer's melt season but has not yet turned to ice (Cogley et al. 2011). Metamorphic processes of heat and refreezing transform the fine-grained seasonal snow into coarse-grained snow, and by early autumn the coarse grains freeze together, forming a hard, pavement-like surface. King essentially accused Muir of mistaking a perennial snowfield for a glacier. Perhaps we just needed to wait a century for King to be right, but not for the reason he stated. With climate warming and the shrinkage of the glaciers, the Lyell Glacier, the one Muir studied, has stopped moving and is technically no longer a glacier (National Park Service 2013).

More Discoveries and the Importance of Outdoor Clubs

In the decades following King and Muir's discoveries, a gold rush of glacier discoveries across the west followed. These included discoveries in the Wind

River Range, Wyoming (Hayden 1878, 1883), Colorado Front Range (Stone 1887), the Lewis Range, Montana—now Glacier National Park (Culver 1891; Cheney 1895)—and the Beartooth Mountains, Montana (Kimball 1899). At that time, interest in glaciers of the American West was geographical, with most attention on their location and physical appearance. After repeated visits it became clear the glaciers were changing and the scientific interest focused on their advance and retreat activity (e.g., Russell 1892; Reid 1906). At the same time, mapping the glacial history of the region and former extent of the ice sheet was of intense interest (Whitney 1869; LeConte 1873).

Prior to World War II, only a small group of scientists, mostly from the U.S. Geological Survey with a few university scientists, were interested in and reported on glacier activity (Russell 1898; Gilbert 1904; Reid 1906). Alpine recreational activity was becoming popular, however, and soon hiking clubs became interested in glaciers. Since the late 1800s, Americans were becoming outdoor recreation enthusiasts (Collingwood 2006). The rapid rise of U.S. industrialization combined with the settlement of the West triggered a sense of a lost American frontier. The passenger pigeon and bison were gone and almost half of the national forests had been cleared. In response, a national conservation movement formed to save what was left, leading to the establishment of the Adirondack forest wilderness area in New York (1872), Yosemite National Park (1890), the National Park System, and the Yellowstone Timberland Reserve (1891)—the first of the national forest system. Outdoor recreation clubs were established to provide exercise and the chance to experience these vanishing landscapes. Clubs included the Appalachian Mountain Club (1876), Sierra Club (1892), Mazamas (Portland, Oregon, 1894), and the Seattle Mountaineers (1906). The most well-known U.S. conservationists date from this period, including John Muir, Gifford Pinchot, and President Theodore Roosevelt. The expansion of the railroads made access to remote landscapes easier than ever before.

Realizing the scientific potential of this alpine activity, scientists such as Reid (1906) and Mazamas and Gilbert (1904) with the Sierra Club encouraged their respective clubs to take repeated photographs of glaciers from established camera locations to track glacier change. This encouragement probably met with little enthusiasm because relatively few photographs in the archives of the hiking clubs date from this period. Perhaps the lack of interest was a response to the expensive, heavy, and delicate camera equipment required. Furthermore, the glaciers were not changing much. During the glacier explorations of the late nineteenth century, glacier change was equivocal (Figure 4; Russell 1892; Basagic and Fountain 2011). Perhaps the combination of the high cost in labor and materials posed by the bulky cameras and the small payoff resulting from little glacier change did not make monitoring the glaciers worthwhile.

The climate was warming, however; the Little Ice Age was past (Mann et al. 2009), and things were about to change. Camera technology was improving with smaller, lighter, less expensive cameras. By the 1930s, the climate was warming rapidly and the glaciers were retreating quickly (Mann et al. 2009; Basagic and Fountain 2011; DeVisser and Fountain 2015). Seeing the obvious signs of rapid glacier recession, and perhaps fearing the loss of the glaciers (e.g., "Glaciers Disappear" 1932), hiking clubs initiated programs of glacier monitoring through photography. The most extensive programs were by the Sierra Club and the Mazamas. In fact, the Mazamas's "Research Committee" designed a field-monitoring program on the Eliot Glacier, Mt. Hood, Oregon, and flew aerial photographic surveys over glaciers in Oregon and Washington (e.g., Phillips 1938). The clubs collected data on glacier change, fulfilling the hopes of Reid and Gilbert. Because of these important activities, members of the hiking clubs were represented on scientific committees constituted by professional scientific societies to track glacier change. The 1939 roster of the American Geophysical Union's Committee on Glaciers included a representative from the Sierra Club's "Committee on Glacier Studies," and the Research Committee of the Mazamas (Matthes 1939). Our understanding today of the rate of glacier retreat during the first half of the twentieth century is based on the studies organized and executed by members of these outdoor clubs.

The Professional Scientists

Glacier studies slowed during World War II as men and material were focused on conflicts overseas. The war invigorated science research in the United States, however, and the surge in science funding came after the Soviet launch of the Sputnik satellite in 1957 (Schweber 1988). The geophysical sciences held the International Geophysical Year in 1957–1958, which coordinated geophysical measurements globally (Collis and Dodds 2008). It was during this time that

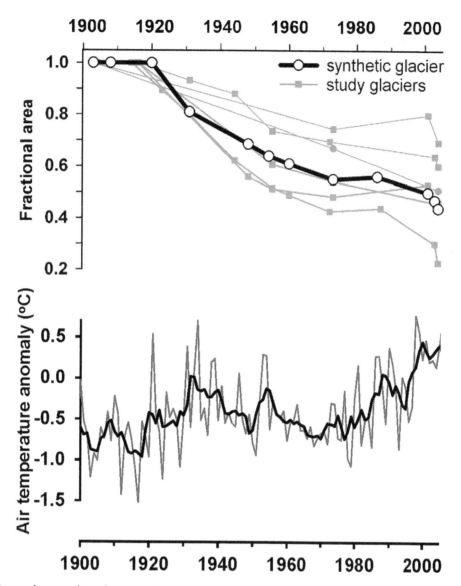

Figure 4. Fractional area changes of six glaciers in the Sierra Nevada and a hypothetical "synthetic" glacier inferred from averaging and interpolating the data from the measured glaciers. Fractional area is the ratio of the glacier area to its initial area. Air temperature anomaly is for the continental United States; the light gray line is annual data and the bold black line is a five-year running mean. *Source:* Adapted from Basagic and Fountain (2011). Data from the National Oceanic and Atmospheric Administration nClimGrid data set (Vose et al. 2014) with a base period of 1981 to 2010. The anomaly data were downloaded from the National Centers for Environmental Information, National Oceanic and Atmospheric Administration (n.d.).

the U.S. Geological Survey initiated a systematic glacier monitoring effort at South Cascade Glacier in the Cascade Range of Washington (Meier 1961). A similar program at Blue Glacier, Olympic Mountains, Washington, by the University of Washington (LaChapelle 1959). The collaboration between these two groups, with the participation of the California Institute of Technology, resulted in a remarkable number of advances in our understanding of glaciers (Kamb 1959; LaChapelle 1960; Post 1960; Meier 1961; Meier and Tangborn 1961). These projects were the vanguard of modern glacier studies in the Western United

States, employing state-of-the-art scientific instrumentation and a geophysical approach toward understanding glacier behavior. The transition from outdoor club–based projects to professionally trained scientists was largely complete by the mid-1960s.

Conclusion

The discovery of glaciers in the American West closely coincided with scientific interest in the geology of the region. Unlike in Europe, where farming communities and villages shared valleys with glaciers, in the

West, European inhabitation was distant from such alpine environments and these regions had yet to be explored by scientists. Although it is clear that the first recorded observations of the glaciers were by amateurs, neither the scientific significance nor the wide dissemination of the discovery occurred until the first scientist, Clarence King, published his findings. He was able to add scientific value to the discovery by providing the critical context for establishing the unique nature of the observation and describing its importance to science. During this time, many professionals in the emerging field of geology dismissed amateur efforts. Ironically, the best glacier science in these early days was done by the amateur John Muir. By the twentieth century, glacier scientists worked in close collaboration with amateurs in making important observations about the advance and retreat activity of the glaciers. This collaboration was motivated by the rapid glacier shrinkage during the warming of the 1930s and potential loss of the alpine glaciers. Many of the early observations and photographs of glaciers were in fact compiled by hiking clubs in the Western states. We see reflections of this today with current rapid glacier shrinkage and the increasing attention and engagement of the public. Since World War II, the approach to glacier studies shifted from observing glacier change to a more geophysical, process-oriented, approach toward understanding why and how they change. Echoing the naturalist to specialist transition of scientific field parties encouraged by King, scientific glacier studies left the realm of the amateur and became the domain of professional scientists.

Acknowledgments

Thanks to Adrian Howkins, who made key suggestions that improved the article significantly, and to the editor and anonymous reviews who helped to clarify the text.

Funding

This work was funded by the U.S. Geological Survey as part of the Western Mountain Initiative.

References

Bahn, P. G. 2007. *Cave art: A guide to the decorated Ice Age caves of Europe.* London: Frances Lincoln.

Basagic, H. J., and A. G. Fountain. 2011. Quantifying twentieth century glacier change in the Sierra Nevada, California. *Arctic, Antarctic, and Alpine Research* 43:317–30.

Becky, F. 2003. *A range of glaciers: The exploration and survey of the Northern Cascade Range.* Portland: Oregon Historical Society Press.

California Academy of Sciences. 1872. Minutes of the regular meeting of March 6th, 1871. *California Academy of Sciences* 4:161–62.

Cheney, L. W., Jr. 1895. A glacier in the Montana Rockies. *Science* 2 (50): 792–96.

Cogley, J. G., R. Hock, L. A. Rasmussen, A. A. Arendt, A. Bauder, R. J. Braithwaite, P. Jansson, G. Kaser, M. Möller, L. Nicholson, and M. Zemp. 2011. Glossary of glacier mass balance and related terms. IHP-VII Technical Documents in Hydrology, 86, IACS Contribution No. 2, UNESCO-IHP, Paris.

Coleman, E. T. 1869. Mountaineering on the Pacific. *Harpers New Monthly Magazine* 39 (234): 793–817.

———. 1871. Glaciers on the Pacific. *Morning Oregonian* 13 February.

———. 1877. Mountains and mountaineering in the far west. *Alpine Journal* 8:233–42.

Collingwood, R. G. 2006. History of Anglo-European outdoor recreation. In *Outdoor recreation in America.* 6th ed., ed. C. R. Jensen and S. P. Guthrie, 19–38. Champaign, IL: Human Kinetics.

Collis, C., and K. Dodds. 2008. Assault on the unknown: The historical and political geographies of the International Geophysical Year (1957–8). *Journal of Historical Geography* 34 (4): 555–73.

Cruikshank, J. 2005. *Do glaciers listen? Local knowledge, colonial encounters, and social imagination.* Vancouver, BC, Canada: UBC Press.

Cuffey, K., and W. S. B. Paterson. 2010. *The physics of glaciers.* 4th ed. Cambridge, UK: Elsevier.

Culver, G. E. 1891. Notes on a little known region in northwestern Montana. *Wisconsin Academy of Sciences, Arts, and Letters* 8:187–205.

DeVisser, M. H., and A. G. Fountain. 2015. A century of glacier change in the Wind River Range, WY. *Geomorphology* 232:103–16.

Discovery of glaciers on the Pacific Coast. 1871a. *Green Mountain Freeman* 28 (5): 1.

Discovery of glaciers on the Pacific Coast. 1871b. *The Oregonian* 11 February:3.

Farquhar, F. P. 1965. *History of the Sierra Nevada.* Berkeley: University of California Press.

Fladmark, K. R. 1979. Routes: Alternate migration corridors for early man in North America. *American Antiquity* 44:55–69.

Gibbs, G. 1873. Physical geography of the north-western boundary of the United States. *Journal of the American Geographical Society of New York* 4:298–392.

Gilbert, G. K. 1904. Variations of Sierra glaciers. *Sierra Club Bulletin* 5 (1): 20–25.

Glaciers disappear. 1932. *Eagle Valley Enterprise* 23 September 1932:6.

The glaciers of the Northwest. 1871. *Saline County Journal* 1 (7): 1.

Goetzmann, W. H. 1966. *Exploration and empire.* New York: Knopf.

Guyton, B. 1998. *Glaciers of California: Modern glaciers, Ice Age glaciers, origin of Yosemite Valley, and a glacier tour in the Sierra Nevada.* Berkeley: University of California Press.

Hayden, F. V. 1878. Discovery of recent glaciers in Wyoming. *American Naturalist* 12:830–31.

————. 1883. *12th annual report of the United States Geological and Geographical Survey of the territories: A report of the progress of the exploration in Wyoming and Idaho for the year 1878.* Washington, DC: U.S. Government Printing Office.

Hopkins, D. M. 1967. *The Bering land bridge.* Vol. 3. Stanford, CA: Stanford University Press.

Kamb, B. W. 1959. Ice petrofabric observation from Blue Glacier, Washington in relation to theory and experiment. *Journal of Geophysical Research* 64 (11): 1891–909.

Kautz, A. V. 1875. Ascent of Mount Rainier. *The Overland Monthly* 14:393–403.

Kimball, J. P. 1899. The granites of Carbon County, Montana: A division and glacier field of the Snowy Range. *Journal of the American Geographical Society of New York* 31:199–216.

King, C. 1871a. Active glaciers with the United States. *Atlantic Monthly* 27 (411): 371–77.

————. 1871b. On the discovery of actual glaciers on the mountains of the Pacific slope. *American Journal of Science and Arts, Third Series* 1 (1): 157–67.

————. 1871c. Shasta. *Atlantic Monthly* 28 (420): 710–20.

King, C., and J. Gardiner. 1878. *Systematic geology: Report of the Geological Exploration of the Fortieth Parallel.* Vol. 1. Washington, DC: U.S. Government Printing Office.

LaChapelle, E. 1959. Annual mass and energy exchange on the Blue Glacier, Mount Olympus (Washington). *Journal of Geophysical Research* 64:585.

————. 1960. Recent glacier variations in western Washington. *Journal of Geophysical Research* 65:2505.

LeConte, J. 1873. On some of the ancient glaciers of the Sierras. *American Journal of Science and Arts, Third Series* 5 (29): 325–42.

Lee, C. M. 2012. Withering snow and ice in the mid-latitudes: A new archaeological and paleobiological record for the Rocky Mountain Region. *Arctic* 65:165–77.

Mann, M. E., Z. Zhang, S. Rutherford, R. S. Bradley, M. K. Hughes, D. Shindell, C. Ammann, G. Faluvegi, and F. Ni. 2009. Global signatures and dynamical origins of the Little Ice Age and Medieval Climate Anomaly. *Science* 326 (5957): 1256–60.

Masson-Delmotte, V., M. Schulz, A. Abe-Ouchi, J. Beer, A. Ganopolski, J. F. González Rouco, E. Jansen, et al. 2013. Information from paleoclimate archives. In *Climate change 2013: The physical science basis. Contribution of Working Group I to the fifth assessment report of the Intergovernmental Panel on Climate Change,* ed. T. F. Stocker, D. Qin, G. K. Plattner, M. Tignor, S. K. Allen, J. Boschung, A. Nauels, Y. Xia, V. Bex, and P. M. Midgley, 383–464. Cambridge, UK: Cambridge University Press.

Matthes, F. E. 1939. Report on the committee of glaciers, April 1939. *Transactions of the American Geophysical Union* 20 (4): 518–23.

Meier, M. F. 1961. Mass budget of South Cascade Glacier, 1957–1960. *Geological Survey Professional Papers* 424-B:206–11.

Meier, M. F., and W. V. Tangborn. 1961. Distinctive characteristics of glacier runoff: U.S. *Geological Survey Professional Papers* 424-B:B14–B16.

Muir, J. 1871. Yosemite glaciers. *New York Tribune* 5 December:5–6.

————. 1872. Living glaciers of California. *Overland Monthly* 9:547–49.

————. 1874. Studies in the Sierra, Mountain sculpture-origin of Yosemite Valley. *Overland Monthly* 12 (6): 489–500.

————. 1875. Living glaciers of California. *Harper's New Monthly Magazine* 51 (306): 769–76.

National Centers for Environmental Information, National Oceanic and Atmospheric Administration. n.d. National temperature index. https://www.ncdc.noaa.gov/temp-and-precip/national-temperature-index/ (last accessed 6 January 2016).

National Park Service. 2013. Yosemite National Park's largest glacier stagnant. Yosemite National Park news release, 4 February. http://www.nps.gov/yose/learn/news/lyellglacier.htm (last accessed 7 December 2016).

O'Connor, J. E. 2013. "Our vanishing glaciers": One hundred years of glacier retreat in Three Sisters Area, Oregon Cascade Range. *Oregon Historical Quarterly* 114 (4): 402–27.

Oregon glaciers. 1871. *Morning Oregonian* 25 March:2.

Phillips, K. N. 1938. Our vanishing glaciers: Observations by the Mazama Research Committee on glaciers of the Cascade Range, in Oregon. *Mazama* 20 (12): 24–41.

Post, A. S. 1960. The exceptional advances of the Muldrow, Black Rapids, and Susitna glaciers. *Journal of Geophysical Research* 65 (11): 3703–12.

Reid, H. F. 1906. Studies of the glaciers of Mt. Hood and Mt. Adams. *Zeitschrift fur Gletscherkunde fur Eiszeitforschung und Geschichte des Klimas* 1:113–32.

Russell, I. C. 1892. Climatic changes indicated by the glaciers of North America. *The American Geologist* 9:322–36.

————. 1898. The glaciers of North America. *The Geographical Journal* 12 (6): 553–64.

Schweber, S. S. 1988. The mutual embrace of science and the military: ONR and the growth of physics in the United States after World War II. *Science, Technology and the Military* 1:1–45.

Silliman's American Journal of Science and Arts. 1871. *Nature* 3:172.

Stevens, H. 1876. The ascent of Takhoma. *Atlantic Monthly* 38 (229): 513–30.

Stone, G. H. 1887. A living glacier on Hague's Peak, Colorado. *Science* 10 (242): 153–54.

Vancouver, G. 1798. *A voyage of discovery to the North Pacific Ocean, and round the world.* Vol. 3. London: G. G. & J. Robinson and J. Edwards.

Vose, R. S., S. Applequist, I. Durre, M. J. Menne, C. N. Williams, C. Fenimore, K. Gleason, and D. Arndt. 2014. Improved historical temperature and precipitation time series for U.S. climate divisions. *Journal of Applied Meteorology and Climatology* 53:1232–51.

Whitney, J. 1869. *The Yosemite book.* New York: Julius Bien.

Wilson, R. 2006. *The explorer king: Adventure, science, and the great diamond hoax—Clarence King in the Old West.* New York: Scribner.

Incorporating Autonomous Sensors and Climate Modeling to Gain Insight into Seasonal Hydrometeorological Processes within a Tropical Glacierized Valley

Robert Åke Hellström, Alfonso Fernández, Bryan Greenwood Mark, Jason Michael Covert, Alejo Cochachín Rapre, and Ricardo Jesús Gomez

Peru is facing imminent water resource issues as glaciers retreat and demand for water increases, yet limited observations and model resolution hamper understanding of hydrometerological processes on local to regional scales. Much of current global and regional climate studies neglect the meteorological forcing of lapse rates (LRs) and valley and slope wind dynamics on critical components of the Peruvian Andes' water cycle, and herein we emphasize the wet season. In 2004 and 2005 we installed an autonomous sensor network (ASN) within the glacierized Llanganuco Valley, Cordillera Blanca (9° S), consisting of discrete, cost-effective, automatic temperature loggers located along the valley axis and anchored by two automatic weather stations. Comparisons of these embedded hydrometeorological measurements from the ASN and climate modeling by dynamical downscaling using the Weather Research and Forecasting model elucidate distinct diurnal and seasonal characteristics of the mountain wind regime and LRs. Wind, temperature, humidity, and cloud simulations suggest that thermally driven up-valley and slope winds converging with easterly flow aloft enhance late afternoon and evening cloud development, which helps explain nocturnal wet season precipitation maxima measured by the ASN. Furthermore, the extreme diurnal variability of along-valley-axis LR and valley wind detected from ground observations and confirmed by dynamical downscaling demonstrate the importance of realistic scale parameterizations of the atmospheric boundary layer to improve regional climate model projections in mountainous regions.

秘鲁因为冰川后退和水资源需求增加的缘故, 正面临迫切的水资源问题, 但有限的观察和模型分辨率, 却阻碍了理解地方和区域尺度的水文气象过程。诸多当前的全球和区域气候研究, 忽略了递减率 (LRs) 的气象驱动以及谷风和坡风动态, 对于秘鲁安第斯山的水循环的关键组成之影响, 因此我们在此强调湿季。我们于 2004 年和 2005 年间, 在布兰卡山脉冰川化的良卡鲁库谷地 (9° S) 植入自动监测网络 (ASN), 该网络包含分离且符合成本效益的自动气温记录器, 随着河谷轴线进行安置, 并以两座自动天气观测站进行锚定。本文比较这些来自 ASN 的植入水文气象测量方法, 以及运用天气研究和预报模型的动态降尺度所进行的气候模式化, 阐明山区风系和递减率的独特每日及每季特徵。风, 气温, 湿度和云量模拟显示, 由暖流所驱动的升谷风与坡风, 与向东的高空流相结合, 增加了傍晚和晚上的云量发展, 并有助于解释由 ASN 所测得的最大湿季夜间降雨量。此外, 沿着河谷轴线的LR的日间极端变量, 以及从地面观察所侦测到的、且由动态降尺度所确认的谷风, 显示出大气边界层的实际尺度参数化之于增进山区的区域气候模型推测的重要性。 *关键词: 关键词·自动监测网络, 气候模式化, 递减率, 冰前河谷, 谷风。*

El Perú enfrenta problemas inminentes con el recurso del agua a medida que los glaciares se retraen y aumenta la demanda de agua, aunque lo limitado de las observaciones y la resolución de los modelos dificultan el entendimiento de los procesos hidrometeorológicos a escalas locales y regionales. La mayor parte de los estudios climáticos globales y regionales actuales desestiman la fuerza del gradiente vertical de la temperatura (LRs) y la dinámica de los vientos de valle y ladera sobre los componentes críticos del ciclo hidrológico de los Andes peruanos, y aquí enfatizamos la estación lluviosa. En 2004 y 2005 instalamos una red de sensores autónomos (ASN) dentro del valle glaciado de Llanganuco, en la Cordillera

Blanca (9°S), consistente en anotadores de temperatura automáticos, discretos y muy efectivos por costo, localizados a lo largo del eje del valle y anclados en dos estaciones meteorológicas automáticas. Las comparaciones de estas medidas hidrometeorológicas incrustadas de los ASN y el modelado climático por reducción dinámica de la escala usando el modelo de Investigación Meteorológica y Pronóstico esclarecen distintas características diurnas y nocturnas del régimen de vientos de montaña y los LRs. Las simulaciones del viento, temperatura, humedad y nubes sugieren que el viento valle arriba y de ladera controlados térmicamente que convergen por arriba con el flujo del este fortalecen el desarrollo de nubosidad bien avanzada la tarde y en la noche, lo cual ayuda a explicar la máxima precipitación nocturna en la estación húmeda, medida por los ASN. Adicionalmente, la variabilidad diurna extrema del LR a lo largo del eje del valle y el viento del valle detectado mediante observaciones sobre el terreno y confirmadas por la reducción dinámica de la escala demuestran la importancia de la parametrización realista de la escala de la capa limítrofe atmosférica para mejorar las proyecciones modelo del clima regional en regiones montañosas.

The tropical Andes is facing critical water resource issues as mountain valleys experience persistent glacier recession (Vuille et al. 2008). Research in this context has focused on Peru's Cordillera Blanca (CB, Figure 1), Earth's most glacierized tropical mountain range. Glacial-fed tributaries flow from proglacial valleys of the CB to supply about two thirds of discharge to the upper Santa River (Mark, McKenzie, and Gomez 2005), whose waters are utilized for municipal supplies, hydroelectric generation, and agricultural irrigation to the Pacific coast (Mark et al. 2010; Bury et al. 2013). Recent satellite image analysis has indicated accelerated recession of CB glaciers and a 25 percent loss of their area between 1987 and 2010 (Burns and Nolin 2014). The rapid loss of glacierized area in the CB has increased the need for greater understanding of the factors controlling the elevation of the freezing level during precipitation events, which has been shown to be crucial for the ablation process (Bradley et al. 2009).

Given evidence showing that the major mass loss of tropical glaciers occurs during the wet season (Wagnon et al. 1999; Francou et al. 2003), a process-based understanding of valley and slope winds and their connection to the freezing level as projected by air temperature lapse rates (LRs) is important for accurately deriving glacier mass balance in proglacial valleys. We postulate that valley-specific processes, such as the valley wind interaction with easterly synoptic flow, influence cloud development and hence rainfall over proglacial valleys. We define the valley atmosphere as the volume of air laterally bounded by steep walls, a narrow outlet (mouth) opening to a plain, with the upper elevation extending to the surrounding mountain ridges (Figure 1B). We hypothesize that coupling observations and modeling over scales that capture both diurnal convection patterns and free-atmosphere meteorological forcing, such as seasonally modulated humidity flux, will give new insights into hydrometeorological processes in glacierized tropical mountains.

Multiscale valley observations elucidate diurnal patterns of convective precipitation that suggest coupling to both local and synoptic processes (Bendix et al. 2009). Tropical precipitation in mountainous regions has a pronounced diurnal cycle with higher potential in the late afternoon or at night (Dai 2001). Decreases in the daily temperature range of the CB from 1983 to 2012 possibly coincide with observed increases in specific humidity or cloud cover (Schauwecker et al. 2014). Furthermore, the thermally induced valley wind circulation results from the amplified temperature range of a column of air within the valley, as compared to the adjacent plain, the topographic amplification factor described by Whiteman (2000). Because the valley atmosphere is more confined than the adjacent plain, it warms by solar heating and cools by infrared radiation release faster, thereby creating a diurnally oscillating horizontal pressure gradient and hence the direction of airflow (Vergeiner 1987). The nocturnal portion of this wind pattern is less pronounced during wet periods, however (Bianco et al. 2006). Assessment of local meteorological forcing over time in periglacial and glacial environments that compare dynamical modeling simulations and ground-based measurements are limited in the Peruvian Andes due in large part to challenges of instrument maintenance, data continuity, and downscaling of climate models (Mölg and Kaser 2011; Hofer, Marzeion, and Mölg 2012).

Here we analyze hourly measurements from an autonomous sensor network (ASN) and output from dynamical downscaling using a climate model to (1) reveal diurnal and seasonal patterns of solar radiation, rainfall, LRs, and wind within and from glacierized

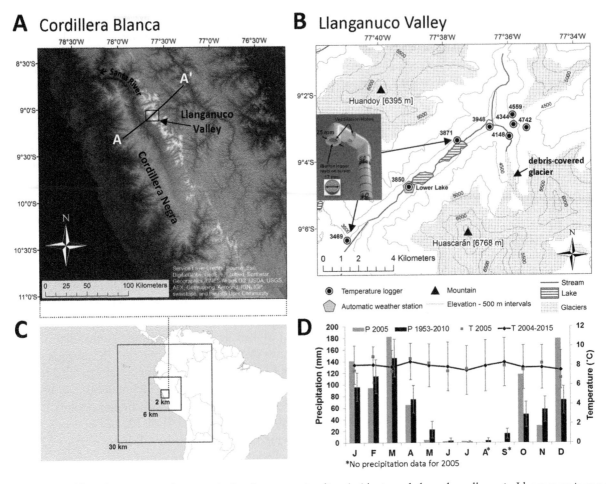

Figure 1. (A) Cordillera Blanca regional map marked with cross-section line A-A′ oriented along the valley axis. Llanganuco is one of several proglacial hanging valleys in the region. (B) Llanganuco Valley contours and location of weather instruments: Note that mountain glaciers flank the valley to the north and south of the southwest–northeast trending valley, including inset of iButton temperature sensor in aspirated PVC enclosure. (C) Advanced Research Weather–Research and Forecasting model nested domains with resolution in kilometers. (D) Llanganuco Valley monthly average precipitation and air temperature showing wet and dry seasons with whiskers representing one standard deviation. (Note: 2005 values are superimposed, but some months have missing data.) PVC = polyvinyl chloride.

mountain peaks surrounding a proglacial valley and (2) identify mechanisms coupling (or decoupling) the valley atmosphere with the overlying free atmosphere, particularly during the wet season. We study the impact of LRs, derived from observations and climate modeling, on freezing level and cloud development, relating it to observed diurnal precipitation patterns during the wet season. We compare sample periods of wet and dry seasons and evaluate the potential for strong stratification between the valley atmosphere and the overlying free air.

Study Area

We study the Llanganuco Valley (LV), a hanging glacier valley in the CB, with about 33 percent glacierized area that drains southwest to the Santa River (Figure 1). Glaciers terminate at a mean elevation of ~4,850 m (all elevations are above mean sea level), with some reaching as low as 4,450 m (Juen, Kaser, and Georges 2007). The valley lost 19.5 percent of glacier extent between 1987 and 2010 (Burns and Nolin 2014). The climate in the CB is semiarid in the valleys and moist in higher elevations with a distinct rainy season between October and April and dry the remaining months (Kaser and Osmaston 2002). Average annual precipitation in Llanganuco ranges from 8 mm ($\sigma = 14$) for June, July, and August of the dry season to 258 mm ($\sigma = 101$) for December, January, and February of the wet season, based on the 1953 to 2010 monthly totals near the lower lake and our 2004 to 2015 air temperature record as depicted by Figure 1D. This climograph illustrates a relatively unchanging annual temperature range and strong seasonal cycle of precipitation with a maximum in December 2005, superimposed with significant interannual variability.

Data and Methods

ASN Measurements

In July 2004, we installed an automatic weather station (AWS; www.onsetcomp.com/HOBO®) near the lower lake in the LV (3,850 m) in collaboration with the Peruvian Institute of Natural Resources, Division of Glaciology and Water Resources, and the Huascaran National Park and UNESCO Biosphere Reserve. The AWS records hourly averages of 10-second samples of air temperature ($\pm0.2°C$), wind speed (±1 m s^{-1}) and direction ($\pm5°$), relative humidity (±2.5 percent), solar radiation (±1.3 W m^{-2}), and precipitation (±0.2 mm). We synthesized all hourly ASN measurements into composite diurnal cycles for twenty-four-day periods representative of the wet (7 to 31 December 2005) and dry (17 June to 11 July 2005) seasons (Figure 1D). These equal-sized intervals contain complete and continuous data sets with sufficiently contrasted humidity and precipitation. Although longer sample periods for each season were desirable, gaps in data prevented inclusion.

We extended near-surface temperature observations to measure near-surface LRs spanning ~1,300 m elevation along the LV axis in June 2005 by installing a cost-effective ASN consisting of eight temperature loggers (iButton Thermochron®, Embedded Data Systems, Lawrenceburg, KY, USA) inconspicuously placed 2 m above the ground in shaded branches of *Polylepis* trees to minimize radiation error and avert possible theft. We embedded the loggers along the main access road for the park at 150 m elevation intervals, transecting the valley axis from the entrance at 3,469 m upward to the Portachuelo pass at 4,742 m (Figure 1B). Each iButton, approximately the size of a U.S. nickel, is powered by an internal battery and can store up to 8,192 measurements on flash memory. The reported iButton resolution is 0.5°C, and accuracy is $\pm1°C$ with a range of $-40°C$ to $+85°C$. Each iButton logger was placed inside a specially designed 2.54-cm inner diameter polyvinyl chloride (PVC) shield (inset, Figure 1B) and programmed to sample air temperature every hour. One iButton was calibrated under actual field conditions

against the shielded air temperature from the AWS, and the R^2 value of the linear regression was in excess of 0.99 with an offset of less than 0.2°C, well within the tolerance of both the iButton and AWS temperature sensors. We expected some impact of radiative warming of the iButtons, particularly during dry season in the afternoon at one of the higher elevation sites (anomalously high temperature at 4,559 m; Figures 2E and 2F), with inconsistent shading under a sparse vegetation canopy. iButton data were recovered bimonthly on site.

We create composite averages of twenty-four hourly measurements for all twenty-four days representing the wet and dry periods—the diurnal cycle. We derive near-surface LRs by calculating the average temperature for each hour during the twenty-four-day period at each elevation and then calculating composite three-hourly intervals from 00 to 21 hours (using local standard time). We applied a linear regression to estimate the lapse rate according to the composite iButton temperatures versus elevation. LR values are positive by convention with temperature decreasing with increasing elevation.

Climate Model Simulations

To develop a mechanistic explanation of the features revealed by the ASN, we performed a dynamical downscaling of reanalysis data for the year 2005 using the Advanced Research Weather–Research and Forecasting (ARW–WRF, henceforth WRF) modeling system (Skamarock et al. 2008) version 3.4.1. WRF is a numerical model developed for weather prediction and atmospheric simulation, widely employed in climatic research and operational meteorology. WRF includes a dynamical solver that integrates the compressible, nonhydrostatic Euler equations. The spatial discretization is composed of an Arakawa C-grid in the horizontal and a terrain-following mass coordinate system for the vertical. The simulation was configured with three domains (Figure 1C). The parent domain extended from south of Colombia to north of Chile (120 × 120 cells, 30 km spatial resolution), with a nested, intermediate domain covering most of central and north Peru (201 × 201 cells, 6-km spatial resolution), and the smallest domain centered on the CB covers a section of north-central Peru at 2-km (150 × 150 cells) spatial resolution. This represents the highest spatial resolution that WRF has been applied to date over the tropical Andes, bearing in mind that the

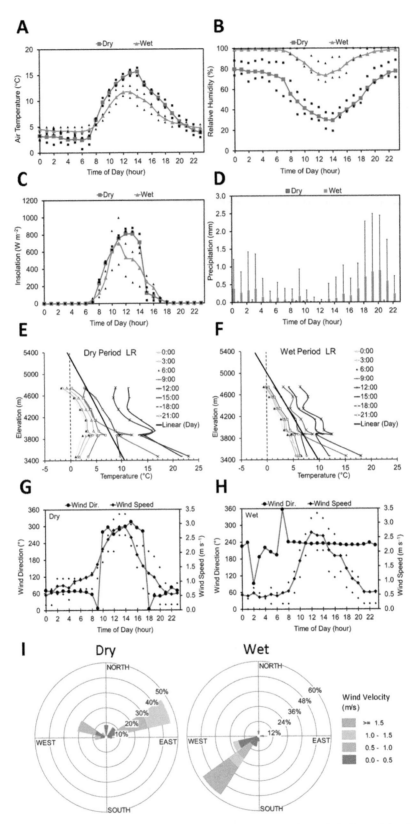

Figure 2. Composite diurnal cycles as recorded by the automatic weather station at the base of the lower lake. Comparisons for the dry and wet seasons include (A) air temperature, (B) relative humidity, (C) incoming global solar radiation, and (D) precipitation. Day-to-day variation is shown by the first and third quartiles (25th and 75th percentile) and is designated by small symbols above or below the mean. (E) Dry and (F) wet season diurnal and average near-surface lapse rates. (G) Dry and (H) wet season wind speed and direction and (I) the wind roses. (Color figure available online.)

Table 1. Configuration of the Advanced Research Weather–Research and Forecasting simulation

Parameterization	Scheme
Cloud microphysics	WSM 6-Class
Longwave radiation	NCAR Community Atmosphere Model
Shortwave radiation	NCAR Community Atmosphere Model
Surface layer	MM5 similarity theory
Land surface model	Noah
Planetary boundary layer	Yonsei University

comparison of the point measurements with grid values introduces inherent uncertainty over the complex terrain geometry (Horvath et al. 2012). The parameterizations selected for cloud microphysics, shortwave radiation, longwave radiation, and surface and boundary layer processes are shown in Table 1. Boundary conditions for the model came from the Climate Forecast System Reanalysis (Saha et al. 2010). Reanalysis fields bounding the largest domain forced the WRF model every six hours.

The output variables we analyzed from WRF included wind speed and direction, vertical wind (positive upward), cloud cover, air temperature, and freezing level every three hours starting at 00 hours each day. For the vertical component of the wind and for cloud cover, we constructed profiles from the Cordillera Negra (to the west of the CB) to the eastern slope of the CB following the main axis of the LV (section

A-A′ in Figure 1A). In the analysis of horizontal wind, we constructed wind maps at two geopotential levels: ~665 mb (~3,500 m following the terrain of the valley approximately 50 to 100 m above ground level) and ~292 mb (~9,000 m). Finally, for temperature we studied the patterns of LRs by determining this gradient using two different approaches: first, from twenty-four-day wet and dry composites of the near-surface temperature of the cells in which the ASN instruments are located and, second, as an average free-air (vertical column) twenty-four-day composite LR following the main axis of the valley, from the mouth at ~720 mb to the overlying air at ~447 mb (~6,000 m). We acknowledge that freezing levels were estimated from near-surface LRs. All analyses considered the same composite period denoted in the previous section.

With respect to precipitation and cloud development, our intention was to simulate timing and location, not the magnitude or type during the wet season. WRF is particularly skilled in this regard when simulating at high spatial resolution, as concluded by Mourre et al. (2016) using a 3-km grid spacing in the smallest domain over the CB.

Results

There are marked contrasts in all variables between wet and dry seasons (Table 2) for the AWS, as shown by the mean and standard deviations of day-to-day

Table 2. Composite daily meteorological forcing for dry and wet seasons for the automatic weather station and the iButton air temperature sensor network

AWS Units below	Dry	Wet	Wet/dry	iButton elevation (m)	Dry (°C)	Wet (°C)	Wet/dry
Air temperature (°C)	7.6 (0.8)	6.6 (0.7)	0.87	4,742	2.84 (0.80)	1.77 (0.78)	0.62
Maximum temperature (°C)	16.0 (1.2)	12.7 (2.2)	0.79	4,559	4.39 (0.80)	3.06 (0.78)	0.70
Minimum temperature (°C)	1.5 (1.2)	3.6 (1.0)	2.40	4,344	5.31 (0.82)	3.56 (0.71)	0.67
Temperature range (°C)	14.5	9.1	0.63	4,148	6.68 (0.79)	5.66 (1.79)	0.85
Relative humidity (%)	59.5 (12.2)	92.4 (8.3)	1.55	3,948	8.12 (0.68)	5.48 (0.94)	0.67
Vapor pressure (kPa)	0.62 (0.12)	0.90 (0.06)	1.45	3,871	8.59 (0.76)	7.45 (0.79)	0.87
Insolation (MJ m^{-2})	16.8 (1.8)	13.3 (4.9)	0.80	3,850	7.49 (1.09)	6.24 (0.95)	0.83
Wind (m s^{-1})	1.42 (0.51)	1.18 (0.51)	0.83	3,469	9.49 (0.58)	9.46 (0.95)	1.00
Wind direction (°)	39	233	N/A				
Precipitation (mm day^{-1})	0.08 (0.09)	7.36 (8.15)	92.0	Lapse rate (°C km^{-1})	5.3 (1.1)	5.8 (0.5)	1.09
				0°C elevation (m)	5,302 (223)	5,010 (118)	0.94

Note: The wet/dry ratio shows the seasonal difference. Standard deviations are shown in parentheses. AWS = automatic weather station.

variations. Comparing the dry season to the wet season reveals a greater range but similar air temperature trend for the dry season (Figure 2A) and surprisingly high insolation during the wet season (Figure 2C). Rainfall is nearly absent during the dry season and strongly nocturnal for the wet season (Figure 2D).

Lapse Rates

Figures 2E and 2F show the diurnal cycle of near-surface LRs as measured by the iButtons of the ASN at different elevations for the composite wet and dry seasons, and Figure 3A compares the diurnal composite LRs with the WRF output. Near-surface LRs for the ASN and WRF are steeper (greater) during the dry season, ASN = 9.05 and WRF = 7.45°C km^{-1}, compared to the wet season, ASN = 6.03 and WRF = 5.81°C km^{-1}. The WRF LRs are comparable to the average of near-surface LRs calculated from ASN temperatures (Figure 3A) for the wet season, but the ASN LR is significantly larger than the WRF LR for the dry season. We attribute this difference to the spatial distribution of the observations in relation to the 2-km spatial resolution of the model; WRF LR is essentially based on air temperature of grid squares at two elevations, 3,747 and about 4,750 m. In part because of less accurate temperature representation in steeper terrain, where five iButtons are clustered in one 2-km grid square, WRF near-surface LRs are lower than those of the ASN (Figure 3A), particularly during times of solar heating, as shown in Figure 2C. Figure 3A shows significant underestimation by WRF from 09 to 15 hours and overestimation from 18 to 00 hours, although the daily composite average LR of WRF and the ASN for the wet season are comparable. WRF simulations did not detect the impact of insolation at the surface and the turbulent sensible heat flux that warms the near-surface valley air measured by the ASN. The freezing level calculated from averaging the composite three-hourly values (Table 3) is lower during the dry season, ASN = 4,817 and WRF = 4,766, compared to the wet season, ASN = 4,913 and WRF = 4,792 (all values in meters above mean sea level). The wet season diurnal range of the freezing level is 734 m for the ASN and 196 m for WRF projections. The seasonal and particularly the diurnal WRF simulations of near-surface LRs and freezing levels are less than those from the ASN.

With regard to the wet season, the free-air WRF LRs calculated from the main axis of the valley indicate a difference in the thermal structure of the valley atmosphere versus the free atmosphere above the CB (Figure 3B). First, the WRF LRs of the two levels closest to the ground show a somewhat opposite behavior with respect to ASN near-surface temperature; that is, increasing LRs between 12 and 18 hours, whereas ASN LRs peak at 12 hours, decreasing thereafter (Figure 3A). Second, the LRs get increasingly steeper (greater) at higher altitudes, suggesting less moisture content in higher layers or that most moisture is concentrated in the lower layers. Subsidence warming, created by downward-directed easterly free air flow into the layer of air below 588 mb, might also contribute to the steeper LRs in the top two layers (Figure 3B), as confirmed by slightly negative vertical wind in the upper portion of the valley and relatively stronger rising air in the slope air layer over the valley floor (Figure 4A, 12–18 hours). Third, there is an opposite daily cycle above and below the highest mountain tops in the study area (~588 mb or ~5,500 m), whereas below that elevation LRs increase from a minimum at 21 hours to a maximum at 15 to 18 hours and in the sections above ~588 mb the minimum occurs between 12 and 18 hours. This pattern differs from the dry season, with relatively higher LRs above the lowest layer under much drier conditions but sharing a similar diurnal cycle at all levels. Specifically, the minimum at around 18 hours appears across all levels, whereas the maximum occurs at 21 hours above ~588 mb (Figure 3C). During the wet season it is much more likely that air parcels will be saturated and hence, if forced to rise above the ~588 mb, will cool at the moist adiabatic LR of 6.0 to 6.5°C km^{-1}, which is significantly less than the 8.0 to 8.25°C km^{-1} in the layer of air above the valley and consequently will destabilize the rising air and enhance convection.

Dynamical Forcing and Wet Season Convection

We observe seasonal contrasts in both near-surface wind speed and direction (Figures 2G, 2H, and 2I). Southwest winds, directed up the valley during daylight hours of the wet season (Figure 2H), correspond to strong insolation (Figure 2C) from 09 hours to 15 hours. On the other hand, the wind direction during the dry season does not show the direct up-valley winds that dominate the wet season (Figures 2G, 2H, and 2I); rather, the stronger wind flows from the west and northwest, plausibly from a curvature-induced secondary circulation as described by Weigel and Rotach

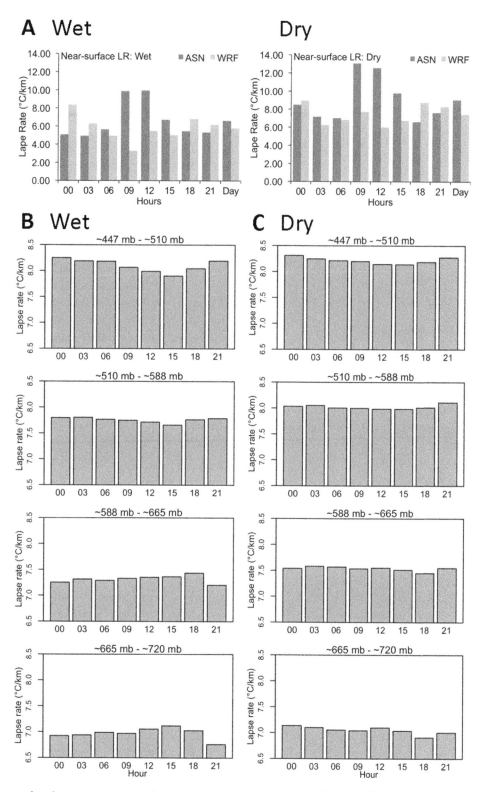

Figure 3. (A) Near-surface lapse rates wet and dry season comparison: Advanced Research Weather–Research and Forecasting versus autonomous sensor network. (B) Simulated vertical free-air lapse rates averaged along the main axis of the valley during the wet season according to four layers delineated by geopotential levels. (C) Same as (B) for dry season. LR = lapse rate; ASN = autonomous sensor network; work; WRF = Advanced Research Weather–Research and Forecasting.

Table 3. Freezing level (m) from the autonomous sensor network and Advanced Research Weather–Research and Forecasting model at every output hour according to grid squares containing the ASN locations

Hour	Dry ASN	Dry WRF	Wet ASN	Wet WRF
00 hours	4,457	4,764	4,760	4,768
03 hours	4,402	4,716	4,714	4,724
06 hours	4,705	4,703	4,612	4,732
09 hours	4,357	4,685	4,843	4,740
12 hours	5,168	4,694	5,155	4,728
15 hours	5,559	4,793	5,346	4,808
18 hours	5,238	4,896	5,053	4,920
21 hours	4,648	4,880	4,822	4,912
M	4,817	4,766	4,913	4,792
Range	1,202	210.9	734	196.1
σ	448	84	249	82

Note. ASN = autonomous sensor network; WRF = Advanced Research Weather–Research and Forecasting.

(2004), but further analysis is outside the scope of this study. After sunset during the dry season, consistent low-speed down-valley winds are probably caused by nocturnal cold air drainage under cloudless skies and partially from forced channeling of prevailing easterly flow aloft as we found from WRF model results for the dry season (not shown in this article). On the other hand, nocturnal down-valley winds are largely undetected during the wet season (Figures 2H and 2I) when clouds enhance downward infrared radiation, as indicated by higher minimum temperatures (Table 2) than the dry season. This agrees with the results of Bianco et al. (2006) showing weak and sporadic nocturnal components of thermally driven winds.

In the analysis of modeled wind patterns, we focus on the wet season to address the observed nocturnal peak of precipitation (Figures 2G and 2I). We acknowledge the limitations of WRF in complex terrain (Jiménez and Dudhia 2013), particularly with respect to uncertainty of wind direction of near-surface winds at particular grid locations. At the free atmosphere level (~292 mb, Figure 5A), the wind vectors indicate divergent flow above the valley atmosphere and a strong southwestward flow through the mouth of the valley along the west slope of the CB, which could be explained by the valley-exit jet (Chrust, Whiteman, and Hoch 2013) typically maximized from 50 to 200 m above the ground and is within the surface wind confines of WRF. Along the valley axis at ~665 mb, particularly near the mouth of the valley (Figure 5B), wind vectors point down-valley

consistently throughout the day. It is important to note that WRF simulates the surface wind approximately 50 to 100 m above the ground following the terrain with hydrostatic pressure coordinates. The converging downslope flow from the two mountain peaks flanking the valley (Figure 5B) air just east of the mouth of the LV is partially responsible for the ascending motion (positive values) along the valley axis during the afternoon (Figure 4A), predominantly between 15 and 21 hours. During all other hours, descending motion prevails with a speed almost an order of magnitude smaller but sufficient to promote subsidence heating, especially at the mouth of the LV.

The diurnal composites of WRF simulations show evidence of convection through cloud profiles (Figure 4B) for the wet season. Most cloud cover over the valley develops after 15 hours, around the time of maximum valley wind (Figure 2H). Cloud thickness increases significantly at 18 hours above the ridge of the CB and then tends to expand westward and down-valley with a peak from 21 to 00 hours. This pattern indicates that the main source of moisture and cloudiness comes from outside the valley, probably of Amazon origin (Figure 4B).

Discussion

Hourly meteorological measurements from the ASN combined with climate model output for the wet and dry seasons in 2005 reveal strongly diurnal meteorological forcing within the LV. With our emphasis on the wet season, we find observational and model evidence of an up-valley wind and strongly stratified LRs that plausibly enhance the potential for local convection, as indicated by persistent up-valley winds during the wet season and nocturnal cloud development. It is important to note, though, that observations from this single annual cycle should not be compared to multi-year studies, given the significant interannual variability observed in some CB valleys, as evident from the large standard deviation of the 1953 to 2010 monthly rainfall record from Llanganuco (Figure 1D), usually attributed to phases of the El Niño–Southern Oscillation (Vuille, Kaser, and Juen 2008).

We summarize our results in a conceptual model (Figure 6). The persistent wet season features of a nocturnal rainfall maximum, high insolation, and up-valley surface winds from 09 to 18 hours observed by the ASN agree with previous studies.

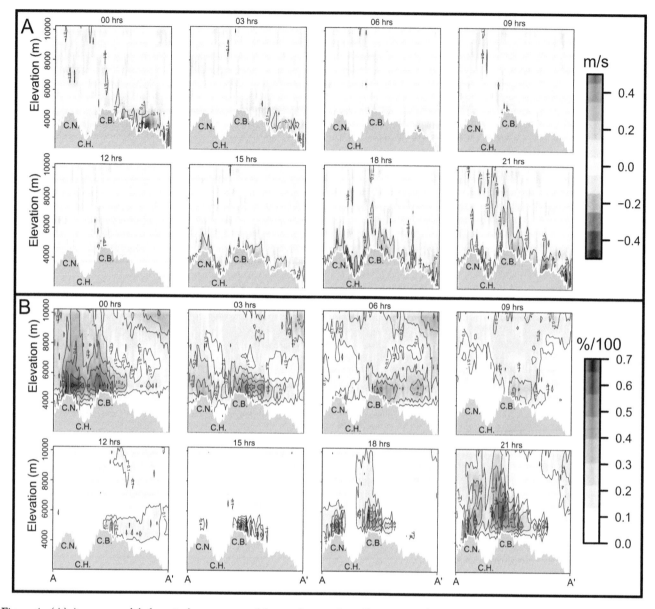

Figure 4. (A) Average modeled vertical component of the wind across the valley axis A-A′ extending from the Cordillera Negra toward the Amazonian (eastern) slope of the Cordillera Blanca. The Llanganuco proglacial valley cuts down the west-facing slope of the Cordillera Blanca. (B) Average modeled cloud cover profile using the same extents as in (A). C.N. = Cordillera Negra; C.B. = Cordillera Blanca; C.H. = Cordillera Huayhuash. (Color figure available online.)

Mapes et al. (2003) observed similar afternoon rainfall maxima over most of South and Central America often caused by relatively small convective cloud systems, and they found a nocturnal maximum of rainfall over some large valleys in the Andes. Bendix, Rollenbeck, and Reudenbach (2006) used K-band rain radar to measure nocturnal and early morning rainfall maxima in a valley with east–west orientation between the southern Ecuadorian Andes and the Amazon basin.

Horizontal and vertical wind during the wet season (Figures 4A, 5A, and 5B) and analysis of vertical

layers of free-air LRs (Figure 3B) suggest decoupling of the stratified valley air and free atmosphere prior to 15 hours when LRs in the valley oppose the behavior of the free atmosphere. We hypothesize that thermally driven down-slope flow from the mountain peaks flanking the valley promotes low-level air convergence, representing a thermally driven mechanism that promotes vertical convection from 18 to 00 hours (Figure 6). Convection is further enhanced by upper air divergence (Figure 5A) over the valley during this time of peak precipitation.

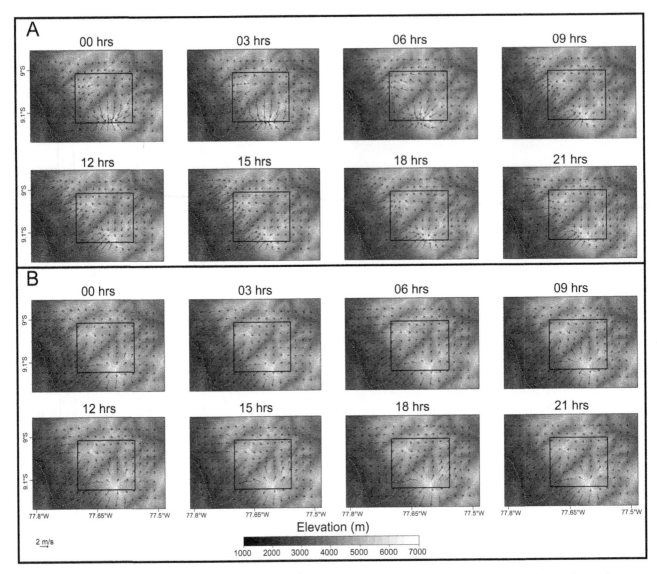

Figure 5. (A) Average modeled horizontal wind vectors (m s^{-1}) indicating direction of airflow overlying the study area during the wet season at ~292 mb (free-air, aloft). (B) Same for ~665 mb (terrain-following about 100 m above ground). Note the elevation shading and locations of the flanking mountain peaks and debris glacier extending north of Huascarán. Elevation data for this figure come from the Shuttle Radar Topographic Mission (Rabus et al. 2003).

Our results suggest that the nocturnal precipitation maximum for the wet season (Figure 2D) could be enhanced by the following factors: (1) strong heating at the mouth of the valley during midday hours as observed by rapid rise in air temperature (Figure 2F) helping drive the valley wind and destabilizing the overlying air column in the later afternoon; (2) simulated persistent down-slope (katabatic) flow off the flanking glacierized peaks leading to convergent flow at about 50 m above ground level in the valley (Figure 5B), some of which rises along the valley axis and diverges aloft while the remainder moves down-valley as an elevated valley-exit jet (Figure 6); and (3) convergence between the valley wind (Figure 5B) and

prevailing easterly free-air flow (Figure 5A) promotes convection (Figure 4A) and partially explains peak cloud development between 21 and 00 hours (Figure 4B). WRF simulations of horizontal wind vectors at ~292 mb (Figure 5B) revealed relatively strong air convergence over the valley from 18 to 00 hours, concurrent with significant cloud production. The diurnal trend of simulated cloud cover over the LV suggests that moisture has an easterly origin, propagates westward after sunset (18 hours), peaks from 19 to 21 hours, and dissipates at sunrise (06 hours), coinciding with precipitation trends observed by the ASN (Figure 2D). This contrasts to the 12 to 17 hours peak and far less diurnal variability of precipitation over

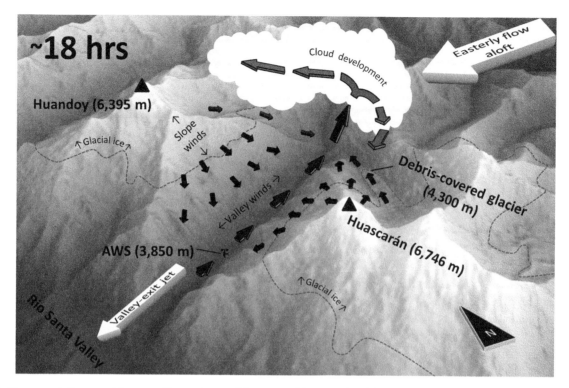

Figure 6. Conceptual diagram with arrows illustrating wind direction of dynamical and thermally driven winds in the Llanganuco Valley at approximately 18 hours. Prevailing easterly flow aloft brings Amazonian moisture and flanking mountain peaks create a persistent thermally driven slope wind that induces convergence near the surface. The valley-exit jet and divergence above the valley atmosphere completes the circulation, which promotes upward convection and cloud development. Note the elevation shading, location of glacier coverage, and location of the weather station. *Source:* ASTER GDEM is a product of METI and NASA (Land Processes Distributed Active Archive Center 2015). AWS = automatic weather station.

the nonmountainous areas of the central and west central Amazon (Angelis, McGregor, and Kidd 2004; Tanaka, Satyamurty, and Machado 2014).

The temperature in the LV tends to be higher than over the lowlands because of the volume effect described by Steinacker (1984) that produces a jump in temperature at the valley's mouth. Figure 2E clearly shows a volume effect jump in temperature at the mouth (~3,400 m) of the LV for both seasons (predominately 09–15 hours), especially in the afternoon with more direct insolation of the southwest downward slope. Southwestward wind vectors in Figure 5B, indicative of a valley-exit jet, likely result from the horizontal pressure gradient acting from the relatively high pressure created by converging downslope flow off Huandoy and Huascarán toward the low pressure created by the warm air in the LV. Our comparisons of measured near-surface versus simulated free-air LRs provide two-dimensional evidence of strong temperature variations that plausibly lead to pressure gradients that create the observed diurnal cycle of the valley wind circulation. Our ASN observations and WRF simulations confirm that

prominent daytime valley winds during the wet season (Figures 2H, 2I, 4A, and 5B) play a critical role in controlling convection.

Implications and Limitations

As demonstrated by Vuille et al. (2003), the western slope of the CB is experiencing a higher rate of warming than the eastern slope based on weather station records with slightly less warming from 1983 to 2012 (Schauwecker et al. 2014). If this warming trend continues, we would expect strengthening of valley winds, particularly during the wet season when they are most prominent (Figures 2H, 4A, and 5). Consequently, this might promote stronger nocturnal convection and raise the freezing level (Table 3) during the wet season due to convergence with moisture-rich easterly synoptic flow, which will reduce solid precipitation and potential for refreezing. We have presented evidence of wet season nocturnal coupling between the air within the LV and the overlying synoptic scale easterly flow. Future work should focus on confirming this mechanism with additional years of data and observations in other valleys where changes in glacial extent have been reported.

We acknowledge limitations of the WRF model with respect to winds and LRs. Even at 2-km resolution, WRF cannot precisely resolve intravalley wind patterns, so we place more emphasis on the general wind pattern of the valley atmosphere versus the free atmosphere during the wet season, which agrees well with the diurnal wind pattern observed at the AWS. We clarify that LR from model output was calculated both for the vertical air column of a 2-km grid square centered in the Valley and that near-surface LR was calculated from two 2-km grid squares, one at about the elevation of the AWS in the lower elevation and one centered at the upper elevation of the LV. Comparing iButton and WRF model LRs has limitations; vertical column LRs are largely created by vertical diffusion of heat, whereas near-surface LRs are closely coupled with the underlying ground. There remain extreme challenges with obtaining observations from and modeling areas of complex terrain as are evident by the limitations to evaluating lapse rates and intravalley winds. We expanded our air-flow analysis beyond intravalley extent to better accommodate the WRF model 2-km spatial resolution. Despite the limitations noted, our approach that combines autonomous sensor observations and climate modeling brings forth valuable insight that advances the understanding of hydrometeorological processes in tropical mountain climates.

Acknowledgments

We are grateful for the collaboration with ANA-UGRH and PNH for permissions and data collection in the National Park.

Funding

Funding for this project was provided by The Ohio State University Climate, Water & Carbon Program, Department of Geography and Office of International Affairs. Additional student involvement and travel was covered under National Science Foundation projects #1010550 and #0752175 (to Brian Greenwood Mark) and the Presidential Fellows Award through Bridgewater State College (Robert Åke Hellström). Alfonso Fernández acknowledges CONICYT Becas-Chile and an American Association of Geographers dissertation research grant.

References

Angelis, C. F., G. R. McGregor, and C. Kidd. 2004. Diurnal cycle of rainfall over the Brazilian Amazon. *Climate Research* 26:139–49.

Bendix, J., R. Rollenbeck, and C. Reudenbach. 2006. Diurnal patterns of rainfall in a tropical Andean valley of southern Ecuador as seen by a vertically pointing K-band Doppler radar. *International Journal of Climatology* 26:829–46.

Bendix, J., K. Trachte, J. Cermak, R. Rollenbeck, and T. Nauß. 2009. Formation of convective clouds at the foothills of the tropical eastern Andes (South Ecuador). *Journal of Applied Meteorology and Climatology* 48:1682–95.

Bianco, L., B. Tomassetti, E. Coppola, A. Fracassi, M. Verdecchia, and G. Visconti. 2006. Thermally driven circulation in a region of complex topography: Comparison of wind-profiling radar measurements and MM5 numerical predictions. *Annales of Geophysicae* 24:1537–49.

Bradley, R. S., F. T. Keimig, H. F. Diaz, and D. R. Hardy. 2009. Recent changes in freezing level heights in the Tropics with implications for the deglacierization of high mountain regions. *Geophysical Research Letters* 36: L17701.

Burns, P., and A. Nolin. 2014. Using atmospherically-corrected Landsat imagery to measure glacier area change in the Cordillera Blanca, Peru from 1987 to 2010. *Remote Sensing of Environment* 140:165–78.

Bury, J., B. G. Mark, M. Carey, K. R. Young, J. M. McKenzie, M. Baraer, A. French, and M. H. Polk. 2013. New geographies of water and climate change in Peru: Coupled natural and social transformations in the Santa River watershed. *Annals of the Association of American Geographers* 103:363–74.

Chrust, M. F., C. D. Whiteman, and S. W. Hoch. 2013. Observations of thermally driven wind jets at the exit of Weber Canyon, Utah. *Journal of Applied Meteorology and Climatology* 52 (5): 1187–1200.

Dai, A. 2001. Global precipitation and thunderstorm frequencies: Part II. Diurnal variations. *Journal of Climate* 14:1112–28.

Francou, B., M. Vuille, P. Wagnon, J. Mendoza, and J. E. Sicart. 2003. Tropical climate change recorded by a glacier in the central Andes during the last decades of the twentieth century: Chacaltaya, Bolivia, 16 degrees S. *Journal of Geophysical Research-Atmospheres* 108:4154.

Hofer, M., B. Marzeion, and T. Mölg. 2012. Comparing the skill of different reanalyses and their ensembles as predictors for daily air temperature on a glaciated mountain (Peru). *Climate Dynamics* 39:1969–80.

Horvath, K. D., D. Koracin, R. Vellore, J. Jiang, and R. Belu. 2012. Sub-kilometer dynamical downscaling of near-surface winds in complex terrain using WRF and MM5 mesoscale models. *Journal of Geophysical Research* 117:D11111.

Jiménez, P. A., and J. Dudhia. 2013. On the ability of the WRF model to reproduce the surface wind direction over complex terrain. *Journal of Applied Meteorology and Climatology* 52 (7): 1610–17.

Juen, I., G. Kaser, and C. Georges. 2007. Modelling observed and future runoff from a glacierized tropical catchment (Cordillera Blanca, Peru). *Global & Planetary Change* 59:37–48.

Kaser, G., and H. A. Osmaston. 2002. *Tropical glaciers.* Cambridge, UK: Cambridge University Press.

Land Processes Distributed Active Archive Center (LP DAAC). 2015. ASTER global digital elevation model. http://lpdaac.usgs.gov/dataset_discovery/aster/aster_products_table/astgtm (last accessed 9 November 2015).

Mapes, B. E., T. T. Warner, M. Xu, and A. J. Negri. 2003. Diurnal patterns of rainfall in northwestern South America: Part I. Observations and context. *Monthly Weather Review* 131:799–812.

Mark, B. G., J. Bury, J. M. McKenzie, A. French, and M. Baraer. 2010. Climate change and tropical Andean glacier recession: Evaluating hydrologic changes and livelihood vulnerability in the Cordillera Blanca, Peru. *Annals of the Association of American Geographers* 100:794–805.

Mark, B. G., J. M. McKenzie, and J. Gomez. 2005. Hydrochemical evaluation of changing glacier meltwater contribution to stream discharge: Callejon de Huaylas, Peru. *Hydrological Sciences Journal-Journal Des Sciences Hydrologiques* 50:975–87.

Mölg, T., and G. Kaser. 2011. A new approach to resolving climate–cryosphere relations: Downscaling climate dynamics to glacier-scale mass and energy balance without statistical scale linking. *Journal of Geophysical Research: Atmospheres* 116 (D16): 101.

Mourre, L., T. Condom, C. Junquas, T. Lebell, J. E. Sicart, R. Figueroa, and A. Cochachin. 2016. Spatio-temporal assessment of WRF, TRMM and in situ precipitation data in tropical mountain environment (Cordillera Blanca, Peru). *Hydrology and Earth System Sciences* 20:125–41.

Rabus, B., M. Eineder, A. Roth, and R. Bamler. 2003. The shuttle radar topography mission—A new class of digital elevation models acquired by spaceborne radar, ISPRS J. *Photogrammetry and Remote Sensing* 57 (4): 241–62.

Saha, S., S. Moorthi, H.-L. Pan, X. Wu, J. Wang, S. Nadiga, P. Tripp, et al. 2010. The NCEP climate forecast system reanalysis. *Bulletin of the American Meteorological Society* 91 (8): 1015–57.

Schauwecker, S., M. Rohrer, D. Acuña, A. Cochachin, L. Dávila, H. Frey, C. Giráldez, et al. 2014. Climate trends and glacier retreat in the Cordillera Blanca, Peru, revisited. *Global and Planetary Change* 119:85–97.

Skamarock, W. C., J. B. Klemp, J. Dudhia, D. O. Gill, D. M. Barker, M. G. Duda, X.-Y. Huang, W. Wang, and J. G. Powers. 2008. *A description of the Advanced Research WRF version 3.* Boulder, CO: National Center for Atmospheric Research.

Steinacker, R. 1984. Area–height distribution of a valley and its relation to the valley wind. *Contributions to Atmospheric Physics* 57:64–71.

Tanaka, L. M., P. Satyamurty, and L. A. T. Machado. 2014. Diurnal variation of precipitation in central Amazon Basin. *International Journal of Climatology* 34 (13): 3574–84.

Vergeiner, I. 1987. An elementary valley wind model. *Meteorology and Atmospheric Physics* 36:255–63.

Vuille, M., R. S. Bradley, M. Werner, and F. Keimig. 2003. 20th century climate change in the tropical Andes: Observations and model results. *Climatic Change* 59:75–99.

Vuille, M., B. Francou, P. Wagnon, I. Juen, G. Kaser, B. G. Mark, and R. S. Bradley. 2008. Climate change and tropical Andean glaciers: Past, present and future. *Earth-Science Reviews* 89:79–96.

Vuille, M., G. Kaser, and I. Juen. 2008. Glacier mass balance variability in the Cordillera Blanca, Peru and its relationship with climate and the large-scale circulation. *Global and Planetary Change* 62:14–28.

Wagnon, P., P. Ribstein, B. Francou, and B. Pouyaud. 1999. Annual cycle of energy balance of Zongo Glacier, Cordillera Real, Bolivia. *Journal of Geophysical Research–Atmospheres* 104:3907–23.

Weigel, A. P., and M. W. Rotach. 2004. Flow structure and turbulence characteristics of the daytime atmosphere in a steep and narrow Alpine valley. *Quarterly Journal of the Royal Meteorological Society* 130:2605–27.

Whiteman, C. D. 2000. *Mountain meteorology: Fundamentals and applications.* New York: Oxford University Press.

How Rivers Get Across Mountains: Transverse Drainages

Phillip H. Larson, Norman Meek, John Douglass, Ronald I. Dorn, and Yeong Bae Seong

Although mountains represent a barrier to the flow of liquid water across our planet and an Earth of impenetrable mountains would have produced a very different geography, many rivers do cross mountain ranges. These transverse drainages cross mountains through one of four general mechanisms: *antecedence*—the river maintains its course during mountain building (orogeny); *superimposition*—a river erodes across buried bedrock atop erodible sediment or sedimentary rock, providing a route across what later becomes an exhumed mountain range; *piracy or capture*—where a steeper gradient path captures a lower gradient drainage across a low relief interfluve; and *overflow*—a basin fills with sediment and water, ultimately breaching the lowest sill to create a new river. This article reviews research that aids in identifying the mechanism responsible for a transverse drainage, notes a major misconception about the power of headward eroding streams that has dogged scholarship, and examines the transverse drainage at the Grand Canyon in Arizona.

尽管山岳对于横越地球的水流带来阻碍，且若地球充满无法穿越的山岳的话将会产生非常不同的地理，但诸多河流的确横跨了山脉。这些横断水系透过四个普遍的机制之一穿越山脉：前行——河流在山岳形成时（造山运动）维持其河道；叠加——河流侵蚀越过在可侵蚀的沉积物或沉积岩上被深埋的基岩，提供越过日后成为崛生山脉之路径；袭夺或捕集——其中一个更为陡峭的倾斜路径捕集了横越低地势起伏的河间地块的较低倾斜度之水系；以及溢流——一个充满沉积与水的流域最终破坏了最低的岩床，创造了新的河流。本文回顾协助指认造成横断水系的机制之研究，指明长期纠缠学术界的有关向源侵蚀河流力量的重大错误概念，并检视亚利桑那州大峡谷中的横断水系。 关键词： 前行，溢流，袭夺，叠加，横断水系。

Aunque las montañas representan una barrera contra el flujo del agua líquida a través del planeta y una Tierra de montañas impenetrables habría producido una geografía muy diferente, la verdad es que muchos ríos se abren camino a través de cadenas montañosas. Estos sistemas de avenamiento transversal logran cruzar las montañas por medio de uno de cuatro mecanismos generales: *antecedencia*—el río conserva su curso durante el proceso de construcción de las montañas (orogenia); *superimposición*—un río erosiona al través de la roca madre sepultada sobre un sedimento erosionable o de roca sedimentaria, generando una ruta a través de lo que luego se convertirá en una cadena montañosa exhumada; *piratería* o *captura*—donde una trayectoria de gradiente más pronunciado captura un gradiente de avenamiento más bajo a través de un inteﬂuvio de bajo relieve; y *desborde*—una cuenca se llena con sedimento y agua, rompiendo en últimas la estructura inferior para crear un nuevo río. Este artículo revisa la investigación que ayuda a identificar el mecanismo responsable de un avenamiento transversal, hace notar una concepción equivocada acerca del poder erosivo contracorriente que ha sido persistente en la erudición, y examina el avenamiento transverso del Gran Cañón en Arizona.

Anew or young hydrogeomorphic landscape is commonly perceived to have a few likely origins. The new or young landscape is either exposed by deglaciation, altered by volcanism or mass wasting processes, or has previously been hydrologically isolated by active tectonics. In North America the characteristics of young landscapes often mark the boundaries between major physiographic regions (e.g., Powell 1896; Fenneman 1931, 1938; Hunt 1967; Graf 1987). Often overlooked in this general physiographic perspective of landscapes are a suite of processes that result in new hydrogeomorphic systems via transverse drainage development (Douglass and Schmeeckle 2007). Transverse drainages are river pathways counterintuitive to the perception that rivers commonly originate in mountain ranges rather than pass completely through them. Ultimately, a transverse drainage connects two or more landscapes by cutting through a mountain range. This article explores specific processes that result in transverse drainages; we then review modeling and criteria to identify the process. Finally, we conclude with a case study highlighting the most visited transverse drainage in the world, the Colorado River at Grand Canyon.

Transverse Drainage Mechanisms

Prior scholarship includes four mechanisms to explain the origin of transverse drainages: antecedence, superimposition, overflow, and piracy or capture (Figure 1). These mechanisms are evoked in isolation (e.g., Stokes and Mather 2003; House et al. 2005) or in combination (e.g., Harvey and Wells 1987; Alvarez 1999; Mayer et al. 2003).

The *antecedence* hypothesis requires that a river existed prior to orogeny that then incises through uplifting mountains. Thus, antecedent streams commonly occur in regions of active tectonism and volcanism. Evidence must exist indicating that the river predates the mountain range. Examples of antecedence include the Columbia River gorge through the Cascade Range (Douglass et al. 2009), drainages exiting the Himalaya (e.g., Wager 1937; Searle and Treloar 1993; Lang and Huntington 2014), drainages of the Umbria–Marchean Apennines (Alvarez 1999), streams exiting the High Atlas Mountains of Morocco (Stokes et al. 2008), rivers of northern Peloponnesus in Greece (Zelilidis 2000), the Hsiukuluan River of eastern Taiwan (Lundberg and Dorsey 1990), and the Santa Ana River in Southern California (Dudley 1936).

The *superimposition* hypothesis posits that mountainous terrain was once buried by unconsolidated or easily erodible material, with the river flowing atop this cover mass. With an increase in stream power, the river first erodes the cover mass and then exhumes the underlying bedrock. Geomorphologists support superimposition when a river crosses a resistant rock layer multiple times or a river flows long distances against the slope of an exhumed bedrock plain or has barbed drainages. The Susquehanna River and others in the northeastern

United States (e.g., W. M. Davis 1909; Johnson 1931a, 1931b), the drainages of the Umbria–Marchean Apennines (Alvarez 1999; Mayer et al. 2003), the Aguas and Feos Rivers of southeastern Spain (Harvey and Wells 1987), streams of the Zagros Mountains of Iran (Oberlander 1965), and the Colorado River across the Marble Platform in Arizona (Babenroth and Strahler 1945) exemplify superimposed drainages.

The *overflow* (or spillover) hypothesis posits that a basin fills with sediment and water. When the water and sediment reach the lowest elevation interfluve, an overflow occurs into an adjacent region. The overflow model has been applied to the lower Colorado River, downstream of the Grand Canyon (House et al. 2005; House, Pearthree, and Perkins 2008); the Mojave River (Meek 1989; Reheis, Miller, and Redwine 2007; Reheis and Redwine 2008); the Salt River of Arizona (Larson et al. 2010; Larson et al. 2014); the Ebro River in Spain (Arche, Evans, and Clavell 2010); the Hutuo River in China (Ren et al. 2014); the Kashmir Valley (Ganjoo 2014); Rio Mimbres and Rio Grande Rivers of the Rio Grande Rift (Mack, Love, and Seager 1997); the Amargosa River (Morrison 1991; Menges and Anderson 2005); and elsewhere.

The *stream piracy* (capture) hypothesis requires a stream to divert its course to flow down a new drainage path that is steeper. Douglass and Schmeeckle (2007) modeled four possible processes that result in piracy (capture): headward erosion, channel aggradation, sapping, and lateral erosion. Locations where piracy has recently been discussed include the Rio Almanzora crossing the Sierra Almagro in Spain (Stokes and Mather 2003), in California's coastal range (Garcia and Stokes 2003; Garcia 2006), the Cahabón River in Guatemala (Brocard et al. 2012), the Osip-cheon River in Korea (Lee et al. 2011), and the Apennines of central Italy (Mayer et al. 2003).

Results of Investigating Biases

Some transverse drainage scholarship contains bias favoring the idea that headward erosion of streams commonly results in stream piracy or capture. In reality, headward erosion is a slow process, even in the softest sediments (Douglass and Schmeeckle 2007). Meek (2009) argued that piracy (via headward erosion) is likely where mass movement processes are active on the topographic barrier separating rivers, similar to explanations of the San Lorenzo River and Pancho Creek, California (Garcia and Stokes 2003).

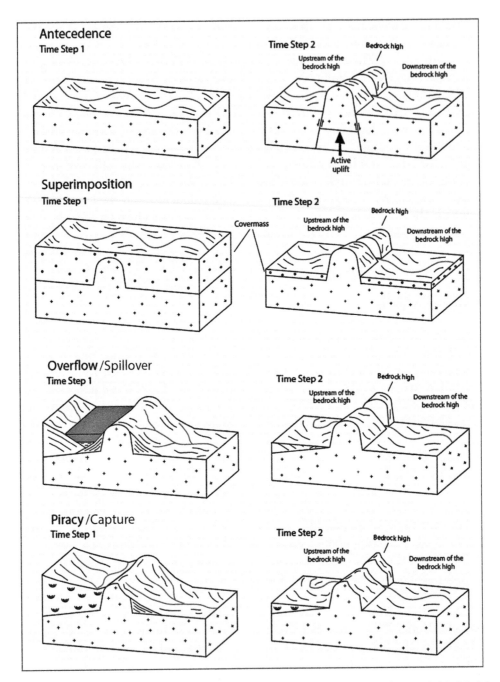

Figure 1. Transverse drainage mechanisms of antecedence, superimposition, overflow and piracy (capture). Modified from Douglass et al. (2009); used with permission.

Both modeling (Douglass and Schmeeckle 2007) and field observations (e.g., Figure 2) reveal that piracy can occur through channel aggradation, sapping undermining the interfluve, or stream lateral erosion removing the topographic barrier—referred to as drainage diversion (Bishop 1995). The lower gradient stream commonly ends up diverting from the top down, into the steeper gradient channel and forming an elbow of capture.

Because piracy or capture via headward erosion has become the prevailing conceptual view for many, Meek (2002) investigated the intellectual upbringing of earth scientists through the treatment of transverse drainage processes in textbooks. The only textbook found discussing overflow was W. M. Davis and Snyder (1898). In 2002, of the nine randomly sampled introductory physical geology texts, eight presented piracy and five antecedence and superimposition. Of the

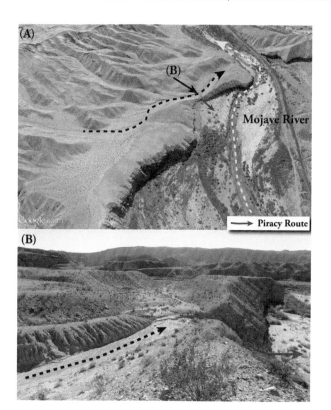

Figure 2. Piracy or capture via drainage diversion (Bishop 1995) occurring near the Mojave River, California, as a result of lateral stream erosion removing the topographic barrier between the streams, 35.03592° N, and 116.37257° W. (A) Oblique Google Earth view of a pirated drainage in development. (B) A photograph at the piracy apex taken by Norman Meek in winter of 2015. (Color figure available online.)

nine introductory physical geography textbooks sampled, only three discussed piracy, antecedence, and superposition. All five geomorphology texts sampled presented piracy, and only four discussed antecedence and superposition. In addition, several textbooks confuse headward erosion with knickpoint recession, leading to the mistaken belief that stream piracy can be caused by vigorous growth of a "precocious gully"—an issue of muddled thinking recognized more than forty-five years ago (Hunt 1969). Thus, it is not surprising that bias is engrained. We agree with Hunt (1969) and Bishop (1995) that piracy is overutilized in academia and comparatively rare in reality.

Criteria to Identify Transverse Drainages and Reduce Bias

Following physical modeling experiments (Douglass and Schmeeckle 2007) and field investigations across the southwestern United States, Douglass et al.

(2009) developed a "decision tree" of criteria to determine the mechanism responsible for a transverse drainage (Figure 3). To show its applicability, we provide an example, the Salt River of south central Arizona. The Salt River forms a bedrock canyon that connects the Tonto basin with the lower Verde River basin and Higley and Paradise basins near Phoenix, Arizona.

Figure 3 plots evidence from the Salt River where Douglass et al. (2009) determined that overflow was likely responsible for this transverse drainage. The process begins by investigating whether the drainage is older or younger than the mountain(s) it crosses. Recent studies (Larson et al. 2010; Larson et al. 2014) and a recent dissertation (Larson 2013) revealed that the Salt River significantly postdates the age of the Mazatzal Mountains the river now crosses. Larson et al. (2010) also identified a new, topographically higher river terrace that contains sedimentary evidence indicating a shifting provenance of the Salt River through time, from local to more distant sources. This shifting provenance represents establishment of the transverse drainage and adjustment of the upstream Tonto basin in response to integration. Earlier, Laney and Hahn (1986) recorded subsurface evidence in deposits under eastern metropolitan Phoenix for the sudden arrival of ancestral Salt River gravels. Larson et al. (2010) concluded that this terrace, the existence of a Pliocene lake in the Tonto basin (Peirce 1984), and the striking similarities to the sedimentologic sequences along the lower Colorado (House et al. 2005; House, Pearthree, and Perkins 2008) support an overflow origin. Despite this evidence, headward erosion was again used to explain the bedrock canyon passages in the Gila River drainage—of which the Salt River is a tributary.

> Bedrock canyon passages (BCP) form where headward erosion from a lowstanding basin breaches the bedrock divide forming a barrier separating the basin from a highstanding basin, commonly leading to sediment aggradation on the lowstanding basin floor and to dissection of the highstanding basin floor. A bedrock canyon passage might also logically arise from erosion downward into bedrock from the level of a sediment ramp transiting the barrier range. (Dickinson 2015, 7)

Unfortunately, the discussion of the Salt River in Dickinson's manuscript did not reference the most recent work or the physical models of Douglass and Schmeeckle (2007). Discussions on the terraces along the Salt did not include the new, higher terrace and

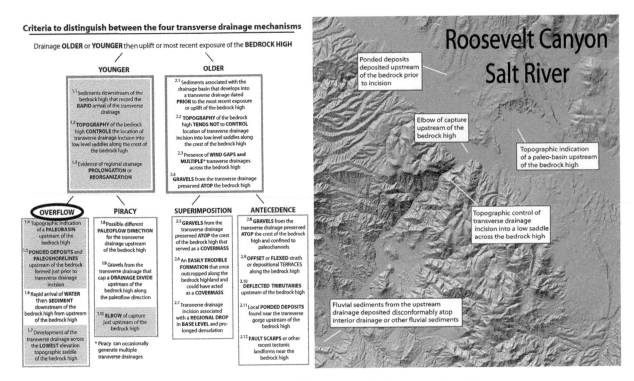

Figure 3. Using the criteria to determine the mechanism responsible for Roosevelt Canyon on the Salt River, Arizona.

did not discuss the criteria-based decision tree or conclusions of Douglass et al. (2009). Out of the three transverse mechanisms suggested by Dickinson in the Gila River drainage, two involved headward erosion of some form (bedrock canyon passages and alluviated gaps), and the last (spillover ramp passages) was discredited by using an argument frequently used to reject overflow:

> But no instances of that behavior have been detected within the Gila River drainage. Nor is there any expectation in the arid to semiarid environment of the Gila River that the surface of any lake occupying a basin of interior drainage ever rose to the level required to overtop a barrier range and initiate the erosion of a bedrock canyon by an outlet stream. (Dickinson 2015, 7)

The misconception invoked is that present-day elevation of lacustrine sediments in a former lake basin indicates the elevation of a paleo-lake surface. In the case of the Salt River, it is likely that adjustment to integration and subsequent erosion of prior lake sediments in the Tonto basin has been ongoing for hundreds of thousands of years, based on cosmogenic nuclide dating of postintegration terraces downstream (Larson et al. forthcoming). In addition, periods of cooler and wetter climatic (pluvial) conditions in the southwestern United States have been well documented throughout

the Pleistocene. Thus, the perception that this region has always been too arid for water to accumulate in a basin and spill over is an unsupported assessment of these transverse systems. Further research will likely investigate linkages between transverse drainage development and paleoclimatic variability.

Grand Canyon

The Grand Canyon is the most visited transverse drainage in the world, as more than 4 million visitors entered Grand Canyon National Park each of the past twenty years to view where the Colorado River cuts through the Kaibab Plateau (Figure 4). One of the longest academic debates in earth science concerns the origin of the Grand Canyon of the Colorado River, which began with the hypothesis of lake overflow (Newberry 1861, 1862). After hiking to the Colorado River in the western Grand Canyon, Newberry documented the existence of lake clays now called the Bidahochi formation along the Little Colorado River. He proposed lake overflow to explain the formation of the Grand Canyon, an idea later elaborated (Blackelder 1934; Gross et al. 2001; Meek and Douglass 2001; Spencer, Smith, and Dowling 2008; Douglass et al. 2009).

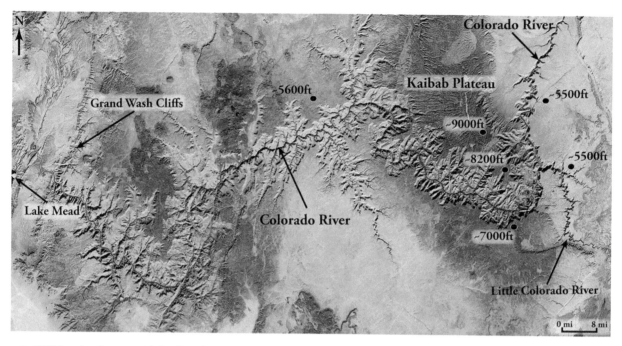

Figure 4. 2013 Landsat 8 mosaic of the Grand Canyon, Arizona. At Grand Canyon the Colorado River becomes a transverse drainage as it crosses the uplifted range of the Kaibab Plateau and exits the Grand Canyon at the Grand Wash Cliffs. *Source:* Modified mosaic from U.S. Geological Survey, http://landsat.usgs.gov/images/gallery/317_L.jpg. (Color figure available online.)

To evaluate Newberry's idea, lake overflow can be ruled out if the river's course is older than the mountains it crosses, and this is not the case. The Kaibab Plateau uplifted during the Laramide Orogeny between ~60 and 40 million years ago (Ma; Yonkee and Weil 2015)—whereas the Colorado River reached its exit from the Grand Canyon at ~5–6 Ma (Roskowski et al. 2010; Spencer et al. 2013; Pearthree and House 2014) and brought in sediment characteristic of the Colorado River to the lower Colorado River basin (Kimbrough et al. 2015).

An age for the Grand Canyon younger than ~6 Ma is consistent with a basic geomorphic understanding of the behavior of rivers (Karlstrom et al. 2008; Darling and Whipple 2015) and with the paleogeography and stratigraphy of the area (Young and Crow 2014). Thus, modern proponents of a Grand Canyon older than 6 Ma (Polyak, Hill, and Asmerom 2008; Wernicke 2011; Sears 2013), including those still supporting Powell's (1875) preference for antecedence, must grapple with overwhelming evidence for a youthful age for the Colorado River's course through the Laramide uplifted Kaibab Plateau.

A younger age for the river also rules out superimposition as considered by Dutton (1882), because an ancestral Colorado could not have flowed on top of slowly eroding Mesozoic layers that were upwarped more than 35 Ma before the canyon was born. The broad curve of the Colorado River to the south might involve some superimposition from an erosional scarp retreating in a circular fashion off the Kaibab Plateau (Strahler 1948; Lucchitta 1984; Douglass 1999)—it is important to note that in our own physical modeling tests we could not replicate what we see at the Grand Canyon using superimposition in this way.

A young age for the Colorado River could be explained by stream piracy first discussed by McKee et al. (1967) and expanded on using groundwater sapping (Hunt 1969; Hill and Polyak 2014; Crossey et al. 2015). No evidence exists, however, to explain how a singular stream could extend 322 km (Spencer and Pearthree 2001) in 6 Ma across the Colorado Plateau as postulated by Crossey et al. (2015)—when insight reveals headward erosion to be slow and relatively ineffective in most situations (Bishop 1995; Douglass and Schmeeckle 2007; Meek 2009). Proponents of headward erosion must present clear geomorphic evidence that a stream working from the Colorado Plateau's rim could erode headward, through the plateau, at a rate averaging ~54 mm per year. Most rates reported in literature are much lower Despite the lack of evidence supporting headward erosion-driven piracy, recent work still argues against overflow (Dickinson 2012) and once again suggests a headward eroding stream piracy or capture model (Pelletier 2010).

The strengths of the lake overflow model are that (1) it has explained the evidence for numerous transverse drainages on the lower Colorado River downstream of the Grand Canyon; (2) the age of existing river sediments from Colorado and Wyoming to the Salton Trough are consistent with a Colorado River extending downstream over time via ponding and overflow; and (3) it would seem to require extraordinary conditions to explain one transverse drainage through a convoluted and poorly constrained form of piracy, where elsewhere along the river the overflow mechanism clearly explains the pathway's transverse drainage (House, Pearthree, and Perkins 2008; Roskowski et al. 2010; Spencer et al. 2013; Pearthree and House 2014).

In the end, this debate will continue and we agree with the perspective that "it is exciting to realize that such a well-known landform like the Grand Canyon still holds an element of mystery" (Dexter 2010, 47).

Conclusion

This article recognizes the importance of mountains and their interaction with river systems. Recent advances in geomorphic and physical geographic theory provide a set of tools and conceptual models by which future geographers and geomorphologists can adjudicate between the different processes that result in rivers crossing mountains resulting in transverse drainages. We hope that this work opens doors for a new generation of research to help resolve ongoing debates over famous landscapes in the world, such as the Grand Canyon.

References

Alvarez, W. 1999. Drainage on evolving fold thrust belts: A study of transverse canyons in the Apennines. *Basin Research* 11:267–84.

Arche, A., G. Evans, and E. Clavell. 2010. Some considerations of the initiation of the present SE Ebro river drainage system: Post- or pre-Messinian? *Journal of Iberian Geology* 36 (1): 73–85.

Babenroth, D. L., and A. N. Strahler. 1945. Geomorphology and structure of the East Kaibab Monocline, Arizona and Utah. *Geological Society of America Bulletin* 56:107–50.

Bishop, P. 1995. Drainage rearrangement by river capture, beheading and diversion. *Progress in Physical Geography* 19 (4): 449–73.

Blackwelder, E. 1934. Origin of the Colorado River. *Geological Society of America Bulletin* 45:551–66.

Bonsall, C. 2008. The Mesolithic of the Iron Gates. In *Mesolithic Europe*, ed. G. Bailey and P. Spikins, 238–79. Cambridge, UK: Cambridge University Press.

Brocard, G., J. Willembring, B. Suski, P. Audra, C. Authemayou, B. Cosenza-Muralles, S. Moran-Cal, F. Demory, P. Rochette, T. Vennemann, and K. Holliger. 2012. Rates and processes of river network rearrangement during incipient faulting: The case of the Cahabon River, Guatemala. *American Journal of Science* 312:449–507.

Crossey, L. C., K. E. Karlstrom, R. Dorsey, J. Pearce, E. Wan, L. S. Beard, Y. Asmerom, et al. 2015. Importance of groundwater in propagating downward integration of the 6–5 Ma Colorado River system: Geochemistry of springs, travertines, and lacustrine carbonates of the Grand Canyon region over the past 12 Ma. *Geosphere* 11:660–82.

Darling, A., and K. Whipple. 2015. Geomorphic constraints on the age of the western Grand Canyon. *Geosphere* 11:958–76.

Davis, W. M. 1909. The rivers and valleys of Pennsylvania. In *Geographical essays*, 413–84. Mineral, NY: Dover Publications, Inc.

Davis, W. M., and W. H. Snyder. 1898. *Physical geography.* Boston, MA: Ginn & Company.

Dexter, L. R. 2010. Grand Canyon: The puzzle of the Colorado River. In *Geomorphological landscapes of the world*, ed. P. Migon, 49–58. New York: Springer.

Dickinson, W. R. 2012. Rejection of the lake spillover model for initial incision of the Grand Canyon, and discussion of alternatives. *Geosphere* 9:1–20.

———. 2015. Integration of the Gila River drainage system through the Basin and Range province of southern Arizona and southwestern New Mexico (USA). *Geomorphology* 236:1–24.

Douglass, J. C. 1999. Late Cenozoic landscape evolution study of the eastern Grand Canyon region. MS thesis, Northern Arizona University, Flagstaff, AZ.

Douglass, J. C., N. Meek, R. I. Dorn, and M. W. Schmeeckle. 2009. A criteria-based methodology for determining the mechanism of transverse drainage development, with application to the southwestern United States. *Geological Society of America Bulletin* 121 (3–4): 586–98.

Douglass, J. C., and M. W. Schmeeckle. 2007. Analogue modeling of transverse drainage mechanisms. *Geomorphology* 84:22–43.

Dudley, P. H. 1936. Physiographic history of a portion of the Perris Block, Southern California. *Journal of Geology* 44:358–78.

Dutton, C. E. 1882. Tertiary history of the Grand Canyon District. *U.S. Geological Survey Annual Report* 2:49–166.

Fenneman, N. M. 1931. *Physiography of the western United States.* New York: McGraw-Hill.

———. 1938. *Physiography of the eastern United States.* New York: McGraw-Hill.

Ganjoo, R. 2014. The vale of Kashmir: Landform evolution and processes. In *Landscapes and landforms of India*, ed. V. S. Kale, 125–33. New York: Springer.

Garcia, A. F. 2006. Thresholds of strath genesis deduced from landscape response to stream piracy by Pancho Rico Creek in the Coast Ranges of California. *American Journal of Science* 306:655–81.

Garcia, A. F., and M. Stokes. 2003. Transverse drainage development along a tectonically active transform plate boundary (San Lorenzo River and Pancho Rico Creek, Monterey County, California). Paper presented at the American Geophysical Union Fall Meeting, San Francisco.

Graf, W. L. 1987. Regional geomorphology of North America. In *Geomorphic systems of North America*, ed. W. L. Graf., 1–4. Boulder, CO: Geological Society of America.

Gross, E. L., P. J. Patchett, T. A. Dallegge, and J. E. Spencer. 2001. The Colorado River system and Neogene sedimentary formations along its course: Apparent Sr isotopic connections. *Journal of Geology* 109:449–61.

Harvey, A. M., and S. G. Wells. 1987. Response of Quaternary fluvial systems to differential epeirogenic uplift: Aguas and Feos river systems, southeast Spain. *Geology* 15:689–93.

Hill, C. A., and V. J. Polyak. 2014. Karst piracy: A mechanism for integrating the Colorado River across the Kaibab uplift, Grand Canyon, Arizona, USA. *Geosphere* 10:627–40.

House, P. K., K. A. Howard, J. W. Bell, M. E. Perkins, J. E. Faulds, and A. L. Brock. 2005. *Geological Society of America field guides*. Washington, DC: Geological Society of America.

House, P. K., P. A. Pearthree, and M. E. Perkins. 2008. *Geological Society of America special papers*. Washington, DC: Geological Society of America.

Hunt, C. B. 1967. *Physiography of the United States*. San Francisco: Freeman.

———. 1969. Geologic history of the Colorado River. U.S. Geological Survey Professional Paper 669-C, 59–130.

Johnson, D. W. 1931a. *Stream sculpture on the Atlantic slope*. New York: Columbia University Press.

———. 1931b. A theory of Appalachian geomorphic evolution. *The Journal of Geology* 39:497–508.

Karlstrom, K. E., R. Crow, L. J. Crossey, D. Coblentz, and J. W. Van Wijk. 2008. Model for tectonically driven incision of the younger than 6 Ma Grand Canyon. *Geology* 36:835–38.

Kimbrough, D. L., M. Grove, G. E. Gehrels, R. J. Dorsey, K. A. Howard, O. Lovera, A. Aslan, P. K. House, and P. A. Pearthree. 2015. Detrital zircon U-Pb provenance record of the Colorado River: Implications for late Cenozoic drainage evolution of the American Southwest. *Geosphere* 11:1–30.

Laney, R. L., and H. E. Hahn. 1986. Hydrogeology of the eastern part of the Salt River Valley area, Maricopa and Pinal Counties, Arizona. U.S. Geological Survey Water-Resources Investigations Report 86–4147.

Lang, K., and K. W. Huntington. 2014. Antecedence of the Yarlung-Siang-Brahmaputra River, eastern Himalaya. *Earth and Planetary Science Letters* 397:145–85.

Larson, P. H. 2013. Desert fluvial terraces and their relationship with basin development in the Sonoran Desert, Basin and Range: Case studies from south-central Arizona. Dissertation, Arizona State University, Tempe, AZ.

Larson, P. H., R. I. Dorn, Z. Bowles, E. J. Harrison, S. Kelley, M. W. Schmeeckle, and J. Douglass. 2014. Pediment response to drainage basin evolution in south-central Arizona. *Physical Geography* 35 (5): 369–89.

Larson, P. H., R. I. Dorn, J. Douglass, B. F. Gootee, and R. Arrowsmith. 2010. Stewart Mountain Terrace: A new Salt River terrace with implications for landscape evolution of the Lower Salt River Valley, Arizona. *Journal of the Arizona-Nevada Academy of Science* 42:26–35.

Larson, P. H., S. B. Kelley, R. I. Dorn, and Y. B. Seong. Forthcoming. Pace of landscape change and pediment development in the northeastern Sonoran Desert, United States. *Annals of the Association of American Geographers*.

Lee, S. Y., Y. B. Seong, Y. Shin, K. H. Choi, H. Kang, and J. Choi. 2011. Cosmogenic 10Be and OSL dating of fluvial strath terraces along the Osip-cheon River, Korea: Tectonic implications. *Geosciences Journal* 15 (4): 359–78.

Lucchitta, I. 1984. Development of landscape in northwestern Arizona; the country of plateaus and canyons. In *Landscapes of Arizona: The Geological Story*, eds. T. L. Smiley, J. D. Nations, T. L. Pewe, and J. P. Schafer, 269–301. Lanham: University Press of America.

Lundberg, N. and R. J. Dorsey. 1990. Rapid Quaternary emergence, uplift, and denudation of the Coastal Range, eastern Taiwan. *Geology* 18:638–41.

Mack, G. H., D. W. Love, and W. R. Seager. 1997. Spillover models for axial rivers in regions of Continental extension: The Rio Mimbres and Rio Grande in the southern Rio Grande rift, USA. *Sedimentology* 44:637–52.

Mayer, L., M. Menichetti, O. Nesci, and D. Savelli. 2003. Morphotectonic approach to the drainage analysis in the North Marche region, central Italy. *Quaternary International* 101–102:157–67.

McKee, E. D., R. F. Wilson, W. J. Breed, and C. S. Breed. 1967. Evolution of the Colorado River in Arizona. *Museum of Northern Arizona Bulletin* 44:1–68.

Meek, N. 1989. Geomorphic and hydrologic implications of the rapid incision of Afton Canyon, Mojave Desert, California. *Geology* 17:7–10.

———. 2002. Ponding and overflow: the forgotten transverse drainage hypothesis. 98th Annual Meeting Abstracts, Association of American Geographers, Los Angeles, CA.

———. 2009. "How (in)effective is the headward erosion process," 105th Annual Meeting Abstracts, Association of American Geographers, Las Vegas, NV.

Meek, N., and J. Douglass. 2001. Lake Overflow: An alternative hypothesis for Grand Canyon incision and development of the Colorado River. In *Colorado River Origin and Evolution*, ed. R. A. Young and E. E. Spamer, 199–204. Grand Canyon, AZ: Grand Canyon Association.

Menges, C. M., and D. E. Anderson. 2005. Late Cenozoic drainage history of the Amargosa River, southwestern Nevada and eastern California. In *Geologic and Biotic Perspectives on Late Cenozoic Drainage History of the Southwestern Great Basin and Lower Colorado River Region*, ed. M. Reheis, 8. U.S. Geological Survey Open-File Report 2005–1404, United States Geological Survey, Washington, DC.

Morrison, R. B. 1991. Quaternary stratigraphic, hydrologic and climatic history of the Great Basin, with emphasis on Lakes Lahontan, Bonneville and Tecopa. In *Quaternary nonglacial geology; coterminous US, the geology of North America*, vol. K–2, ed. R. B. Morrison, 283–320. Boulder, CO: Geological Society of America.

Newberry, J. S. 1861. *Report upon the Colorado River of the West, Part III.* Washington, DC: U.S. Government Printing Office.

———. 1862. Colorado River of the west. *American Journal of Science* 33:376–403.

Oberlander, T. 1965. *The Zagros streams.* Syracuse, NY: Syracuse University Press.

Ollier, C., and C. Pain.. 2000. *The origin of mountains.* London and New York: Routledge.

Pearthree, P. A., and P. K. House. 2014. Paleogeomorphology and evolution of the early Colorado River inferred from relationships in Mohave and Cottonwood valleys, Arizona, California, and Nevada. *Geosphere* 10:1139–60.

Peirce, W. H. 1984. Some late Cenozoic basins and basin deposits of southern and western Arizona. In *Landscapes of Arizona. The geological story,* ed. T. L. Smiley, J. D. Nations, T. L. Pewe, and J. P. Schafer, 207–27. Lanham, MD: University Press of America.

Pelletier, J. D. 2010. Numerical modeling of the late cenozoic geomorphic evolution of Grand Canyon, Arizona. *Geological Society of America Bulletin* 122:595–608.

Phillips, R. L., and I. Berczi. 1985. Processes and depositional environments of Neogene deltaic—lacustrine sediments, Pannonian basin, southeast Hungary; core investigation summary. USGS Open-File Report 85–360.

Polyak, V., C. Hill, and Y. Asmerom. 2008. Age and evolution of the Grand Canyon revealed by U-Pb dating of water table–type speleothemetry. *Science* 319:1377–80.

Powell, J. W. 1875. *Exploration of the Colorado River of the West and its tributaries: Explored in 1869, 1870, and 1872 under the direction of the secretary of the Smithsonian Institution.* Washington, DC: U.S. Government Printing Office.

———. 1896. *The physiography of the United States.* New York: American Book Company.

Reheis, M. C., D. M. Miller, and J. L. Redwine. 2007. Quaternary stratigraphy, drainage-basin development and geomorphology of the Lake Manix Basin, Mojave Desert. Guidebook for the fall fieldtrip, Friends of the Pleistocene, Pacific Cell. USGS Open-File Report 2007-1281. Reston, VA: U.S. Geological Survey.

Reheis, M. C., and J. L. Redwine. 2008. Lake Manix shorelines and Afton Canyon terraces: Implications for incision of the Afton Canyon. *Geological Society of America Special Papers* 439:227–59.

Ren, J., S. Zhang, A. J. Meigs, R. S. Yeats, R. Ding, and X. Shen. 2014. Tectonic controls for transverse drainage and timing of the Xin-Ding paleolake breach in the upper reach of the Hutuo River, north China. *Geomorphology* 206:452–67.

Roskowski, J. A., P. J. Patchett, J. E. Spencer, P. A. Pearthree, D. L. Dettman, J. E. Faulds, and A. C. Reynolds. 2010. A late Miocene–early Pliocene chain of lakes fed by the Colorado River: Evidence from Sr, C, and O isotopes of the Bouse Formation and related units between Grand Canyon and the Gulf of California. *Geological Society of America Bulletin* 122:1625–36.

Searle, M. P., and P. J. Treloar. 1993. Himalayan tectonics—An introduction. *Geological Society, London, Special Publications* 74:1–7.

Sears, J. W. 2013. Late Oligocene–early Miocene Grand Canyon: A Canadian connection? *GSA Today* 25 (11): 1–10.

Spencer, J. E., P. J. Patchett, P. A. Pearthree, P. K. House, A. M. Sarna-Wojcicki, E. Wan, J. A. Roskowski, and J. E. Faulds. 2013. Review and analysis of the age and origin of the Pliocene Bouse Formation, lower Colorado River Valley, southwestern USA. *Geosphere* 9 (3): 444–59.

Spencer, J. E., and P. A. Pearthree. 2001. Headward erosion versus closed basin spillover as alternative causes of Neogene capture of the ancestral Colorado River by the Gulf of California. In *Colorado River origin and evolution: Grand Canyon,* ed. R. A. Young and E. E. Spamer, 215–19. Grand Canyon, AZ: Grand Canyon Association.

Spencer, J. E., G. R. Smith, and T. E. Dowling. 2008. Middle to late Cenozoic geology, hydrography, and fish evolution in the American Southwest. *Geological Society of America Special Papers* 439:279–99.

Stokes, M. and A. E. Mather. 2003. Tectonic origin and evolution of a transverse drainage: The Rıo Almanzora, Betic Cordillera, Southeast Spain *Geomorphology* 50:59–81.

Stokes, M., A. E. Mather, A. Belfoul, and F. Farik. 2008. Active and passive tectonic controls for transverse drainage and river gorge development in a collisional mountain belt (Dades Gorges, High Atlas Mountains, Morocco). *Geomorphology* 102:2–20.

Strahler, A. N. 1948. Geomorphology and strucutre of the West Kaibab fault zone and Kaibab Plateau, Arizona. *Geological Society of America Bulletin* 59:513–40.

Wager, L. R. 1937. The Arun River drainage pattern and the rise of the Himalaya. *The Geographical Journal* 89:239–50.

Wernicke, B. 2011. The California River and its role in carving the Grand Canyon. *Geological Society of America Bulletin* 123:1288–316.

Yonkee, W. A., and A. B. Weil. 2015. Tectonic evolution of the Sevier and Laramide belts within the North American Cordillera orogenic system. *Earth Science Reviews* 150:531–93.

Young, R. A., and Y. Crow. 2014. Paleogene Grand Canyon incompatible with Tertiary paleogeography and stratigraphy. *Geosphere* 10:664–79.

Zelilidis, A. 2000. Drainage evolution in a rifted basin, Corinth graben, Greece. *Geomorphology* 35:69–85.

Geomorphometric Controls on Mountain Glacier Changes Since the Little Ice Age in the Eastern Tien Shan, Central Asia

Yanan Li, Yingkui Li, Xiaoyu Lu, and Jon Harbor

The linkage between glacier change and climate has garnered significant attention in recent decades, but little is known about the role of local geomorphometric factors on glacier changes since the Little Ice Age (LIA), approximately 100 to 700 years ago. This study examines the spatial pattern of changes in glacier area in the eastern Tien Shan based on geomorphological mapping of LIA glacial extents and contemporary glaciers from the Second Glacier Inventory of China. Partial least squares regression was applied to examine the correlations between geomorphometric factors, including glacier area, slope, aspect, shape, elevation, and hypsometry, and relative glacier area loss, both in the whole area and in three subregions (the Boro-Eren Range, the Bogda Range, and the Karlik Range). Our results show that the area of 640 mapped LIA glaciers decreased from 791.6 ± 18.7 km^2 to 483.9 ± 31.2 km^2 between 2006 and 2010, a loss of 38.9 ± 2.7 percent. The losses for three subregions are 43.4 ± 3.2 percent, 35.9 ± 2.4 percent, and 30.2 ± 1.8 percent, respectively. Elevation, slope, and area of a glacier are the three most significant geomorphometric factors to glacier area change, at both regional and subregional scales. The west–east decreasing trend of glacier retreat and different variances explained in subregional regressions might reflect the influence from the shifting dominance of the westerlies and the Siberian High.

冰川改变和气候之间的关联, 在近数十年来获得了显着的关注, 但我们对于大约一百至七百年前的小冰期(LIA)以降, 冰川改变的在地地形计测系数所扮演的角色却所知甚少。本研究根据中国第二次冰川编目中的小冰期冰河范围和目前冰川的地形计测製图, 检视天山东部冰川地区的空间模式变化。本文应用篇最小平方迴归, 检视在整个地区和在三大次区域中(博罗—依连山脉、博格达山脉、喀尔里克山脉²), 包含冰川面积、坡度、坡向、形态、海拔和测高法的地形计测系数与冰川面积的相对丧失之间的相互关系。我们的研究结果显示, 六百四十个绘製的LIA冰川面积, 在 2006 年至 2010 年间, 从 791.6 ± 18.7 平方公里, 降至 483.9 ± 31.2 平方公里, 一共损失百分之38.9 ± 2.7。三大次区域的丧失各为百分之 43.4 ± 3.2, 百分之 35.9 ± 2.4, 以及百分之 30.2 ± 1.8。冰川的海拔、坡度和面积, 是区域及次区域尺度中冰川面积改变的三个最重要的地形计测系数。冰川后退的东西向减少趋势, 以及次区域迴归所解释的不同变异, 可能反映出西风带和西伯利亚高压支配转变的影响。 *关键词:* *天山东部, 地形计测系数, 冰川改变, 小冰期。*

El nexo entre cambios de los glaciares y clima ha logrado atraer notable atención en décadas recientes, pero poco se sabe acerca del papel de factores geomorfométricos locales sobre los cambios glaciarios desde la Pequeña Edad del Hielo (PEH), aproximadamente entre 100 y 700 años atrás. Este estudio examina los cambios del patrón espacial en el área glaciada de la parte oriental del Tien Shan con base en cartografía geomorfológica de las extensiones glaciares de la PEH y de los glaciares contemporáneos del Segundo Inventario Glaciar de China. Se aplicó una regresión parcial de mínimos cuadrados para examinar las correlaciones entre factores geomorfométricos, incluyendo el área de los glaciares, la inclinación, aspecto, forma, elevación e hipsometría, y la pérdida relativa de área glaciaria, tanto en el área total como en tres subregiones (la Cordillera Boro-Eren, la Cordillera Bogda y la Cordillera Karlik). Nuestros resultados muestran que el área cartografiada de 640 glaciares de la PEH disminuyó de 791.6 ± 18.7 km2 a 483.9 ± 31.2 km2 entre 2006 y 2010, una pérdida de 38.9 ± 2.7 por ciento. Las pérdidas para las tres subregiones son 43.4 ± 3.2 por ciento, 35.9 ± 2.4 por ciento y 30.2% ± 1.8 por ciento, respectivamente. La elevación, la inclinación y el área de un glaciar son los tres factores geomorfométricos más significativos para el cambio de área del glaciar, tanto a escala regional como subregional. La tendencia oeste–este de recesión del glaciar y las diferentes varianzas explicadas en regresiones subregionales podrían reflejar la influencia del cambio de predominio de los vientos oestes y de la zona de Alta Presión de Siberia.

Glacier change is one of the manifestations of global climate change, especially in mountainous regions where meteorological data are scarce. Although glacier mass balances are mainly affected by the dominance of shifting climate controls, geomorphometric and glaciological factors also play significant roles in forming the spatial pattern of glacier changes (Furbish and Andrews 1984; DeBeer and Sharp 2007; K. Li et al. 2011; Delmas, Gunnell, and Calvet 2014). In particular, the difference in glacier size, aspect, shape, slope, elevation, and hypsometry could cause the heterogeneous response of glaciers to climate change in different regions. Assuming that the same climate forcing occurred during a climatic event at a regional scale, different responses of glaciers are more likely driven by microclimates surrounding individual glaciers caused by geomorphometric characteristics (Furbish and Andrews 1984). Therefore, it is of great importance to investigate the geomorphometric controls on variations in glacial extents.

The Tien Shan, a large mountain range in arid and semiarid regions of central Asia, is well known as the "Water Tower of Central Asia" (Sorg et al. 2012). The large inventory of mountain glaciers in this mountain range provides vital water resources to the local population and ecosystems (Bishop et al. 2011; Gao et al. 2013; Farinotti et al. 2015). Since the end of the Little Ice Age (LIA) around the 1890s, most glaciers in the Tien Shan have been retreating extensively with an accelerated recession rate that has occurred notably in the last few decades (Bolch 2007; Narama et al. 2010; Sorg et al. 2012). Changes in climate conditions are the main causes of the glacier recession. V. B. Aizen et al. (2007) noted that an increase in air temperatures and changes in precipitation partitioning (rain replacing snow at high elevations) have led to negative mass balance of glaciers and changed river runoff regimes in central Asia. Located at the confluence of the midlatitude westerlies and the Siberian high-pressure system, glaciers and rivers in the Tien Shan are sensitive to the temporal variations in the dominance of these systems (Benn and Owen 1998; E. M. Aizen et al. 2001; F. H. Chen et al. 2008). As much effort has been devoted to understanding the linkage between glacier change and climate change, studies have also investigated how glacier changes are affected by geomorphometric factors. For example, studies around the world have shown a consistent pattern that small glaciers have experienced relatively larger proportional changes than large glaciers in the past decades (e.g., S. Y. Liu et al. 2003; Ye et al. 2003; Bhambri et al. 2011; K. Li et al. 2011; Cogley

2014; Y. N. Li and Li 2014; Q. Liu et al. 2015). The fact that more glaciers exist with poleward aspects indicates that equator-ward facing slopes receive more solar radiation to discourage glacier preservation (Evans 2006b). In the Tien Shan, we previously examined the effect of local factors on glacier area and equilibrium-line altitude (ELA) changes within a small region and found that the glacier area and mean elevation are the two main factors in the relative area changes, whereas ELA changes are not strongly associated with local factors (Y. N. Li and Li 2014). Detailed studies on geomorphometric controls in glacier change are still limited in the Tien Shan, although a glacier inventory (e.g., Glacier Inventory of China [GIC]) has been updated and is freely available (Y. F. Shi, Liu, and Kang 2009; S. Y. Liu et al. 2015).

This study focuses on the glacier change since the LIA, a cold era during the last millennium, approximately between AD 1350 and 1890 (Grove 2004). Most studies in the Tien Shan have been conducted for recent decades based on remotely sensed data, but little is known about glacial recession since the LIA. Here, we aim to answer two research questions: (1) How much glacier area has been reduced across three mountain ranges in the eastern Tien Shan from the LIA to present (2006–2010)? and (2) What geomorphometric factors, including glacier area, slope, aspect, shape, elevation, and hypsometry, can be used to explain the variance in glacier area changes?

Study Area

Formed by the collision of the Indian and Eurasian continental plates about 40 to 50 million years ago (Yin and Harrison 2000), the Tien Shan is an ~2,500-km-long mountain series stretching from the western boundary of Kyrgyzstan across the Xinjiang Uyghur Autonomous region in China. The relatively higher, western part of the Tien Shan is recognized as the Kyrgyz Tien Shan, and to the east, approximately two thirds of the total length (~1,700 km) lies within Xinjiang, China. Our study area includes the mountain ranges of the eastern Tien Shan in China, ranging approximately from 85°40′E to 94°50′E and from 42°40′N to 44°00′N (Figure 1). This area includes many individual ranges, most of which reach about 4,000 m above sea level (a.s.l.), and is characterized with a sharp contrast of landscapes of high mountains, intervening valleys, and desert basins (V. B. Aizen et al. 1997; Ye et al. 2005). Glaciers are developed at high elevations and play an important role in the hydrologic

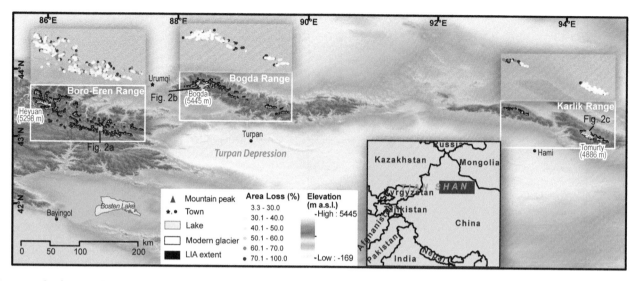

Figure 1. Study area of the eastern Tien Shan (red block on the inset map) and three subregions (white-outlined boxes). The spatial pattern of relative area change is shown using color-graded dots in gray boxes corresponding to each subregion. m a.s.l. = meters above sea level. (Color figure available online.)

cycle, water balance, ecosystems, and socioeconomic development of central Asia. Most glaciers in the study area are clean-ice glaciers, with less than ten glaciers identified as debris-covered glaciers in the GIC (Guo et al. 2014). Glacier types include valley glaciers, hanging glaciers, and cirque glaciers. Three subregions were defined based on the clustering of glacier distribution, separated by two passes of 1,100 m a.s.l. and 860 m a.s.l. in elevation (Figure 1). From west to east, the first subregion is named the Boro-Eren Range (coded BE), which include the eastern Borohoro Mountains and the Eren Habirga Mountains; its highest peak is Heyuan (5,298 m a.s.l.). The second subregion is the Bogda Range (coded BG), located adjacent to the regional capital city Urumqi, and its highest peak is Bogda (5,445 m a.s.l.). The third subregion, the Karlik Range (coded KL), is at the eastern end of the Tien Shan, with relatively fewer glaciers, and the highest peak is Tomurty (4,886 m a.s.l.).

Data and Methods

Data Sets and Data Processing

The outlines of contemporary glaciers were obtained from the Second GIC (Guo et al. 2014), which provides detailed information for each glacier in China. Glacier coverage in the Second GIC was derived from multiple satellite images (Landsat TM/ETM+, ASTER, and SPOT imagery) that were acquired between 2005 and 2010 (S. Y. Liu et al.

2015). The boundaries of contemporary glaciers in our study area (Figure 1) were solely delineated from Landsat TM/ETM+ scenes between 2006 and 2010. The horizontal error of these glacier boundaries is ±30 m (Guo et al. 2014; S. Y. Liu et al. 2015).

The LIA glacial extents were manually delineated in Google Earth (Figure 1; Y. N. Li, Li, Chen, et al. 2016). We defined the outermost fresh and unvegetated moraine in front of glaciers as the maximum extent of LIA glaciers and mapped them in the three mountain ranges of the eastern Tien Shan based on the geomorphic location and relationship, morphology, vegetation cover, and weathering characteristics. Numerical dating of the LIA moraines has been conducted at a few sites in the eastern Tien Shan. For example, the fresh-looking moraine in front of Glacier No. 1 in the BE subregion is well constrained to LIA ages (~430 ± 100 years) using cosmogenic [10]Be exposure dating recently (Y. K. Li et al. 2014; Y. N. Li, Li, Harbor, et al. 2016; Figure 2A). Previous studies using other methods, such as lichenometry and radiocarbon dating, also assigned LIA ages to the outermost moraine at this site. J. Chen (1989) dated three moraine ridges at this Glacier No. 1 site to 472 ± 20 years (outermost), 233 ± 20 years, and 139 ± 20 years (innermost) using lichenometry, and Yi et al. (2004) used radiocarbon dating of organic calcium oxalate coatings from the outermost moraine and obtained ages of 460 ± 120 years and 490 ± 120 years (recalculated in Xu and Yi 2014). In the Karlik Range, Y. X. Chen et al. (2015) reported an older moraine

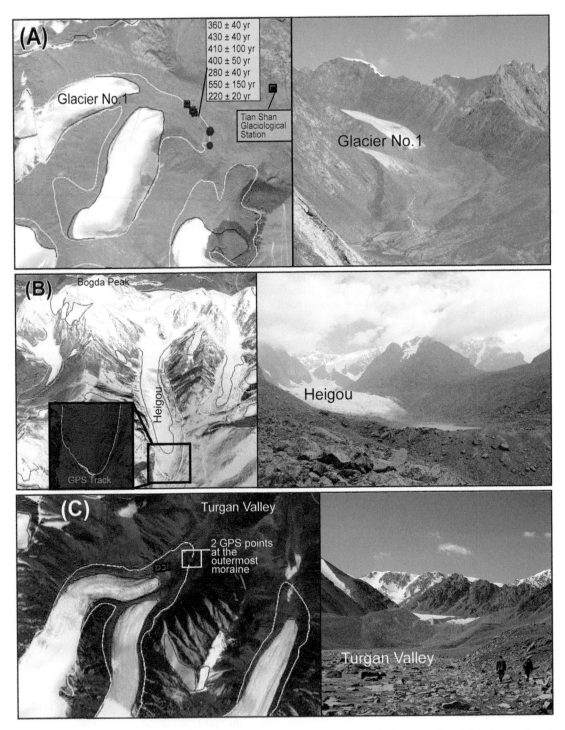

Figure 2. Examples of contemporary glaciers and delineated LIA moraines in Google Earth and in field photos: (A) The Urumqi River headwaters area in the Boro-Eren subregion, with seven cosmogenic [10]Be ages indicating the LIA moraine age (Y. K. Li et al. 2014; Y. N. Li, Li, Harbor, et al. 2016). (B) The Heigou Valley in the Bogda subregion, with a GPS receiver track route to validate the delineation of LIA moraine. (C) The Turgan Valley in the Karlik subregion, with two GPS points collected at the outermost fresh moraine (modified from Figures 3 and 4 in Y. N. Li, Li, Chen, et al. 2016). LIA = Little Ice Age; GPS = Global Positioning System. (Color figure available online.)

age of 790 ± 300 years (recalculated in Y. N. Li, Li, Harbor, et al. 2016) based on eleven [10]Be exposure ages from the fresh moraines at the Turgan Valley.

The manually delineated LIA moraines were saved as Keyhole Markup Language files in Google Earth and converted into shapefiles for further processes in

ArcGIS. The detailed delineation procedure can be found in Y. N. Li, Li, Chen, et al. (2016). To assess the accuracy of our delineated LIA extent of glaciers, we applied a revised automated proximity and conformity analysis (R-APCA; Y. K. Li, Napieralski, and Harbor 2008) to quantify the offset between our field Global Positioning System (GPS) measurements and delineations from Google Earth (Figure 2). The average offsets were 10.5 m, 9.4 m, and 9.1 m at sampled sites in the three subregions, respectively. Hence, the horizontal error of our delineated LIA glacial extents is about ±10 m. This error estimate does not include the potential identification error that occurred in the boundary extraction processes.

We used the Shuttle Radar Topography Mission (SRTM) digital elevation model (DEM; 30 m) downloaded from USGS EarthExplorer (earthexplorer.usgs. gov) to calculate the geomorphometric factors for each glacier. We applied a cubic convolution algorithm to fill some void cells with no data in the original 30-m SRTM DEM. Glacier area was derived directly for both the contemporary and LIA glacier layers in ArcGIS. The area change was then calculated as the difference between glacier areas at these two times. In case of an LIA glacier that had split into individual contemporary glaciers, the change was based on the difference between the total area of individual contemporary glaciers and their corresponding LIA glacier. The change in glacier area might not reflect the ice volume loss but is still a good indicator of glacier change, especially when no ice thickness data are available. To accommodate varied sizes of glaciers, we calculated the relative change of glacier area (in percentage) using the following equation:

$$\Delta A = \left(1 - \frac{\sum A_M}{A_{LIA}}\right) * 100, \quad (1)$$

where ΔA is the relative area change (%); A_M is the area of contemporary glacier(s), and A_{LIA} is the glacier area during the LIA.

We used the method described in Bolch, Menounos, and Wheate (2010), Wei et al. (2014), and S. Y. Liu et al. (2015) for error analysis of glacier area changes. The error of glacier area ε_i is defined as

$$\varepsilon_i = \frac{N * \lambda_i^2}{2}, \quad (2)$$

where N is the number of cells intersecting with the glacier's outline, λ_i is the line pixel error, and i refers to the contemporary or LIA time period. The error in absolute area change $\varepsilon_{(A_{LIA} - A_M)}$ from LIA to present is defined as

$$\varepsilon_{(A_{LIA} - A_M)} = \sqrt{\varepsilon_M^2 + \varepsilon_{LIA}^2} \quad (3)$$

and the error in relative area change $\varepsilon_{\Delta A}$ is defined as

$$\varepsilon_{\Delta A} = \left(1 - \frac{A_M}{A_{LIA}}\right) * \sqrt{\left(\frac{\varepsilon_{LIA}}{A_{LIA}}\right)^2 + \left(\frac{\varepsilon_M}{A_M}\right)^2}$$
$$* 100. \quad (4)$$

Geomorphometric Factors and Statistical Analysis

In addition to glacier area, we derived aspect, slope, shape, median elevation, and hypsometry, based on the LIA glacier polygons and the DEM in ArcGIS. To incorporate the spatial difference in glacier area loss, the longitude of the centroid ("Easting") of each glacier was added as another factor (Table 1). To examine the statistical relationships between these seven factors and relative glacier area change, we used partial least squares regression (PLSR) because this model accounts for multicollinearity between independent variables and has shown stronger explanatory capacity, compared to principal components analysis and multiple regression (Kemsley 1996; Carrascal, Galván, and Gordo 2009; Yan et al. 2013; Z. Shi et al. 2014). PLSR uses the most important linear combinations (components) in the regression equation, which is achieved by maximizing the covariance between the dependent variable and all possible linear functions of independent variables (Abdi 2003). In our model, the relative glacier area change (ΔA) is the dependent variable (Y), and seven factors are independent variables (X). Glacier area data are highly skewed toward small values (Figure 3A), so they were logarithmic transformed for normal distribution and ln(Area) was used in the regression. Given the circular and continuous nature of the aspect factor, we adopted the method described in Evans (2006a, 2006b, 2011) to transform it into cosine and sine using the first harmonic of a Fourier regression.

Several parameters were used to measure the explanatory and predictive ability of the PLSR model (Abdi 2003; Wold et al. 2004). R^2 is the coefficient of determination and represents the amount explained. Q^2 is the cross-validated R^2 and represents the amount "predicted," defined as $(1 - PRESS/SS)$, where $PRESS$ is predictive residual sum of squares and SS is the sum

Table 1. Information of geomorphometric variables derived for the LIA glacial extents

Factor	Variable name	Unit	Summary statistics	Calculation source and method	Note
Glacier area	Area	m^2	—	Calculated using glacier polygons, logarithmic transformed for normal distribution.	It reflects glacier size.
Median elevation	Elevation	M	Median	Derived based on DEM	It reflects glacier's altitudinal location.
Surface slope	Slope	°	Mean	Derived based on DEM	Slope can range from 0° to 90°.
Surface facing	Aspect	°	Directional mean	Derived based on DEM	Aspect spans clockwise from 0° (due north) to 360° (again due north). We calculated the cosine and sine of aspect using Fourier regression.
Distance east	Easting	°	—	Longitude of the centroid of glaciers	It records the longitudinal locations of glaciers in our study area.
Hypsometric integral	HI	N/A	Mean	$\frac{Elev_{mean} - Elev_{min}}{Elev_{max} - Elev_{min}}$	Hypsometry is a simplified factor representing mass balance distribution, ranging from 0 to 1. Small values indicate more area or mass distributed at low elevations and vice versa.
Shape index	Shape	N/A	—	$\frac{D_{min}}{D_{max}}$	Shape is the ratio of largest diameter to the smallest diameter orthogonal to it. It ranges from 0 (extremely elongated) to 1 (equiaxed).

Note: LIA = Little Ice Age; DEM = digital elevation model.

of squares of Y corrected for the mean (Wold et al. 2004). R^2_{cum} is the cumulative R^2 over the selected X variables, and Q^2_{cum} is the cumulative Q^2 over all of the selected PLSR components. The root mean squared error of prediction (RMSEP) is another parameter that contains useful information to calibrate and develop the regression model and is calculated as

$$RMSEP = \sqrt{\frac{\sum_{i=1}^{n} \left(\bar{y}_{predicted} - y_{observed} \right)^2}{n}}. \quad (5)$$

In a PLSR model, the relative importance of each variable is given by the variable importance for the projection (VIP). Variables with larger VIP values are the most relevant for explaining the dependent variable. The regression coefficients reveal the direction and strength of the impact of each variable in the PLSR model. PLSR was performed for glaciers in the entire study area, as well as in each of the three subregions, to examine whether different dominating factors exist in explaining local glacier change. All statistical analyses were performed using R (Ihaka and Gentleman 1996; Björn-Helge and Wehrens 2007).

Results and Discussion

Geomorphometry and Glacier Changes Since the LIA

The total number of delineated LIA glacial extents is 640, corresponding to 865 contemporary glaciers in the Second GIC in the eastern Tien Shan (Table 2). The summary statistics and frequency distribution of geomorphometric features of the 640 LIA glaciers are illustrated in Figure 3. Glacier area has a left-skewed distribution due to the domination of glaciers <2.0 km² (550 out of 640, 86 percent), although the range of glacier area is from 0.1 km² to 17.4 km² (both minimum and maximum values are from the BE subregion; Figure 3A). The mean slope values of these 640 LIA glaciers are likely normally distributed with a mean of 26.6° and a standard deviation of 4.6° (Figure 3B). The median elevations show a slightly left-skewed but close to normal distribution, with a mean of 3,826 m a.s.l. and a standard deviation of 172 m (Figure 3C). The lowest median glacier elevation (3,369 m a.s.l.) is more than 1,000 m lower than the highest one (4,382 m a.s.l.), and both are in the BE range. The shape index ranges from 0.17 to 0.90, with a mean and median of ∼0.54

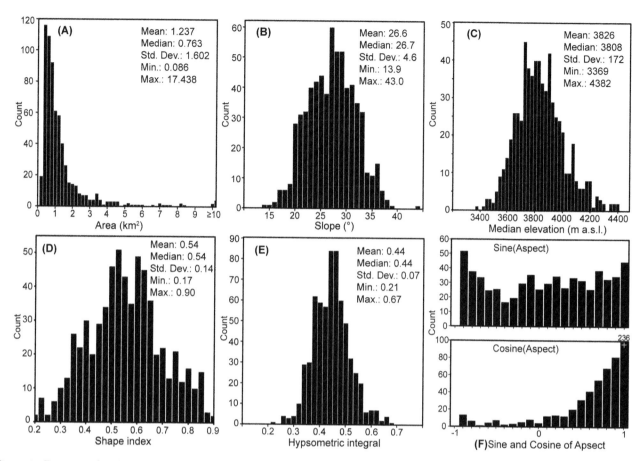

Figure 3. Frequency distribution of six geomorphometric factors: (A) area, (B) slope, (C) median elevation, (D) shape index, (E) hypsometric integral, and (F) sine and cosine of aspect. The sample size is 640.

(Figure 3D). The hypsometric integral (HI) values show a narrow bell-shaped distribution with the same mean and median value of 0.44 (Figure 3E). This might be attributed to the similar geologic background of these alpine glaciers in the study region, but such hypsometry of Tien Shan glaciers could be distinct from that of other glaciated landscapes (Brocklehurst and Whipple 2004). The aspect of glaciers was transformed to Sin(Aspect) and Cos(Aspect), two variables, and the sine term measures east–west differences, whereas the cosine term measures north–south differences (Evans

2006b). As shown in Figure 3F, our glaciers show no apparent east–west contrast, whereas there is a big difference in amount of north-facing and south-facing glaciers. Most glaciers (519 out of 640, 81 percent) are with north-, northeast-, and north-west-facing directions (Figure 4A).

The total area of these glaciers decreased from 791.6 ± 18.7 km² during the LIA to 483.9 ± 31.2 km² in 2006 to 2010, with a reduction of 38.9 ± 2.7 percent. This reduction percentage represents a minimum estimate of glacier area loss because some LIA glaciers that might have completely disappeared are not included in

Table 2. Comparison of the LIA glacial extents and contemporary glaciers in the eastern Tien Shan

	LIA		Present-day			
	No.	Area (km²)	No.	Area (km²)	Absolute area loss (km²)	Relative area loss (%)
Boro-Eren	392	432.1 ± 10.6	541	244.4 ± 16.9	187.7 ± 20.0	43.4 ± 3.2
Bogda	168	201.1 ± 4.7	202	129.0 ± 8.1	72.1 ± 9.4	35.9 ± 2.4
Karlik	80	158.4 ± 3.3	122	110.6 ± 6.2	47.8 ± 7.0	30.2 ± 1.8
Total	640	791.6 ± 18.7	865	483.9 ± 31.2	307.7 ± 36.4	38.9 ± 2.7

Note: LIA = Little Ice Age.

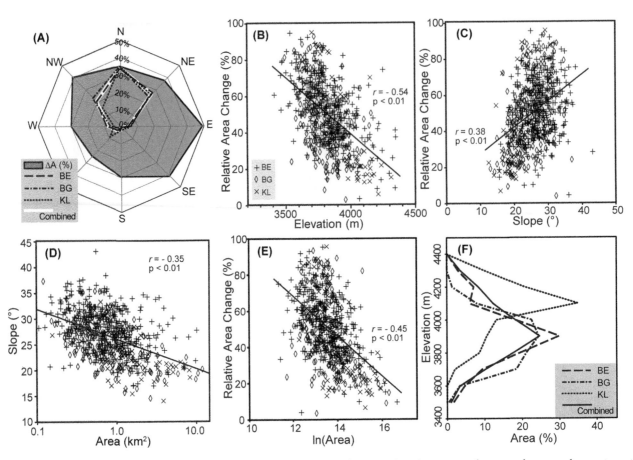

Figure 4. Diagrams and scatter plots to show some important relationships between the relative area change and geomorphometric variables: (A) △A vs. aspect; (B) △A vs. elevation; (C) △A vs. slope; (D) slope vs. area; (E) △A vs. ln(area); and (F) area distribution along elevation bins. In (A), both percentages of glaciers at eight directions and the relative area change share the same scale from 0 to 50 percent. *Note:* BE = Boro-Eren Range; BG = Bogda Range; KL = Karlik Range.

the calculation. For the three subregions, BE contains 541 contemporary glaciers that correspond to 392 LIA glaciers, more than half of the total numbers. BE contemporary glaciers have an area of 244.4 ± 16.9 km², accounting for about half of the total area. BG is a smaller range compared to BE, and it has 202 glaciers with an area of 129.0 ± 8.1 km², corresponding to 168 LIA glaciers. KL is the smallest region with 122 contemporary glaciers (corresponding to 80 LIA glaciers) and a total area (110.6 ± 6.2 km²) similar to the BG glacier area (Table 2). A west–east decreasing trend is observed in relative area loss in three subregions, ranging from 43.4 ± 3.2 percent, 35.9 ± 2.4 percent, to 30.2 ± 1.8 percent in BE, BG, and KL, respectively.

Important Geomorphometric Factors

Different settings of area, slope, aspect, elevation, shape, and hypsometry of a glacier could influence its response sensitivity to climate change. Pearson's correlation analysis shows pairwise correlations among these independent variables and relative area change (Table 3). For 640 LIA glaciers, the relative area change (△A) is correlated with six variables (area, elevation, slope, cosine A, easting, and HI) at the 0.05 significance level. A positive correlation is observed between relative area change and slope, and negative correlations exist between relative area change and area, elevation, cosine A, easting, and HI.

To deduce the relative importance of these geomorphometric variables on glacier changes, we performed the PLSR model on 640 glaciers over the study region (Table 4). Results show that the minimum RMSEP and maximum Q^2 were obtained with the first four components, and the addition of any more components produced an insignificant increase in amount explained because the other components were not strongly correlated with the residuals of the dependent variable. The first component explained 29.22 percent of the total variance, and the addition of the other three components led to a cumulative explained variance of 49.47 percent and a maximum Q^2 of 0.481 (Table 4).

Table 3. Correlation matrix of the relative area change and eight independent variables

	$\triangle A$	Area	Elevation	Slope	Cosine A	Sine A	Easting	HI	Shape
$\triangle A$	1								
Area	−0.45**	1							
Elevation	−0.54**	0.22**	1						
Slope	0.38**	−0.35**	−0.07	1					
Cosine A	−0.09*	0.11**	0.04	−0.05	1				
Sine A	0.05	−0.02	−0.02	−0.02	−0.03	1			
Easting	−0.19**	0.15**	0.23**	−0.38**	0.01	−0.04	1		
HI	−0.20**	−0.02	0.44**	0.21**	−0.03	−0.02	0.07	1	
Shape	0.05	0.02	0.07	0.16**	0.04	−0.02	0.04	0.10**	1

Note: HI = hypsometric integral.
*Indicates significance at the 0.05 level ($p < 0.05$)
**Indicates significance at the 0.01 level ($p < 0.01$).

Table 5 shows the composition of only the components that entered the model when minimum RMSEP and maximum Q^2 were reached, in the form of linear combinations of selected predictor variables (geomorphometric factors). In the model of 640 glaciers from the entire study area, the first component is elevation, with a PLSR weight of –1.0; the second component is dominated by slope, with a positive weight; the third component is mainly on easting with a positive weight; and the fourth component is mostly loaded with area and easting, both with negative weights (Table 5). The weight values are indicators for how

contributive individual variables are in each component in prediction of relative area change. The variable with the highest VIP value is median elevation (VIP = 2.174), followed by slope (VIP = 1.370). Although some research suggested an arbitrary value of 1 as the threshold to select important variables (Yan et al. 2013; Z. Shi et al. 2014), we feel that it is necessary to include area (VIP = 0.811) and easting (VIP = 0.835) in our regression model for their comparative importance to the rest factors. As the final step in the PLSR model, we only used the four most important variables to generate the regression for the

Table 4. Summary of the partial least squares regression model of relative area change

Response variable	Region	R^2	Q^2	Component	% of explained variance	Cumulative % of explained variance	RMSEP	Q^2_{cum}
Relative area change ($\triangle A$)	Combined (n = 640)	0.49	0.48	1	29.22	29.22	15.87	0.288
				2	12.02	41.24	14.51	0.405
				3	3.10	44.34	14.14	0.435
				4	**5.13**	**49.47**	**13.54**	**0.481**
				5	0.05	49.52	13.55	0.480
	BE (n = 392)	0.49	0.47	1	29.17	29.17	13.99	0.280
				2	7.10	36.27	13.31	0.347
				3	**12.57**	**48.84**	**12.02**	**0.467**
				4	0.21	49.05	12.05	0.465
	BG (n = 168)	0.43	0.37	1	19.12	19.12	18.00	0.171
				2	18.64	37.76	16.02	0.342
				3	**5.55**	**43.31**	**15.64**	**0.372**
				4	4.09	43.65	15.69	0.367
	KL (n = 80)	0.77	0.74	1	58.09	58.09	15.48	0.560
				2	14.59	72.68	12.60	0.708
				3	**4.24**	**76.92**	**11.99**	**0.735**
				4	0.42	77.30	12.28	0.721

Note: RMSEP = root mean squared error of prediction; BE = Boro-Eren Range; BG = Bogda Range; KL = Karlik Range. Values in bold indicate the number of components to reach minimum RMSEP and maximum Q^2.

Table 5. Variable importance for the projection values and partial least squares regression weights for the relative area change model in the whole study area

Region		Area	Elevation	Slope	Cosine A	Sine A	Easting	HI	Shape
Combined	Coefficient	−6.918	−0.056	0.900	−1.070	1.397	0.496	−0.128	0.175
	VIP	0.811	2.174	1.370	0.208[a]		0.835	0.018	0.021
	W[1]		**−1.000**						
	W[2]			**0.987**			−0.217		
	W[3]	−0.188					**1.310**		
	W[4]	**−0.606**		−0.218	−0.120	0.145	**−0.742**		
BG	Coefficient	−5.101	−0.070	1.376	−4.345	0.263	−0.811	−0.221	0.390
	VIP	0.797	1.872	1.825	0.684[a]		0.189	0.076	0.155
	W[1]		**−1.000**						
	W[2]	−0.102		**0.998**					
	W[3]	**−0.739**		−0.173	−0.654	0.136			
KL	Coefficient	−6.413	−0.073	1.787	−0.613	2.496	−1.222	−0.135	0.213
	VIP	0.643	2.452	1.207	0.260[a]		0.214	0.049	0.066
	W[1]		**−1.000**						
	W[2]	−0.117		**0.999**					
	W[3]	**−0.898**		−0.217	−0.109	0.394			
BE	Coefficient	−7.756	−0.052	0.374	0.216	1.371	0.693	−0.139	0.067
	VIP	1.370	2.181	1.119	0.275[a]		0.188	0.041	0.039
	W[1]		**−1.000**						
	W[2]			**1.028**					
	W[3]	**−0.926**		−0.324		0.145	0.151		

Note: HI = hypsometric integral; BG = Bogda Range; KL = Karlik Range; BE = Boro-Eren Range. The values shown in bold indicate that the PLSR components are mainly loaded on these corresponding variables. [1], [2], [3], and [4] indicate partial least squares regression components 1, 2, 3, and 4, respectively.
[a] The variable importance for the projection of the aspect is combined from the variable importance for the projection of CosA and SinA as $\sqrt{VIP_{CosA}^2 + VIP_{SinA}^2}$.

640 LIA glaciers in the study area:

$$\Delta A = 297.383 - 7.024 * \ln(Area) - 0.057 * Elev. + 0.883 * Slope + 0.477 * Easting \quad (6)$$

The Q^2 and the R^2 of the model are 0.483 and 0.491, respectively.

The positive coefficient of easting is contrary to our observed pattern of declining loss from west to east (Table 2). This is likely due to the highly clustered (thus nonnormal) distribution of glaciers in the study area. After excluding the easting variable in the model, the final regression is

$$\Delta A = 336.202 - 7.046 * \ln(Area) - 0.055 * Elev. + 0.788 * Slope \quad (7)$$

The Q^2 and the R^2 of the model are 0.479 and 0.487, respectively. Compared to Equation 6, the Q^2 and the R^2 only suffered a minuscule decrease, but the variables with the strongest explanatory and predictive abilities are kept. Further removal of variables would

significantly reduce the explanatory amount and thus no stepwise regression was conducted after this.

Elevation indirectly affects the mass balance of glaciers as the change in elevation is proportional to the change in temperature (Glickman 2000). The negative correlation and negative regression coefficient indicate that glaciers located at higher elevations tend to have less glacier shrinkage compared with those at lower elevations (Figure 4B). Such a finding is in accordance with the general situation that low temperatures at high elevations help retain ice, whereas relatively higher temperatures at low elevations increase ablation. Our previous test in the Boro-Eren range also found that elevation is one of the key factors to glacier changes (Y. N. Li and Li 2014). The role of elevation in determining glacial melt pattern is also emphasized when modeling future change patterns. Hall and Fagre (2003) simulated future melting as influenced by topography for glaciers in Glacier National Park, Montana, and concluded that among three derived factors, "elevation, which corresponds to temperature, was at least twice as powerful a predictor as either aspect or slope in explaining which cells had melted" (137). Our VIP values show a consistency with their finding.

Slope, as the second important factor in our regression, exhibits a positive relationship with the relative area change (Figure 4C), indicating that glaciers on steeper slopes tend to suffer larger fractions of area loss in the eastern Tien Shan. This is contradictory to the well-recognized relationship between slope and the sensitivity of a glacier to climate change. The general situation is that the same upward shift of the ELA will substantially increase the ablation area when the slope is gentle. Some previous studies have discussed this situation with examples of the Nigardsbreen Glacier in Norway (Oerlemans 1992), the Franz Josef and Fox glaciers in New Zealand (Chinn 1996), and 286 Himalayan glaciers (Scherler, Bookhagen, and Strecker 2011). Some studies also found no correlations between magnitude of glacier shrinkage and slope, such as for 321 glaciers in the North Cascades National Park Complex, Washington (Granshaw and Fountain 2006), and for 489 glaciers in the Svartisen region, northern Norway (Paul and Andreassen 2009). The slope effect on glacier dynamics could show substantial variations due to different evolving glacial landscapes of arbitrary regions. In the eastern Tien Shan, the mean slope exhibits a statistically significant dependence on glacier area ($r = -0.35$, $p < 0.01$) as the greater the slope, the smaller the glacier (Figure 4D). Many small glaciers, located at mountain ridges or cirque headwalls, as remnant patches of former valley glaciers, experience more and accelerated glacier change because the entire glacier might exist below the snow line. For example, Dong et al. (2012) reported that in 2008, the ELA of the Urumqi Glacier No. 1 reached an altitude of 4,168 m a.s.l., close to the glacier summit, implying that almost the whole glacier was ablating.

Many studies have shown that small glaciers tend to lose area relatively faster than large glaciers (e.g., Ye et al. 2003; Paul et al. 2004; Bhambri et al. 2011; K. Li et al. 2011). In the high mountains of central Asia, Bolch (2007) found a strong dependence of the glacier retreat (relative area change from 1955 to 1999) on glacier size in the northern Kyrgyz Tien Shan; Bhambri et al. (2011) reported that smaller glaciers (<1 km^2) lost proportionately about six times more of their ice than larger glaciers (>50 km^2) in the central Himalaya from 1968 to 2006. Although glacier sizes are overall much smaller in the eastern Tien Shan compared to these regions, a similar trend is clearly shown here (Figure 4E), as well as in previous studies of Boro-Eren glaciers (Y. N. Li and Li 2014), Bogda glaciers (K. Li et al. 2011), and Karlik glaciers (Wang, Li, and Gao 2011). It is suggested that small glaciers are more sensitive to changes in climate conditions, whereas large compound glaciers tend to respond more slowly (Bahr et al. 1998; Bolch 2007).

Other geomorphometric factors, including aspect, hypsometry, and shape, are not statistically significant factors to relative area changes in our models but could still play a role in the mechanism of glacier dynamics. Although aspect might affect glaciers by creating variations in solar radiation incidence, temperature, wind, and cloudiness (Evans 2006b), our results show that east- and northwest-facing glaciers have higher area loss than others, whereas north-facing glaciers lost less (Figure 4A). Shape could relate to the confinement and potential mass inputs from avalanching from surrounding slopes (DeBeer and Sharp 2007). It could enhance the sensitivity of a glacier but might not be a direct factor. Hypsometry is a simplified factor representing mass balance distribution (Furbish and Andrews 1984). Our narrowly distributed HI values indicate similar mass balance distribution of most glaciers, which have all ranges of relative area change.

Differences among Three Subregions

For all PLSR models of three subregions, the minimum RMSEP and maximum Q^2 were obtained with the first three components (Table 4). The model for the subregion KL produced an R^2 of 0.77 and Q^2 of 0.74, much higher than these in the other two subregion models (BE: $R^2 = 0.49$, $Q^2 = 0.47$; BG: $R^2 = 0.43$, $Q^2 = 0.37$; see Table 4). The selected geomorphometric factors explained the most variance in relative area change in KL, whereas in BE and BG, less than 50 percent of the variance can be explained. One reason for such subregional differences is the smaller number of glaciers in KL ($n = 80$), whereas the ability of factors to explain variances was dampened by relatively large numbers of glaciers in BE ($n = 392$) and BG ($n = 168$). The importance order of the geomorphometric factors was assessed by VIP values in three subregions, as shown in Table 5. Similar to the entire study area, median elevation is the first in all subregions, and the slope and area are the following important factors. In the BE subregion, median elevation and area are the two factors ranked highest, which is consistent with the findings in our previous test (Y. N. Li and Li 2014). Slope is the second highest ranking factor in both BG and KL, but in BG it is almost equally as important as the median elevation, whereas in KL it is only half of the VIP value of the median elevation. Based on the hypsometries as depicted in Figure 4F, KL glaciers show a distinctive curve from other subregions and the entire area, indicating more glaciers distributed at higher altitudes.

The variation in responses of glaciers implies influences from geomorphometric factors as well as other features that could cause local climate differences. The unexplained portions in the models are possibly subject to differences in shade, wind redistribution, avalanching, and other processes, which could modify incoming solar radiation and snow accumulation of a glacier. Unfortunately, these contributors to glacier changes are difficult to quantify, especially for large amounts of glaciers, and were not analyzed in this study.

Other than different roles of important geomorphometric factors within subregions, the easting factor did not show significance in statistical models. Small VIP values (<0.215; Table 5) in the BE, BG, and KL subregions indicate a limited effect of longitudinal location at a local level. The overall decline of relative glacier area changes from west to east was observed (Table 2), which might suggest a spatial pattern of climatic controls across the eastern Tien Shan. In general, the westerlies transport moisture from large water bodies such as the Caspian Sea and Aral Sea in the west, and the orographic effect of the Tien Shan reduces the moisture amount and forms a west–east gradient of precipitation across these mountain ranges (Sorg et al. 2012). Several studies have noted that during past glacial stages, the extent of glaciers in the central Asian mountains is influenced by shifting dominance of the westerlies (Kreutz et al. 1997; Benn and Owen 1998; Gong and Ho 2002; V. B. Aizen et al. 2006). For example, significant glacier advances during Marine Oxygen Isotope Stages (MIS) 4 and MIS 3 in central Asia were accounted for by abundant precipitation carried by the westerlies (Zech 2012; Y. K. Li et al. 2014), whereas arid conditions during the global last glacial maximum (~MIS 2) in the Tien Shan helped to explain the relatively restricted glacial advances (Abramowski et al. 2006; Narama et al. 2007; Koppes et al. 2008). As the westerlies become less powerful eastwards, the Siberian High dominates and delivers cold air masses to the eastern end of Tien Shan. A generally stronger Siberian High since the LIA (Gong and Ho 2002; D'Arrigo et al. 2005) counteracted the effect of limited precipitation; hence, it is suggested that in the Karlik Range, the lower temperature and relatively higher altitudes resulted in less glacier retreats than in other parts of study region (Y. X. Chen et al. 2015; Y. N. Li, Li, Harbor, et al. 2016). Knowledge of climate conditions in

three subregions during the LIA is still limited, so it prevents us from investigating how glaciers behaved differently in relation to the in situ climate conditions. To better quantify climate controls on glacier changes across the Tien Shan, we need further work and more information on past climate and robust numerical models at a detailed subregional spatial scale.

Conclusions

This study investigated the glacier changes since the LIA in the eastern Tien Shan based on the Second GIC data, delineations of LIA extents from Google Earth, and a 30-m SRTM DEM. The total area of glaciers decreased from 791.6 ± 18.7 km^2 during the LIA to 483.9 ± 31.2 km^2 during 2006 to 2010, a loss of 38.9 ± 2.7 percent. Within the whole study area and each subregion, large variability in relative glacier area change suggests that local geomorphometric settings are important in controlling the behavior of glaciers. To quantify the importance of such local factors, glacier area, aspect, slope, shape, median elevation, hypsometry, as well as the longitude were extracted for each LIA glacier and used as predictor variables, whereas the relative area change of each glacier from LIA to present was used as the response variable. Although Pearson's correlation test showed the significance of each factor in relation to relative area change, the PLSR model revealed that the variance in the response variable was attributed mainly to three factors: glacier elevation, slope, and area. Models for three subregions showed similar results to the whole area. The predictive ability is relatively higher in the Karlik, the easternmost subregion. Overall, from the western to the eastern subregions, the area loss exhibits a decreasing trend. The influence of climatic changes on glacier changes might vary spatially: The western part is more influenced by the westerlies, whereas the eastern regions are more under the impact of the Siberian High in the past few hundreds of years. To better understand the linkage among glacier, geomorphometry, and climate, we need more data and further modeling work in the eastern Tien Shan.

References

Abdi, H. 2003. *Partial least square regression (PLS regression)*. Thousand Oaks, CA: Sage.

Abramowski, U., A. Bergau, D. Seebach, R. Zech, B. Glaser, P. Sosin, P. W. Kubik, and W. Zech. 2006. Pleistocene glaciations of central Asia: Results from

[10]Be surface exposure ages of erratic boulders from the Pamir (Tajikistan), and the Alay-Turkestan range (Kyrgyzstan). *Quaternary Science Reviews* 25 (9–10): 1080–96.

Aizen, E. M., V. B. Aizen, J. M. Melack, T. Nakamura, and T. Ohta. 2001. Precipitation and atmospheric circulation patterns at mid-latitudes of Asia. *International Journal of Climatology* 21 (5): 535–56.

Aizen, V. B., E. M. Aizen, J. M. Melack, and J. Dozier. 1997. Climatic and hydrologic changes in the Tien Shan, Central Asia. *Journal of Climate* 10 (6): 1393–1404.

Aizen, V. B., E. M. Aizen, A. B. Surazakov, and V. A. Kuzmichenok. 2006. Assessment of glacial area and volume change in Tien Shan (central Asia) during the last 150 years using geodetic, aerial photo, ASTER and SRTM data. *Annals of Glaciology* 43:202–13.

Aizen, V. B., V. A. Kuzmichenok, A. B. Surazakov, and E. M. Aizen. 2007. Glacier changes in the Tien Shan as determined from topographic and remotely sensed data. *Global and Planetary Change* 56 (3): 328–40.

ArcGIS. Version 10.1. Redlands, CA: Esri.

Bahr, D. B., W. T. Pfeffer, C. Sassolas, and M. F. Meier. 1998. Response time of glaciers as a function of size and mass balance: 1. Theory. *Journal of Geophysical Research: Solid Earth* 103 (B5): 9777–82.

Benn, D., and L. Owen. 1998. The role of the Indian summer monsoon and the mid-latitude westerlies in Himalayan glaciation: Review and speculative discussion. *Journal of the Geological Society* 155 (2): 353–63.

Bhambri, R., T. Bolch, R. K. Chaujar, and S. C. Kulshreshtha. 2011. Glacier changes in the Garhwal Himalaya, India, from 1968 to 2006 based on remote sensing. *Journal of Glaciology* 57 (203): 543–56.

Bishop, M. P., H. Björnsson, W. Haeberli, J. Oerlemans, J. F. Shroder, M. Tranter, V. P. Singh, P. Singh, and U. K. Haritashya. 2011. *Encyclopedia of snow, ice and glaciers*. Berlin: Springer Science & Business Media.

Björn-Helge, M., and R. Wehrens. 2007. The PLS package: Principal component and partial least squares regression in R. *Journal of Statistical Software* 18 (2): 1–24.

Bolch, T. 2007. Climate change and glacier retreat in northern Tien Shan (Kazakhstan/Kyrgyzstan) using remote sensing data. *Global and Planetary Change* 56 (1): 1–12.

Bolch, T., B. Menounos, and R. Wheate. 2010. Landsat-based inventory of glaciers in western Canada, 1985–2005. *Remote Sensing of Environment* 114 (1): 127–37.

Brocklehurst, S. H., and K. X. Whipple. 2004. Hypsometry of glaciated landscapes. *Earth Surface Processes and Landforms* 29:907–26.

Carrascal, L. M., I. Galván, and O. Gordo. 2009. Partial least squares regression as an alternative to current regression methods used in ecology. *Oikos* 118 (5): 681–90.

Chen, F. H., Z. C. Yu, M. L. Yang, E. Ito, S. M. Wang, D. B. Madsen, X. Z. Huang, et al. 2008. Holocene moisture evolution in arid central Asia and its out-of-phase relationship with Asian monsoon history. *Quaternary Science Reviews* 27 (3–4): 351–64.

Chen, J. 1989. Preliminary researches on lichenometric chronology of Holocene glacial fluctuations and on other topics in the headwater of Urumqi River, Tianshan Mountains. *Science in China Series B: Chemistry* 32 (12): 1487–1500.

Chen, Y. X., Y. K. Li, Y. Y. Wang, M. Zhang, Z. J. Cui, C. L. Yi, and G. N. Liu. 2015. Late Quaternary glacial history of the Karlik Range, easternmost Tian Shan, derived from [10]Be surface exposure and optically stimulated luminescence datings. *Quaternary Science Reviews* 115:17–27.

Chinn, T. J. 1996. New Zealand glacier responses to climate change of the past century. *New Zealand Journal of Geology and Geophysics* 39 (3): 415–28.

Cogley, J. G. 2014. Glacier shrinkage across High Mountain Asia. *Annals of Glaciology* 57 (71): 41–49.

D'Arrigo, R., G. Jacoby, R. Wilson, and F. Panagiotopoulos. 2005. A reconstructed Siberian High index since A.D. 1599 from Eurasian and North American tree rings. *Geophysical Research Letters* 32 (5): L05705.

DeBeer, C. M., and M. J. Sharp. 2007. Recent changes in glacier area and volume within the southern Canadian Cordillera. *Annals of Glaciology* 46 (1): 215–21.

Delmas, M., Y. Gunnell, and M. Calvet. 2014. Environmental controls on alpine cirque size. *Geomorphology* 206:318–29.

Dong, Z. W., D. H. Qin, J. W. Ren, K. M. Li, and Z. Q. Li. 2012. Variations in the equilibrium line altitude of Urumqi Glacier No. 1, Tianshan Mountains, over the past 50 years. *Chinese Science Bulletin* 57:4776–83.

Evans, I. S. 2006a. Glacier distribution in the Alps: Statistical modelling of altitude and aspect. *Geografiska Annaler: Series A-Physical Geography* 88 (2): 115–33.

———. 2006b. Local aspect asymmetry of mountain glaciation: A global survey of consistency of favoured directions for glacier numbers and altitudes. *Geomorphology* 73 (1–2): 166–84.

———. 2011. Glacier distribution and direction in Svalbard, Axel Heiberg Island and throughout the Arctic: General northward tendencies. *Polish Polar Research* 32 (3): 199–238.

Farinotti, D., L. Longuevergne, G. Moholdt, D. Duethmann, T. Mölg, T. Bolch, S. Vorogushyn, and A. Güntner. 2015. Substantial glacier mass loss in the Tien Shan over the past 50 years. *Nature Geoscience* 8:716–22.

Furbish, D., and J. Andrews. 1984. The use of hypsometry to indicate long-term stability and response of valley glaciers to changes in mass transfer. *Journal of Glaciology* 30 (105): 199–211.

Gao, M., T. Han, B. Ye, and K. Jiao. 2013. Characteristics of melt water discharge in the Glacier No. 1 basin, headwater of Urumqi River. *Journal of Hydrology* 489:180–88.

Glickman, T. S. 2000. *Glossary of meteorology*. Boston: American Meteorological Society.

Gong, D. Y., and C. H. Ho. 2002. The Siberian High and climate change over middle to high latitude Asia. *Theoretical and Applied Climatology* 72 (1–2): 1–9.

Granshaw, F. D., and A. G. Fountain. 2006. Glacier change (1958–1998) in the north Cascades national park complex, Washington, USA. *Journal of Glaciology* 52 (177): 251–56.

Grove, J. M. 2004. *Little ice ages: Ancient and modern*. 2 vols. London and New York: Routledge.

Guo, W., J. Xu, S. Liu, D. Shangguan, L. Wu, X. Yao, J. Zhao, Q. Liu, Z. Jiang, and P. Li. 2014. *The second glacier inventory dataset of China* (Version 1.0). Lanzhou, China: Cold and Arid Regions Science Data Center at Lanzhou.

Hall, M. H., and D. B. Fagre. 2003. Modeled climate-induced glacier change in Glacier National Park, 1850–2100. *BioScience* 53 (2): 131–40.

Ihaka, R., and R. Gentleman. 1996. R: A language for data analysis and graphics. *Journal of Computational and Graphical Statistics* 5 (3): 299–314.

Kemsley, E. 1996. Discriminant analysis of high-dimensional data: A comparison of principal components analysis and partial least squares data reduction methods. *Chemometrics and Intelligent Laboratory Systems* 33 (1): 47–61.

Koppes, M., A. R. Gillespie, R. M. Burke, S. C. Thompson, and J. Stone. 2008. Late Quaternary glaciation in the Kyrgyz Tien Shan. *Quaternary Science Reviews* 27 (7–8): 846–66.

Kreutz, K., P. Mayewski, L. Meeker, M. Twickler, S. Whitlow, and I. Pittalwala. 1997. Bipolar changes in atmospheric circulation during the Little Ice Age. *Science* 277:1294–96.

Li, K., H. Li, L. Wang, and W. Gao. 2011. On the relationship between local topography and small glacier change under climatic warming on Mt. Bogda, eastern Tian Shan, China. *Journal of Earth Science* 22 (4): 515–27.

Li, Y. K., G. N. Liu, Y. X. Chen, Y. N. Li, J. Harbor, A. P. Stroeven, M. Caffee, M. Zhang, C. C. Li, and Z. J. Cui. 2014. Timing and extent of quaternary glaciations in the Tianger Range, eastern Tian Shan, China, investigated using [10]Be surface exposure dating. *Quaternary Science Reviews* 98:7–23.

Li, Y. K., J. Napieralski, and J. Harbor. 2008. A revised automated proximity and conformity analysis method to compare predicted and observed spatial boundaries of geologic phenomena. *Computers & Geosciences* 34 (12): 1806–14.

Li, Y. N., and Y. K. Li. 2014. Topographic and geometric controls on glacier changes in the central Tien Shan, China, since the Little Ice Age. *Annals of Glaciology* 55 (66): 177–86.

Li, Y. N., Y. K. Li, Y. X. Chen, and X. Y. Lu. 2016. Presumed Little Ice Age glacial extent in the eastern Tian Shan, China. *Journal of Maps*. Advance online publication. http://dx.doi.org/10.1080/17445647.2016.1158595

Li, Y. N., Y. K. Li, J. Harbor, G. N. Liu, C. L. Yi, and M. Caffee. 2016. Cosmogenic [10]Be constraints on Little Ice Age glacial advances in the eastern Tian Shan, China. *Quaternary Science Reviews* 138:105–18.

Liu, Q., S. Y. Liu, W. Q. Guo, Y. Nie, D. H. Shangguan, J. L. Xu, and X. J. Yao. 2015. Glacier changes in the Lancang River Basin, China, between 1968–1975 and 2005–2010. *Arctic, Antarctic, and Alpine Research* 47 (2): 335–44.

Liu, S. Y., W. X. Sun, Y. P. Shen, and G. Li. 2003. Glacier changes since the Little Ice Age maximum in the western Qilian Shan, northwest China, and consequences of glacier runoff for water supply. *Journal of Glaciology* 49 (164): 117–24.

Liu, S. Y., X. J. Yao, W. Q. Guo, J. L. Xu, D. H. Shangguan, J. F. Wei, W. J. Bao, and L. Z. Wu. 2015. The contemporary glaciers in China based on the Second Chinese Glacier Inventory. *Acta Geographica Sinica* 70 (1): 3–16.

Narama, C., A. Kääb, M. Duishonakunov, and K. Abdrakhmatov. 2010. Spatial variability of recent glacier area changes in the Tien Shan Mountains, Central Asia, using Corona (~1970), Landsat (~2000), and ALOS (~2007) satellite data. *Global and Planetary Change* 71 (1): 42–54.

Narama, C., R. Kondo, S. Tsukamoto, T. Kajiura, C. Ormukov, and K. Abdrakhmatov. 2007. OSL dating of glacial deposits during the last glacial in the Terskey-Alatoo Range, Kyrgyz Republic. *Quaternary Geochronology* 2 (1–4): 249–54.

Oerlemans, J. 1992. Climate sensitivity of glaciers in southern Norway: Application of an energy-balance model to Nigardsbreen, Hellstugbreen and Alfotbreen. *Journal of Glaciology* 38:223–32.

Paul, F., and L. M. Andreassen. 2009. A new glacier inventory for the Svartisen region, Norway, from Landsat ETM+ data: Challenges and change assessment. *Journal of Glaciology* 55 (192): 607–18.

Paul, F., A. Kääb, M. Maisch, T. Kellenberger, and W. Haeberli. 2004. Rapid disintegration of Alpine glaciers observed with satellite data. *Geophysical Research Letters* 31 (21): L21402.

R Development Core Team. *R*. Version 3.2.2. Vienna, Austria: R Development Core Team.

Scherler, D., B. Bookhagen, and M. R. Strecker. 2011. Spatially variable response of Himalayan glaciers to climate change affected by debris cover. *Nature Geoscience* 4 (3): 156–59.

Shi, Y. F., C. H. Liu, and E. Kang. 2009. The Glacier Inventory of China. *Annals of Glaciology* 50 (53): 1–4.

Shi, Z., X. Huang, L. Ai, N. Fang, and G. Wu. 2014. Quantitative analysis of factors controlling sediment yield in mountainous watersheds. *Geomorphology* 226:193–201.

Sorg, A., T. Bolch, M. Stoffel, O. Solomina, and M. Beniston. 2012. Climate change impacts on glaciers and runoff in Tien Shan (Central Asia). *Nature Climate Change* 2 (10): 725–31.

Wang, W., K. Li, and J. Gao. 2011. Monitoring glacial shrinkage using remote sensing and site-observation method on southern slope of Kalik Mountain, eastern Tian Shan, China. *Journal of Earth Science* 22:503–14.

Wei, J., S. Liu, W. Guo, X. Yao, J. Xu, W. Bao, and Z. Jiang. 2014. Surface-area changes of glaciers in the Tibetan Plateau interior area since the 1970s using recent Landsat images and historical maps. *Annals of Glaciology* 55 (66): 213–22.

Wold, S., L. Eriksson, J. Trygg, and N. Kettaneh. 2004. *The PLS method—partial least squares projections to latent structures—and its applications in industrial RDP (research, development, and production)*. Umeå, Sweden: Umeå University.

Xu, X., and C. Yi. 2014. Little Ice Age on the Tibetan Plateau and its bordering mountains: Evidence from moraine chronologies. *Global and Planetary Change* 116:41–53.

Yan, B., N. Fang, P. Zhang, and Z. Shi. 2013. Impacts of land use change on watershed streamflow and sediment yield: An assessment using hydrologic modelling and partial least squares regression. *Journal of Hydrology* 484:26–37.

Ye, B., Y. Ding, F. Liu, and C. Liu. 2003. Responses of various-sized alpine glaciers and runoff to climatic change. *Journal of Glaciology* 49 (164): 1–7.

Ye, B., D. Yang, K. Jiao, T. Han, Z. Jin, H. Yang, and Z. Li. 2005. The Urumqi River source Glacier No. 1, Tianshan, China: Changes over the past 45 years. *Geophysical Research Letters* 32 (21): L21504.

Yi, C., K. Liu, Z. Cui, K. Jiao, T. Yao, and Y. He. 2004. AMS radiocarbon dating of late Quaternary glacial landforms, source of the Urumqi River, Tien Shan—A pilot study of ^{14}C dating on inorganic carbon. *Quaternary International* 121 (1): 99–107.

Yin, A., and T. M. Harrison. 2000. Geologic evolution of the Himalayan–Tibetan orogen. *Annual Review of Earth and Planetary Sciences* 28 (1): 211–80.

Zech, R. 2012. A late Pleistocene glacial chronology from the Kitschi-Kurumdu Valley, Tien Shan (Kyrgyzstan), based on 10Be surface exposure dating. *Quaternary Research* 77 (2): 281–88.

Some Perspectives on Avalanche Climatology

Cary J. Mock, Kristy C. Carter, and Karl W. Birkeland

Avalanche climatology is defined as the study of the relationships between climate and snow avalanches, and it contributes in aiding avalanche hazard mitigation efforts. The field has evolved over the past six decades concerning methodology, data monitoring and field collection, and interdisciplinary linkages. Avalanche climate research directions are also expanding concerning treatment in both spatial scale and temporal timescales. This article provides an overview of the main themes of avalanche climate research in issues of scale from local to global, its expanding interdisciplinary nature, as well as its future challenges and directions. The growth of avalanche climatology includes themes such as its transformation from being mostly descriptive to innovative statistical methods and modeling techniques, new challenges in microscale efforts that include depth hoar aspects and increased field studies, expanding synoptic climatology applications on studying avalanche variations, efforts to reconstruct past avalanches and relate them to climatic change, and research on potential avalanche responses to recent twentieth-century and future global warming. Some suggestions on future avalanche climatology research directions include the expansion of data networks and studies that include lesser developed countries, stronger linkages of avalanche climate studies with GIScience and remote sensing applications, more innovative linkages of avalanches with climate and societal applications, and increased emphases on modeling and process-oriented approaches.

雪崩气候学定义为气候和雪崩之间的关系之研究，并对于增进雪崩灾害减轻的努力做出贡献。该领域中的方法论、数据监控、田野搜集和跨领域连结，在过去六十年来已有所进展。雪崩气候研究的方向，在处理空间尺度和暂时的时间尺度上亦有所扩张。本文提供了雪崩气候研究在从地方到全球尺度，其扩张中的跨领域本质，及其未来的挑战和方向等重大议题之概要。雪崩气候学的成长，包含了诸如从多半为描述性到创造性的统计方法和模式化技术之变革，包含雪中白霜面和增加的田野研究之微观尺度努力的挑战，扩张的天气气候学在研究雪崩变异性方面的应用，重建过往雪崩并将其连结至气候变迁的努力，以及对于晚近二十世纪和未来全球暖化的潜在雪崩回应之研究等主题。对于未来雪崩气候学研究方向的若干建议，包含扩大数据网络与含纳第三世界国家的研究，雪崩气候研究和 GIS 科学与遥测应用之间更为强大的连结，雪崩与气候和社会应用之间更为创新的连结，以及更为强调模式化和以过程为导向的方法。 关键词： 雪崩气候学，微气候学，雪科学，空间尺度，天气气候学。

La climatología de avalancha se define como el estudio de las relaciones entre el clima y las avalanchas de nieve, y contribuye en los esfuerzos de mitigación por los riesgos de aludes. Esta especialidad ha evolucionado durante las pasadas seis décadas en lo que se refiere a metodología, monitoreo y recolección de datos de campo, y vínculos interdisciplinarios. Las direcciones de la investigación sobre clima de avalancha también se están ampliando en lo que concierne al tratamiento del problema a escalas espacial y temporal. Este artículo suministra una visión de conjunto de los temas principales de investigación sobre clima de avalancha en aspectos relacionados con la variación de escala de lo local a lo global, su naturaleza interdisciplinaria en expansión, lo mismo que sus retos y direcciones futuras. El crecimiento de la climatología de avalancha incluye temas como su transformación de ser principalmente descriptiva al uso de métodos estadísticos innovadores y técnicas de modelado, nuevos desafíos en los esfuerzos a microescala que incluyen aspectos de escarcha de profundidad y más estudios de campo, expandiendo las aplicaciones de climatología sinóptica al estudio de variaciones de las avalanchas, esfuerzos para reconstruir avalanchas pasadas para relacionarlas con cambio climático, e investigación sobre avalanchas potenciales en respuesta al reciente calentamiento global del siglo XX y hacia el futuro. Algunas sugerencias sobre direcciones futuras de investigación en la climatología de avalancha incluyen la expansión de redes de datos y estudios que incluyan a los países menos desarrollados, vínculos más fuertes de los estudios de clima de avalancha con la ciencia de los SIG y aplicaciones de percepción remota, vínculos más innovadores de los estudios de avalanchas con aplicaciones climáticas y sociales, y un mayor énfasis en el diseño de modelos y en enfoques orientados a proceso.

Avalanches are a dangerous natural hazard in mountainous regions worldwide, killing hundreds of people annually. Avalanche victims in less developed countries tend to be killed by large avalanches descending into villages or onto roads. Exact fatality numbers are hard to ascertain in these areas, but fatalities from well-documented avalanche cycles in these countries occasionally number in the hundreds. Although less frequent, such avalanche deaths also occur in developed countries such as in the European Alps, occasionally involving accidents with many dozens of deaths. Zoning efforts over most developed areas ensure that such incidents are relatively rare. Some areas, though, like Juneau, Alaska, still have high population densities living in avalanche-prone areas and are threatened by nearby large avalanche paths.

Avalanche victims in developed countries are primarily people recreating on skis, snowboards, snowmobiles, and snowshoes, and they typically trigger the avalanches that kill them. Europe and Canada have about fifty and fifteen victims annually, respectively. Twenty-five to thirty people are killed in avalanches annually in the United States, a number that has stayed consistent since the early 2000s (Figure 1). Dangerous avalanche conditions routinely close highways, causing sizable economic impacts. For example, an avalanche closure of Little Cottonwood Canyon near Salt Lake City, Utah, affected approximately 10,000 automobiles per day in the early 1990s (Blattenberger and Fowles 1995; Figure 2), and the number of affected automobiles and associated economic costs are much higher today.

Climate acts as a background condition that characterizes snowpack conditions that can cause different types of avalanching, and seasonal-to-daily weather conditions are the primary drivers for large avalanche cycles. The application of climate and weather data as a first trigger for understanding snow avalanche activity has been routinely conducted by scientists dating back to the earlier twentieth century (Seligman 1936) and was coined *avalanche climatology* initially by R. L. Armstrong and Armstrong (1987) and popularized by Mock and Kay (1992). Avalanche climatology deals with a range of timescales from daily to decadal. In recent decades, climatologists have studied climatic processes related to various environmental responses in more detail, linking information from various space scales to gain a more holistic knowledge of climate system components. This article describes the historical developments of avalanche climatology in recent decades, emphasizing issues of spatial scale from local to regional, growing interdisciplinary aspects, and some suggestions for future directions.

Avalanche Climatology at Local and Microscales

Prior to the mid-1900s, research involving weather as an important control on avalanche activity was mostly locally applied with a forecast emphasis. This research was basically descriptive, supplemented by some field observations. As longer and more numerous mountain weather and avalanche records became available in the 1970s for many mountain settings around the world (Beniston, Diaz, and Bradley 1997), statistical quantification became more common. These methods initially focused on basic bivariate analyses to identify important weather variables associated with avalanching at certain locations, like heavy precipitation and wind speed (Perla 1970). The statistical techniques employed for local avalanche forecasting eventually evolved to more multivariate and later more nonparametric approaches (e.g., Föhn et al. 1977; Davis et al. 1999; McCollister et al. 2003; Hendrikx, Murphy, and Onslow 2014). Supplemental nonstatistical approaches on avalanche forecasting, using increased understanding of snowpack processes gathered from field data, continued as demonstrated by Fitzharris (1976) for Mt. Cook in New Zealand and Conway and Raymond (1993) for the Pacific Northwest of the United States. Despite advances in statistical techniques, avalanche forecasting today continues to rely largely on conventional techniques (LaChapelle 1980), although methods of remote data collection and data sharing (Stethem et al. 2003) have expanded dramatically in the past thirty-five years. Climatologically, coastal locations might rely more heavily on weather data for avalanche forecasting, whereas intermountain and continental zones require careful analysis of existing snow structure (LaChapelle 1966, 1980). Ultimately, conventional avalanche forecasting requires the consideration of all factors, regardless of the climate zone.

The formation of snow is influenced by many meteorological variables that are changing constantly in time and space (Nakaya 1954; Libbrecht 2005). Each storm is unique in its temperature, humidity, wind speed, turbulence, and other factors; thus, each snowfall deposits a unique layer. Once on the ground, these layers further differentiate due to rapid snow

Fatalities in the United States, 1951-2014

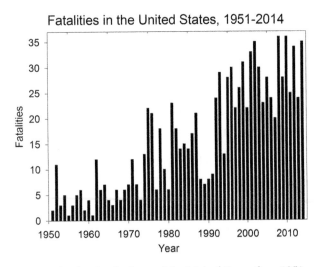

Figure 1. Avalanche fatalities of the United States from 1951 to 2014.

metamorphism (Fierz et al. 2009), and the nature of these layers ultimately determines the avalanche danger. When snow falls in mountainous terrain, micrometeorological processes significantly influence both the snow surface and the subsurface snow. This emphasizes the importance of addressing different spatial scales when studying avalanche climatology. Limited work has been completed to address the micrometeorological influences on avalanche activity, due to the challenging nature of researching constantly changing snow packs. The research that has been done is also geographically limited to British Columbia (Colbeck, Jamieson, and Crowe 2008; Bellaire and Jamieson 2013; Horton, Bellaire, and Jamieson 2014), Davos, Switzerland (Stossel et al. 2010; Reuter et al. 2015), and northern Wyoming and western Montana (Birkeland 2001; Kozak et al. 2003; Slaughter et al. 2009) and not necessarily representative of all avalanche-prone areas.

One of the most important formations on top of the snow surface from an avalanche perspective is that of surface hoar (Figure 3). The winter equivalent of dew, surface hoar is feathery or solid crystals that grow upward with formation typically taking place overnight during clear, calm, and relatively humid conditions (Colbeck 1988). Once buried by subsequent snowfall, surface hoar creates a thin, dangerous, and persistent weak layer. Indeed, Jamieson and Johnston (1992) reported that surface hoar was present in 50 percent of avalanche accidents involving professionals. The difficult assessment of surface hoar instability is amplified further by its variable distribution pattern across an already inconsistent mountain terrain, in terms of both surface hoar formation and its persistence until burial (Lutz and Birkeland 2011; Helbig and van Herwijnen 2012).

Figure 2. A large avalanche descends Mount Superior in Utah's Little Cottonwood Canyon. (Color figure available online.)

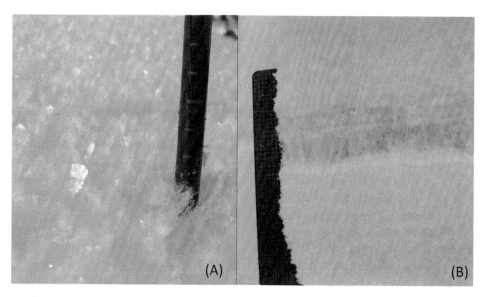

Figure 3. Surface hoar (A) is typically composed of feathery crystals (probe markings are 1 cm) and (B) creates a dangerous weak layer when buried (grid size is 2 mm). (Color figure available online.)

Spatial variations of surface hoar growth have been modeled based on terrain characteristics (Helbig and van Herwijnen 2012), weather models (Bellaire and Jamieson 2013; Horton, Bellaire, and Jamieson 2014), and laboratory experiments (Slaughter et al. 2009). Other surface hoar research includes tracking the hardness of snow layers (Kozak et al. 2003), a direct comparison between eddy flux measurements of moisture transport and surface mass changes (Stossel et al. 2010), and examining turbulence, vapor pressure, and solar radiation influences of surface hoar growth at the microlevel (Colbeck 1988). The lack of fine-scale micrometeorology measurements on the snow surface challenges our ability to track surface hoar growth and persistence over a wider spatial extent.

Regional Avalanche Climatology and Synoptic Climatology

Some research focuses on surface avalanche and weather conditions at longer regional and temporal scales. Compared to conventional climatology research, avalanche climatology studies are mostly restricted to time frames of no more than a few decades per location, given the paucity of upper elevation data near avalanche paths. Climate is generally viewed as statistical mean characteristics of weather series under steady-state conditions, and many studies use this steady-state concept. These steady-state characteristics could vary within longer timescales, however, thus placing the concept of avalanche climatology quite differently when assessing consequences of climate change on avalanche activity.

Regionalization of three main avalanche climates of the western United States, initially based on field observations, have long been recognized to exist, representing generally a west–east gradient: the coastal, intermountain, and continental (LaChapelle 1966). This zonation provides a general framework for understanding how climatic variables influence avalanche activity. The coastal zone is characterized by abundant snowfall, higher snow densities, warmer temperatures, and a higher frequency of avalanches. The continental zone is characterized by cooler temperatures, lower snowfall, lower snow densities, and extensive faceted crystal growth resulting from high temperature gradients in the snowpack. Relatively lower numbers of avalanches occur in the continental zone, but this zone has a higher avalanche hazard potential due to difficulties in predicting the behavior of buried faceted crystals (Birkeland 1998). The intermountain zone is intermediate between the other two types. R. L. Armstrong and Armstrong (1987), Fitzharris (1981), McClung and Tweedy (1993), Mock and Kay (1992), Mock (1995), Mock and Birkeland (2000), and Haegeli and McClung (2003) further documented similar climate zones over western North America. Some studies used multivariate statistics for classification (e.g., Mock 1995), whereas some had more of a qualitative–quantitative mixture such as a decision trees approach (Mock and Birkeland

2000). Hackett and Santeford (1980) mapped Alaska in a similar avalanche climate classification, based mostly on field observations. Recently, Ikeda et al. (2009) extended Mock and Birkeland's (2000) classification approach for analyzing the avalanche climate of Japan.

Operational avalanche forecasting has a long traditional synoptic meteorological component dating back to the advent of detailed daily synoptic weather maps in the mid-twentieth century. This component links daily synoptic-scale circulation patterns to big avalanche events, some of which are in governmental reports. Published studies on this approach include those by Birkeland and Mock (1996) for Bridger Bowl, Montana; Fitzharris (1981) for Rogers Pass, British Columbia; Calonder (1986) and Hächler (1987) for the Swiss Alps, which involve the application of daily Grosswetterlagen synoptic types to avalanche activity; Rangachary and Bandyopadhyay (1987) for the Himalayas; and Gürer et al. (1995) on an avalanche case study for northwest Anatolia, Turkey.

An avalanche climate provides an idea of the avalanche problems faced in a certain area. The concept of avalanche climate is especially important during exceptional years or when climate changes. As discussed in Mock and Birkeland (2000), an area in a coastal avalanche climate might occasionally have a year with a continental avalanche climate (drier and colder than normal) and during that year they would anticipate having avalanche problems characteristic of more continental locations such as full-depth avalanches failing on depth hoar. Most avalanche and synoptic climate studies conducted to date are based on using avalanche extremes to study synoptic patterns (surface-to-circulation approach), although Mock and Birkeland (2000) and Birkeland and Mock (2001), in a few examples, used a circulation-to-surface approach to map surface avalanche climate variations for selected years. Fitzharris and Bakkehøi (1986) mapped synoptic patterns and used circulation indexes to explain major avalanche winters in Norway, and Fitzharris (1987) employed a similar approach to study major avalanche winters at Rogers Pass, British Columbia. Mock and Kay (1992) and Mock (1995, 1996) defined avalanche extremes at the surface using multivariate statistical methods on surface avalanche and weather data to then employ synoptic composite anomaly maps linked to avalanche extremes for areas in the western United States and southern Alaska. Birkeland, Mock, and Shinker (2001) expanded on this type of approach for the

intermountain west but based surface avalanche extremes from a daily avalanche index based largely on size and compositing NCEP-Reanalysis data products for studying synoptic patterns. Hansen and Underwood (2012) demonstrated a comprehensive synoptic approach as related to large avalanches at Mt. Shasta, California, using synoptic composites, soundings, and trajectory models.

Some papers have addressed possible linkages between teleconnections and avalanche activity. Keylock (2003) illustrated that positive phases in the North Atlantic Oscillation can lead to increased storms and major avalanche events in Iceland, but this association is not always consistent. McClung (2013) and Dixon et al. (1999) suggested linkages between La Niña and El Niño with avalanche activity in parts of northwestern North America, with La Niña causing more snow and avalanches. McClung's (2013) results, however, are marginally statistically significant via t tests and failed to closely address different strengths and character of individual El Niño Southern Oscillation (ENSO) events, as well as weekly to daily weather variations within individual ENSO winters. Reardon et al. (2008) suggested possible Pacific Decadal Oscillation (PDO) signals from avalanche data at longer timescales, with linkages with decadal climate and avalanches that remain weak.

More Interdisciplinary Aspects

Proxy data, as commonly used in paleoclimatology, have similarly been used for reconstructing avalanche activity prior to modern records. Historical documents have been used to compile avalanche histories, although the subjectivity and sporadic availability of historical data through time have made reliable continuous reconstructions difficult to relate with climate. B. R. Armstrong (1977) used historical data to compile a continuous time series of avalanche fatalities for Ouray County, Colorado, back to the late nineteenth century and made broad inferences on an active avalanche period around the 1880s that was potentially linked to climate change. Butler (1986) used newspapers from Glacier National Park, Montana, to reconstruct avalanche events and attributed them to large-scale atmospheric circulation patterns. The most comprehensive historical avalanche reconstructions come from Europe, which has longer written histories. Laternser and Pfister (1997) demonstrated that severe avalanche winters can be reconstructed from documentary data extending back 500 years. They also attributed many of these extremes to synoptic patterns that similarly appear in the more recent modern record. Hétu, Fortin, and Brown (2015) analyzed a mixture of documentary and meteorological data as well as maps and photographs to compile an avalanche history of the Québec City region dating back to the early 1800s. They attributed increased avalanche activity at the end of the nineteenth century to increased snowfall, but land use and reforestation also played important roles variably with time.

Some research has reconstructed avalanche activity from tree rings sampled in avalanche paths and cross-dated, assuming that suppressed tree growth immediately following avalanche occurrence can be reliably assessed (e.g., Muntán et al. 2009). These potential avalanche responses can be clarified through comparisons with tree-ring records outside avalanche paths. Viewpoints on how these tree ring reconstructions of avalanches specifically relate to climate varies among different environments, as tree-ring response is limited to within the avalanche path and also relates largely to additional nonclimatic factors. For example, Rayback's (1998) tree-ring study on avalanches for the Colorado Front Range indicated avalanche magnitude in relation to climate is limited mostly to large avalanche extremes, such as the severe avalanche winter of 1985–1986. Hebertson and Jenkins (2003) described similar results through their tree-ring reconstruction from south-central Utah. Historical data have also been used for verifying tree-ring avalanche reconstructions, as demonstrated by Carrara (1979) for Ophir, Colorado, and Butler (1986) for northwest Montana. Corona et al. (2012) provided a tree-ring reconstruction of avalanches from the French Alps back to 1338 AD and attributed some linkages of avalanche activity with annual-to-decadal temperature variability through the Little Ice Age. Schläppy et al. (2016) further studied the French Alps record via logistic regression approaches, demonstrating that their tree rings can reveal some climate signals.

Few studies have addressed snow avalanche variations related to recent climatic changes at regional to hemispheric scales, including forcing mechanisms such as prominent large-scale cycles and those related to potential global warming. Fitzharris (1981), building on previous work by Touhinsky (1966), synthesized long-term avalanche records dating back to the early twentieth century spread out in the Northern Hemisphere. He found no clear linkages of avalanche activity with solar cycles. Glazovskaya (1998)

suggested linkages of climatic change governing spatial patterns of avalanche variability to the global scale, urging the need to incorporate variables of snowfall, snow days, and snow depth, but ideas were generalized.

Schneebeli, Laternser, and Ammann (1997) addressed varying long-term avalanche data quality for Switzerland as hampering efforts to detect signals related to recent large-scale warming trends. Eckert, Baya, and Deschatres (2010), later expanded by Eckert et al. (2010), applied a hierarchical modeling framework to study the spatiotemporal variability of avalanche occurrences in the northern French Alps and found no clear relation to climate trends but did find a decreasing trend probably related to avalanche runout for recent decades. Castebrunet, Eckert, and Giraud (2012) used a more detailed northern French Alps data set from 1958 to 2009 and a time-explicit model to specify climate signals. They found that avalanches at low frequencies were weakly related to temperature increases. They also found a second avalanche signal that relates to decadal variations of cold and snowier conditions. Lazar and Williams (2008) provided a local example for Aspen Mountain, Colorado, applying general circulation and regional model output to snowmelt runoff and snow quality models in forecasting wet avalanche activity. They suggested that by 2030, wet avalanches at higher elevations will occur from two to nineteen days earlier than today's climate and sixteen to twenty-seven days earlier at lower and middle elevations. Castebrunet et al. (2014) expanded on their previous work on the French Alps, conducting statistical downscaling techniques to predict avalanche activity for the 2020 to 2050 and 2070 to 2100 periods. They imply a general decrease of 20 to 30 percent of avalanche activity as well as interannual variability in the future but an increase of wet snow avalanches at higher altitudes in midwinter.

Conclusions

Several important developments on avalanche climatology are apparent (Figure 4). Due to increased data collection, technology, and data sharing, there has been increased sophistication in research concerning statistics and other methods. An increase in field and local studies concerning snow science is apparent, and this is being supplemented with more studies at larger spatial scales. Avalanche climate and related research is only prominent at a small number of academic and research centers; it also largely involves the

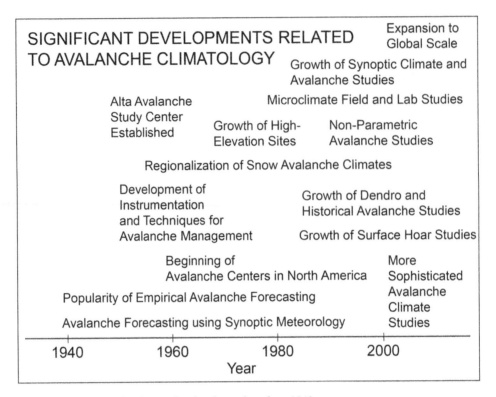

Figure 4. Timeline of important events related to avalanche climatology from 1940 to present.

applied community of avalanche practitioners such as avalanche forecasters, ski patrollers, and guides.

Several avenues exist for expanding future research involving climatology and avalanches. First, avalanche climate research to date has rarely been largely involved in GIScience applications, particularly considering climate variability. Most GIScience studies on avalanches use digital elevation models, vegetation, and basic snowfall parameters for mapping avalanche hazard zones. Some studies, particularly those from Switzerland, clearly demonstrate GIScience potential for incorporating more of a climate component (e.g., Schweizer, Mitterer, and Stoffel 2009), and this might be more important in future developments to link dynamical modeling of avalanches with climate variability and perhaps climate model output. Remote sensing might eventually become a major tool in avalanche climatology. Such applications have been useful in detecting avalanche occurrence (Eckerstorfer et al. 2016) and for surface hoar, which might relate to triggering of weak slab avalanches (Bühler, Meier, and Ginzler 2015).

Furthermore, future directions involve more research studying avalanches and climates that also includes lesser developed countries with a high avalanche hazard. Continued improvement of high-elevation networks of climate and avalanche data,

including field aspects and automated data networks, is imperative. These data sets can be used to improve and verify the continued use of statistical and dynamical modeling perspectives. Longer temporal avalanche climate perspectives can be expanded for prior to the mid-twentieth century, combining paleoenvironmental and historical research with newer developments from the twentieth century reanalysis project and modeling. Another interesting area for future avalanche climate research would be to integrate climate and avalanches with society. Podolskiy et al. (2014) provided an excellent example on integrating avalanche fatalities of the Sakhalin and Kuril Islands for the last century with changing land use, politics, and climate.

Avalanche climatological research has become increasingly complex, similar to many other fields of applied climatology. Future successful avalanche research should not be based solely in climate and weather determinism approaches. Avalanche climatology is an important theme for studying future mountain environments and climatic change, and this work will continue to involve scholars ranging from atmospheric scientists to geographers to snow scientists. Given the complexities on linking climate, weather, and avalanche processes at different temporal and spatial scales, successful future research must emphasize more process-oriented and modeling approaches as opposed to being engrained in basic empiricism. Supplementing process-oriented approaches with statistical techniques and mixed-method quantitative and qualitative approaches is still vital, but geographers involved in avalanche climatology should possess substantial training spanning from meteorology to snow science to successfully tackle future challenging process-oriented approaches.

Acknowledgments

We gratefully thank two anonymous reviewers for their very thoughtful comments and suggestions.

References

Armstrong, B. R. 1977. Avalanche hazard in Ouray County, Colorado 1877–1976. Boulder, CO: Institute of Arctic and Alpine Research Occasional Paper 24, University of Colorado.

Armstrong, R. L., and B. R. Armstrong. 1987. Snow and avalanche climates of the western United States: A comparison of maritime, intermountain and continental conditions. *International Association of Hydrological Sciences Publications* 162:281–94.

Bellaire, S., and B. Jamieson. 2013. Forecasting the formation of critical snow layers using a coupled snow cover and weather model. *Cold Regions Science and Technology* 94:37–44.

Beniston, M., H. F. Diaz, and R. S. Bradley. 1997. Climatic change at high-elevation sites: An overview. *Climatic Change* 36 (3–4): 233–51.

Birkeland, K. W. 1998. Terminology and predominant processes associated with the formation of weak layers of near-surface faceted crystals in the mountain snowpack. *Arctic and Alpine Research* 30 (2): 193–99.

———. 2001. Spatial patterns of snow stability throughout a small mountain range. *Journal of Glaciology* 47 (157): 176–86.

Birkeland, K. W., and C. J. Mock. 1996. Atmospheric circulation patterns associated with heavy snowfall events, Bridger Bowl, Montana, U.S.A. *Mountain Research and Development* 16:281–86.

———. 2001. The major snow avalanche cycle of February 1986 in the western United States. *Natural Hazards* 24:75–95.

Birkeland, K. W., C. J. Mock, and J. J. Shinker. 2001. Extreme avalanche hazard events and atmospheric circulation patterns. *Annals of Glaciology* 32:135–40.

Blattenberger, G., and R. Fowles. 1995. Road closure to mitigate avalanche danger: A case study for Little Cottonwood Canyon. *International Journal of Forecasting* 11 (1): 159–74.

Bühler, Y., L. Meier, and C. Ginzler. 2015. Potential of operational high spatial resolution near-infrared remote sensing instruments for snow surface type mapping. *IEEE Geoscience and Remote Sensing Letters* 12 (4): 821–25.

Butler, D. R. 1986. Snow-avalanche hazards in Glacier National Park, Montana, meteorologic and climatologic aspects. *Physical Geography* 7:72–87.

Calonder, G. P. 1986. *Ursachen, Wahrschlinlichkeit und Intensität von Lawinenkatastrophen in den Schweizer Alpen* [Causes, probability and intensity of avalanche disasters in the Swiss Alps, diploma thesis, University of Zurich]. Zürich, Switzerland: Diplomarbeit, University of Zürich.

Carrara, P. E. 1979. The determination of snow avalanche frequency through tree-ring analysis and historical records at Ophir, Colorado. *Geological Society of America Bulletin* 90:773–80.

Castebrunet, H., N. Eckert, and G. Giraud. 2012. Snow and weather climatic control on snow avalanche occurrence fluctuations over 50 yr in the French Alps. *Climate of the Past* 8 (2): 855–75.

Castebrunet, H., N. Eckert, G. Giraud, Y. Durand, and S. Morin. 2014. Projected changes of snow conditions and avalanche activity in a warming climate: The French Alps over the 2020–2050 and 2070–2100 periods. *The Cryosphere* 8:1673–97.

Colbeck, S. C. 1988. On the micrometeorology of surface hoar growth on snow in mountainous areas. *Boundary Layer Meteorology* 44:1–12.

Colbeck, S. C., B. Jamieson, and S. Crowe. 2008. An attempt to describe the mechanism of surface hoar growth from valley clouds. *Cold Regions Science and Technology* 54 (2): 83–88.

Conway, H., and C. F. Raymond. 1993. Snow stability during rain. *Journal of Glaciology* 39 (133): 635–42.

Corona, C., J. L. Saez, M. Stoffel, G. Rovera, J. L. Edouard, and F. Berger. 2012. Seven centuries of avalanche activity at Echalp (Queyras massif, southern French Alps) as inferred from tree rings. *The Holocene* 23 (2): 292–304.

Davis, R. E., K. Elder, D. Howlett, and E. Bouzaglou. 1999. Relating storm and weather factors to dry slab avalanche activity at Alta, Utah, and Mammoth Mountain, California, using classification and regression trees. *Cold Regions Science and Technology* 30 (1–3): 79–89.

Dixon, R. W., D. R. Butler, L. M. Dechano, and J. A. Henry. 1999. Avalanche hazard in Glacier National Park: An El Niño connection? *Physical Geography* 20 (6): 461–67.

Eckerstorfer, M., Y. Bühler, R. Frauenfelder, and E. Malnes. 2016. Remote sensing of snow avalanches: Recent advances, potential, and limitations. *Cold Regions Science and Technology* 121 (February): 126–40.

Eckert, N., H. Baya, and M. Deschatres. 2010. Assessing the response of snow avalanche runout altitudes to climate fluctuations using hierarchical modeling: Application to 61 winters of data in France. *Journal of Climate* 23 (12): 3157–80.

Eckert, N., E. Parent, R. Kies, and H. Baya. 2010. A spatiotemporal modelling framework for assessing the fluctuations of avalanche occurrence resulting from climate change: Application to 60 years of data in the northern French Alps. *Climatic Change* 101 (3): 515–53.

Fierz, C., R. Armstrong, Y. Durand, P. Etchevers, E. Greene, D. McClung, K. Nishimura, P. K. Satyawali, and S. Sokratov. 2009. The international classification for seasonal snow on the ground. Paris: IHP-VII Technical Documents in Hydrology, No. 83, IACS Contribution No. 1, UNESCO-IHP.

Fitzharris, B. B. 1976. An avalanche event in the seasonal zone of the Mount Cook region, New Zealand. *New Zealand Journal of Geology and Geophysics* 19:449–62.

———. 1981. Frequency and climatology of major avalanches at Rogers Pass, 1909–1977. Ottawa, Canada: National Research Council, Canadian Association Committee on Geotechnical Research, DBR Paper No. 956.

———. 1987. A climatology of major avalanche winters in western Canada. *Atmosphere-Ocean* 25 (2): 115–36.

Fitzharris, B. B., and S. Bakkehøi. 1986. A synoptic climatology of major avalanche winters in Norway. *Journal of Climatology* 6:431–46.

Föhn, P. M., B. W. Good, P. Bois, and C. Obled. 1977. Evaluation and comparison of statistical and conventional methods of forecasting avalanche hazard. *Journal of Glaciology* 19 (81): 375–87.

Glazovskaya, T. G. 1998. Global distribution of snow avalanches and changing activity in the Northern Hemisphere due to climate change. *Annals of Glaciology* 26:337–42.

Gürer, I., O. Murat Yavas, T. Erenbilge, and A. Sayin. 1995. Snow avalanche incidents in north-western Anatolia, Turkey during December 1992. *Natural Hazards* 11:1–16.

Hächler, P. 1987. Analysis of the weather situations leading to severe and extraordinary avalanche situations. *International Association of Hydrological Sciences Publications* 162:295–304.

Hackett, S. W., and H. S. Santeford. 1980. Avalanche zoning in Alaska, U.S.A. *Journal of Glaciology* 26:377–92.

Haegeli, P., and D. M. McClung. 2003. Avalanche characteristics of a transitional snow climate—Columbia Mountains, British Columbia, Canada. *Cold Regions Science and Technology* 37:255–76.

Hansen, C., and S. J. Underwood. 2012. Synoptic scale weather-patterns and size-5 avalanches on Mt. Shasta, California. *Northwest Science* 86 (4): 329–41.

Hebertson, E. G., and M. J. Jenkins. 2003. Historic climate factors associated with major avalanche years on the Wasatch Plateau, Utah. *Cold Regions Science and Technology* 37 (3): 315–32.

Helbig, N., and A. van Herwijnen. 2012. Modeling the spatial distribution of surface hoar in complex topography. *Cold Regions Science and Technology* 82:68–74.

Hendrikx, J., M. Murphy, and T. Onslow. 2014. Classification trees as a tool for operational avalanche forecasting on the Seward Highway, Alaska. *Cold Regions Science and Technology* 97:113–20.

Hétu, B., G. Fortin, and K. Brown. 2015. Climat hivernal, aménagement du territoire et dynamique des avalanches au Québec méridional: Une analyse à partir des accidents connus depuis 1825 [Winter climate, territory/land management and avalanche dynamics of southern Quebec: An analysis from known accidents since 1825]. *Canadian Journal of Earth Sciences* 52:307–21.

Horton, S., S. Bellaire, and B. Jamieson. 2014. Modelling the formation of surface hoar layers and tracking post-burial changes for avalanche forecasting. *Cold Regions Science and Technology* 97:81–89.

Ikeda, S., R. Wakabayashi, K. Izumi, and K. Kawashima. 2009. Study of snow climate in the Japanese Alps: Comparison to snow climate in North America. *Cold Regions Science and Technology* 59 (2–3): 119–25.

Jamieson, J. B., and C. Johnston. 1992. Snowpack characteristics associated with avalanche accidents. *Canadian Geotechnical Journal* 29:862–66.

Keylock, C. J. 2003. The North Atlantic Oscillation and snow avalanching in Iceland. *Geophysical Research Letters* 30 (5): 1254.

Kozak, M. C., K. Elder, K. Birkeland, and P. Chapman. 2003. Variability of snow layer hardness by aspect and prediction using meteorological factors. *Cold Regions Science and Technology* 37 (3): 357–71.

LaChapelle, E. R. 1966. Avalanche forecasting—A modern synthesis. *International Association of Hydrological Sciences Publication* 69:350–56.

———. 1980. The fundamental processes in conventional avalanche forecasting. *Journal of Glaciology* 26 (94): 75–84.

Laternser, M., and C. Pfister. 1997. Avalanches in Switzerland 1500–1900. In *Rapid mass movement as a source of climatic evidence for the Holocene*, ed. B. Frenzel, J. Matthews, A. Gläser, and M. Weiss. 239–66. Stuttgart, Germany: Gustav Fischer Verlag.

Lazar, B., and M. Williams. 2008. Climate change in western ski areas: Potential changes in the timing of wet avalanches and snow quality for the Aspen ski area in the years 2030 and 2100. *Cold Regions Science and Technology* 51 (2–3): 219–28.

Libbrecht, K. G. 2005. The physics of snow crystals. *Reports on Progress in Physics* 68:855–95.

Lutz, E. R., and K. W. Birkeland. 2011. Spatial patterns of surface hoar properties and incoming radiation on an inclined forest opening. *Journal of Glaciology* 57 (202): 355–66.

McClung, D. M. 2013. The effects of El Niño and La Niña on snow and avalanche patterns in British Columbia, Canada, and central Chile. *Journal of Glaciology* 59 (216): 783–92.

McClung, D. M., and J. Tweedy. 1993. Characteristics of avalanching: Kootenay Pass, British Columbia, Canada. *Journal of Glaciology* 39:316–22.

McCollister, C., K. Birkeland, K. Hansen, R. Aspinall, and R. Comey. 2003. Exploring multi-scale spatial patterns in historical avalanche data, Jackson Hole Mountain Resort, Wyoming. *Cold Regions Science and Technology* 37 (3): 299–313.

Mock, C. J. 1995. Avalanche climatology of the continental zone in the southern Rocky Mountains. *Physical Geography* 16:165–87.

———. 1996. Avalanche climatology of Alyeska, Alaska, U.S.A. *Arctic and Alpine Research* 28 (4): 502–08.

Mock, C. J., and K. W. Birkeland. 2000. Snow avalanche climatology of the western United States mountain ranges. *Bulletin of the American Meteorological Society* 81 (10): 2367–92.

Mock, C. J., and P. A. Kay. 1992. Avalanche climatology of the Western United States, with an emphasis on Alta, Utah. *The Professional Geographer* 44:307–18.

Muntán, E., C. García, P. Oller, G. Martí, A. García, and E. Gutiérrez. 2009. Reconstructing snow avalanches in the southeastern Pyrenees. *Natural Hazards and Earth System Science* 9 (5): 1599–1612.

Nakaya, U. 1954. *Snow crystals: Natural and artificial.* Cambridge, MA: Harvard University Press.

Perla, R. 1970. On contributory factors in avalanche hazard evaluation. *Canadian Geotechnical Journal* 7 (4): 414–19.

Podolskiy, E. A., K. Izumi, V. E. Suchkov, and N. Eckert. 2014. Physical and societal statistics for a century of snow-avalanche hazards on Sakhalin and the Kuril Islands (1910–2010). *Journal of Glaciology* 60 (221): 409–30.

Rangachary, N., and B. K. Bandyopadhyay. 1987. An analysis of the synoptic weather pattern associated with extensive avalanching in western Himalaya. *International Association of Hydrological Sciences Publication* 162:311–16.

Rayback, S. A. 1998. A dendrogeomorphological analysis of snow avalanches in the Colorado Front Range, USA. *Physical Geography* 19 (6): 502–15.

Reardon, B. A., G. T. Pederson, C. J. Caruso, and D. B. Fagre. 2008. Spatial reconstructions and comparisons of historic snow avalanche frequency and extent using tree rings in Glacier National Park, Montana, U.S.A. *Arctic, Antarctic, and Alpine Research* 40 (1): 148–60.

Reuter, B., A. Van Herwijnen, J. Veitinger, and J. Schweizer. 2015. Relating simple drivers to snow instability. *Cold Regions Science and Technology* 1220:168–78.

Schläppy, R., V. Jomelli, N. Eckert, M. Stoffel, D. Grancher, D. Brunstein, C. Corona, and M. Deschartres. 2016. Can we infer avalanche–climate relations using tree-ring data? Case studies in the French Alps. *Regional Environmental Change* 16:629–42.

Schneebeli, M., M. Laternser, and W. Ammann. 1997. Destructive snow avalanches and climate change in the Swiss Alps. *Eclogae Geologicae Helvetiae* 90 (3): 457–61.

Schweizer, J., C. Mitterer, and L. Stoffel. 2009. On forecasting large and infrequent avalanches. *Cold Regions Science and Technology* 59:234–41.

Seligman, G. 1936. *Snow structure and ski fields.* Cambridge, UK: International Glaciology Society.

Slaughter, A. E., D. McCabe, H. Munter, P. J. Staron, E. E. Adams, D. Catherine, I. Henninger, M. Cooperstein, and T. Leonard. 2009. An investigation of radiation-recrystallization coupling laboratory and field studies. *Cold Regions Science and Technology* 59 (2–3): 126–32.

Stethem, C., B. Jamieson, P. Schaerer, D. Liverman, D. Germain, and S. Walker. 2003. Snow avalanche hazard in Canada—A review. *Natural Hazards* 28 (2–3): 487–515.

Stossel, F., M. Guala, C. Fierz, C. Manes, and M. Lehning. 2010. Micrometeorological and morphological observations of surface hoar dynamics on a mountain snow cover. *Water Resources Research* 46 (4): WO4511.

Touhinsky, G. K. 1966. Avalanche classification, and rhythms in snow cover and glaciation of the Northern Hemisphere in historical times. *International Association of Hydrological Sciences Publication* 69:382–93.

Characteristics of Precipitating Storms in Glacierized Tropical Andean Cordilleras of Peru and Bolivia

L. Baker Perry ⓘ, Anton Seimon, Marcos F. Andrade-Flores, Jason L. Endries, Sandra E. Yuter, Fernando Velarde, Sandro Arias, Marti Bonshoms, Eric J. Burton, I. Ronald Winkelmann, Courtney M. Cooper, Guido Mamani, Maxwell Rado, Nilton Montoya, and Nelson Quispe

Precipitation variability in tropical high mountains is a fundamental yet poorly understood factor influencing local climatic expression and a variety of environmental processes, including glacier behavior and water resources. Precipitation type, diurnality, frequency, and amount influence hydrological runoff, surface albedo, and soil moisture, whereas cloud cover associated with precipitation events reduces solar irradiance at the surface. Considerable uncertainty remains in the multiscale atmospheric processes influencing precipitation patterns and their associated regional variability in the tropical Andes—particularly related to precipitation phase, timing, and vertical structure. Using data from a variety of sources—including new citizen science precipitation stations; new high-elevation comprehensive precipitation monitoring stations at Chacaltaya, Bolivia, and the Quelccaya Ice Cap, Peru; and a vertically pointing Micro Rain Radar—this article synthesizes findings from interdisciplinary research activities in the Cordillera Real of Bolivia and the Cordillera Vilcanota of Peru related to the following two research questions: (1) How do the temporal patterns, moisture source regions, and El Niño-Southern Oscillation relationships with precipitation occurrence vary? (2) What is the vertical structure (e.g., reflectivity, Doppler velocity, melting layer heights) of tropical Andean precipitation and how does it evolve temporally? Results indicate that much of the heavy precipitation occurs at night, is stratiform rather than convective in structure, and is associated with Amazonian moisture influx from the north and northwest. Improving scientific understanding of tropical Andean precipitation is of considerable importance to assessing climate variability and change, glacier behavior, hydrology, agriculture, ecosystems, and paleoclimatic reconstructions.

热带高山的降水变异, 是影响在地气候表现以及包括冰川行为和水资源的多样环境过程的关键因素, 但却未能受到良好的理解。降雨类型, 日行性, 频率及雨量, 影响着水文径流, 地表反照率及土壤湿度, 而与降雨事件有关的云层覆盖, 则降低了地表的太阳辐射。在影响降雨模式的多重尺度大气过程, 及其在热带安第斯地区中的相关区域变异中, 仍然持续有着大量的不确定性, 特别是有关降雨时期, 时机与垂直结构。本文运用来自多样资源的数据——包括崭新的公民科学雨量站; 在查卡塔雅, 玻利维亚和秘鲁魁尔克亚的冰冠的崭新高海拔综合雨量监控站; 以及垂直观测的微观降雨雷达, 综合从玻利维亚的雷亚尔山脉与秘鲁韦尔卡努塔山脉的跨领域研究活动中有关下列两大问题的发现: (1) 时间模式, 湿度来源区域和圣婴—南方振荡现象, 与降雨发生之间的关系为何有所变异? (2) 什麼是热带安第斯山降雨的垂直结构 (例如反射率, 多普勒速度, 融解层高度), 及其如何随着时间变化? 研究结果指出, 大量降雨多半发生在夜间, 在结构上是层状而非对流的, 并与亚马逊湿气从北方与西北方流入有关。增进对於热带安第斯山降雨的科学性理解, 将对评估气候变异与变迁, 冰川行为, 水文, 农业, 生态系统与古气候的再结构具有重要影响。 关键词: 水文气象学, 融解层高度, 降水, 热带安第斯山。

La variabilidad en precipitaciones de las altas montañas tropicales es un factor fundamental, pero todavía pobremente entendido, que influye en la expresión climática local y en una variedad de procesos ambientales, incluyendo el comportamiento de los glaciares y los recursos hídricos. El tipo de precipitación, el carácter diurno, frecuencia y cantidad influyen la escorrentía hidrológica, el albedo de la superficie y la humedad del suelo, mientras que la cubierta de nubes asociada con los eventos de la precipitación reduce la irradiación solar de la superficie. Una considerable incertidumbre subsiste en los procesos atmosféricos de multiescala que influencian los patrones de precipitación y su asociada variabilidad regional en los Andes tropicales—en particular lo relacionado con la fase de precipitación, tipo y estructura vertical. Con el uso de datos de una variedad de fuentes—incluyendo estaciones de precipitación de la nueva ciencia ciudadana; las nuevas estaciones de altura que monitorean la precipitación en todos sus aspectos en Chacaltaya, Bolivia, y en el casquete nevado de Quelccaya, Perú; y un Micro Rain Radar orientado verticalmente—este artículo sintetiza los hallazgos de las actividades de investigación interdisciplinaria en la Cordillera Real de Bolivia y en la Cordillera Vilcanota del Perú, en relación con las siguientes dos preguntas de investigación: (1) ¿Cómo varían las relaciones de los patrones temporales, humedad de las regiones fuente y la Oscilación Meridional de El Niño con la ocurrencia de la precipitación? y (2) ¿Cuál es la estructura vertical (e.g., reflectividad, velocidad Doppler, alturas de la capa de fusión) de la precipitación andina tropical y cómo evoluciona ésta temporalmente? Los resultados indican que gran parte de la alta precipitación ocurre durante la noche, es más de estructura estratiforme que convectiva y está asociada con el influjo de humedad amazónica del norte y noroeste. La mejora en el entendimiento de la precipitación andina tropical es de gran importancia para evaluar la variabilidad y cambio climáticos, el comportamiento de los glaciares, la hidrología, la agricultura, los ecosistemas y las reconstrucciones paleoclimáticas.

Mountains represent approximately 24 percent of the land surface of the Earth and play a vital role in sustaining ecosystems and humanity (Marston 2008). Precipitation is the critical freshwater input to the hydrological system in mountain regions of the world (e.g., Singh, Singh, and Haritashya 2011), is a fundamental influence on mountain glaciers and ecosystems (e.g., Francou et al. 2003; Barry 2008; Kaser et al. 2010; Vuille 2011), and is the primary parameter preserved in ice cores obtained from mountain glaciers and ice caps (e.g., Thompson 2000). Precipitation type (e.g., rain vs. snow), timing, frequency, and amount control surface albedo in mountain environments, whereas cloud cover associated with precipitation events reduces solar irradiance at the surface, together resulting in a major influence on climate.

In the outer tropical Andes Mountains of southern Peru and Bolivia (12–16°S; Figure 1), a region where glacier meltwater is critical for buffering water supplies during the dry season (Chevallier et al. 2011), precipitation is of added significance in influencing glacier mass balance and surface albedo (Francou et al. 2003). Precipitation is the primary influence on oxygen stable isotope ratios ($\delta^{18}O$, hydrogen-deuterium oxide) preserved in tropical ice cores (Vimeux et al. 2009). Precipitation also provides the critical freshwater inputs to hydrological resources (e.g., irrigation,

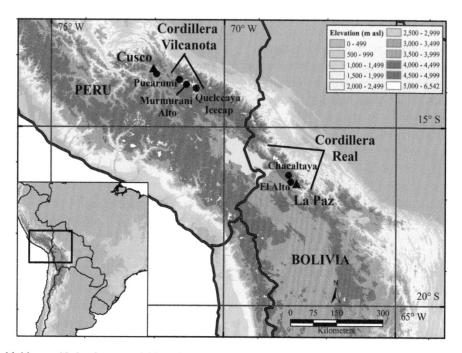

Figure 1. Location of field sites. (Color figure available online.)

hydroelectricity, water reservoirs) and is a major influence on ecosystems and agriculture (Bush, Hanselman, and Gosling 2010; Yager and Meneses 2010).

Previous investigations of tropical Andean precipitation have focused primarily on (1) the large-scale atmospheric circulation on seasonal (wet vs. dry) and interannual timescales (e.g., Garreaud 1999; Vuille 1999; Garreaud, Vuille, and Clement 2003; Vuille and Keimig 2004); (2) precipitation–climate–glacier interactions in the inner and outer tropics (e.g., Hardy et al. 1998; Francou et al. 2003; Favier, Wagnon, and Ribstein 2004; Salzmann et al. 2013); and (3) atmospheric influences on $\delta^{18}O$ values recorded in precipitation in the context of paleoclimatic interpretation (e.g., Vimeux et al. 2005). The outer tropical Andes (e.g., 12–16°S) are characterized by distinct seasonality of precipitation, with a wet season from November to March (shorter in the south and west) and a dry season from April through October. Moisture transport, as inferred from backward air trajectories, is almost exclusively from the Amazon basin (Vimeux et al. 2005; Perry, Seimon, and Kelly 2014), although the trajectories vary depending on regional topographic setting. Hurley et al. (2015) even suggested that heavy snowfall events on the Quelccaya Ice Cap in the Cordillera Vilcanota might be tied to extratropical cold surges in southeastern South America.

Tropical Pacific sea-surface temperatures (SSTs) and the associated El Niño-Southern Oscillation (ENSO) provide a strong influence on the interannual variability of precipitation in the outer tropical Andes. According to Vuille (1999), easterly flow and the advection of Amazon moisture is enhanced in the wet season during the cold phase of ENSO (La Niña), resulting in above-normal precipitation, greater cloud cover, and lower temperatures in the outer tropical Andes. In contrast, during the warm phase of ENSO (El Niño), enhanced upper level westerly flow presumably leads to a delay in the onset of the wet season. Additionally, much of the tropical Andes experiences below-normal precipitation and associated reductions in cloud cover (Vuille and Keimig 2004) and higher temperatures (Vuille et al. 2003). Nonetheless, the spatial variability in the ENSO response is complex (e.g., Vuille and Keimig 2004; Ronchail and Gallaire 2006), and precipitation totals are higher in the Cordillera Vilcanota during El Niño years (Perry, Seimon, and Kelly 2014). Several studies (e.g., Wagnon et al. 2001; Francou et al. 2003) have reported rapid glacier ablation and highly negative mass balance

in the Cordillera Real of Bolivia during strong El Niño events, such as 1997–1998.

Prior to studies specifically addressing diurnality, tropical Andean precipitation was described simply as the late-day convective response to daytime solar heating in the presence of adequate boundary layer moisture (Johnson 1976; Aceituno 1997; Garreaud, Vuille, and Clement 2003; Vuille and Keimig 2004). New quantitative studies reveal that in reality, nighttime stratiform precipitation contributes a majority of the total annual precipitation in many locations (e.g., Biasutti et al. 2012; Romatschke and Houze 2013; Mohr, Slayback, and Yager 2014; Perry, Seimon, and Kelly 2014). A substantial fraction of satellite-borne reflectivity signatures across portions of the tropical Andes, including the Cordillera Vilcanota and Cordillera Real, are indicative of stratiform precipitation (Romatschke and Houze 2013) and yield a nighttime maximum in the timing of precipitation (Mohr, Slayback, and Yager 2014). This is in agreement with precipitation gauge observations from Cusco and the Cordillera Vilcanota that exhibit a nighttime precipitation maximum, which are inferred to be stratiform in character (Perry, Seimon, and Kelly 2014).

With the exception of simple classifications based solely on an automated system using temperature in the Cordillera Real, Bolivia (L'hôte et al. 2005), and on Antisana Volcano, Ecuador (Favier, Wagnon, and Ribstein 2004), studies discriminating precipitation phase in the tropical Andes using high–temporal resolution present weather sensors (e.g., Yuter et al. 2006) have been limited. Likewise, no studies exist of ground-based radar-derived melting layer heights or their associated tropospheric conditions in the glacierized regions of the outer tropical Andes.

The limited understanding of tropical Andean precipitation patterns and processes introduces considerable uncertainty in interpreting precipitation–glacier–hydroclimate interactions, conducting paleoclimatic reconstructions, and predicting future climate scenarios in the region (Vuille et al. 2008; Perry, Seimon, and Kelly 2014). Taken together, these uncertainties reduce the climatological inference that can be derived from ice cores in the region (e.g., Thompson et al. 1985; Thompson et al. 1986; Ramirez et al. 2003; Thompson et al. 2006; Kellerhals et al. 2010; Thompson et al. 2013). Furthermore, understanding regional variability in climate and precipitation is particularly important to interpreting local implications of precipitation changes projected by climate models

Table 1. Summary of data sources

Location	Elevation (m asl)	Temporal resolution	Period	Source
1. Precipitation				
Cusco	3,350	24 hr	1963–2010	UNSAAC Observatory
La Paz	4,038	24 hr	1980–2010	SENAMHI Bolivia
2. Precipitation, snowfall, and snow depth				
Murmurani Alto	5,050	24 hr	2010–2015	Citizen Science Observers
Pucarumi	4,100	24 hr	2011–2015	Citizen Science Observers
Quelccaya Base	4,950	24 hr	2014–2015	Citizen Science Observers
3. Present weather				
Cusco/SPZO	3,310	1 hr	2011–2015	Aviation METARs
La Paz/SLLP	4,038	1 hr	2011–2015	Aviation METARs
Murmurani Alto	5,050	1 hr	2012–2015	Parsivel Disdrometer
Quelccaya Icecap	5,650	1 hr	2014–2015	Parsivel Disdrometer
Nevado Chacaltaya	5,160	1 hr	2014–2015	Parsivel Disdrometer
4. Liquid equivalent precipitation				
Quelccaya Icecap	5,650	1 hr	2014–2015	Pluvio^2 weighing gauge
Nevado Chacaltaya	5,160	1 hr	2014–2015	Pluvio^2 weighing gauge
5. Backward air trajectories				
South America		1 hr	2014–2015	GDAS 0.5° data
6. Radar reflectivity, Doppler velocity, melting layer height				
Cusco	3,350	1 min	2014–2015	Micro Rain Radar
La Paz	3,440	1 min	2015–2016	Micro Rain Radar

incorporating anthropogenic forcings and multidecadal teleconnections and oscillations (Urrutia and Vuille 2009).

This article is guided by the following research questions:

1. How do the temporal patterns, moisture source regions, and ENSO relationships with precipitation occurrence vary in the outer tropical Andes?
2. What is the vertical structure (e.g., reflectivity, Doppler velocity, melting layer heights) of tropical Andean precipitation and how does it evolve temporally?

The results presented here, although particular to the Cordillera Vilcanota and Cordillera Real, contribute more broadly to the understanding of the meteorological factors that influence precipitation patterns and hydroclimate in the tropical high Andes.

Data and Methods

Precipitation Data Sources and Time Periods

Table 1 summarizes the data sources and time periods used for this study. In situ manual precipitation data were obtained from (1) the Universidad Nacional de San Antonio de Abád de Cusco (UNSAAC, 1963–2010) and (2) the La Paz/El Alto JFK International Airport (SLLP, 1980–2010). We obtained hourly present weather observations from the Cusco International Airport (SPZO) and SLLP Meterological Terminal Aviation Routine (METAR) weather reports for the period 2011 to 2015. Manual observations from three citizen science stations located in the Cordillera Vilcanota also provided a valuable source of data for this study. The citizen science stations are all located above 4,000 m above sea level (asl; all elevations hereafter are asl). Manual liquid equivalent precipitation measurements were taken each morning at the citizen science stations at approximately 0700 LST (1200 UTC) by a trained observer using established protocols developed for measuring snowfall (Doesken and Judson 1996) and for citizen scientist precipitation observers as part of the Collaborative Rain, Hail, and Snow (CoCoRaHS) network in the United States and Canada (Cifelli et al. 2005). Hourly present weather observations were also obtained from three automated high-elevation precipitation monitoring stations: Murmurani Alto and Quelccaya in the Cordillera Vilcanota and Chacaltaya in the Cordillera Real. We used the OTT PARSIVEL present weather sensors (Löffler-Mang and Joss 2000; Löffler-Mang and Blahak 2001) at all three stations to classify the timing, type, and intensity of precipitation.

National Oceanic and Atmospheric Administration's Hybrid Single Particle Lagrangian Integrated Trajectory Trajectory Tool

The National Oceanic and Atmospheric Administration's (NOAA) Hybrid Single Particle Lagrangian Integrated Trajectory (HYSPLIT) model (Draxler and Rolph 2015) was used to simulate seventy-two-hour backward air trajectories ending at the date and time of the maturation hours of precipitation events at Cusco, Quelccaya Ice Cap, and Chacaltaya. We used ending heights of 4,000 m and 6,000 m for Cusco and 5,000 m for both Quelccaya and Chacaltaya. These levels are typically within the lower cloud layer during precipitation events and are therefore used as a proxy for low-level moisture transport. HYSPLIT backward trajectories were derived using four-dimensional (x, y, z, t) meteorological fields from the Global Data Analysis System (GDAS) data set. GDAS data are available from 2007 at three-hourly temporal resolution and 0.5° (latitude/longitude grid) spatial resolution with twenty-three vertical levels. We performed a cluster analysis of the HYSPLIT backward air trajectories for Cusco, Quelccaya, and Chacaltaya. This method groups or clusters similar trajectories and maximizes the differences in the clusters of trajectories (e.g., Taubman et al. 2006; Kelly et al. 2013).

Micro Rain Radar

We deployed a 24-GHz vertically pointing Micro Rain Radar (MRR; e.g., Peters, Fischer, and Andersson 2002) at the SENAMHI Peru office in Cusco from August 2014 to February 2015 and at the UMSA Cota Cota campus in La Paz from October to December 2015. The MRR provided continuous profiles of hydrometeor reflectivity (dBZ) and Doppler velocity from 3,350 m (3,440 m) in Cusco (La Paz) up to 7,850 m (7,940 m) using thirty 150-m gates. MRR data were postprocessed using the technique developed by Maahn and Kollias (2012) to remove noise, improve data quality, and improve data sensitivity in snow. Distinct melting layers are present in vertically pointing radar profiles when vertical air motions are weak, which usually corresponds to periods of stratiform precipitation (Houze 1997). The top of the melting layer observed in radar reflectivity corresponds to the highest altitude 0°C level height during periods of precipitation (Minder and Kingsmill 2013). Particle fall speed increases as snow fully melts into rain at the bottom of the melting layer and is

detectable in vertically pointing Doppler velocity (Houze 1993). In our analysis, the bottom of the melting layer is identified as the most negative gradient in Doppler velocity in the profile and the top of the melting layer as the most negative gradient in reflectivity. To determine the top of the melting layer, the average hourly melting layer thickness (top minus bottom) was combined with the 1-minute values of melting layer in that hour. We elected to discard any melting layer height values greater than one standard deviation of the hourly mean. Eight heavy nighttime stratiform events in Cusco and four in La Paz (1) had MRR data available, (2) occurred primarily at nighttime (between 0000 and 1200 UTC), (3) had an hour or more of continuous stratiform precipitation characterized by a well-defined melting layer, (4) had a duration of longer than four hours, and (5) deposited liquid equivalent precipitation of more than ~10 mm at Cusco (SPZO) or Chacaltaya. For these events, HYSPLIT backward air trajectories were also calculated ending at 4,000 m, 6,000 m, and 8,000 m over Cusco and La Paz.

Results and Discussion

2014–2015 Precipitation Totals

Precipitation totals for the 2014–2015 hydrological year for stations analyzed in this article (Table 2) ranged from 633 mm at El Alto to 1,143 mm in Pucarumi, with totals slightly above the long-term mean annual values at most locations. A bimodal diurnal distribution of frequency of precipitation occurrence was evident across all stations, consisting of afternoon and nighttime maxima (Figure 2). The afternoon maximum occurred around 2000 UTC (1500 LST) in Cusco and the Cordillera Vilcanota but nearly two hours earlier, around 1700 UTC (1300 LST), in the Cordillera Real. The timing of the nighttime maximum was very similar across all stations, with a broad peak centered between 0400 and 0500 UTC (midnight LST). An afternoon maximum in frequency of precipitation occurrence existed for the higher elevation stations of Quelccaya, Murmurani Alto, and Chacaltaya, whereas this pattern reversed for the lower elevation stations of Cusco and El Alto, where the nighttime maximum was clearly dominant. The bimodal distribution of precipitation is consistent with the results of Perry, Seimon, and Kelly (2014) and confirms the importance of widespread nighttime precipitation across the region.

Table 2. Precipitation characteristics from 2014–2015 hydrological year

Location	Mean annual precipitation (mm)	2014–2015 hydrologic year precipitation totals (mm)												
		Jul	Aug	Sep	Oct	Nov	Dec	Jan	Feb	Mar	Apr	May	Jun	Total
Peru														
Cusco	820	3	13	15	79	18	171	167	146	87	73	19	4	794
Pucarumi	1,111	1	10	63	69	78	205	236	178	170	118	11	6	1,143
Murmurani Alto	732	−99	8	39	66	67	132	125	121	89	109	34	18	809
Quelccaya Base	−99	−99	12	22	30	61	89	155	89	59	110	85	18	730
Bolivia														
El Alto	618	4	17	46	29	26	115	135	86	96	76	4	0	633
Chacaltaya	−99	−99	−99	−99	−99	51	144	196	110	161	102	11	0	779

Note: −99 denotes missing data. See Table 1 for years used for the calculation of the mean annual precipitation.

Precipitation Type

Observations of precipitation type at the three stations located above 5,000 m indicated that the frequency of precipitation was greatest in the austral summer (December–February) and that solid precipitation dominated, representing more than 95 percent of all precipitation hours (Figure 3). Snow was the primary precipitation type, but graupel (*phati* in the native Quechua language) also constituted a significant fraction of precipitation hours, particularly on the Quelccaya Ice Cap and at Murmurani Alto during the austral winter. Detailed field observations of snow particle type and degree of riming during research expeditions over a five-year period confirm that the particles are indeed graupel, consisting of a heavily rimed snow crystal whose original structure is unrecognizable (e.g., Mosimann, Weingartner, and Waldvogel 1994). Although liquid precipitation (rain and freezing rain) was infrequent (5 percent of precipitation hours; range of 0–18 percent of precipitation hours per month), its presence at elevations above 5,000 m is

noteworthy, as many glacier termini in the Cordillera Vilcanota (e.g., Salzmann et al. 2013) and Cordillera Real are found below this elevation. As freezing levels continue to rise globally in the tropics (Diaz et al. 2003) and in the tropical Andes (Bradley et al. 2009), it is not surprising that liquid precipitation occurs in 5 percent of our observations at Murmurani Alto at 5,050 m. This suggests that rain occurs on occasion in the ablation zone of nearby glaciers, promoting enhanced melting of the glacier surface and reducing surface albedo and likely influencing glacier mass balance (e.g., Francou et al. 2003). High-altitude liquid precipitation occurrence might therefore be a strong indicator of climate change in the region.

Backward Air Trajectories

The seventy-two-hour backward air trajectories for precipitation events in Cusco were primarily from the northwest (Figures 4A, 4B), in general agreement with the results of Perry, Seimon, and

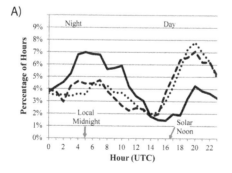
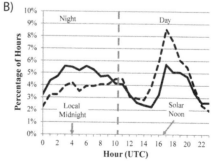

Figure 2. Precipitation timing at (A) Cusco (solid), Murmurani Alto (dotted), and Quelccaya Ice Cap (dashed); and (B) El Alto (solid) and Chacaltaya (dashed). Data time period for Cusco is 11 August 2011 to 30 June 2015; Murmurani Alto is 4 April 2012 to 13 January 2015; Quelccaya Ice Cap is 4 November 2014 to 30 June 2015; El Alto is August 2011 to 30 June 2015; and Chacaltaya is 1 November 2014 to 30 June 2015.

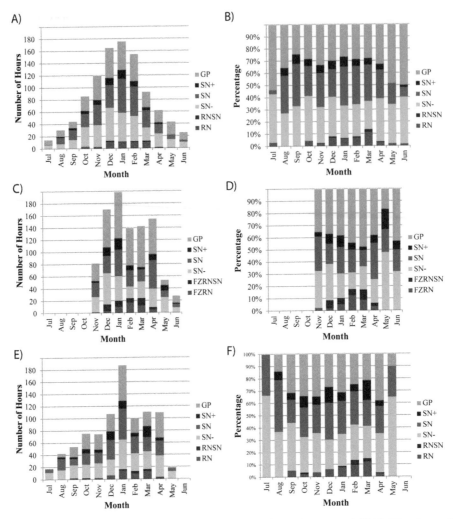

Figure 3. All precipitation hours by month at (A, B) Murmurani Alto, (C, D), Quelccaya, and (E, F) Chacaltaya. GP = graupel; SN+ = heavy snow; SN = moderate snow; SN− = light snow; RN/SN = rain and snow mix; RN = rain. Data time period for Murmurani Alto is 1 July 2012 to 30 July 2014 and monthly averages are plotted; Quelccaya is 4 November 2014 to 30 June 2015; and Chacaltaya is 1 December 2014 to 30 November 2015. (Color figure available online.)

Kelly (2014). The clusters comprising the largest percentage of events produced by HYSPLIT at 4,000 and 6,000 m both had trajectories from this direction. There were also smaller easterly clusters that originated in the Amazon at both ending heights as well as one southerly cluster at both heights from the Pacific. The northwest clusters had the shortest mean trajectories (i.e., weakest flow) and traveled closer to the ground. In general, the trajectory analyses for both Quelccaya and Chacaltaya are similar (Figures 4C, 4D) and suggest that Amazon air parcels originating to the northwest and north of the Cordillera Vilcanota and Cordillera Real were associated with the majority of the precipitation events during the 2014–2015 wet season. These trajectories are consistent with the climatologically favored backward trajectories

for the Zongo Valley of the Cordillera Real calculated by Vimeux et al. (2005), which were predominantly out of the north and north-northwest during the austral wet season. Nonetheless, these trajectories represent precipitation events for only one hydrological year and a longer time series will increase confidence of the climatological relationships.

Multidecadal ENSO–Precipitation Relationships

Long-term (greater than twenty-five years) hydrological year annual precipitation totals at Cusco (UNSAAC) and La Paz (SLLP) plotted according to ENSO phase (Figures 5A, 5B) highlight the complexity of ENSO–precipitation relationships in the outer tropical Andes. In Cusco, strong El Niño years are

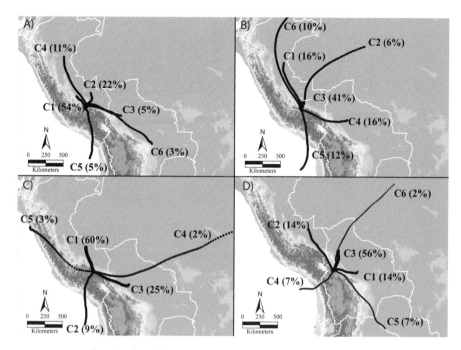

Figure 4. Clustered mean seventy-two-hour backward air trajectories for 2014–2015 precipitation events at (A) Cusco (ending at 4,000 m asl), (B) Cusco (ending at 6,000 m asl), (C) Quelccaya (ending at 6,000 m asl), and (D) Chacaltaya (ending at 6,000 m asl). (Color figure available online.)

characterized by positive precipitation anomalies, in contrast to strong La Niña years, which are associated with negative anomalies. In La Paz, both strong El Niño and strong La Niña years exhibit negative precipitation anomalies, with all other years much wetter. The positive (negative) anomalies associated with El Niño (La Niña) in Cusco are generally consistent throughout the hydrological year, whereas in La Paz precipitation with both El Niño and La Niña is close to normal until the austral fall (March–May), when negative anomalies predominate.

Vertical Structure of Precipitation

Observations of the vertical structure of precipitation from the MRR in the Cordillera Vilcanota and Cordillera Real indicate that the afternoon precipitation maxima was primarily convective, whereas the nighttime precipitation was largely stratiform in character. A fifteen-hour period in Cusco on 11 and 12 February 2015 (Figure 6) illustrates the discontinuous and cellular nature, widely varied reflectivity signatures, and lack of a well-defined melting layer typical

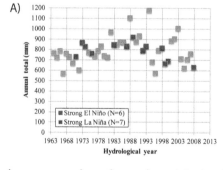
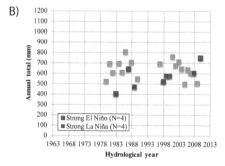

Figure 5. Hydrological year scatterplots of annual precipitation by El Niño-Southern Oscillation phase for (A) Cusco–UNSAAC (3,365 m) 1963–2009 and (B) La Paz–El Alto (4,038 m) 1979–2009 according to December to March Multivariate El Niño-Southern Oscillation Index means. The characterization of strong events is determined by the mean December to March MEI, with values > +1.00 used to identify strong El Niño events and values < −1.00 used for strong La Niñas; given that MEI is assessed based on bimonthly statistics this means average December to January, January to February, and February to March MEI values. All other "regular" years fall between these two thresholds; that is, MEI values < +1.00 and > −1.00. Source data for MEI are from National Oceanic and Atmospheric Administration–Earth System Research Laboratory, Physical Sciences Division (2017). MEI = Multivariate El Niño-Southern Oscillation. (Color figure available online.)

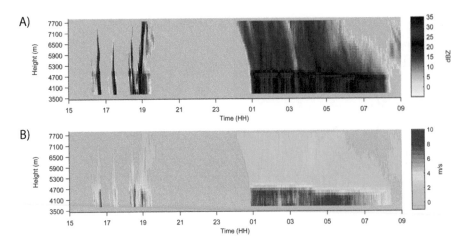

Figure 6. Summary of Micro Rain Radar (A) radar reflectivity (dBZ) and (B) Doppler velocity (m s^{-1}) for 1500 UTC 11 February 2015 to 0900 UTC 12 February 2015 in Cusco. (Color figure available online.)

of convective precipitation between 1500 and 2100 UTC (1000–1600 LST). In contrast, the more continuous nature, longer duration, and presence of a well-defined melting layer between 0100 and 0800 UTC (2100–0300 LST) indicate largely uniform layers of stratiform precipitation. These cases, along with many others at both Cusco and La Paz (not shown), confirm previous inferences by Perry, Seimon, and Kelly (2014) that afternoon precipitation is largely convective and nighttime precipitation is primarily stratiform. Considerable uncertainty remains as to the origin of the nighttime stratiform events and will require detailed investigations of individual events.

Frequency distributions of melting layer heights for precipitation observed by the MRR in Cusco and La Paz (Figure 7) share common characteristics. Median values (standard deviations) for melting layer heights at Cusco were 4,810 m (264 m) and 4,786 m (255 m) for La Paz, with most melting layer heights between 4,400 and 5,100 m. Although these median values are still below the ablation zones of most glaciers in the Cordillera Vilcanota and Cordillera Real, an increase in median melting layer heights of 200 m would result in

rain falling during 49 percent (54 percent) of all precipitating hours at 5,000 m in Cusco (La Paz), compared to 14 percent (16 percent) during the periods of study.

Case Studies

An analysis of eight heavy (liquid equivalent precipitation > ~10 mm) nighttime stratiform precipitation events in Cusco (Table 3) and four in La Paz (Table 4) highlights numerous consistencies in their timing (peak around 0400 to 0500 UTC [~midnight LST]), total precipitation (~10–20 mm), maximum column reflectivity (40–45 dBZ), mean column reflectivity (18–22 dBZ), and mean melting layer heights (4,500–4,700 m). The seventy-two-hour backward trajectories ending at 4,000 m for the Cusco precipitation events all originated in the Amazon basin to the northwest, whereas the trajectories ending at 6,000 m and 8,000 m were more varied, with several originating over the Pacific Ocean. The seventy-two-hour backward trajectories for the four events in La Paz were all characterized by north or north-northwest flow that originated in the Amazon basin at all levels.

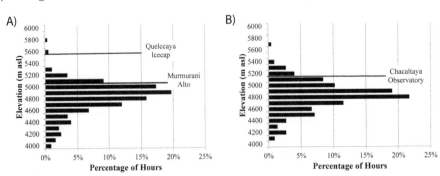

Figure 7. Frequency distribution of melting layer heights for (A) Cusco and (B) La Paz. Cusco data are for August 2014 to February 2015 and La Paz data are for October to December 2015. (Color figure available online.)

Table 3. Characteristics of heavy nighttime stratiform precipitation events in Cusco

Date(s)	Time range (UTC)		Total precipitation (mm)	Maximum reflectivity (dBZ)	Mean column reflectivity (dBZ)	Mean melting layer height (m asl)	Trajectory origin		
	Start	End					4,000 m	6,000 m	8,000 m
8 Oct 2014	0100	0700	16.6	42.2	21.4	4,545	N/Amazon	NW/Amazon	E/Amazon
9 Dec 2014	0300	0700	15.2	41.2	22.5	4,719	NW/Amazon	S/Pacific	SW/Pacific
17 Dec 2014	0000	0900	23.0	38.6	19.2	4,571	NW/Amazon	S/Pacific	E/Amazon
19–20 Dec 2014	2300	0600	15.6	40.7	18.2	4,688	NW/Amazon	S/Pacific	S/Pacific
19 Jan 2015	0100	0700	14.6	44.0	20.0	4,550	NW/Amazon	S/Andes	W/Andes
26–27 Jan 2015	2200	0800	9.8	43.6	17.3	4,639	NW/Amazon	E/Amazon	SE/Amazon
10 Feb 2015	0800	1300	13.0	39.2	19.8	4,550	NW/Amazon	NW/Amazon	NW/Amazon
11 Feb 2015	0100	1000	14.4	39.7	18.9	4,704	NW/Amazon	NW/Amazon	NW/Amazon

Table 4. Characteristics of heavy nighttime stratiform precipitation events in La Paz

Date(s)	Time range (UTC)		Total precipitation (mm)	Maximum reflectivity (dBZ)	Mean column reflectivity (dBZ)	Mean melting layer height (m asl)	Trajectory origin		
	Start	End					4,000 m	6,000 m	8,000 m
3 Oct 2015	0700	1100	11.4	32.0	15.0	−99	N/Amazon	N/Amazon	N/Amazon
29 Oct 2015	0100	0600	21.3	39.0	18.0	4,578	N/Amazon	N/Amazon	N/Amazon
12 Nov 2015	0100	0700	10.8	36.0	18.0	4,715	NNW/Amazon	NNW/Amazon	NNW/Amazon
19 Dec 2015	0500	1200	9.7	39.9	15.0	4,640	N/Amazon	N/Amazon	N/Amazon

These results suggest that all of the analyzed heavy nighttime stratiform events were associated with low-level moisture in association with north or northwest flow originating in the Amazon basin and mid- and upper level moisture primarily originating in the Amazon basin but with greater variability in the backward trajectories for Cusco.

Summary and Conclusions

In this article, we have investigated the spatiotemporal patterns, moisture source regions, ENSO relationships, and vertical structure of precipitation in the Cordillera Vilcanota of southern Peru and the Cordillera Real of Bolivia. Results indicate that much of the heavy precipitation occurs at night, is stratiform rather than convective in structure, and is associated with Amazonian moisture influx from the north and northwest. Our findings of positive (negative) precipitation anomalies during El Niño (La Niña) in Cusco and negative anomalies during both El Niño and La Niña in La Paz also highlight the complex ENSO–precipitation relationships in the region. Melting layer heights for most precipitation hours are currently beneath the levels of glacier termini in the Cordillera Vilcanota and Cordillera Real, but in situ observations of precipitation type in the vicinity of ablation zones of glaciers in the region indicate that liquid precipitation accounts for 5 percent of all precipitation hours. Melting layer height increases of only 200 m, however, could result in approximately half of the total precipitation falling as rain on tropical Andean glaciers. Such height increases, which equate to approximately a 1.2°C mid-tropospheric temperature increase (which is close to the midcentury projected value according to the Intergovernmental Panel on Climate Change 5th Assessment Report; Cubasch et al. 2013), would act to augment the rapid ablation of regional glaciers already exhibiting strongly negative mass balance under the current warming regime. As such, the capacity of tropical Andean glaciers to serve as reliable buffers for water supply to human and natural systems, already in decline, looks to dwindle even more rapidly as rainfall encroaches steadily higher into the Andean nival zone.

Acknowledgments

The authors gratefully acknowledge NOAA's Air Resources Laboratory for the use of the HYSPLIT model; citizen science observers Pedro Godofredo, Felipe Quispe, and Mateo Condori for taking daily precipitation measurements; Pedro Godofredo for permission to install a meteorological station; the Crispin family and Adrián Ccahuana for logistical support; and Appalachian State University summer study abroad students (2012–2014) for help servicing the Murmurani Alto station. Fabricio Avila provided technical support for instrumentation in Bolivia. The authors are grateful for the comments of two anonymous reviewers and for discussions with Ronnie Ascarza, Rimort Chavez, Doug Hardy, Christian Huggel, Paul Mayewski, Richard Poremba, Charles Rodda, and Simone Schwauwecker.

Funding

Funding supporting this research was provided by the Justin Brooks Fisher Foundation for field support in the Cordillera Vilcanota in 2012 and the National Science Foundation through Grant AGS-1347179 (CAREER: Multiscale Investigations of Tropical Andean Precipitation).

ORCID

L. Baker Perry http://orcid.org/0000-0003-0598-6393

References

Aceituno, P. 1997. Climate elements of the South American Altiplano. *Revista Geofísica* 44:37–55.

Barry, R. G. 2008. *Mountain weather and climate*. 3rd ed. Cambridge, UK: Cambridge University Press.

Biasutti, M., S. E. Yuter, C. D. Burleyson, and A. H. Sobel. 2012. Very high resolution rainfall patterns measured by TRMM precipitation radar: Seasonal and diurnal cycles. *Climate Dynamics* 39:239–58.

Bradley, R. S., F. Keimig, H. F. Diaz, and D. R. Hardy. 2009. Recent changes in freezing level heights in the tropics with implications for the deglacierization of high mountain regions. *Geophysical Research Letters* 36: L17701.

Bush, M. B., J. A. Hanselman, and W. D. Gosling. 2010. Nonlinear climate change and Andean feedbacks: An imminent turning point? *Global Change Biology* 16:3223–32.

Chevallier, P., B. Pouyaud, W. Suarez, and T. Condom. 2011. Climate change threats to environment in the tropical Andes: Glaciers and water resources. *Regional Environmental Change* 11 (Suppl. 1): S179–S187.

Cifelli, R., N. Doesken, P. Kennedy, L. D. Carey, S. A. Rutledge, C. Gimmestad, and T. Depue. 2005. The Community Collaborative Rain, Hail, and Snow Network:

Informal education for scientists and citizens. *Bulletin of the American Meteorological Society* 86:1069–77.

Cubasch, U., D. Wuebbles, D. Chen, M. C. Facchini, D. Frame, N. Mahowald, and J.-G. Winther. 2013: Introduction. In *Climate change 2013: The physical science basis. Contribution of Working Group I to the Fifth Assessment Report of the Intergovernmental Panel on Climate Change*, ed. T. F. Stocker, D. Qin, G.-K. Plattner, M. Tignor, S. K. Allen, J. Boschung, A. Nauels, Y. Xia, V. Bex, and P. M. Midgley, 119–58. Cambridge, UK: Cambridge University Press.

Diaz, H. F., J. K. Eischeid, C. Duncan, and R. S. Bradley. 2003. Variability of freezing levels, melting season indicators, and snow cover for selected high-elevation and continental regions in the last 50 years. *Climatic Change* 59:33–52.

Doesken, N. J., and A. Judson. 1996. *The snow booklet: A guide to the science, climatology, and measurement of snow in the United States*. Fort Collins, CO: Colorado Climate Center.

Draxler, R. R., and G. D. Rolph. 2015. HYSPLIT (HYbrid Single-Particle Lagrangian Integrated Trajectory) model access. http://ready.arl.noaa.gov/HYSPLIT.php (last accessed 10 December 2015).

Favier, V., P. Wagnon, and P. Ribstein. 2004. Glaciers of the outer and inner tropics: A different behaviour but a common response to climatic forcing. *Geophysical Research Letters* 31:L16403.

Francou, B., M. Vuille, P. Wagnon, J. Mendoza, and J.-E. Sicart. 2003. Tropical climate change recorded by a glacier in the central Andes during the last decades of the twentieth century: Chacaltaya, Bolivia, 16°S. *Journal of Geophysical Research* 108:D18106.

Garreaud, R. D. 1999. Multiscale analysis of the summertime precipitation over the central Andes. *Monthly Weather Review* 127:901–21.

Garreaud, R. D., M. Vuille, and A. C. Clement. 2003. The climate of the Altiplano: Observed current conditions and mechanisms of past changes. *Palaeogeography, Palaeoclimatology, Palaeoecology* 194:5–22.

Hardy, D., M. Vuille, C. Braun, F. Keimig, and R. S. Bradley. 1998. Annual and daily meteorological cycles at high altitude on a tropical mountain. *Bulletin of the American Meteorological Society* 79:1899–1913.

Houze, R. A., Jr. 1993. *Cloud dynamics*. San Diego, CA: Academic.

———. 1997. Stratiform precipitation in regions of convection: A meteorological paradox? *Bulletin of the American Meteorological Society* 78:2179–96.

Hurley, J. V., M. Vuille, D. R. Hardy, S. J. Burns, and L. G. Thompson. 2015. Cold air incursions, $\delta^{18}O$ variability, and monsoon dynamics associated with snow days at Quelccaya Icecap, Peru. *Journal of Geophysical Research: Atmospheres* 120.

Johnson, A. M. 1976. The climate of Peru, Bolivia, and Ecuador. In *World survey of climatology*, ed. W. Schwerdtfeger, 147–218. New York: Elsevier.

Kaser, G., T. Molg, N. J. Cullen, D. R. Hardy, and M. Winkler. 2010. Is the decline of ice on Kilimanjaro unprecedented in the Holocene? *The Holocene* 20:1079–91.

Kellerhals, T., S. Brutsch, M. Sigl, S. Knusel, H. W. Gaggeler, and M. Schwikowski. 2010. Ammonium concentration in ice cores: A new proxy for regional temperature reconstruction? *Journal of Geophysical Research* 115:D16123.

Kelly, G. M., B. F. Taubman, L. B. Perry, P. T. Soulé, J. P. Sherman, and P. Sheridan. 2013. Relationships between aerosols and precipitation in the southern Appalachian Mountains. *International Journal of Climatology* 33 (14): 3016–28.

L'hôte, Y., P. Chevallier, A. Coudrain, Y. Lejeune, and P. Etchevers. 2005. Relationship between precipitation phase and air temperature: Comparison between the Bolivian Andes and the Swiss Alps. *Hydrological Sciences Journal* 50:989–97.

Löffler-Mang, M., and U. Blahak. 2001. Estimation of the equivalent radar reflectivity factor from measured snow size spectra. *Journal of Applied Meteorology* 40:843–49.

Löffler-Mang, M., and J. Joss. 2000. An optical disdrometer for measuring size and velocity of hydrometeors. *Journal of Atmospheric and Oceanic Technology* 17:130–39.

Maahn, M., and P. Kollias. 2012. Improved micro rain radar snow measurements using Doppler spectra post processing. *Atmospheric Measurement Techniques* 5:2661–73.

Marston, R. A. 2008. Presidential address: Land, life, and environmental change in mountains. *Annals of the Association of American Geographers* 98:507–20.

Minder, J. R., and D. E. Kingsmill. 2013. Mesoscale variations of the atmospheric snow line over the northern Sierra Nevada: Multiyear statistics, case study, and mechanisms. *Journal of the Atmospheric Sciences* 70:916–38.

Mohr, K. I., D. Slayback, and K. Yager. 2014. Characteristics of precipitation features and annual rainfall during the TRMM era in the central Andes. *Journal of Climate* 27 (11): 3982–4001.

Mosimann, L., E. Weingartner, and A. Waldvogel. 1994. An analysis of accreted drop sizes and mass on rimed snow crystals. *Journal of the Atmospheric Sciences* 51:1548–58.

National Oceanic and Atmospheric Administration–Earth System Research Laboratory, Physical Sciences Division. 2017. Multivariate ENSO Index (MEI). Boulder, CO: NOAA/ESRL/PSD. https://www.esrl.noaa.gov/psd/enso/mei/ (last accessed 6 January 2017).

Perry, L. B., A. Seimon, and G. M. Kelly. 2014. Precipitation delivery in the tropical high Andes of southern Peru: New findings and paleoclimatic implications. *International Journal of Climatology* 34:197–215.

Peters, G., B. Fischer, and T. Andersson. 2002. Rain observations with a vertically looking Micro Rain Radar (MRR). *Boreal Environment Research* 7:353–62.

Ramirez, E., G. Hoffmann, J. D. Taupin, B. Francou, P. Ribstein, N. Caillon, F. A. Ferron, et al. 2003. A new Andean deep ice core from Nevado Illimani (6350 m), Bolivia. *Earth and Planetary Science Letters* 212:337–50.

Romatschke, U., and R. A. Houze. 2013. Characteristics of precipitating convective systems accounting for the summer rainfall of tropical and subtropical South America. *Journal of Hydrometeorology* 14:25–46.

Ronchail, J., and R. Gallaire. 2006. ENSO and rainfall along the Zongo Valley (Bolivia) from the Altiplano to the Amazon Basin. *International Journal of Climatology* 26:1223–36.

Salzmann, N., C. Huggel, M. Rohrer, W. Silverio, B. G. Mark, P. Burns, and C. Portocarrero. 2013. Glacier changes and climate trends derived from multiple sources in the data scarce Cordillera Vilcanota region, Southern Peruvian Andes. *The Cryosphere* 7:103–18.

Singh, V. P., P. Singh, and U. K. Haritashya, eds. 2011. *Encyclopedia of snow, ice, and glaciers.* Dordrecht, The Netherlands: Springer.

Taubman, B. F., J. D. Hains, A. M. Thompson, L. T. Marufu, B. G. Doddridge, J. W. Stehr, C. A. Piety, and R. R. Dickerson. 2006. Aircraft vertical profiles of trace gas and aerosol pollution over the mid-Atlantic U.S.: Statistics and meteorological cluster analysis. *Journal of Geophysical Research* 111:D10S07.

Thompson, L. G. 2000. Ice core evidence for climate change in the Tropics: Implications for our future. *Quaternary Science Reviews* 19:19–35.

Thompson, L. G., E. Mosley-Thompson, J. F. Bolzan, and B. R. Koci. 1985. A 1500 year record of tropical precipitation in ice cores from the Quelccaya Ice Cap, Peru. *Science* 229:971–73.

Thompson, L. G., E. Mosley-Thompson, W. Dansgaard, and P. M. Grootes. 1986. The "Little Ice Age" as recorded in the stratigraphy of the tropical Quelccaya ice cap. *Science* 234:361–64.

Thompson, L. G., E. Mosley-Thompson, M. E. Davis, P.-N. Lin, T. Mashiotta, and K. Mountain. 2006. Abrupt tropical climate change: Past and present. *Proceedings of the National Academy of Science* 103:10536–43.

Thompson, L. G., E. Mosley-Thompson, M. E. Davis, V. S. Zagorodnov, I. M. Howat, V. N. Mikhalenko, and P.-N. Lin. 2013. Annually resolved ice core records of tropical climate variability over the past ~1800 years. *Science* 340:945–50.

Urrutia, R., and M. Vuille. 2009. Climate change projections for the tropical Andes using a regional climate model: Temperature and precipitation simulations for the end of the 21st century. *Journal of Geophysical Research* 114:D02108.

Vimeux, F., R. Gallaire, S. Bony, G. Hoffman, and J. Chiang. 2005. What are the climatic controls on delta D in precipitation in the Zongo Valley (Bolivia)? Implications for the Illimani ice core interpretation. *Earth and Planetary Science Letters* 240:205–20.

Vimeux, F., P. Ginot, M. Schwikowski, M. Vuille, G. Hoffmann, L. G. Thompson, and U. Schotterer. 2009. Climate variability during the last 1000 years inferred from Andean ice cores: A review of methodology and recent results. *Palaeogeography, Paleoclimatology, Paleoecology* 281:229–41.

Vuille, M. 1999. Atmospheric circulation over the Bolivian Altiplano during dry and wet periods and extreme phases of the Southern Oscillation. *International Journal of Climatology* 19:1579–600.

———. 2011. Climate variability and high altitude temperature and precipitation. In *Encyclopedia of snow, ice, and glaciers*, ed. V. P. Singh, P. Singh, and U. K. Haritashya, 153–56. Dordrecht, The Netherlands: Springer.

Vuille, M., R. S. Bradley, M. Werner, and F. Keimig. 2003. 20th century climate change in the tropical Andes: Observations and model results. *Climatic Change* 59:75–99.

Vuille, M., B. Francou, P. Wagnon, I. Juen, G. Kaser, B. G. Mark, and R. S. Bradley. 2008. Climate change and tropical Andean glaciers: Past, present, and future. *Earth-Science Reviews* 89:79–96.

Vuille, M., and F. Keimig. 2004. Interannual variability of summertime convective cloudiness and precipitation in the central Andes derived from ISCCP-B3 data. *Journal of Climate* 17:3334–48.

Wagnon, P., P. Ribstein, B. Francou, and J. E. Sicart. 2001. Anomalous heat and mass budget of Glaciar Zongo, Bolivia, during the 1997–98 El Niño year. *Journal of Glaciology* 47:21–28.

Yager, K., and R. I. Meneses. 2010. Interpreting and monitoring Andean peatbogs. Presentation at Global Change and the World's Mountains conference, Perth, Australia.

Yuter, S. E., D. Kingsmill, L. B. Nance, and M. Löffler-Mang. 2006. Observations of precipitation size and fall speed characteristics within coexisting rain and wet snow. *Journal of Applied Meteorology and Climatology* 45:1450–64.

On the Production of Climate Information in the High Mountain Forests of Guatemala

Diego Pons, Matthew J. Taylor, Daniel Griffin, Edwin J. Castellanos, and Kevin J. Anchukaitis

Guatemala's population is dependent on cash crops and subsistence agriculture, the yield of which depends on both the timing and quantity of rainfall. Detailed knowledge about Guatemala's past, current, and future climate is therefore critical to the well-being of a country so reliant on agriculture. Relatively little information about Guatemala's climate exists, though, due to sparse instrumental records and limited high-resolution paleoclimate data. Given this situation, the development of climate data is the necessary first step toward facilitating improved decision making and robust adaptation in the face of predicted future climate change. Here we document how we successfully used tree rings to produce an annually resolved paleoclimate record from Guatemala stretching back to the late seventeenth century. These data provide a more comprehensive understanding of the range of natural variability in local and regional hydroclimate. This increased understanding could then be used to generate locally relevant climate information, to assist in planning, and toward reducing climate-related vulnerability at regional to local scales in agriculturally dependent communities. Our goal herein is to begin to close the gap between climate data generation and the use of relevant agrometeorological information in Guatemala by identifying key participants, decision makers, and modes of stakeholder engagement that are critical to coproduce climate information in the mountain regions of Guatemala.

危地马拉的人口依赖经济作物和生计农业，而其产出同时取决于降雨的时机和雨量。因此有关危地马拉过去、现在及未来的气候之详尽知识，对于此一极度依赖农业的国家福祉而言相当关键。尽管如此，危地马拉的气候信息却相对而言鲜少存在，因为仪器记录的数量零星，且高度辨识率的古气候数据相当有限。有鉴于此一情况，气候数据的建立，是在面对预测的未来气候变迁时，迈向促进改进的决策制定和强健的调适的必要第一步。我们于此记录我们如何成功地运用树轮来生产回溯至十七世纪晚期以年进行分辨的危地马拉古气候记录。这些数据，对于在地方和区域的水文气候中的自然变化范围，提供了更为综合性的理解。增进的理解，随后能够进行运用，在依赖农业的社区中生产与地方相关的气候信息、协助规划，并迈向减少从区域到地方尺度的与气候相关的脆弱性。因此我们的目标是透过指认对危地马拉山区气候信息的共同生产而言，重要的关键参与者、决策制定者和利害关系人参与的模式，着手弥合危地马拉的气候数据生产和相关农业气象信息运用之间的落差。 关键词： 气候变迁，年轮气候学，危地马拉，气候知识生产。

La población de Guatemala depende de cultivos comerciales y de agricultura de subsistencia, cuyo rendimiento lo determinan la coordinación del tiempo y la cantidad de lluvias. Por ello, el conocimiento detallado del pasado de Guatemala, el clima actual y futuro son críticos para el bienestar de un país que tanto depende de la agricultura. Sin embargo, la información existente sobre el clima de Guatemala es relativamente escasa, debido a lo limitado de los registros instrumentales de datos paleoclimáticos a alta resolución. Tomando en cuenta esta situación, la producción de datos sobre clima es el primer paso necesario que facilite una mejor toma de decisiones y una adaptación robusta frente a un futuro cambio climático pronosticado. Aquí documentamos cómo utilizamos con éxito los anillos del tronco arbóreo para producir un registro paleoclimático de resolución anual para Guatemala que se remonte hasta finales del siglo XVII. Estos datos proporcionan un entendimiento más amplio del ámbito de la variabilidad natural en hidroclima local y regional. Este entendimiento mejorado puede usarse luego para generar información climática de relevancia local que ayude en planificación y a reducir la vulnerabilidad relacionada con clima, de escalas regionales a locales, en comunidades que dependen de la agricultura. En este contexto, nuestro propósito es empezar a cerrar la brecha entre la generación de datos climáticos y el uso de información agrometeorológica relevante en Guatemala, identificando los participantes claves, los tomadores de decisiones y los modos de involucramiento de interesados, los cuales son cruciales para ayudar a producir información climática en las regiones montañosas de Guatemala.

Guatemala is among the most vulnerable countries in the world to climate change, primarily because half of its population lives in poverty in rural areas and relies on rain-fed agriculture. In addition, 70 percent of the territory is at risk of extreme weather (United Nations Development Programme 2009). Textiles, coffee, sugar, and bananas are the most important pillars of Guatemala's economy (Banco de Guatemala 2014). Most of these important export crops, as well as the maize and beans on which most of Guatemala's population subsists, are all grown in rain-fed systems. Guatemala's terrain is predominantly mountainous, and it is in these mountainous regions that the majority of Guatemala's population resides and tends fields for subsistence (maize and beans) and export (coffee and vegetables). In 2015 and 2016 much of Guatemala was affected by insufficient and erratic rainfall associated with a strong El Niño event (United Nations Office for the Coordination of Humanitarian Affairs 2016). Drought conditions reduced crop production for small-scale producers of maize and beans in several areas of Central America between 50 and 100 percent (Famine Early Warning System 2015). Interannual climate variability already presents hazards for agriculturalists, but projections of future climate change for Guatemala suggest further impending challenges for agricultural productivity. For instance, climate projections consistently simulate a reduction of precipitation in both dry and wet seasons and an increase in temperature (Neelin et al. 2006; Karmalkar, Bradley, and Díaz 2011; Magrin et al. 2014). Because of the rain-fed dependency of agriculture in Guatemala, the reduction in precipitation could significantly affect this sector. The overall consequence of decreased precipitation and increased temperatures would be an increase in the evaporative demand and reduced available soil moisture (Shelton 2012). Such changes could imply potentially severe consequences for a country reliant on agriculture. A greater understanding, then, of Guatemala's past, present, and future climate is imperative to assist in mitigating the consequences of climate change.

Guatemala lacks sufficient climate information for robust monitoring and climate model validation, and the limited temporal and spatial resolution of the current network of weather stations impedes a better understanding of climate processes, particularly at local and regional scales. The number of weather stations in Central America has declined in recent years (Giannini, Cane, and Kushnir 2001). The average density of weather stations in the country is currently around one station per 1,000 km^2 with a limited temporal resolution extending back to 1970 in most cases (Georgiou 2014). Moreover, the Global Climate Observing System (GCOS) installed only two weather stations in Central America for long-term surface measurements, neither of which are above 2,000 m (GCOS 2013). At the same time, the relatively coarse spatial resolution of general circulation models—typically on a scale of hundreds of kilometers—reduces their use and relevance for local and regional planning. It is virtually impossible to derive reliable climate information from limited climate data currently existing for Guatemala. Uncertainty about the future of Guatemala's climate, and therefore the fate of its agriculture and economy, not only jeopardizes development objectives for the region but could also lead to civil unrest, migration, and changes in critical ecosystem services (Milan and Ruano 2014). The climate data that are available come predominantly from stations in urban centers and commercially important agricultural regions. This means that smallholder, agriculture-dependent communities in rural areas are often without access to information relevant to their crops.

To begin to address the lack of climate data, we turned to Guatemala's high mountains to reconstruct past climate variability using tree rings. Longer climate records from Guatemala's mountain forests could be used both to test climate model skill through validation against hindcast simulations and to place current observations in the context of recent and decadal-scale climate variability. This is critical to assist in vulnerability reduction through risk management at regional and local scales in agriculturally dependent locations (e.g., coffee growing areas; Figure 1) because it can add to our knowledge on the magnitude, duration, frequency, and impact of the physical exposure to extreme weather events. Integrating this climate knowledge with the appropriate social research could transform these forecasts into useful information for farmers (Bell et al. 2011). Yet, the overall generation of locally relevant climate information applicable at the farm level can only be achieved by engaging the communities in a more participatory and interactive approach (Lemos and Morehouse 2005). We are striving toward a model of knowledge coproduction, wherein researchers interact closely with regional stakeholders over long periods of time to develop iterative and collaborative relationships. This should engage communities of researchers and decision makers toward a collective goal of shaping feasible research questions that could lead to information production that is relevant to climate-vulnerable groups and that generates some action toward improved resilience across space and through time (Cash et al. 2003).

Figure 1. Topographic map of Guatemala showing coffee-growing areas and tree-ring research sites sampled since 2008 by the authors. *Source:* Authors, with information from Asociación Nacional del Café (n.d.). (Color figure available online.)

Here, we present a summary of the results of tree-ring research in Guatemala's high mountains and describe how this research on climate science during the last decade in Guatemala has helped us identify some of the barriers that limit the generation and communication of climate knowledge. We then identify strategies, key stakeholders, and decision makers that can, in the near term, engage these actors in producing usable climate data. We also discuss the ways in which different modes of stakeholder engagement have been successfully used throughout our research and how the incorporation of the other principles of coproduction can be used in Guatemala to begin to close the gap

between climate knowledge and locally relevant climate information. This article also adds to the literature of successful empirical work that could lead to interactive research among scientists and decision makers (Lemos and Morehouse 2005).

Geography of the Mountains of Guatemala

Two thirds of Guatemala is mountainous and the steep slopes of volcanic and nonvolcanic mountain ranges contribute, in part, to low agricultural productivity (Pope et al. 2015). Torrential downpours and

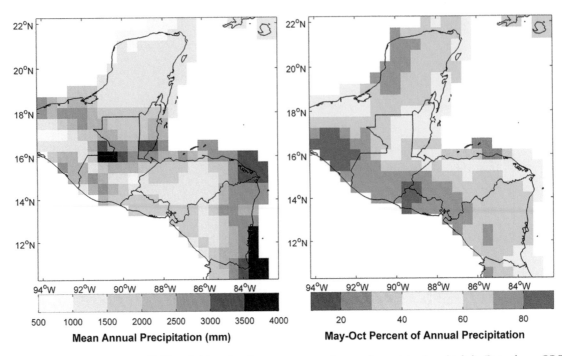

Figure 2. Mean annual precipitation (left) and May–October percentage of annual precipitation (right). Data from GPCCv6 0.5°, 1901–2010 (Schneider et al. 2011). (Color figure available online.)

landslides are a common threat to communities located in mountain areas (Sutton and Restrepo 2013). Flooding and landslides associated with Hurricane Stan in 2005 caused more than 600 deaths and affected almost half a million people (United Nations Environment Programme/United Nations Office for the Coordination of Humanitarian Affairs 2005). The mountainous regions are also considered high-risk areas because of the limited access to water and health services and the limited options available to the inhabitants of these montane regions to manage natural disaster risk (World Bank 2011). In terms of water access, 80 percent to 90 percent of Guatemala's population relies on groundwater for drinking—water that is ultimately derived from precipitation. Yet, approximately 30 percent of the rural population does not have household water connections, making these groups particularly vulnerable to meteorological and hydrological droughts (World Health Organization 2015). Furthermore, the magnitude and socioeconomic consequences of these conditions become clear once we take into account that half of the population lives in rural areas and that 76 percent of those rural residents live in poverty (World Bank 2015). Rural residents are often excluded from access to the political and social mainstream—including access to climate information—making this group especially vulnerable to climate change and natural disasters that affect the region.

Similar to much of Central America, Guatemala experiences a seasonal precipitation climate, with a boreal summer wet season and a winter dry season. The country is influenced by the northern extent of the Intertropical Convergence Zone during the boreal summer. The rainy season, dominated by spatially complex convective thunderstorms, typically begins in April or early May, although in dry years the onset of the rainy season has been delayed. It lasts until October in most of the country but persists through early to mid-December in the northern and eastern extent of the country. The dry season extends from November through April. Topographic complexity of Guatemala's mountains and volcanoes imparts important spatial heterogeneity in mean climate conditions. Even in gridded data products such as the Global Precipitation Climatology Centre 0.5° data set, the scale of which underrepresents the spatial complexity of regional precipitation, remarkable variation is evident (Figure 2). Annual precipitation in parts of the northwestern lowlands exceeds 4,000 mm, whereas in the so-called dry corridor, near the triple junction of Guatemala, Honduras, and El Salvador, annual precipitation is less than 1,000 mm. These complex hydroclimate conditions also interact with the diversity of the land use in the Guatemalan highlands to create an array of local microclimates (Manoharan, Welch, and Lawton 2009). Generally, it is wetter and cooler at higher elevation and on northern aspects, whereas it is drier and warmer at lower elevation and on southern exposures. These details translate to spatial complexity in soil water balance. Interannual soil moisture variation is particularly important for agriculture. If climate model projections of increased temperature and reduced precipitation are accurate, the amount of semiarid land in Guatemala would increase, limiting agricultural production in those regions (Shelton 2012).

Recent Climate Reports from Central America and the Caribbean

Although the scarcity of weather station networks in mountain regions limits our capacity to identify climate trends in these locations, many mountain regions are seemingly experiencing seasonal warming at a greater rate than the global average (e.g., Rangwala and Miller 2012; Pepin et al. 2015). According to Aguilar et al. (2005), there was a general warming trend in the Central American region with both maximum and minimum temperatures increasing for the period from 1961 to 2003. Recent infestations and outbreaks of coffee pests in Guatemala at elevations higher than historical records might also be an indicator of increased minimum temperatures in mountainous regions (Georgiou 2014). Precipitation in the Central American and Caribbean region, on the other hand, has exhibited more complex spatial and temporal patterns that require additional analysis (Jones et al. 2016). Nevertheless, impacts on subsistence annual crops like maize and beans have been associated with recent droughts that are probably associated with the onset of El Niño (United Nations Office for the Coordination of Humanitarian Affairs 2016). This impact can be particularly dire for the dry corridor region of eastern Guatemala. In 2009 this area experienced its worst drought in thirty years, reducing the harvest of maize and beans by half and affecting 2.5 million Guatemalans (World Bank 2015). Perennial crop yields (e.g., coffee) have also decreased, which has been associated with temporal variations in precipitation (Castellanos et al. 2013). The coffee sector of Guatemala was heavily affected by droughts that same year (Eakin, Tucker, and Castellanos 2005), in addition to a serious decline in world coffee prices in 2001. Not only droughts affect coffee plantations; heavy rains caused by hurricanes have often led to substantial losses—stressing the impact of precipitation

anomalies in the country and their different impacts on crops (Tucker, Eakin, and Castellanos 2010).

Due to Guatemala's reliance on rain-fed agriculture, a better understanding of local and regional climate in the past, present, and future is necessary. This knowledge is a prerequisite for monitoring current climate change, reconstructing past variability, testing the skill of climate models at regional scales, and eventually mitigating the adverse effects of climate variability (Mengistu 2011). Yet, the expansion of climate science independent of its application has contributed to a gap between the production of climate change knowledge and its subsequent transformation into locally relevant climate information and subsequent policy initiatives (Lynch, Tryhorn, and Abramson 2008). Nonetheless, the disparities in the production of these two forms of knowledge need to be understood and addressed if sustainable development is truly to be achieved (Gunasekera 2010). Because of the urgency to mitigate the adverse effect of climate variability, more attention has been placed on finding effective ways of production, dissemination, and communication of climate information (Tarhule and Lamb 2003; Sivakumar 2006; Brondizio and Moran 2008). This call to practice within climate change science is persistent and is the next required step, especially in countries most vulnerable to climate change (Moser 2010; Klenk et al. 2015).

Generation of Climate Information in Mountainous Guatemala Using Tree Rings

The processes of production, dissemination, and subsequent use of climate information relevant to communities at the local scale are critical in mitigating the adverse effects of climate variability (Mengistu 2011). Climate information needs to be created and disseminated in a manner that is both useful and accessible to stakeholders. For example, finding ways to communicate the probabilistic framework on which forecasts and predictions of climate are created and communicated has proven to be challenging (Hammer et al. 2001). Differences in spatial and temporal scales between scientists and farmers also play a role in the utility and application of climate information and have been recognized as important barriers (Lynch, Tryhorn, and Abramson 2008).

In the absence of instrumental records, scientists investigating the causes and consequences of climate variability and change often depend on natural proxy records that can be used to reconstruct past conditions.

Tree-ring chronologies form the bulk of the available high-resolution terrestrial proxy records covering the last several centuries, but they are still largely absent from the tropical Americas south of Mexico. As a consequence, for instance, the North American Drought Atlas (Cook et al. 2004), which provides a spatiotemporal perspective on the frequency, magnitude, and severity of drought covering the last millennium, ends in southern Mexico, limited by a lack of tree-ring chronologies collected further south from Central America. Although some areas of the tropics lack the necessary seasonal climate variability to induce the formation of annual tree rings, many tropical regions do have a highly seasonal precipitation regime, with a distinct dry season of several months. It has been conclusively demonstrated that many tropical trees do form annual rings (e.g., Stahle et al. 1998; Speer et al. 2004; Brienen and Zuidema 2006; D'Arrigo et al. 2006; Therrell et al. 2006), although the development of long and precisely dated annual time series remains more challenging than in temperate regions due to irregular growth patterns, high species diversity, forest

disturbance, and relatively short individual tree life spans.

Tropical dendrochronology presents technical and methodological challenges that have long been recognized (Schulman 1944). Even once progress has been made in overcoming issues related to chronology, however, new challenges arise connected to both climate as well as human factors. Guatemala's mountain landscapes have been the site of human settlement for more than 10,000 years. This period of uninterrupted occupancy (Lovell 2003) implies a continual pressure on forest resources. Long-term reliance on trees for building, high population densities, and continuing dependence on firewood for cooking and heating have, based on our last decade of experience and exploration, led to a paucity of old trees in the mountains of Guatemala.

Despite these challenges, we have made significant progress in Guatemala. Extensive reconnaissance and fieldwork over a decade led to the discovery of old stands of trees and allowed us to reconstruct interannual precipitation back to the

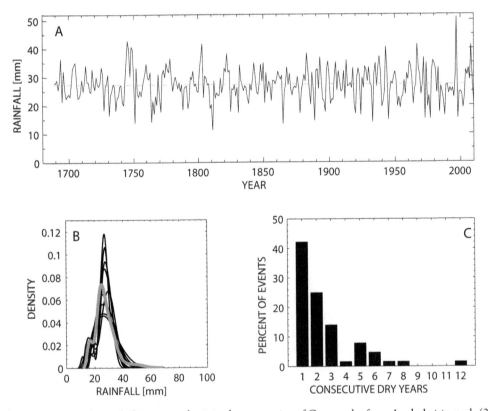

Figure 3. Results from tree-ring analysis of *Abies guatemalensis* in the mountains of Guatemala, from Anchukaitis et al. (2015). (A) Reconstructed January through March total precipitation for the Altos Cuchumatanes (Anchukaitis et al. 2015). (B) Kernel density estimates for reconstructed January through March total for all nonoverlapping thirty-year periods between 1710 and 2009. The most recent thirty-year period (1980–2009) is shown in a gray line and is not exceptional in either mean or distribution (Anchukaitis et al. 2015). (C) Occurrence of droughts as a function of their duration, expressed as the percentage of drought events that persist for a given number of consecutive years. Note that the event occurring for twelve consecutive years is in a poorly replicated part of the chronology.

late seventeenth century using the precisely dated annual growth rings of Guatemalan fir—*Abies guatemalensis* (Anchukaitis et al. 2013; Anchukaitis et al. 2015). We are also developing chronologies from *Pinus hartwegeii*. Our initial reconstruction from the Cuchumatanes (Figure 3) reveals that recent trends in winter–spring rainfall are not yet outside the range of the natural variability of the last several centuries. Indeed, our research shows a significant yet variable decadal component to regional rainfall, associated in part with both the El Niño Southern Oscillation and the North Atlantic atmospheric circulation. The tree rings reveal the existence of past multiyear droughts, including events where up to eight consecutive dry seasons experienced below-average rainfall. Although declining precipitation in Guatemala is a robust prediction of the current generation of climate models, the long-term perspective provided in our chronology suggests that recent rainfall patterns are not extraordinary or unusual. To generate this record it was necessary to overcome multiple challenges associated with the process of locating a climate-sensitive species, accessing the forests, collecting the samples, and proving that growth rings formed annually could be dated according to the rigorous methodological standard of dendrochronology. None of this production of climate data would have been possible without the involvement of local key stakeholders from the very beginning.

Stakeholder Engagement and the Process of Climate Knowledge Production

We now illustrate the different modes of stakeholder engagement and approaches that were used in the different stages of production of climate knowledge in the mountain regions of Guatemala. We frame our analysis under the structure described by Meadow et al. (2015). We discuss the different climate knowledge production pathways used with different stakeholders and decision makers throughout a decade of research as well as the potential of some of these approaches for future research, especially in the context of climate change science in mountain regions. We discuss the potential of coproducing climate records using dendrochronology to provide local records of the climate and a better understanding of the overall climate system at different scales as well as its potential to connect traditionally excluded communities to climate information, hence reducing the overall vulnerability of these distant mountain communities.

The Complex Task of Sampling Trees

In the case of Guatemala, the search for old trees can be complicated. In addition to the challenges associated with locating a climate-sensitive species in the tropics, fieldwork is complicated by Guatemala's linguistic diversity—twenty-three different languages are spoken in Guatemala (Grandin, Levenson-Estrada, and Oglesby 2011), and most of this linguistic and cultural diversity is located in the Western Highlands. Despite our knowledge of Spanish, we were cognizant of the need to communicate in indigenous languages, making the use of a translator necessary. The past and present repression of indigenous communities in the Guatemalan highlands has undermined the trust that these communities have of foreign individuals and nonindigenous Guatemalans. Land distribution inequalities and ambiguous rights over communal territories have created tension over the entitlement and overall use of communal forests where old trees might be found (Elías and Wittman 2005). The legal definitions of the land and subsequent right-of-use are often susceptible to changing political circumstances and changes in the central power in Guatemala City and hence alienated from local realities. In addition to these circumstances, the often unwanted presence of foreign extractive industries in some of these territories complicates access to these forests by researchers (LaPlante and Nolin 2014). Nevertheless, these communities bring an immense and invaluable perspective to the table, one that should be more central to Western models of research programs (Clark et al. 2016). During the search for old trees in their territories, community leaders often referred to them as "the wise ancient trees," a term that we adopted and used in following visits to the communities to highlight how, from our own perspective, these trees were also considered archives of ancient climate information. Despite an apparent disconnection from global scientific issues, these communities have a local empirical knowledge that is irreplaceable and that should be taken into account when trying to understand local impacts of global phenomena. An appreciation for and inclusion of multiple perspectives, including local and qualitative knowledge, can lead to practical outcomes on adaptation (Lynch, Tryhorn, and Abramson 2008).

Figure 4. (A) Collecting increment core samples with forest ranger Don Doroteo from the Todos Santos Municipality, Huehuetenango. (B) Don Geronimo, a forest guard and community leader from the Todos Santos Municipality, prepares to take a sample from *Abies guatemalensis*. Don Geronimo has enabled much of our research in the Cucuhumantan region. Don Doroteo and Don Geronimo are Mam Maya and speak Mam. In this region they cultivate maize and fava beans but mainly potatoes because much of their lands are above the elevation where maize can be cultivated. Sheep are also an important part of the household economy in this region (Steinberg and Taylor 2008). (Color figure available online.)

Gaining access to the forests in mountainous regions of Guatemala is possible but requires substantial effort, including the appropriate communication of the project expectations, methods, and potential outcomes with central and local government entities and nongovernmental power structures months in advance to build trust. In Guatemala, the National System of Development Council represents one of the most important channels for participation of the rural population. It includes the Community Development Council as the most elemental component of the system, which is made up of civil society organizations. It is through the appropriate communication of the project with these organizations that the production of climate information begins. During meetings with these entities, the outputs of research can be discussed and the usefulness of them assessed. It is only after these meetings and discussions that participatory sampling with community leaders can be planned (Figure 4). Because the distribution of tree species in Guatemala is not well documented, a first exploratory field trip with community leaders can help identify forests with dendrochronological potential. Once this has been accomplished, the project needs to be presented to the National Council for Protected Areas in Guatemala City. The involvement of this entity was not only a legal requirement due to the protected status of *Abies guatemalensis* but it reinforced the legitimacy of the project to local authorities (both governmental and nongovernmental) although not necessarily with the communities. The politicization of formal knowledge, as described by Clark et al. (2016), is often perceived by these communities as a way to control the activities of rural land users and an intrusion into their worldviews. Hence, it becomes an important barrier to overcome when working on rural development projects. Due to the time that it takes to build a sense of legitimacy of the project in these communities, the process of production of climate data often takes longer than the standard funding cycles of agencies like the National Science Foundation (NSF).

This first step in finding useful trees for dendroclimatological purposes could be framed as a combination between a contractual and a collaborative mode of stakeholder engagement. The former has been described as a research process in which scientists make use of the land, services, and resources to test experimental technology and the latter a mode that requires stakeholder input, such as the location of old trees and undisturbed areas (Meadow et al. 2015). The approach used here to gain collaboration in this first step was a rapid assessment process as described by Meadow et al. (2015) in which the local situation was assessed from the perspective of the stakeholders to understand and fit local knowledge frameworks. One of the most challenging barriers to overcome while using this approach was the lack of interest in projects that generated information as an outcome (our goal) as opposed to infrastructure projects (a desire of several of the communities in which we worked).

Select studies have reported meaningful and positive outcomes from tailoring tree-ring reconstructions of climate to stakeholder interests (e.g., Woodhouse and Lukas 2006; Rice, Woodhouse, and Lukas 2009), but the value of projects that culminate in the production of knowledge versus the construction of tangible infrastructure outcomes like clinics or wells can be difficult to explain or justify to local communities. The expectation of outcomes that include infrastructure is a result, in part, of the large number of nongovernmental projects in the Western Highlands after the signing of the peace accords in 1996 and the implementation of stipulations for development set out in those accords (Moran-Taylor 2008). The understandable pragmatism of infrastructural development projects can obscure the potential value of information—information that could then be used to improve planning or implementation of infrastructure projects. For instance, knowing the magnitude, severity, and return interval of historical drought cycles was not necessarily perceived as useful by all the community members. Locals often only saw the relevance of climate information after we illustrated how it could be applied to physical infrastructure projects; for example, in the location and sizing of microhydropower installations.

Analyzing Tree Samples, Creating Local Capacities

The production of climate information through dendrochronology requires the analysis of the samples in a laboratory to derive the climate information. Yet, it is hard to involve agriculturists in this endeavor because it requires a substantial amount of time and preparation that would take them away from their agricultural livelihoods. This leaves a dilemma about how to avoid the pitfall of outside experts monopolizing the creation and analysis of local climate proxies. One of the most urgent calls in the literature of human dimensions of climate change is facilitating a sustainable system for access by vulnerable communities to relevant climate information. It is necessary that climate information remains available for much longer than the life span of any single project. To achieve this important component of production of climate information, part of the funding for our research was directed to local capacity building. A fully functional dendrochronology lab was established at Del Valle

University in Guatemala City in 2012 and training was provided to Guatemalan students.

NSF funding also enabled us to run the first ever Central American Dendroecological Fieldweek in Guatemala in March 2015, which was attended by Central American scholars as well as climate and forestry practitioners. Three researchers from Del Valle University in Guatemala City have continued to develop dendrochronology projects in the country. Two of them recently completed a dendrochronology workshop in Mexico with internationally recognized tree-ring researchers. These two researchers are currently writing a proposal to investigate the potential of pine species to reconstruct severe frost events in the Guatemalan highlands. A third student (the first author) is currently studying for a PhD in geography and working on dendrochronology as a tool to reconstruct precipitation in coffee regions of the country. In addition to funding availability, this cohort of Guatemalan researchers represents a promising pathway to ensure that relevant climate data will be produced, analyzed, and disseminated to appropriate communities in Guatemala by Guatemalans. Due to the nature of this process in which local institutions (stakeholders) and scientists are partners in the research process, we identify this stage of climate knowledge production as a collaborative process. The ongoing applied research at the local institution by trained research team members can be acknowledged as successful stakeholder engagement. Moreover, according to Meadow et al. (2015), because these stakeholders are now conducting their own research and solving real-life problems with enhanced capacity, this stage could be defined as collegial because local capacities have been strengthened with support from international research.

Returning to the Communities to Disseminate Results

As part of our commitment to the communities with which we work, the dissemination of the results has always been the most important part of the process of production of climate information. Yet, this step has proven difficult. This is due to language barriers, literacy levels, the communication of the probabilistic scenarios of our reconstructions, and the difficulties associated with previous knowledge systems like different world visions. Although we have been able to communicate some insights on the frequency of droughts and placed recent trends in the context of long-term variability in the communities where we worked, this newly generated information has not yet been transformed into agrometeorological information. We have translated academic publications into Spanish, but these publications are of little use to farmers because they are just that—academic publications. Although the literacy rate reported for adults (age fifteen and older) in Guatemala is 77 percent (World Bank 2015), in some cases these communities fall within the 23 percent of the adult population that cannot read in Spanish. To overcome this challenge, we turned to local community leaders to communicate our findings. Usually, these community leaders speak Spanish and play important decision-making roles. In the case of our ongoing research in the mountain region of Kanchej, Quetzaltenango, we shared the preliminary results with the Office of Forest Affairs representative of the Cantel municipality and with two high school teachers from the same area who took samples of tree cores to show their students how tree rings can contain climate information.

Future Research and Conclusion

The critical challenge that remains now is to work to collaborate with climate modelers and Guatemalan communities (scientific and nonscientific) to create links between the abiotic (climate) and biotic (crop) components in a manner that generates sustainable and actionable engagement. Only when climate data have been generated (e.g., through dendrochronology) and the component of climate variability relevant to the particular crop has been identified by the local stakeholders (e.g., precipitation variability) can the information help adaptation efforts. As stated by Hammer et al. (2001), the predictions are useless unless some changes take place based on them. An interdisciplinary approach where descriptive (using standard social science techniques) and quantitative analyses should come together as an inclusive discussion toward the identification of a realistic set of strategies that can be put in place by smallholder farmers is the next required step (Hammer et al. 2001). Although a multidisciplinary approach has been recently used in Guatemala to work with smallholder farmers on climate variability adaptation (see Castellanos et al. [2008]), we suggest following the approach proposed by Meadow et al. (2015) in which a transdisciplinary approach integrates academic and practitioner knowledge to coproduce climate information that has the

potential to (1) help solve socially relevant problems (e.g., information of environmental hazards like droughts), (2) reconcile the social demand for academic research in excluded communities, and (3) democratize science in countries vulnerable to climate changes.

Guatemala's mountain regions are certain to be challenged by imminent anthropogenic climate change, and this vulnerability is exacerbated by the paucity of climate information. The montane forests of Guatemala, however, preserve a valuable record of the climate conditions locked within the annual rings of trees that cover their slopes. Although we have shown the necessary steps we have taken to demonstrate that the production of climate information is possible using dendrochronology in highland Guatemala, it will be the coproduction of actionable, relevant, and comprehensible climate knowledge with the inhabitants of these forests that will bridge the gap between reducing vulnerability to climate change and these new climate data.

Acknowledgments

We thank our collaborators in Guatemala for making the research possible in their communities: Don Gerónimo Ramirez and Don Doroteo from Todos Santos, Huehuetenango, and Doña Enriqueta Salanic and Don Ramón Rixquicche from the Cantel municipality, Quetzaltenango. Pauline Decamps and Fernando Mejilla made it all possible.

Funding

This research was enabled with funding from National Science Foundation grants BCS 1263609 and 0852652, National Geographic, Conference of Latin Americanist Geographers, and Latin American Specialty Group Field Study Awards; and the Lawrence Herold Fund at the University of Denver.

References

Aguilar, E., T. C. Peterson, P. R. Obando, R. Frutos, J. A. Retana, M. Solera, J. Soley, et al. 2005. Changes in precipitation and temperature extremes in Central America and northern South America, 1961–2003. *Journal of Geophysical Research: Atmospheres* 110:D23.

Anchukaitis, K. J., M. J. Taylor, C. Leland, D. Pons, J. Martin-Fernandez, and E. Castellanos. 2015. Tree-ring reconstructed dry season rainfall in Guatemala. *Climate Dynamics* 45:1537–46.

Anchukaitis, K. J., M. J. Taylor, J. Martin-Fernandez, D. Pons, M. Dell, C. Chop, and E. Castellanos. 2013. Annual chronology and climate response in *Abies guatemalensis rehder* (pinaceae) in Central America. *The Holocene* 23 (2): 270–77.

Asociación Nacional del Café (ANACAFE). n.d. Red de estaciones meteorológicas. http://www.anacafe.org/glifos/index.php/22GIS:Red-de-estaciones-meteorologicastemp/comer/estadisticas_temp.htm (last accessed 19 October 2016).

Banco de Guatemala. 2014. Guatemala en Cifras. http://www.banguat.gob.gt/publica/guatemala_en_cifras_2013.pdf (last accessed 28 November 2015).

Bell, A. R., B. I. Cook, K. J. Anchukaitis, B. M. Buckley, and E. R. Cook. 2011. Repurposing climate reconstructions for drought prediction in Southeast Asia. *Climatic Change* 106 (4): 691–98.

Beniston, M. 2002. Climate modeling at various spatial and temporal scales: Where can dendrochronology help? *Dendrochronologia* 20 (1): 117–31.

Brondizio, E. S., and E. F. Moran. 2008. Human dimensions of climate change: The vulnerability of small farmers in the amazon. *Philosophical Transactions of the Royal Society B: Biological Sciences* 363 (1498): 1803–9.

Cash, D. W., W. C. Clark, F. Alcock, N. M. Dickson, N. Eckley, D. H. Guston, J. Jager, and R. B. Mitchell. 2003. Knowledge systems for sustainable development. *Proceedings of the National Academy of Sciences of the United States of America* 100 (14): 8086–91.

Castellanos, E., R. Díaz, H. Eakin, and Y. G. Jiménez. 2008. Understanding the resources of small coffee growers within the global coffee chain through a livelihood analysis approach. In *Applying ecological knowledge to ecological decisions*, ed. H. Tiessen and J. Stewart, 34–41. The Scientific Committee on Problems of the Environment, SCOPE.

Castellanos, E., C. Tucker, H. Eakin, H. Morales, J. Barrera, and R. Díaz. 2013. Assessing the adaptation strategies of farmers facing multiple stressors: Lessons from the Coffee and Global Changes Project in Mesoamerica. *Environmental Science and Policy* 26:19–28.

Clark, W. C., T. P. Tomich, M. Noordwijk, D. Guston, D. Catacutan, M. Dickson, and E. McNie. 2016. Boundary work for sustainable development: Natural resource management at the Consultative Group on International Agricultural Research (CGIAR). *Proceedings of the National Academy of Sciences* 113 (17): 4615.

Cook, E. R., C. A. Woodhouse, C. M. Eakin, D. M. Meko, and D. Stahle. 2004. Long-term aridity changes in the Western United States. *Science* 306 (5698): 1015–18.

Cutter, S. L., L. Barnes, M. Berry, C. Burton, E. Evans, E. Tate, and J. Webb. 2008. Community and regional resilience: Perspectives from hazards, disasters, and emergency management. CARRI Research Report 1, Community and Regional Resilience Initiative, Oak Ridge National Laboratory, Oak Ridge, TN.

D'Arrigo, R., R. Wilson, J. Palmer, P. Krusic, A. Curtis, J. Sakulich, S. Bijaksana, S. Zulaikah, and L. O. Ngkoimani. 2006. Monsoon drought over Java, Indonesia, during the past two centuries. *Geophysical Research Letters* 33:4709.

Eakin, H., C. M. Tucker, and E. Castellanos. 2005. Market shocks and climate variability: The coffee crisis in Mexico, Guatemala, and Honduras. *Mountain Research and Development* 25 (4): 304–9.

Elías, S., and H. Wittman. 2005. State, forest and community: Decentralization of forest administration in Guatemala. In *The politics of decentralization*, ed. C. J. Pierce and C. D. Capistrano, 282–96. London: Earthscan.

Famine Early Warning System. 2015. Continuing El Niño drives increased food insecurity across many regions. http://www.fews.net/global/alert/october-8-2015 (last accessed 28 November 2015).

Fernández-Llamazares, Á., M. E. Méndez-López, I. Díaz-Reviriego, M. F. McBride, A. Pyhälä, A. Rosell-Melé, and V. Reyes-García. 2015. Links between media communication and local perceptions of climate change in an indigenous society. *Climatic Change* 131 (2): 307–20.

Georgiou, S. 2014. An analysis of the weather and climate conditions related to the 2012 epidemic of coffee rust in Guatemala: Technical report. Costa Rica: CATIE.

Giannini, A., M. Cane, and Y. Kushnir. 2001. Interdecadal changes in the ENSO teleconnection to the Caribbean region and the North Atlantic Oscillation. *Journal of Climate* 14 (13): 2867–79.

Global Climate Observing System (GCOS). 2013. GSN station summary by region. https://www.ncdc.noaa.gov/sites/default/files/attachments/GSN_Stations_by_Region_2013.pdf (last accessed 14 July 2016).

Grandin, G., D. Levenson-Estrada, and E. Oglesby. 2011. *The Guatemala reader: History, culture, politics.* Durham, NC: Duke University Press.

Gunasekera, D. 2010. Use of climate information for socioeconomic benefits. *Procedia Environmental Sciences* 1:384–86.

Hammer, G. L., J. W. Hansen, J. G. Phillips, J. W. Mjelde, H. Hill, A. Love, and A. Potgieter. 2001. Advances in application of climate prediction in agriculture. *Agricultural Systems* 70 (2): 515–53.

Jones, P., C. Harpham, I. Harris, C. M. Goodess, A. Burton, A. Centella–Artola, M. A. Taylor, et al. 2016. Longterm trends in precipitation and temperature across the Caribbean. *International Journal of Climatology* 36 (9): 3314–33.

Karmalkar, A. V., R. S. Bradley, and H. F. Díaz. 2011. Climate change in Central America and Mexico: Regional climate model validation and climate change projections. *Climate Dynamics* 37:605–29.

Klenk, N., K. Meehan, S. Lee, F. Mendez, P. Torres, and D. Kammen. 2015. Stakeholders in climate science: Beyond lip service? *Science* 350:6262.

LaPlante, J. P., and C. Nolin. 2014. *Consultas* and socially responsible investing in Guatemala: A case study examining Maya perspectives on the Indigenous right to free, prior and informed consent. *Society & Natural Resources: An International Journal* 27 (3): 231–48.

Lemos, M. C., and B. J. Morehouse. 2005. The co-production of science and policy in integrated climate assessments. *Global Environmental Change* 15 (1): 57–68.

Lovell, G. 2003. *Conquest and survival in colonial Guatemala: A historical geography of the Cuchumatán highlands, 1500–1821.* Montreal, Canada: McGill-Queen's University Press.

Lynch, A. H., L. Tryhorn, and R. Abramson. 2008. Working at the boundary: Facilitating interdisciplinarity in climate change adaptation research. *Bulletin of the American Meteorological Society* 89 (2): 140.

Magrin, G., J. A. Marengo, J.-P. Boulanger, M. S. Buckeridge, E. Castellanos, G. Poveda, F. R. Scarano, and S. Vicuña. 2014. Central and South America. In *Climate change 2014: Impacts, adaptation, and vulnerability. Part B: Regional aspects. Contribution of Working Group II to the Fifth Assessment Report of the Intergovernmental Panel on Climate Change*, ed. V. R. Barros, C. B. Field, D. J. Dokken, M. D. Mastrandrea, K. J. Mach, T. E. Bilir, M. Chatterjee, et al., chap. 27, pp. 1502. Cambridge, UK: Cambridge University Press.

Manoharan, V. S., R. M. Welch, and R. O. Lawton. 2009. Impact of deforestation on regional surface temperatures and moisture in the Maya lowlands of Guatemala. *Geophysical Research Letters* 36 (21).

Meadow, A. M., D. B. Ferguson, Z. Guido, A. Horangic, G. Owen, and T. Wall. 2015. Moving toward the deliberate coproduction of climate science knowledge. *Weather, Climate, and Society* 7 (2): 179–91.

Mengistu, D. K. 2011. Farmers' perception and knowledge of climate change and their coping strategies to the related hazards: Case study from Adiha, Central Tigray, Ethiopia. *Agricultural Sciences* 2 (2): 138–45.

Milan, A., and S. Ruano. 2014. Rainfall variability, food insecurity and migration in Cabricán, Guatemala. *Climate and Development* 6 (1): 61–68.

Moran-Taylor, M. J. 2008. Guatemala's ladino and Maya migra landscape: The tangible and intangible outcomes of migration. *Human Organization* 67 (2): 111–24.

Moser, S. C. 2010. Now more than ever: The need for more societally relevant research on vulnerability and adaptation to climate change. *Applied Geography* 30 (4): 464–74.

Neelin, J. D., M. Münnich, H. Su, J. E. Meyerson, and C. E. Holloway. 2006. Tropical drying trends in global warming models and observations. *Proceedings of the National Academy of Sciences* 103:6110–15.

Pepin, N., R. Bradley, H. Diaz, M. Baraer, E. Caceres, N. Forsythe, H. Fowler, et al. 2015. Elevation-dependent warming in mountain regions of the world. *Nature Climate Change* 5 (5): 424–30.

Pope, I., D. Bowen, J. Harbor, G. Shao, L. Zanotti, and G. Burniske. 2015. Deforestation of montane cloud forest in the central highlands of Guatemala: Contributing factors and implications for sustainability in Q'eqchi' communities. *International Journal of Sustainable Development & World Ecology* 22 (3): 201–12.

Rangwala, I., and J. R. Miller. 2012. Climate change in mountains: A review of elevation-dependent warming and its possible causes. *Climatic Change* 114 (3): 527–47.

Rice, J., C. Woodhouse, and J. Lukas. 2009. Science and decision-making: Water management and tree-ring data in the Western United States. *Journal of the American Water Resources Association* 45 (5): 1248–59.

Schneider, U., A. Becker, P. Finger, A. Meyer-Christoffer, B. Rudolf, and M. Ziese. 2011. GPCC full data reanalysis version 6.0 at 0.5°: Monthly land-surface precipitation from rain-gauges built on GTS-based and historic data. Deutscher Wetterdienst: Global Precipitation Climatology Centre.

Schulman, E. 1944. Dendrochronology in Mexico. *Tree-Ring Bulletin* 10 (3): 18–24.

Shelton, M. 2012. Land-use changes in southwestern Guatemala: Assessment of their effects and sustainability. ProQuest Dissertations 3563093.

Sivakumar, M. V. K. 2006. Dissemination and communication of agrometeorological information-global perspectives. *Meteorological Applications* 13 (S1): 21–30.

Speer, J. H., K. H. Orvis, H. D. Grissino-Mayer, L. M. Kennedy, and S. P. Horn. 2004. Assessing the dendrochronological potential of *Pinus occidentalis* Swartz in the Cordillera Central of the Dominican Republic. *The Holocene* 14 (4): 563–69.

Stahle, D. W., R. D. D'Arrigo, P. J. Krusic, M. K. Cleaveland, E. R. Cook, R. J. Allen, J. E. Cole, et al. 1998. Experimental dendroclimatic reconstruction of the Southern Oscillation. *Bulletin of the American Meteorological Society* 79 (10): 2137–52.

Stahle, D. W., D. K. Stahle, D. J. Burnette, J. Villanueva, J. Cerano, F. K. Fye, R. D. Griffin, et al. 2012. Tree-ring analysis of ancient baldcypress trees and subfossil wood. *Quaternary Science Reviews* 34:1–15.

Steinberg, M. K., and M. J. Taylor. 2008. Guatemala's Altos de Cuchumatán: Landscape changes on the high frontier. *Mountain Research and Development* 28 (3–4): 255–62.

Sutton, L., and C. Restrepo. 2013. Natural hazards, diverse economy and livelihoods in the Sierra de las Minas, Guatemala. *Journal of Latin American Geography* 12 (3): 137–64.

Tarhule, A., and P. J. Lamb. 2003. Climate research and seasonal forecasting for West Africans: Perceptions, dissemination, and use? *Bulletin of the American Meteorological Society* 84 (12): 1741–59.

Therrell, M. D., D. W. Stahle, L. P. Ries, and H. H. Shugart. 2006. Tree-ring reconstructed rainfall variability in Zimbabwe. *Climate Dynamics* 26 (7): 677–85.

Tucker, C., H. Eakin, and E. Castellanos. 2010. Perceptions of risk and adaptation: Coffee producers, market shocks, and extreme weather in Mexico and Central America. *Global Environmental Change* 20:23–32.

United Nations Development Programme. 2009. *Climate change and agriculture: Guatemala country note.* http://www.adaptationlearning.net/country-profiles/GT (last accessed 1 March 2016).

United Nations Environment Programme/United Nations Office for the Coordination of Humanitarian Affairs Environment Unit. 2005. *Hurricane Stan: Environmental impacts from floods and mudslides in Guatemala. Results from a rapid environmental assessment in Guatemala.* Geneva, Switzerland: UNEP/UNOCHA.

United Nations Office for the Coordination of Humanitarian Affairs (UNOCHA). 2016. Humanitarian needs overview: Central America. https://www.humanitarianresponse.info/en/system/files/documents/files/2016-hno-centralamerica-7jan.pdf (last accessed 14 July 2016).

Woodhouse, C. A., and J. J. Lukas. 2006. Drought, tree rings, and water resource management. *Canadian Water Resources Journal* 31:297–310.

World Bank. 2011. Climate risk and adaptation country profile, Guatemala. http://sdwebx.worldbank.org/clima teportalb/doc/GFDRRCountryProfiles/wb_gfdrr_clima te_change_country_profile_for_GTM.pdf (last accessed 14 July 2016).

———. 2015. Global poverty working group: Annual poverty rates, country profile, Guatemala. http://data.world bank.org/indicator/SI.POV.RUHC/countries/GT?dis play=graph (last accessed 14 July 2016).

World Health Organization. 2015.WHO/UNICEF water supply statistics. https://knoema.com/WHOWSS 2014/who-unicef-water-supply-statistics-2015?location= 1000970-guatemala (last accessed 14 July 2016).

Retreating Glaciers, Incipient Soils, Emerging Forests: 100 Years of Landscape Change on Mount Baker, Washington, USA

Paul Whelan and Andrew J. Bach

Glacial forelands are harsh environments where incipient pedogenesis provides the basis for vegetation establishment and succession. The Easton Glacier foreland on Mount Baker, Washington, has till deposited during five time intervals over the last 100 years as determined from historic ground and air photos. A soil chronosequence was established on the different age surfaces to assess rates of pedogenesis. As hypothesized, all soil variables, except pH, showed increasing values on progressively older surfaces, with several orders of magnitude increase between the active till and the 100-year surface. Till on ice showed no vegetation cover, low organic matter (0.4 percent), little to no nitrogen content (maximum 0.001 percent), minimal carbon (maximum 0.0083 percent), and a carbon/nitrogen (C/N) ratio of 5.9. The 100-year-old surface has continuous vegetation cover, high organic matter (12.6 percent), 0.67 percent nitrogen, and 9.47 percent carbon, and the C/N ratio was at its highest (22.6). Organic matter content started higher than expected in fresh till and gradually increased before vegetation became established, suggesting aeolian deposition of detritus built soil fertility. We estimate that after about sixty years of exposure, till surfaces became fully covered with vegetation and soil organic matter increased by almost 2,800 percent (0.4–12.6 percent). This rapid rate of soil development, given a short growing season, is hypothesized to be related to several edaphic conditions (topographic setting relative to established vegetation, aspect, and andesitic parent material), rather than a normal condition for the Cascades Range as a whole, demonstrating that ongoing climate change is affecting many environmental processes.

冰川前沿是坚困的环境，其中初期的土壤化育提供了植被建立和演替的基础。依据历史基础和空照图的测定，华盛顿贝克山的伊斯顿冰川前沿，在过去一百年来的五次间隔中产生了冰川沉积物。本研究根据不同年代的表层，建立土壤年代序列，以评估土壤化育的速度。如同假说所推定，所有的土壤变异数，除了pH值之外，显示出在日渐老旧的表层中质量逐渐增加，其中在活跃的冰川沉积和一百年之久的表层之间，增加了若干量级。冰上的冰川沉积显示不具有植被覆盖，较低的有机物质（千分之四），少量或不具有氮含量（最多十万分之一），极少的碳（最多十万分之八点三），以及碳氮比（C/N）为五点九。一百年的旧表层，持续具有植被覆盖，高度的有机物质（百分之十二点六），千分之六点七的氮，以及百分之九点四七的碳，且有最高的C/N比（二十二点六）。有机物质含量在新的冰川沉积物中，一开始较预期为高，并在植被形成之前逐渐增加，显示腐植质的风成沉积物建立了土壤的肥沃度。我们评估，在曝露约六十年之后，冰川沉积表面被植被完全覆盖，而土壤有机物质增加了近乎百分之两千八百（从百分之零点四到百分之十二点六）。此一在短期成长季中的快速土壤发展速率，假定与若干土壤条件有关（与已建立的植被、坡向和安山岩母质相关的地形环境），而非喀斯喀特山脉整体的一般条件，证实持续不断的气候变迁，正影响着诸多环境过程。关键词：气候变迁冲击，冰川地形学，山区地理，太平洋西北部，土壤发展。

Las tierras glaciares frontales son entornos ásperos donde la pedogénesis incipiente provee la base para la colonización vegetal y para la sucesión. La parte terminal del Glaciar Easton del Monte Baker, en Washington, tiene till glaciario depositado en cinco intervalos de tiempo durante los pasados 100 años, según puede determinarse a partir de fotos históricas aéreas y de superficie. Se estableció una cronosecuencia del suelo en las diferentes superficies de edad para evaluar las ratas de pedogénesis. En acuerdo con nuestras hipótesis, todas las variables del suelo, excepto el pH, mostraron valores en incremento sobre superficies progresivamente más viejas, con incrementos de varios órdenes de magnitud entre el till activo y la superficie de los 100 años. El till sobre hielo no mostró cobertura vegetal, bajo contenido de materia orgánica (0.4 por ciento), poco o ningún contenido de nitrógeno (máximo de 0.001 por ciento), mínimo de carbono (máximo 0.0083 por ciento) y una razón de carbono/nitrógeno (C/N) de 5.9. La superficie de 100 años de edad tiene una cubierta continua de vegetación, alto contenido de materia orgánica (12.6 por ciento), 0.67 por ciento de nitrógeno y 9.47 por ciento de carbono, y la razón C/N alcanzó su máximo (22.6). El contenido de materia orgánica empezó más alto de lo esperado en till reciente y se incrementó gradualmente antes de que la vegetación se afincase, lo cual sugiere una fertilidad del suelo generada en detritus de deposición eólica. Calculamos que después de alrededor de sesenta años de exposición las superficies cubiertas de till estarían enteramente cubiertas de vegetación y la materia orgánica del suelo incrementada en cerca del 2.800 por ciento (0.4–12.6 por ciento). Nuestra hipótesis es que esta rápida rata de desarrollo del suelo, dada una corta estación de crecimiento, debe estar relacionada con condiciones edáficas (un escenario topográfico relativo a la vegetación establecida, aspecto y material parental andesítico), mejor que una condición normal para la Cadena de las Cascadas, en conjunto, demostrando que el cambio climático en desarrollo está afectando muchos procesos ambientales.

As alpine glaciers recede, new surfaces are exposed that are subsequently modified by soil development and vegetation succession. As surface age increases (i.e., time since exposure via deglaciation), a pattern of development emerges that intertwines incipient stages of soil genesis with colonization by pioneer species that, over time, are outcompeted by late-successional plant life (Burga et al. 2010). This transformation from a barren expanse of unconsolidated glacial till to a productive forest was examined in the Easton foreland—the region located directly below the Easton Glacier terminus on the southern side of Mount Baker, Washington (Figure 1).

Climate change has resulted in rapid glacial recession over the last 100 years, leading to questions about the fate and longevity of alpine and subnival ecosystems (Meier, Dyurgerov, and McCabe 2003; Erschbamer 2007; Intergovernmental Panel on Climate Change [IPCC] 2007; Vos et al. 2008). Depending on how rapid recession occurs, rates and patterns of soil development and vegetation establishment might change (e.g., U. M. Huber, Bugmann, and Reasoner 2005; Erschbamer 2007). The Easton foreland is defined here as the most recently deglaciated trough from the 1912 glacial terminus to the 2012 terminus. Seed rain and detritus input from mature forests atop adjacent Holocene moraines are evident. As retreat continues, however, observed facilitative succession might become untenable as the glacier terminus retreats above the treeline.

Few studies have been conducted in the Cascade Range, resulting in a deficient understanding of how Cascadian forelands develop (e.g., Jumpponen et al. 1998). Vegetation diversity is low in the incipient Easton forefield, leading to species-rich patches of early establishment. Nutrient accumulation in forelands that do not benefit from nearby sources of seed rain and detritus (e.g., Glacier Bay, Alaska) typically amalgamate nutrients incrementally with time (Chapline et al. 1994). In the Easton foreland, surrounding vegetation provides some soil nutrient input in early stages and is increasingly influential in later successional stages.

This research aims to discern the soil processes of a Cascadian foreland. One of the primary objectives is to better understand and capture the characteristics of the Easton foreland using pH, vegetation cover, and nutrients (organic matter [OM], carbon [C], nitrogen [N], and the related carbon/nitrogen [C/N] ratio). We hypothesize that these soil variables will increase in value with increasing age, except pH, which is expected to decrease following normal pedogenesis trajectories (Jenny 1958). From this information, we seek an understanding of what the best indicator of development is—elevation, time, or successional stage. After establishing the strongest indicator of soil genesis, an inference is presented as to how this observed trend might change in the future.

The primary purpose of this study is to contribute to our understanding of foreland pedogenesis. Due to the synergetic and tightly intertwined relationship between plants and soil, however, discussion of plant communities will also be included, as they are the basis for the vegetation zone determination (Legros 1992; Matthews 1992; Jumpponen et al. 1998). Understanding foreland pedogenesis informs many disciplines, as the processes that transform barren surfaces amidst harsh climatic conditions into productive ecosystems are highly complex. Implications remain uncertain of what and how severe the effects of global warming will be on high-elevation ecosystems. It has been shown that high-elevation species' ranges are increasingly fragmented and shifting into so-called new climate space where suitable conditions still exist and of which recently deglaciated valleys can be viable climatic havens (Thuiller 2004; Vos et al. 2008). For subalpine, alpine, and subnival ecosystems, ascension into these new landscapes will be requisite to outpace the expanding elevation range of vegetation communities better suited to warmer conditions (i.e., the montane forest). Due to altitudinal zonality—the relationship between elevation and temperature via the normal lapse rate—this upslope migration will be necessary to maintain the current gradient of primary succession. Many organisms are specifically adapted to these climatic conditions and rarely exist in other environments. Therefore,

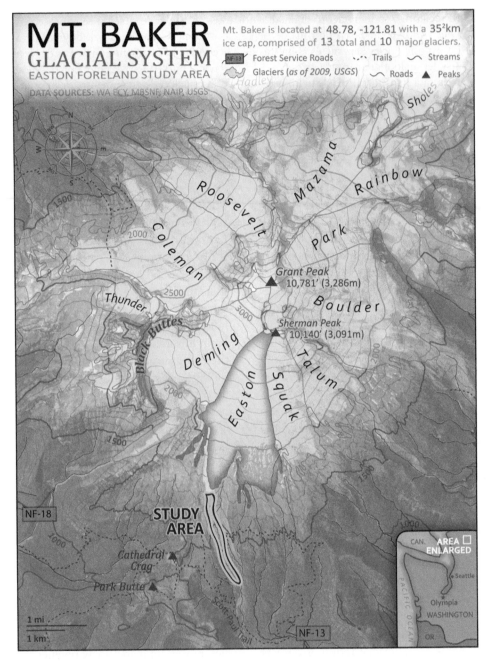

Figure 1. Mount Baker glacial system (ice extent as of 2009) and location of the study area in the Easton Glacier Valley. USGS = U.S. Geological Survey. (Color figure available online.)

the rate and properties of soil development in recently deglaciated valleys are critical to understanding how soils alter the landscape for establishment of high-elevation ecosystems (e.g., abiotic crust, cryogenic particle translocation; Frenot, Van Vliet-Lanoë, and Gloaguen 1995; Frenot, Gloaguen, and Bellido 1998). Overall, understanding shifting environmental conditions and habitat fragmentation in the alpine is severely limited without elucidating the underpinning pedogenic processes (Thuiller 2004).

Study Area

Mount Baker (Figure 1) is a heavily glaciated stratovolcano and the highest peak (3,285 m) in the North Cascades Range. Mount Baker resides in a west coast maritime climate where its high elevation results in heavy winter precipitation (Mass 2008). The glaciers that make up the ice cap, though, are comparatively thin given their planar dimension and have experienced significant recession over the last century (Heikkinen 1984; Harper 1993;

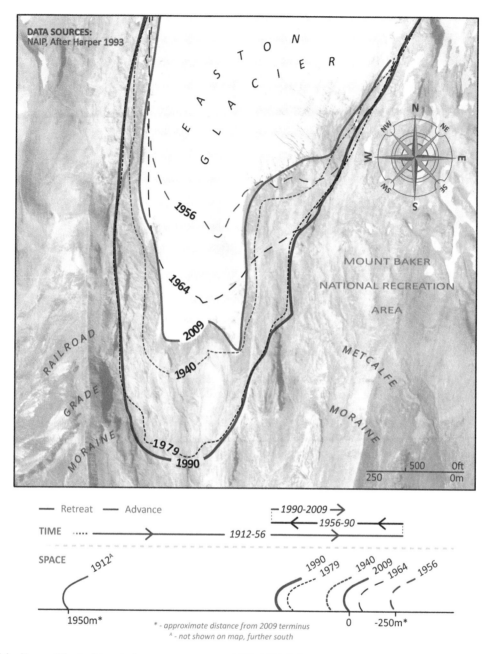

Figure 2. Map of the Easton Glacier historical terminus positions, 1912–2009 (after Harper 1993). The diagram at the bottom offers a visualization of the glacial dynamics over time. (Color figure available online.)

Pelto 2015). This research investigated the Easton foreland, a southern aspect valley spanning 1,250 to 1,650 m above sea level. The nearby town of Glacier, Washington (285 m above sea level), has a long-term weather station, where temperatures average 16°C in summer and 2°C in the winter (Bach 2003). Based on the Cascade lapse rate, temperatures in the Easton foreland are roughly 6.5°C cooler than Glacier, Washington (Minder, Mote, and Lundquist 2010). The dry summer months receive ~5 cm per month in precipitation, whereas winter precipitation exceeds 40 cm snow water equivalent per

month (Bach 2003). Secular variation in precipitation and temperature results in multiple stages of glacial advance and retreat, with roughly ten-year transition periods (Harper 1993; Kovanen 2003; Pelto 2008).

The Easton Glacier retreated rapidly from 1940 to 1956, readvanced until 1990, and has been steadily retreating since (Figure 2; Harper 1993; Pelto 2015). To further extend this timeline, a historic ground photograph of Mount Baker taken in 1912 was employed (Mount Baker Volcano Research Center [MBVRC] 2012). The photograph, taken from nearby Loomis Mountain,

Figure 3. Historic ground photographs taken from Mount Loomis looking north at the south flank of Mount Baker with the Easton Glacier located in the center. The red lines indicate the extent of Easton Glacier in 2012 and the blue lines represent the extent of Easton Glacier in 1912. Note that there is a ridgeline that obscures part of the lower portion of the Easton Glacier valley. (A) Photo taken in 1912 by W. D. Welsh. (B) Repeat photo taken in 2012 by the Mount Baker Volcano Research Center. (C) Enlargement of the 1912 photo to show the location of the Easton Glacier terminus. (D) Enlargement of the 2012 photo to show the retreat of the Easton Glacier over the last hundred years. Photographs used with permission of the Mount Baker Volcano Research Center. (Color figure available online.)

depicts the Easton Glacier to be much more massive than today; a 2012 repeat photograph by the MBVRC helps contextualize its retreat (Figure 3). Analysis of the skyline, ridgelines, and isolated snow patches on the Black Buttes confirms that they were taken from the same vantage point and in similar seasonal ablation conditions. Using a 3D geographic information system (GIS), the approximate location of the 1912 terminus was determined and was used as the lower bound for the study area. The snout elevation was noted during fieldwork that, in aggregate, created a 100-year spatiotemporal sequence.

Due to extreme climatic conditions, the forefield showed little to no evidence of influence attributed to animals. Vegetation is generally described as heath and ground cover with discontinuous Mountain hemlock (*Tsuga mertensiana*) that pervades the entire study area. Alders (*Alnus tenuifolia*) are notably absent in the Easton foreland, with conditions better suited for heather (*Phyllodoce empetriformis*), partridge foot (*Luetkea pectinata*), and bird's beak lousewort (*Pedicularis ornithorhyncha*). The Easton foreland is flanked by 60-m-tall Holocene-age moraines with mature Mt. Hemlock/Silver Fir forests (Figure 2).

Parent material in the Easton foreland varies subtly throughout the study area as a result of many active geomorphic processes, but bedrock and sediment lithology is consistently andesite of various ages and has not experienced tephra deposition over the 100 years (Kovanen, Easterbrook, and Thomas 2001). Variability pervades, as surfaces of deposition versus ablation that populate the hummocky microtopography of forelands produce different soil forming environments. Many qualitative surface types were observed, such as redeposited (flow till), older deposits that are overridden (deformation tills), glacio-fluvial sediments, and colluvium, among others (Lundqvist 1988; Matthews 1992). This study, however, did not consider local parent material as a variable because the sampling strategy attempted to remove variation by avoiding colluvial, fluvial, and other identifiable nonglacial influences. Therefore, for the purpose of this study, parent material was not assessed but is largely considered to be basal till. Aside from till, the primary constituents of local source material are Mount Baker's geologic composition, aeolian (windblown) inputs, and eruptive history (both lava flows and ashfall events; Coombs 1939; Bockheim and Ballard 1975; Kovanen, Easterbrook, and Thomas 2001).

Windblown deposition of fine particles and detritus from the adjacent ridgelines above the study area and more distant sources was difficult to quantify. Observations during the field season revealed windblown conifer needles and other detritus near and on the glacial ice, the source of which was likely the adjacent Holocene moraine. In addition, prevailing westerly winds are bifurcated around the Olympic Range, converge through the Puget Sound, and arrive with a strong southerly component (Mass 2008). These dominant south–southwesterly winds bring detritus, fine-grain material, and other elements of the aeolian biome to the Easton foreland (Litaor 1987). Up-valley, southerly winds were commonly observed during the field season, bringing organics and other fine-grain material from the older forest to younger surfaces. Conversely, northerly katabatic winds flow down the Easton Glacier, stripping away fine-grain material. Ultimately we hypothesized that lateral moraine sources provide more input than katabatic winds remove.

The Easton foreland is located within the Mount Baker National Recreation Area (MBNRA), between the Mount Baker Wilderness to the east and west, and within the larger Mount Baker–Snoqualmie National Forest complex. Multiple uses are permitted, including camping, hiking, horseback riding, and mountaineering. Snowmobile use is permitted after the snow level exceeds two feet at the end of the main access road (NF Road #13). Several conifers were either snow stunted from compaction or had sheared crowns due to snowmobile use. In addition, the litter and trash that pervade the valley present the potential for localized soil pollution (e.g., leaking engine fluids), which, unless obviously observed and avoided, could not be controlled for in this study.

Methods

The delineation of vegetation zones was initially informed by aerial imagery and later ground-truthed during fieldwork. The Easton foreland was mapped into four major successional stages, in addition to samples taken on glacial ice (Figure 4). Zone 4 is the zone of continuous vegetation cover, dominated by conifers

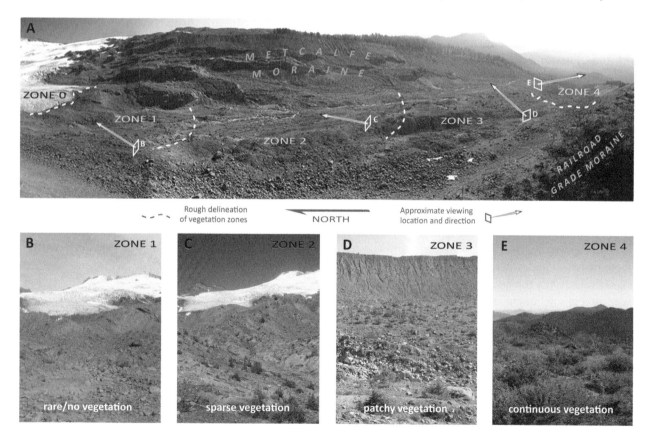

Figure 4. (A) Panoramic photo looking east taken from the Railroad Grade (Figure 2) of the Easton Foreland, showing the vegetation zones. (B) Vegetation Zone 1 was deglaciated by 1990 and has no vegetation. (C) Vegetation Zone 2 was deglaciated about eighty years ago and has sparse vegetation. (D) Vegetation Zone 3 is more than eighty years old and has patches of vegetation. (E) Vegetation Zone 4 was deglaciated about 100 years ago and has continuous vegetation, with trees increasing in size down valley. (Color figure available online.)

and supporting a dense understory. The transition to patchy vegetation cover is represented by Zone 3, which forms a constellation of more productive conditions. In Zone 2 vegetation is sparse and in isolated clumps, where over time surviving individuals become loci for further plant establishment. Sites exposed less than twenty years ago are within Zone 1, where pioneer vegetation is extremely rare and conditions are most harsh. In Zone 0, initial state samples were taken from till on active glacial ice. These successional stages represent the independent variable vegetation zone (Matthews 1992).

Sample locations were randomly generated in a GIS using a coarse estimate of the most recently deglaciated trough and field adapted during survey (Carter 1993). Sites that fell beyond the bounds of this geomorphic zone were relocated during fieldwork. In addition, sites that appeared heavily influenced by either colluvial or fluvial processes were relocated. For example, talus fields and steep slopes were avoided. In general, samples were taken within 2 m of the randomly generated coordinates but no more than 25 m due to the aforementioned considerations.

At each sample site, a one-half-meter quadrant was placed for reference to estimate both percentage vegetation and clasts and fine-sediment cover. Slope was measured using an inclinometer. Aspect was recorded using a compass. Elevation was recorded using a handheld Global Positioning System unit. The top 2.5 cm of soil was collected and no horizon differentiation was applied, as horizons were hardly perceivable or present (Matthews 1992).

Analytical particle size was determined by sieving samples at 2 mm—differentiating fine and coarse grains (Carter 1993). Due to the sampling strategy, some coarse organics were included in the sample (twigs, needles, etc.). After sieving, coarse fractions were visually assessed for these materials and, where present, separated out into coarse organics. Only the fine-grain soil and coarse organics were used in subsequent analyses.

Soil acidity was assessed using the fine-grain (<2 mm) subsample. A digital Oakton Acorn Series: pH5 reader was calibrated daily at pH 4.0 and 7.0. A 10-mL sample was combined with 20 mL of DIH_2O and left for one hour to allow ion equilibration before measurements were taken (Carter 1993).

OM content was determined using loss on ignition (LOI; Carter 1993). The percentage OM recorded was adjusted by weight to calculate the concentration of the entire sample. These values can be artificially high due to dewatering clay minerals or metal oxides, loss of volatile salts, or loss of inorganic carbons (Heiri, Lotter, and Lemcke 2001). Our results, however, were consistent with more sophisticated C/N content analyses.

A Thermo Flash EA 112 Series NC Soil Analyzer (CE Elantech, Lakewood, NJ) calibrated with an atropine standard was used to measure carbon and nitrogen content. Soil samples were first ground to a fine powder using a SPEX Mixer Mill to the consistency of chalk and then placed in a drying oven at 70°C overnight. After drying, the samples were placed in a desiccator and an ~100 mg subsample was taken. Before each batch run, calibration protocols were followed. Based on high and low values indicated in the preceding LOI analysis, a standard curve was constructed. This curve supplies a guideline of values for the instrument to expect, ensuring precise results.

Statistical tests compared independent variables—vegetation zone, elevation, and surface age—to dependent variables—pH, OM, carbon content, nitrogen content, C/N ratio, and vegetation cover. All tests were run at 95 percent confidence. Normality tests determined nonparametric tests to be appropriate. Kolomogorov–Smirnov was used on all of the data and Shapiro–Wilk on each vegetation zone (sample size < 50; McGrew and Monroe 2000). Kruskal–Wallace tests were performed to assess whether the differences in vegetation zones were statistically significant. Spearman's rank-order correlation (rho) was used, and data sets were adjusted using a log transformation to meet the linear relationship requirement (McGrew and Monroe 2000). Descriptive language is employed when discussing correlation results, after Quinnipiac University (2013). A rho value indicated the strength of the relationship as >0.70 = very strong; 0.40 to 0.69 = strong; 0.30 to 0.39 = moderate; 0.20 to 0.29 = weak; and <0.20 = negligible. Additionally, an asterisk denotes normal or significant results.

Results

All soil variables, except pH, related to vegetation following hypothesized trends (Table 1 and Figure 5). Soil acidity began surprisingly acidic on till samples (pH = 4.19), became more neutral in Zones 2 and 3 (pH = 5.43 and 5.51), followed by an increase in acidity in Zone 4 (pH = 5.04). It is suspected that the initial acidity is due to local geothermal deposits that discolored the rocks in Zone 0 (Bockheim and Ballard 1975). On the younger surfaces this trend was

Table 1. Summary results of descriptive statistics: Mean values ± standard deviation

	N	pH	%OM	%C	%N	C/N	%VEG
Average		5.2 ± 0.50	4.3 ± 6.4	2.27 ± 5.53	0.16 ± 0.53	15.3 ± 8.88	31.4 ± 36.5
Zone 0	4	4.19 ± 0.77	0.4 ± 0.1	0.01 ± 0.01	0.00 ± 0.00	5.9 ± 1.9	0 ± 0
Zone 1	16	5.05 ± 0.32	1.5 ± 0.2	0.02 ± 0.01	0.00 ± 0.00	8.2 ± 1.0	4.9 ± 8.6
Zone 2	17	5.43 ± 0.26	1.6 ± 0.6	0.08 ± 0.04	0.01 ± 0.00	12.0 ± 3.5	20.9 ± 32.6
Zone 3	15	5.51 ± 0.32	2.5 ± 0.9	0.40 ± 0.47	0.02 ± 0.01	16.8 ± 6.5	42.7 ± 37.0
Zone 4	17	5.04 ± 0.48	12.6 ± 8.8	9.47 ± 8.45	0.67 ± 0.99	22.6 ± 11.3	64.7 ± 31.1

Note: OM = organic matter; C = carbon; N = nitrogen; VEG = vegetation.
*Significant result at 95% confidence.

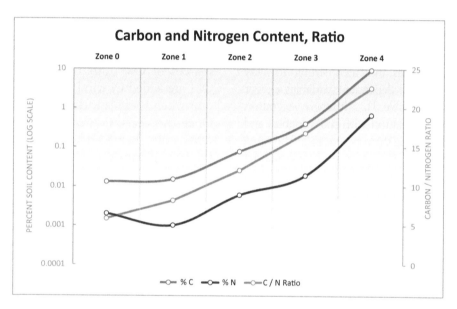

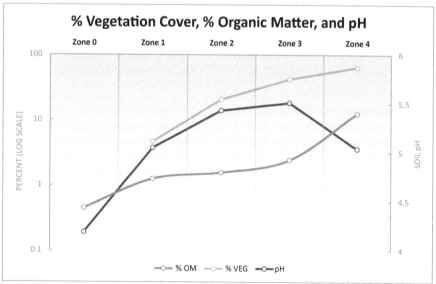

Figure 5. Trends in soil variables. (A) Average percentage carbon, percentage nitrogen, and carbon/nitrogen ratio measured in soils from each vegetation zone. (B) Average soil pH, percentage soil organic matter, and percentage vegetation cover measured in each vegetation zone. (Color figure available online.)

unexpected; the dominance of conifers in Zone 4 causing a drop in pH was an expected outcome (Table 1).

OM showed a small increase from Zone 0 to Zone 2 (0.4–1.6 percent) and then increased to 2.5 percent in Zone 3. In Zone 4, OM rose sharply to 12.6 percent, an increase of 2,793 percent from initial conditions (Table 1). Carbon, nitrogen, and N/C ratio all show the same general patterns of change as OM (Table 1). Finally, percentage vegetation cover very slowly increases from zero in Zone 0 and peaks in Zone 4 at 64.7 percent; it should be noted, however, that these values had high standard deviations (Table 1).

Kruskal–Wallace tests were performed on all variables to determine whether changes were significant between each vegetation zone. Results showed that at 95 percent confidence, differences between zones are significant for all variables (Table 2).

Toposequence rho values indicated strong to very strong relationships with all dependent variables except pH (−0.466 to −0.868; Table 3). Although vegetation cover had the highest correlation with soil variables and increased incrementally along the toposequence, the more categorical view of stages of vegetation succession had the strongest overall correlation to soil variables (Table 3). Deglaciation age had the weakest overall relationship but did show the strongest relationship with soil pH, which highlights a notable change in acidity from till on ice, to the ≤20 surface, to the oldest sites (≤100 years). Glacial advance from the 1950s through 1990s covered the deposits from the period ~1934 through 1990, however, by obliterative overlap, limiting the usefulness of this variable (Gibbons, Megeah, and Pierce 1984).

Vegetation zones showed the strongest overall relationship to soil characteristics (Table 3). The relationship of vegetation zones to OM, carbon, nitrogen, and C/N ratio was very strong, all key indicators of soil development. Soil pH was found not to have significant relationship to vegetation zones, though, suggesting a nontemporal process at work (e.g., hydrothermal activity).

Correlation results demonstrate hypothesized relationships and confirm that soil development is occurring along expected temporal trajectories. Soil pH increased over time and distance, which is the reverse of the expected trend due to unusually acidic nascent till. The build-up of organics strongly covaried with all independent variables (Table 3). It was shown that nutrients build up slowly and then rapidly increase as patches of vegetation become contiguous.

It is also instructive to assess the cross-correlations between dependent variables to evaluate the strength of intervariable relationships. As expected, soil OM, carbon, nitrogen, and C/N ratio all had strong cross-correlations (0.760–0.970; Table 4). This is expected, as OM and %C are closely related (Birkeland 1999), and as these four variables typically change in concert it is difficult to tease them apart. The correlation of all dependent variables to pH was negligible, again suggesting that pH has a nontemporal influence (Table 4). Percentage vegetation cover also showed a negligible correlation to carbon and nitrogen.

Discussion

The relationship between the measures of OM in the soil (%OM, %C, %N, C/N) and vegetation cover between the youngest vegetation zones (0–1) and the older vegetation zones (2–4) is due to two sources of OM into the soils. Small quantities of OM are added by wind at all sites but is most important when there is little OM in the soil (Zones 0–1). Once the vegetation cover is established in Zone 2, OM is also being added to the soil from leaf litter from the plant cover. Additionally, there is a significant jump in age between Zones 1 and 2, from twenty years to more than eighty years, where that additional sixty years of aeolian inputs has increased the OM content and development of the soil biota has set the stage for continuous vegetation cover (Jumpponen et al. 1998).

Our findings and challenges were similar to those of other glacial foreland studies; we found significant trends but also had difficulty distinguishing between the effects of time and vegetation on soil formation, because the time factor also affects successional stages and vegetation colonization (Jumpponen et al. 1998). Our results are discussed in the context of other foreland and nival studies, examining relationships with the Kerguelen Islands, the Alps, the Caucasus Mountains, Iceland, Alaska's Glacier Bay, and the Coleman foreland of Mount Baker. Although the comparability of these studies to the Easton foreland is at times tenuous, they all inform one of the principal research questions: How does this Cascadian foreland relate to others around the world?

At roughly the same latitude as the Easton but in the Southern Hemisphere, the Ampère forefield developed in a fundamentally different way in terms of its geoecology (Frenot, Van Vliet-Lanoë, and Gloaguen 1995; Frenot, Gloaguen, and Bellido 1998). The role of plants, soil OM, and carbon was minimal, suggesting that succession and facilitation are inconsequential.

Table 2. Summary results Kruskal–Wallace test if variables statistically change between vegetation zones

Variable	Statistic	Kruskal–Wallace
pH	χ^2	25.58*
	p value	0.000
OM	χ^2	49.34*
	p value	0.000
%C	χ^2	36.87*
	p value	0.000
%N	χ^2	35.87*
	p value	0.000
C/N ratio	χ^2	17.89*
	p value	0.001
% VEG	χ^2	31.20*
	p value	0.000

Note: OM = organic matter; C = carbon; N = nitrogen; VEG = vegetation.
*Significant result at 95% confidence.

Table 4. Cross-correlation of dependent variables

	pH	OM	%C	%N	C/N
OM	0.137				
%C	0.171	0.908*			
%N	−0.008	0.878*	0.970*		
C/N	0.198	0.800*	0.852*	0.760*	
%VEG	0.052	0.390*	0.280	0.145	0.495*

Note: OM = organic matter; C = carbon; N = nitrogen; VEG = vegetation.
*Significant result.

Instead, abiotic soil genesis and the biological traits of pioneer species are the major drivers of development (Frenot, Gloaguen, and Bellido 1998). The glacial recession in that forefield is much more dramatic than in the Easton, and in the absence of nearby seed rain and detritus inputs, the sequence of plant establishment was observed as gradual. Katabatic winds in the Ampère foreland were documented as inhibitory, stripping away what little OM was present on the nascent till (Frenot, Van Vliet-Lanoë, and Gloaguen 1995; Frenot, Gloaguen, and Bellido 1998).

Studies in the Swiss and Austrian Alps describe sequences of development in forelands and alpine and nival ecotones at comparable latitudes (46–47° N) and slightly higher elevations (2,000–3,100 m above sea level; Conen et al. 2007; E. Huber et al. 2007; Burga et al. 2010). Conen and colleagues (2007) looked at two comparatively short chronosequences (about fifty years), whereas E. Huber and colleagues (2007) investigated an altitudinal and environmental gradient. The former showed a more gradual buildup of soil carbon, whereas in the Easton accumulation is minimal until the late successional stages are reached (Conen et al. 2007). Burga and colleagues (2010) returned to the Morteratsch foreland and studied a longer, 134-year chronosequence. On this larger temporal scale, a similar pattern of soil OM accumulation was observed and is comparable to the Easton trend (Burga et al. 2010). The latter study by E. Huber and colleagues (2007) presented a 200-m toposequence at much higher elevation (~3,000 m above sea level vs. ~1,600 m above sea level) but similar orientation that captured the establishment of early pioneer communities, namely, grasslands and meadows. Once again, this gradual development is unlike the patchy, heterogeneous colonization of the Easton foreland (E. Huber et al. 2007). In contrast, the pattern of establishment in the Swiss and Austrian Alps appears much more linear and gradual, possibly due to aspect or limited sources of aeolian influence.

Near sea level, an Icelandic foreland experiencing cooler temperatures and parent material influenced by tepha has similar trends in OM content and pH (Vilmundardóttir, Gisladóttir, and Lal 2014). Even with scant vegetation, the Icelandic surfaces younger than sixty-five years accumulated OM at a faster rate than we found, attributed to the andic nature of the parent material. The 100- to 120-year surfaces at Easton, however, had over five times more OM than found in Iceland, suggesting that once established, Cascadian vegetation is more productive.

In south central Norway, eighteen forelands were lichenometrically dated, rising from sea level to 2,000 m (Messer 1988). These forelands represent 231 years of glacial retreat under slightly cooler conditions than at the Easton (Messer 1988). Increasing OM content and decreasing pH were principal findings that agree with the Easton sequence. Loffler

Table 3. Summary results of Spearman's correlation analyses between soil variables and site characteristics

	Elevation		Deglaciation age		Vegetation zone	
Log (Var.)	p value	Rho	p value	Rho	p value	Rho
Soil pH	0.124	−0.196	0.000	0.491*a	0.017	0.264*
%OM	0.000	−0.774*	0.000	0.616*	0.000	0.843*a
%C	0.000	−0.868*	0.000	0.777*	0.000	0.919*a
%N	0.000	−0.809*	0.001	0.649*	0.000	0.888*a
C/N ratio	0.000	−0.638*	0.000	0.655*	0.000	0.813*a
%VEG	0.001	−0.466*a	0.000	0.464*	0.001	0.453*

Note: OM = organic matter; C = carbon; N = nitrogen; VEG = vegetation.
[a]Greatest rho coefficient of the three independent variables.
*Significant result at 95% confidence.

(2007) and Messer (1988) pointed to climatic factors to explain major shifts in soil properties, namely, cation exchange capacity and soil depth, as external flux potentials (Jenny 1958). Due to harsh climate, after more than 200 years the soil carbon accumulates to levels found in the Easton after only 100 years.

Further east in the Caucasus Mountains, Makarov and colleagues (2003) studied a 50-m toposequence. As also seen in the Alps, this sequence terminates at the subnival alpine meadow; therefore, it is only applicable to the lower reaches of the Easton sequence. Soil organic carbon and nitrogen, as well as C/N ratio, all increased at a rate similar to that in the Easton valley. The study site elevation is also closer to the Easton at 2,700 m above sea level, albeit at a slightly lower latitude (Makarov et al. 2003). The longer growing season seemed to accelerate development in this foreland, as opposed to the first 50 m of the Easton sequence, which is only sparsely populated by vegetation.

The forelands of Glacier Bay, Alaska, are instructive for their wide breadth of study (Crocker and Dickson 1957; Reiners, Worley, and Lawrence 1971; Bormann and Sidle 1990; Chapline et al. 1994; Crocker and Major 1995; Williamson et al. 2001; Bardgett and Walker 2004). The higher latitude and harsh climate in the Alaskan Gulf make for short and tough growing seasons. There the recession is much more dramatic than at the Easton; after 230 years a fjord almost 100 km has been exposed (Chapline et al. 1994). Therefore, it is important to take scale into consideration as the magnitude of retreat in Glacier Bay is many times greater than the narrow 100-year Easton foreland. The rate of vegetation establishment is more inhibited in the Alaskan foreland. In addition, the wide gulf between glacial snout and mature forest limits seed rain and aeolian detritus to areas immediately up-valley of established vegetation. As a result, development is slow through the valley and various stages of pioneer colonization take place before the late-successional Sitka spruce (*Picea sitchensis*) becomes dominant. The oldest sites show higher carbon and nitrogen contents than the Easton foreland, but twice as much time has elapsed in Glacier Bay.

The gradual progression of successional stages in Glacier Bay stands in contrast to the heterogeneous patches seen in the Easton. This could be attributed to the seed rain and detritus input from the mature forests on the Little Ice Age (LIA) moraines that flank the Easton forefield as opposed to the Glacier Bay sequence where such inputs are limited to the leading edge of the developing foreland and the growing season is

much longer (Jones and del Moral 2005). Also worth noting is the relative absence of alders (*Alnus tenuifolia*) in the Easton foreland, whereas their presence in Glacier Bay is argued to be requisite for late-successional establishment (Lawrence 1953; Crocker and Major 1955). Instead, it is hypothesized that abundant lupine (*Lupinus latifolius*) helps fix soil nitrogen as no alder were observed, even in the oldest area (Zone 4).

Research in the Coleman foreland, located on the northwest flank of Mount Baker, is most applicable, as the physical conditions of both valleys are very similar, only differing in aspect and slope (Heikkinen 1984; Jones and del Moral 2005). Studies in this forefield have documented the glacial advance, which peaked in 1823, and the dramatic glacial history since. Stages of primary succession appear similar to those identified in the Easton sequence, beginning with low shrubs and ending with Mountain hemlock dominance; however, it does contain abundant alder. Ground cover and heath communities seemed better suited to the Easton, perhaps due to the southerly aspect. Although no explicit soil study has been carried out in the Coleman foreland, researchers have noted a comparable accumulation of OM, up to 25 percent in the oldest Coleman sites (~180 years), similar to the observed trajectory of 12.6 percent peak value in the Easton sites after over 100 years (Jones and del Moral 2005). Likewise, the vegetation communities in the Lyman Glacier foreland, located 70 km away, are different from Easton and are composed of different species, yet have similar vegetation zones (Jumpponen et al. 1998).

To forecast likely conditions in the Easton foreland as climate change continues, it is instructive to look to Glacier Bay (Jones and del Moral 2005; IPCC 2007). Seed and detritus sources are increasingly remote from nascent surfaces, thereby delaying colonization in Glacier Bay. This widening gulf between vegetation communities and retreating ice results in slower, gradual successional advance. Depending on the future ablation rate of the Easton, facilitative sources might become increasingly scarce as the foreland advances north of LIA forests (Jones and del Moral 2005).

Conclusions

The glacial history of the Easton foreland does not provide a smooth, linear change in surface age. Instead, a dramatic and deflationary recession that began in the early 1800s was interrupted by a period of advance from the 1950s to the 1980s

(Harper 1993; MBVRC 2012). This readvance diminishes the utility of a chronosequence (or deglaciation age) as only three ages are present (Gibbons, Megeah, and Pierce 1984). A complementary and continuous independent variable was therefore employed: a toposequence. The Easton foreland gains elevation fairly consistently (~0.2 m gain/m); therefore, elevation is closely intertwined with both deglaciation age and vegetation zones. Furthermore, it is assumed that deglaciation from 1912 to 1956 was relatively consistent, although data for this period are sparse and decadal variability could have caused periods of rapid or slowed retreat (MBVRC 2012; Pelto 2015). Vegetation zones were also designated that represent stages of glacial deposition and subsequent vegetation establishment and were found to have the strongest covariance with soil variables.

Significant changes in soil were observed. Till on ice showed little OM (~0.4 percent), little to no nitrogen content (maximum 0.0010 percent), minimal carbon (maximum 0.0083 percent), and high acidity (pH = 3.23–4.02). In the first vegetation zone (Zone 1, ≤20 years), rare vegetation was evident and few signs of soil development were observed. Vegetation cover increased to an average of 4.9 percent, pH became more neutral (5.05), and whereas OM increased to an average of 1.5 percent, carbon and nitrogen contents were similar to till on ice. In the second vegetation zone, average vegetation cover increased to 20.9 percent and the C/N ratio doubled the on-ice average, but carbon, OM, and pH were largely unchanged. Vegetation Zone 3 is the transition to patchy vegetation, where average vegetation cover increased to 42.7 percent. In this stage, nitrogen content increased by an order of magnitude; OM and carbon content also increased to 2.5 percent and 0.40 percent, respectively; and pH remained essentially constant. In Zone 4 vegetation became continuous, acidity increased, the C/N ratio was at its highest, and nitrogen, carbon, and OM contents all increased by an order of magnitude. Binning by vegetation zones showed the strongest relationship to soil genesis, despite variant and confounding environmental influences.

Organics, carbon, and nitrogen build up very slowly in early stages and then rapidly as vegetation cover becomes less fragmented. Furthermore, C/N and percentage carbon both increase throughout the study area, indicating that in the lower reaches organic input outpaces decomposition, setting the stage for O-horizon formation. Similarly, the continual increase in nitrogen despite increasing plant demand indicates

that lupine is fixing nitrogen faster than plants are accessing it. Therefore, soil OM, carbon, nitrogen, and the C/N ratio are strongly indicated by vegetation zones in lieu of consistent and contiguous surface age.

Uncertainty remains regarding whether the Easton and other Cascadian forelands will continue the same successional trajectory as climate change continues or whether climatic shifts will force new regimes of development. Across the North Cascade Range, glacier volumes have fluctuated very similarly, especially the current trend of increased ablation (Pelto 2008). Therefore, with consideration for variation in development due to physical setting (e.g., aspect, slope, elevation), an inference can be made for most Cascadian forelands based on the processes studied in the Easton. The interpretations presented herein provide speculation of conceivable pathways for future soil and vegetation development in the North Cascades. In the Easton foreland, the 2012 snout is lower in elevation than the mature forests on the adjacent Holocene moraines, providing seed rain and detritus input that facilitate species-rich, patchy development of the incipient foreland. As retreat continues, though, the terminus will move above the treeline, and these organic and fine-grain inputs will become increasingly distant (Jones and del Moral 2005). In such a situation, the aeolian biome is diminished and source material might be limited to the leading edge of the successional sequence. Such a reduction of current facilitative succession might cause a transition to processes akin to Glacier Bay, where development is slower and pioneer communities are more gradual and banded, less patchy, and less species diverse (Jones and del Moral 2005).

Albeit dynamic and heterogeneous, the Easton foreland presents a measureable system of development and succession that is slightly different from other postglacial environments. The Easton sequence showed that OM, carbon, and nitrogen content increase along expected trajectories, typically by two to three orders of magnitude with increasing surface age. Regardless of the discontinuous spatial timeline of the Easton's recession, an identifiable sequence was evident and suggested that surface age is not always the principal indicator of soil development. Instead, stages of succession were the strongest indicator of soil genesis. The development of soils and plants represented components of a larger shift in the landscape, and such a perspective that focuses on the plant factor and stages of succession is instructive for foreland studies (Jenny 1958; Matthews 1992). Worldwide and with no exception for the North Cascades, glacial recession is accelerating (Pelto 2008), and understanding the impacts on soil genesis is difficult (IPCC 2007). This study and others like it aim to better understand the dynamic landscapes of glacial forelands, how their soils develop, and how that genesis interacts with vegetation succession. In this way, we can more accurately forecast how they are likely to adapt to the changing conditions of the Anthropocene.

References

Bach, A. J. 2003. Showshed contributions to the Nooksack River watershed, North Cascade Range, Washington. *The Geographical Review* 92 (2): 192–212.

Bardgett, R. D., and L. R. Walker. 2004. Impact of coloniser plant species on the development of decomposer microbial communities following deglaciation. *Soil Biology and Biochemistry* 36:555–59.

Birkeland, P. W. 1999. *Soils and geomorphology*. New York: Oxford University Press.

Bockheim, J. G., and T. M. Ballard. 1975. Hydrothermal soils of the crater of Mt. Baker, Washington. *Soil Science Society of America* 39:997–1001.

Bormann, B., and R. Sidle. 1990. Changes in productivity and distribution of nutrients in a chronosequence at Glacier Bay National Park, Alaska. *Journal of Ecology* 78:561–78.

Burga, C. A., B. Krüsi, M. Egli, M. Wernli, S. Elsener, M. Ziefle, T. Fischer, and C. Mavris. 2010. Plant succession and soil development on the foreland of the Morteratsch glacier (Pontresina, Switzerland): Straightforward or chaotic? *Flora* 205:561–76.

Carter, M. R. 1993. *Soil sampling and methods of analysis*. Boca Raton, FL: CRC.

Chapline, F., L. R. Walker, C. L. Fastie, and L. C. Sharman. 1994. Mechanisims of primary succession following deglaciation at Glacier Bay, Alaska. *Ecological Monographs* 64 (2): 252–84.

Conen, F., M. V. Yakutin, T. Zumrunn, and J. Leifeld. 2007. Organic carbon and microbial biomass in two soil development chronosequences following glacial retreat. *European Journal of Soil Science* 58:758–62.

Coombs, H. A. 1939. Mount Baker, a cascade volcano. *Geological Society of America Bulletin* 50:1493–1509.

Crocker, R. L., and B. A. Dickson. 1957. Soil development on the recessional moraines of the Herbert and Mendenhall Glaciers, South-Eastern Alaska. *Journal of Ecology* 45 (1): 169–85.

Crocker, R. L., and J. Major. 1955. Soil development in relation to vegetation and surface age at Glacier Bay, Alaska. *The Journal of Ecology* 43 (2): 427–48.

Erschbamer, B. 2007. Winners and losers of climate change in a central alpine glacier foreland. *Arctic, Antarctic, and Alpine Research* 39:237–44.

Frenot, Y., J. C. Gloaguen, and A. Bellido. 1998. Primary succession on glacial forelands in the subantarctic Kerguelen Islands. *Journal of Vegetation Science* 9 (1): 75–84.

Frenot, Y., B. Van Vliet-Lanoë, and J. Gloaguen. 1995. Particle translocation and initial soil development on a glacier foreland, Kerguelen Islands. *Arctic and Alpine Research* 27 (2): 107–15.

Gibbons, A. B., J. D. Megeah, and K. L. Pierce. 1984. Probability of moraine survival in a succession of glacial advances. *Geology* 12:327–30.

Harper, J. T. 1993. Glacier terminus fluctuations on Mt. Baker, Washington, U.S.A., 1940–1990, and climate variations. *Arctic and Alpine Research* 25 (4): 332–40.

Heikkinen, O. 1984. Dendrochronological evidence of variations of Coleman Glacier, Mt. Baker, Washington, U.S.A. *Arctic and Alpine Research* 16 (1): 53–64.

Heiri, O., A. F. Lotter, and G. Lemcke. 2001. Loss on ignition as a method for estimating organic and carbonate content in sediments: Reproducibility and comparability of results. *Journal of Paleolimnology* 25:101–10.

Huber, E., W. Wanek, M. Gottifried, H. Pauli, P. Schweiger, S. K. Arndt, K. Reiter, and A. Richter. 2007. Shift in soil-plant N dynamics of an alpine-nival ecotone. *Plant Soil* 301:65–76.

Huber, U. M., H. K. Bugmann, and M. A. Reasoner, eds. 2005. *Global change and mountain regions: An overview of current knowledge*. Dordrecht, The Netherlands: Springer.

Intergovernmental Panel on Climate Change. 2007. *Climate change 2007: The physical science basis. Contribution of Working Group I to the Fourth Assessment Report of the Intergovernmental Panel on Climate Change*. Cambridge, UK: Cambridge University Press.

Jenny, H. 1958. Role of the plant factor in pedogenic functions. *Ecology* 39:5–16.

Jones, C. C., and R. del Moral. 2005. Patterns of primary succesion on the foreland of Coleman Glacier, Washington, U.S.A. *Plant Ecology* 180:105–16.

Jumpponen, A., K. Mattson, J. Trappe, and R. Ohtonen. 1998. Effects of established willows on primary succession on Lyman Glacier forefront, North Cascade Range, Washington, U.S.A.: Evidence for simultaneous canopy inhibition and soil facilitation. *Arctic and Alpine Research* 30 (1): 31–39.

Kovanen, D. 2003. Decadal variability in climate and glacier fluctuations on Mt. Baker, Washington, U.S.A. *Geografiska Annaler* 85A (1): 43–55.

Kovanen, D., D. Easterbrook, and P. Thomas. 2001. Holocene eruptive history of Mount Baker, Washington. *Canadian Journal of Earth Science* 38:1355–66.

Lawrence, D. B. 1953. Development of vegetation and soil in south-eastern Alaska, with special reference to the accumulation of nitrogen. Final Office of Naval Research, Project NR 160-183.

Legros, J. P. 1992. Soils of alpine mountains. In *Weathering, soils and paleosols*, ed. I. P. Martini and W. Chosworth, 155–81. Amsterdam: Elsevier.

Litaor, M. I. 1987. The influence of eolian dust on the genesis of alpine soils in the Front Range, Colorado. *Soil Science of America Journal* 51:142–47.

Loffler, J. 2007. The influence of micro-climate, snow cover, and soil moisture on ecosystem functioning in high mountains. *Journal of Geographical Sciences* 17:3–19.

Lundqvist, J. 1988. Glacigenic processes, deposits and landforms. In *Genetic classification of glacigenic deposits*, ed. R. Goldthwait and C. Matsch, 3–16. Rotterdam, The Netherlands: Balkema.

Makarov, M. I., B. Glaser, W. Zech, T. I. Malysheva, I. V. Bulatnikova, and A. V. Volkov. 2003. Nitrogen dynamics in alpine ecosystems of the northern Caucasus. *Plant and Soil* 256:389–402.

Mass, C. 2008. *The weather of the Pacific Northwest.* Seattle: University of Washington Press.

Matthews, J. 1992. *The ecology of recently deglaciated terrain: A geoecological approach to glacier forelands and primary succession.* Cambridge, UK: Cambridge University Press.

McGrew, J., and C. B. Monroe. 2000. *An introduction to statistical problem solving in geography.* 2nd ed. Long Grove, IL: Waveland.

Meier, M. F., M. B. Dyurgerov, and G. J. McCabe. 2003. The health of glaciers: Recent changes in glacier regime. *Climate Change* 59:123–35.

Messer, A. C. 1988. Regional variations in rates of pedogenesis and the influence of climatic factors on marine chronosequences, southern Norway. *Arctic and Alpine Research* 20:31–39.

Minder, J. R., P. W. Mote, and J. D. Lundquist. 2010. Surface temperature lapse rates over complex terrain: Lessons from the Cascade Mountains. *Journal of Geophysical Research* 115:D14122.

Mount Baker Volcano Research Center. 2012. *Loomis Point photographs.* Mount Baker Volcano Research Center. https://mbvrc.wordpress.com/ (last accessed 1 October 2016).

Pelto, M. S. 2008. Glacier annual balance measurement, forecast and climate correlations, North Cascades, Washington 1984–2006. *The Cryosphere* 2:13–21.

———. 2015. *Climate driven retreat of Mount Baker glaciers and changing water resources.* New York: Springer.

Quinnipiac University. 2013. *Instructor's resource guide for the text—QUC Crude estimates for interpreting correlation strength.* http://faculty.quinnipiac.edu/librarts/polsci/Statistics.html (last accessed 1 June 2013).

Reiners, W. A., I. A. Worley, and D. B. Lawrence. 1971. Plant diversity in a chronosequence at Glacier Bay, Alaska. *Ecology* 52 (1): 55–69.

Thuiller, W. 2004. Patterns and uncertainties of species' range shifts under climate change. *Global Change Biology* 10:2020–27.

Vilmundardóttir, O. K., G. Gisladóttir, and R. Lal. 2014. Early stage development of selected soil properties along the proglacial moraines of Skaftafelsjökull glacier, SE—Iceland. *Catena* 121:142–50.

Vos, C. C., P. Berry, P. Opdam, H. Baveco, B. Nijhof, J. O'Hanley, C. Bell, and H. Kuipers. 2008. Adapting landscapes to climate change: Examples of climate-proof ecosystem networks and priority adaptation zones. *Journal of Applied Ecology* 45:1722–31.

Williamson, C. E., O. G. Olson, S. E. Lott, N. D. Walker, D. R. Engstrom, and B. R. Hargreaves. 2001. Ultraviolet radiation and zooplankton community structure following deglaciation in Glacier Bay, Alaska. *Ecology* 82 (6): 1748–60.

Impacts of Glacier Recession and Declining Meltwater on Mountain Societies

Mark Carey, Olivia C. Molden, Mattias Borg Rasmussen, M Jackson, Anne W. Nolin, and Bryan G. Mark

Glacierized mountains are often referred to as our world's water towers because glaciers both store water over time and regulate seasonal stream flow, releasing runoff during dry seasons when societies most need water. Ice loss thus has the potential to affect human societies in diverse ways, including irrigation, agriculture, hydropower, potable water, livelihoods, recreation, spirituality, and demography. Unfortunately, research focusing on the human impacts of glacier runoff variability in mountain regions remains limited, and studies often rely on assumptions rather than concrete evidence about the effects of shrinking glaciers on mountain hydrology and societies. This article provides a systematic review of international research on human impacts of glacier meltwater variability in mountain ranges worldwide, including the Andes, Alps, greater Himalayan region, Cascades, and Alaska. It identifies four main areas of existing research: (1) socioeconomic impacts; (2) hydropower; (3) agriculture, irrigation, and food security; and (4) cultural impacts. The article also suggests paths forward for social sciences, humanities, and natural sciences research that could more accurately detect and attribute glacier runoff and human impacts, grapple with complex and intersecting spatial and temporal scales, and implement transdisciplinary research approaches to study glacier runoff. The objective is ultimately to redefine and reorient the glacier-water problem around human societies rather than simply around ice and climate. By systematically evaluating human impacts in different mountain regions, the article strives to stimulate cross-regional thinking and inspire new studies on glaciers, hydrology, risk, adaptation, and human–environment interactions in mountain regions.

冰川化的山区, 经常指称为世界的水塔, 因为冰河同时长期储水并调节季节性河流, 并在社会最需要水的旱季释放径流。冰的流失因此有可能会以多样的方式影响着人类社会, 包括灌溉, 农业, 水力发电, 饮用水, 生计, 休憩, 精神灵性, 以及人口。但不幸的是, 聚焦人类对于山区冰河径流变异的影响之研究仍然相当有限, 且该研究经常倚赖有关缩减的冰川对于山区水文与社会的影响之假设, 而非具体的证据。本文对于人类影响全世界山脉的冰河径流变异之国际研究, 提供系统性的回顾, 包括安第斯山, 阿尔卑斯山, 大喜马拉雅地区, 喀斯开山脉, 以及阿拉斯加。本文指认四大主要的既存研究领域: (1) 社会经济冲击; (2) 水力发电; (3) 农业, 灌溉与粮食安全; 以及 (4) 文化冲击。本文同时指出社会科学、人文和自然科学研究能够更准确侦测并归因冰河径流与人类冲击的未来方向, 应对复杂且相互交织的空间与时间尺度, 以及实施跨领域研究方法来研究冰河径流。本文的最终目标在于重新定义冰河水问题, 并将之重新导向人类社会, 而非单纯围绕着冰与气候。本文透过系统化地评估人类在不同山区带来的冲击, 力图刺激跨区域思考, 并激发对于山区中的冰河, 水文, 风险, 调适与人类—环境互动的崭新研究。关键词: 气候变迁冲击, 冰河径流变化, 冰河—社会互动, 水文社会循环, 水资源可持续性。

A menudo las montañas glaciadas son referidas como las torres hídricas de nuestro mundo porque los glaciares a la vez almacenan agua a través del tiempo y regulan el flujo estacional de las corrientes, liberando escorrentía durante las estaciones secas cuando más necesitan agua las sociedades. Por eso la pérdida del hielo tiene el potencial de afectar las sociedades humanas en diversos frentes, incluyendo irrigación, agricultura, energía hidroeléctrica, agua potable, medios de vida, recreación, espiritualidad y demografía. Infortunadamente, la investigación enfocada en los impactos humanos que tiene la variabilidad de la escorrentía glaciaria en las regiones montañosas sigue siendo limitada, y los estudios frecuentemente descansan en suposiciones más que sobre evidencia concreta acerca de los efectos que tiene la recesión de los glaciares sobre la hidrología de montaña y sobre las sociedades. Este artículo suministra una revisión sistemática de la investigación internacional sobre los impactos humanos de la variabilidad del agua de deshielo de los glaciares en las cadenas montañosas de todo el mundo, incluyendo los Andes, Alpes, la gran región de los Himalayas, las Cascadas y Alaska. El artículo identifica cuatro áreas principales de la investigación existente: (1) impactos socioeconómicos; (2) energía hidroeléctrica; (3) agricultura, irrigación y seguridad alimentaria; y (4) impactos culturales. El artículo sugiere también avenidas de avanzada para las ciencias sociales, las humanidades e investigación de las ciencias naturales, que podrían detectar y atribuir con mayor exactitud la escorrentía de los glaciares y los impactos humanos, lidiar con escalas espaciales y temporales complejas y entrecruzadas, e implementar enfoques de investigación transdisciplinaria para estudiar la escorrentía glaciaria. En últimas, el objetivo es redefinir y reorientar el problema del agua de glaciar alrededor de las sociedades humanas en vez de simplemente hacerlo alrededor de hielo y clima. Evaluando sistemáticamente los impactos humanos en diferentes regiones montañosas, el artículo se esfuerza en estimular el pensamiento interregional y en inspirar nuevos estudios sobre glaciares, hidrología, riesgo, adaptación e interacciones humano-ambientales en regiones montañosas.

Glacier retreat generates far-reaching concerns about water supplies for communities in and around the world's glacierized mountains (Intergovernmental Panel on Climate Change 2014). Despite the possible human impacts from shrinking glaciers, few studies focus primarily on the people and societies in glacierized watersheds. There is thus an urgent need to understand the human impacts of glacier runoff variability in mountain regions—not only for local communities, policy makers, and social science and humanities researchers who focus on human societies but also for natural scientists studying glaciers. Natural scientists frequently frame and conclude their scientific publications with claims about the hydrologic effects of glacier change on downstream societies, often making assertions about human impacts that transcend the scope of the scientific research. Bold impact statements also appear in the media, with *The New York Times*, for example, writing that "the 46,000 glaciers of the Third Pole [greater Himalayan] region help sustain 1.5 billion people in 10 countries. ... Scattered across nearly two million square miles, these glaciers are receding at an ever-quickening pace, producing a rise in levels of rivers and lakes in the short term and threatening Asia's water supply in the long run" (Wong 2015).

Such accounts—and academic research as well (National Academy of Sciences 2012)—usually do not explain key issues, though. These include which portion of the population would be influenced, how the number of affected people was determined; when people would be affected; how much glacier runoff is truly accessed for human water use; and what other factors—such as water rights, hydropower development, social conflicts, or agricultural practices—also affect water supplies and management in glacier-fed watersheds. Research on glacier runoff impacts might

overemphasize the role of ice, hold human societies static in hydrologic models, not acknowledge social drivers of change, or not differentiate among biophysical and societal forces of hydrologic transformation in glacier-fed watersheds.

Most glacier runoff studies divorce water supply (the hydrological dimensions of water) from water access and allocation (the social dimensions of water management)—despite existing literature on social–ecological systems (Folke 2006), the hydrosocial cycle (Swyngedouw 2009), sociohydrology (Sivapalan, Savenije, and Blöschl 2012), and hydrosocial approaches to glacierized watersheds (Nüsser, Schmidt, and Dame 2012; Carey et al. 2014). The common tendency is to identify glacier loss and then assume inevitable negative downstream human impacts or social conflicts over water, which is actually a form of environmental determinism (Hulme 2011) and the construction of "apocalyptic imaginaries" (Swyngedouw 2010). Human forces in mountain regions—from tourism and demographic changes to cultural beliefs, race relations, and government laws—are rarely identified and analyzed in glacier runoff research. Instead, the issue of mountain water is reduced to a problem of climate change and glacier shrinkage, an environmental problem driven by global temperatures and the shifting mass balance of glaciers that conceals other drivers of water scarcity (human beings) and thus fails to point toward effective adaptation strategies.

This article evaluates and synthesizes literature examining societal impacts of glacier runoff variability in mountain regions worldwide, simultaneously charting new paths forward to refine and reorient mountain research. Five representative regions serve as case studies for this analysis: the Andes, Alps, greater Himalayan region, North Cascades, and Alaska. The article is organized by impacts rather than by specific mountain range or political boundary. Four impacts predominate in the existing literature: (1) socioeconomic impacts; (2) hydropower; (3) irrigation, agriculture, and food security; and (4) cultural aspects. After synthesizing these societal impacts, three directions for future research are discussed: detection and attribution of glacier runoff impacts; consideration of intersecting spatial and temporal scales; and transdisciplinary research to redefine runoff.

Methods

Research involved a systematic review of existing literature on the societal impacts of glacier runoff variability in the five case study regions, following principles outlined for literature reviews in the social sciences (e.g., Petticrew and Roberts 2006; Rodela 2011). Systematic reviews are essential to digest large and disparate literatures, identify gaps in knowledge, and discover areas of uncertainty in scholarship. They also "flag up areas where spurious certainty abounds. These are areas where we think we know more than we do, but where in reality there is little convincing evidence to support our beliefs" (Petticrew and Roberts 2006, 2).

This systematic review was guided by this overarching research question: What knowledge exists within the published literature regarding the nature and scope of societal impacts of glacier runoff variability? The authors conducted searches for publications published from 1980 to November 2015 using Web of Science, Google Scholar, and Science Direct. Search terms were used for all fields (title, abstract, keywords, full text), such as "glac* runoff," "glacial discharge society," or "glacier water irrigation"—and included mountain ranges (e.g., Cascades), geographic regions (e.g., Himalaya), and mountain and glacier names (e.g., Huascarán). Database filters were used to narrow results to the social sciences and humanities when possible. Publications were included if they (1) addressed societal impacts of glacier runoff variability in a substantive way and (2) examined water flow or hydrological resources from glaciers or in glacier-fed watersheds. The review excluded non-peer-reviewed publications, media articles, organization reports, theses and dissertations, and gray literature. Natural science publications were excluded when they lacked a human component or did not focus on glacier runoff. Glacial lake outburst floods (GLOFs) are a form of hydrologic variability in glacier-fed watersheds, but they were excluded from this review because they do

not necessarily affect water use and are influenced more by moraine and ice dam stability, among other factors, than glacier runoff variability.

More than 500 publications were initially identified (approximately 200 for the Andean region, 200 for the Alps, 120 for the greater Himalayan region, and 50 for the Cascades and Alaska), but only a subset (one quarter to one third) focused on human dynamics of glacier meltwater. This study analyzed the smaller human-focused subset. Coverage was uneven within regions; for example, in the Andes, Peru's Cordillera Blanca has disproportionately high coverage compared with Bolivia, Ecuador, or Patagonia. Analysis also focused on populations affected, geographical scale studied, time range covered, evidence provided, topics studied, methods used, conclusions reached, and other trends. Results showed existing research concentrated in four main categories: socioeconomics, hydropower, agriculture, and cultural aspects. Many studies overlapped in two or more of these categories. Here, we reference only a limited, representative sample of publications analyzed in the research.

Impacts

Socioeconomic Impacts

Research suggests that glacier runoff variability will generate significant direct and indirect socioeconomic impacts. Community livelihoods in rural mountain areas can be vulnerable to small-scale environmental changes due to the interdependence among water, biodiversity, and livelihoods (Chaudhary et al. 2011; Sherpa 2014; Konchar et al. 2015). Glacier runoff variability affects aspects of local livelihoods: livestock production, irrigated agriculture, tourism income, social conflicts, and political struggles for power over water allocation practices (Mark et al. 2010; Bury et al. 2011; French, Barandiarán, and Rampini 2015).

Changes in glacier runoff are likely to affect urban water supplies in some regions. Vergara et al. (2007) reported that reduced glacier runoff will influence La Paz and El Alto, Bolivia, where 2.3 million residents receive 30 to 40 percent of their potable water from Cordillera Real glaciers. Studies suggest future policy implications of glacier runoff variability in Bogotá (Lampis 2013), and Ioris (2012) analyzed the uneven distribution of water in Lima's urban areas. Changes to glacier runoff might also affect mountain tourism, although research on the topic remains limited (Rhoades, Ríos, and Ochoa 2008; Bury et al. 2011). As climate change affects glaciers, local residents could see changes to tourism assets and natural resources (Becken, Lama, and Espiner 2013). Amidst insecurities about glacier runoff variability, increased tourism can escalate water use and exacerbate water competition among locals and tourists (Nyaupane and Chhetri 2009; G. McDowell et al. 2013). Tourism in the Alps might be negatively affected as glacial runoff decreases, influencing both the appearance of the landscape and the associated tourism activities (Beniston 2012).

Water quality is also of critical importance to human livelihoods in glacier-fed watersheds. Glacier retreat can warm downstream water (Mantua, Tohver, and Hamlet 2010; Pelto 2011), with potential implications for fish populations that societies depend on for food security, fishing markets, and cultural aspects (Lackey 2000; Grah and Beaulieu 2013). Mining (Bury 2015) and changes in sediment load resulting from increased glacier melting (Riedel et al. 2015) can both affect water quality, which could be exacerbated by glacier shrinkage and declining runoff.

Assertions about the impacts of glacier runoff variability often lack evidence to substantiate claims, even though some studies identify multiple drivers of socioeconomic changes, such as weather extremes and changing precipitation regimes (Chaudhary et al. 2011; G. McDowell et al. 2013; Sherpa 2014). Nüsser, Schmidt, and Dame (2012) and Sherpa (2014) illustrated that the influence of climate change, glacier shrinkage, and runoff variability on local people is not always distinguished from social, political, economic, and infrastructural inequities. For human migration along glacier-fed waterways, evidence points less to issues of water supplies than to a range of other societal factors (Wrathall et al. 2014; Raoul 2015), which is consistent with the environmental migration literature that identifies multiple drivers of migration (e.g., Black et al. 2011).

Hydropower

Research on glacier runoff indicates diverse effects on the generation of hydroelectricity (Terrier et al. 2011; Bavay, Grünewald, and Lehning 2013; Beniston and Stoffel 2014). Glacier runoff variability in the greater Himalayan region, for instance, will likely have far-reaching effects on hydropower, water storage, dry season flows, and related geopolitical and economic concerns (Qureshi 2011; Tiwari and Joshi 2012; Molden et al. 2014). Impacts could be particularly potent in

regions dependent on hydropower such as Nepal and the Andes, in areas without access to other energy sources, and in countries diversifying energy sources for economic productivity or pursuing sustainable energy (Gardarsson and Eliasson 2006; Einarsson and Jónsson 2010; Bliss, Hock, and Radić 2014). In the Alps, however, studies point to varying outcomes. Terrier et al. (2011) suggested that there will be higher summer water volume and more storage capacity for energy production as glaciers shrink. Schaefli, Hingray, and Musy (2007) and Beniston and Stoffel (2014), on the other hand, indicated a likely decrease in hydropower production due to decreased glacier runoff. Such discrepancies can emerge due to regional variation or from timescales studied. Short-term runoff from shrinking glaciers will often increase, whereas a reduction in runoff occurs below larger glaciers or more glacierized basins on long-term scales. Impacts of glacier runoff variation on hydropower needs further documentation, and societies might not be as negatively affected as commonly asserted, especially in the near future (Mark et al. 2015).

Predictions of future hydroelectric impacts have limitations. For example, Vergara et al. (2007) estimated that glacier retreat will reduce Perú's Cañón del Pato hydropower output. Although this study calculates economic losses and the reduction of gigawatt hours produced in glacier-fed rivers, it does not analyze the dynamic societal factors that influence hydropower generation or energy distribution, usage, and costs. Yet tensions over hydroelectric water management are affected by sociopolitical conflicts, water competition, new water technologies, and economic development (Carey, French, and O'Brien 2012; Lynch 2012; Bury et al. 2013). Further, reservoir construction and management to compensate for glacier runoff reduction can also affect hydropower generation and society more broadly (Rasmussen 2016).

Irrigation, Agriculture, and Food Security

Research on glacier runoff often argues that glacier loss will reduce irrigation water, diminish agricultural productivity, and threaten food security in glacier-fed watersheds. In the Oregon Cascades, glacier meltwater contributions to stream flow are negligible in the lower Hood River watershed but comprise 41 to 73 percent of the upper watershed flow (Nolin et al. 2010), thus suggesting potential impacts on irrigation and agriculture. Quantitative studies in the greater Himalayan region estimate that food security for 4.5 percent of the population living in the Brahmaputra, Indus,

Yangtze, and Ganges basins will be affected by reduced water availability due to the effects of climate change on glaciers and snowmelt (Immerzeel, van Beek, and Bierkens 2010). As glacier and snow meltwater account for approximately 60 percent of upper Indus River stream flow, glacier retreat will likely generate economic and food losses for Pakistan's agriculture (Akhtar, Ahmad, and Booij 2008; Immerzeel, van Beek, and Bierkens 2010). In Peru's Santa River basin, up to 66 percent of dry season flow consists of glacier runoff, which provides most of the water for the large-scale Chavimochic irrigation and agriculture project (Mark et al. 2010; Bury et al. 2013; Carey et al. 2014). Diminishing glacier runoff is likely to affect subsistence agriculture and produce water stress elsewhere in the Andes (Young and Lipton 2006; J. Z. McDowell and Hess 2012). Although highland pastoralists might experience initial benefits from increased runoff, studies claim that water stress will increase in the future with climatic uncertainty (Postigo, Young, and Crews 2008; López-i-Gelats et al. 2015).

Many runoff and agriculture studies highlight factors intersecting with glacier hydrology such as changing precipitation, drying water sources, and warming weather (Chaudhary et al. 2011; J. Z. McDowell and Hess 2012). In Cotacachi, Ecuador, researchers suggest that glacier loss has heightened competition for irrigation water, but they note that precipitation, government policies, water management practices, and the social organization of water also affect agriculture and irrigation practices (Rhoades, Ríos, and Ochoa 2008; Skarbø and VanderMolen 2014). In Bolivia, agriculture is influenced by factors beyond glaciers or water supply, including patterns of production, household assets, and institutional arrangements (J. Z. McDowell and Hess 2012). Other studies suggest an impact of diminishing glacier runoff on agriculture but do not differentiate among the various processes affecting food production, such as crops planted, local knowledge, or cultural influences (Bavay, Grünewald, and Lehning 2013; Beniston and Stoffel 2014).

In some cases, research reveals perceived and observed impacts of diminished glacier runoff. In northern India, residents report crop loss, decline in crop productivity, and reduced livestock productivity, as well as depletion of wells and groundwater sources, cultivable land, forest resources, and soil fertility (Bhadwal et al. 2013). In Peru's Cordillera Blanca, household studies in the Yanamarey Valley showed that 93 percent of agriculture- and livestock-

dependent respondents observed decadal decreases in dry season water supplies alongside a steady decline in glacier coverage (Bury et al. 2011). Other studies in the region suggest that locals perceive vanishing glaciers to be critical for agriculture (Young and Lipton 2006; Jurt et al. 2015). Although individuals might observe retreating glaciers and note less water, they also acknowledge that the current water realities are shaped by water governance (Rasmussen 2015). Water scarcity is as much a matter of equity as it is contingent on environmental conditions (Jaeger et al. 2013).

Cultural Impacts

Mountain residents worldwide are now questioning and, in some cases, transforming their beliefs about which forces affect regional hydrology—whether global economic changes, new power (political) dynamics, or local people's behavior and culture (Paerregaard 2013; Drenkhan et al. 2015; Jurt et al. 2015). The ways in which people imagine causality shapes their responses to water scarcity, and these understandings, local knowledges, and cultural beliefs can differ markedly from scientific depictions of the runoff problem (Williams and Golovnev 2015). Globally, glaciers and the mountains that sustain them have spiritual and cultural value for societies, such as the sacred Khawa Karpo mountain for Tibetan Buddhists, where the upper Yangtse, upper Mekong, upper Salween, and Dulong Rivers flow through the mountains (Allison 2015). Even though few studies discuss spiritual impacts associated with glacier runoff variability, the changes to glaciers, snow, lakes, and rivers could affect spiritual beliefs and practices (Drew 2012; Salick, Byg, and Bauer 2012; Becken, Lama, and Espiner 2013; Konchar et al. 2015).

Boelens (2014) explained how social groups see glacier runoff in the Andes differently: Engineers see it as a technical issue with biophysical characteristics; nongovernmental organizations work through sociolegal frameworks; and local residents understand it through, among other ways, historical and cultural perspectives. Elsewhere in Peru, climate change and glacier runoff variability have caused some local residents to adjust spiritual relationships because they believe that their religious offerings no longer assure adequate water supplies (Paerregaard 2013), and others report that glacier and snow loss illustrates reduced power of their *Apus* (deities). More research is needed to discern how cultural values, narratives, discourse, and perceptions of glaciers and hydrology are changing as glaciers retreat—and how these cultural factors affect water use and management.

Refining Runoff Research: Future Directions

Detection and Attribution

Explicit detection of global environmental changes, as well as the attribution of those observed changes to particular human and environmental drivers (Stone et al. 2013), is essential for glacier runoff research to avoid simplistic assumptions about glaciers. As glaciers shrink, they initially generate more meltwater; subsequently, watersheds experience "peak water," the point after which stream flow declines (National Academy of Sciences 2012). The runoff reduction usually manifests in the dry season, so impacts would not be constant or year-round. Further, even when glacier runoff declines at the glacier terminus, other variables influence downstream hydrology such as wetlands and human land use practices.

Studies detecting glacier runoff quantities, peak water, and downstream contributions of glacier meltwater from ice to ocean are rare in the natural sciences (see Baraer et al. 2012). Researchers usually only assume that glacier runoff has declined or will decline in the near future, without adequate consideration. Many social scientists reproduce a common misunderstanding that a shrinking glacier will cause an immediate downstream water flow reduction. Yet for the greater Himalayan region, for example, Immerzeel, Pellicciotti, and Bierkens (2013) indicated that future changes in monsoon precipitation will likely compensate for glacier runoff reduction in the Indus and Ganges Rivers. Social scientists do not conduct studies to detect glacier runoff variability because it is not their focus or expertise; however, they should analyze hydrologic studies before drawing conclusions about downstream impacts.

Societal changes in glacierized watersheds are often attributed ambiguously or unconvincingly to glacier runoff variability. The role of glaciers versus human forces of change (e.g., institutional change, reallocation of water rights, demographic shifts, or new land and water use practices) is, in other words, often unacknowledged. Studies concluding that water scarcity is a result of glacier loss—without attention to human variables driving water competition or restricting water access—often do not attribute water variability or human impacts to specific drivers. They are thus problematic for several reasons: They are based on speculation and tangential evidence; they do not differentiate between water availability and water allocation; they lack robust consideration of stakeholder competition for water; and they fail

to consider the ways in which glacial ice, snowpack, precipitation, and groundwater interact to affect stream flow throughout glacier-fed watersheds. A central challenge for glacier runoff research is to integrate the diversity of biophysical and social processes, simultaneously disentangling them to understand which forces should be addressed in water management practices.

Spatial and Temporal Scales

Research on glacier runoff variability frequently neglects explicit analysis of temporal and spatial scales, as well as recognition that scale in and of itself is socially and historically constructed (Neumann 2015; Margulies et al. 2016). Claims about water scarcity resulting from glacier retreat generally fail to identify when a reduction in glacier runoff has occurred, or will occur, and to what effect. Further, social scientists often draw conclusions about short-term or present-day human impacts of glacier runoff variability, whereas natural scientists project long-term impacts in future decades. Identifying when glacier-fed watersheds will reach peak water—that is, an explicit consideration of temporal scales and runoff regimes over time—is critical for evaluation of hydrologic risk or promotion of specific adaptation strategies.

Moreover, greater recognition of seasonally variable impacts is essential because glacier shrinkage will have the most pronounced effect during dry seasons and summers. During winters or wet seasons, precipitation markedly increases stream flow, often above what surrounding societies use. Studies of glacier runoff variability must thus identify when the downstream effects are most potent—and how specific water demands such as crops grown, annual timing of tourism cycles, and even time of day for hydroelectricity generation shape watershed hydrology over different timescales.

More place-based studies are also needed to understand diverse hydrologic processes and human water use dynamics across a range of watersheds, mountain ranges, countries, and cultures. Gender relations, ethnic and political inequities, community-level adaptations, spiritual relations with glaciers, access to alternative water sources, and differences in local knowledge vary from one region to another (Drew 2012; Salick, Byg, and Bauer 2012; Becken, Lama, and Espiner 2013; Gagné, Rasmussen, and Orlove 2014; Rasmussen 2015; Williams and Golovnev 2015). Impacts cannot be homogenized or universalized (scaled up) even when studies show consistent glacier shrinkage across mountain ranges. Careful citation practices can help to avoid cross-regional application of claims that have not been validated.

Transdisciplinarity and Redefining the Runoff Problem

Transdisciplinary research on water in glacierized basins is essential but must be approached carefully. Transdisciplinarity consists of interaction and collaboration across academic disciplines and also the integration and acceptance of a diverse range of knowledges, including indigenous knowledge, women's voices, local farmers' observations, and other stakeholder knowledge alongside the natural sciences (Wainwright 2010; Klenk and Meehan 2015; Carey et al. 2016). In transdisciplinary research on glacier runoff, the natural sciences and social sciences should work in "conversation" with each other—on equal footing—to detect glacier runoff variability and attribute human impacts to social–ecological drivers. Bridge building across disciplines and between researchers and nonacademic stakeholders—without one subsuming or appropriating the other—is crucial (Lave 2015).

Transdisciplinary collaboration is not enough, though, and we argue that the entire glacier runoff problem needs to be redefined—potentially in ways that transcend academic disciplinary structures (Wainwright 2010). Currently, most studies construct the problem as a glacier, water, and climate change issue. Ultimately, though, water scarcity is a socioeconomic issue (Jaeger et al. 2013). Human beings determine how much water to use and how (and to whom) it gets distributed. The glacier runoff problem needs to undergo a transformation in conceptualization, discourse, narratives—and research. We need to investigate questions of glacier runoff impacts through approaches that link biophysical and human systems, placing local livelihoods within a context of multiple forces influencing water availability and use, from changing precipitation patterns and glacier runoff rates to government policies, race relations, inequality, spirituality, ethics, and poverty (Drew 2012; Salick, Byg, and Bauer 2012; Bury et al. 2013; Carey et al. 2014; Gagné, Rasmussen, and Orlove 2014; Allison 2015).

Conclusions

Although this article addresses glacier recession and mountain hydrology specifically, it offers insights for human and physical geographers more broadly. It argues for the critical importance of detection and attribution of climate change impacts across all disciplines, topics, and world regions. The glacier case also demonstrates why careful analysis of spatial and temporal scales is vital. Additionally, the article challenges a fundamental paradigm driving much of today's glacier runoff research: that shrinking glaciers lead inevitably and immediately to water scarcity for societies—an underlying assumption that hinges on environmental determinism. Glacier (and other) research must avoid such generalizations, speculative assertions, and apocalyptic claims that lack supporting evidence. The article also reinforces the need for more social science and humanities voices in global environmental change research. The dynamic interactions between water supply and water management justify transdisciplinary collaborations, which are lacking despite decades of calls for more coupled natural–human systems research. The need to reorient glacier runoff scholarship around human dynamics—as previously done with natural disaster and environmental migration scholarship—is acute. Yet folding social sciences and humanities into the natural sciences is simply not enough when it comes to the lives and livelihoods of people living in the world's mountains. Hydrologic problems in glacierized basins cannot be resolved, water cannot be equitably distributed, hazard risk reduction cannot occur, and effective climate change adaptation will not occur without the integrated glacier runoff approach we are proposing for mountain regions.

Funding

Mark Carey's work is supported by the U.S. National Science Foundation under Grant #1253779. M Jackson's work is supported by a U.S. Fulbright–National Science Foundation Arctic Research Grant. Anne W. Nolin's work is partially supported by the U.S. National Science Foundation under Grant #1414106.

References

Akhtar, M., N. Ahmad, and M. J. Booij. 2008. The impact of climate change on the water resources of Hindukush-Karakorum-Himalaya region under different glacier coverage scenarios. *Journal of Hydrology* 355:148–63.

Allison, E. A. 2015. The spiritual significance of glaciers in an age of climate change. *WIREs Climate Change* 6 (5): 493–508.

Baraer, M., B. G. Mark, J. McKenzie, T. Condom, J. Bury, K. I. Huh, C. Portocarrero, J. Gómez, and S. Rathay. 2012. Glacier recession and water resources in Peru's Cordillera Blanca. *Journal of Glaciology* 58 (207): 134–50.

Bavay, M., T. Grünewald, and M. Lehning. 2013. Response of snow cover and runoff to climate change in high Alpine catchments of Eastern Switzerland. *Advances in Water Resources* 55:4–16.

Becken, S., A. K. Lama, and S. Espiner. 2013. The cultural context of climate change impacts: Perceptions among community members in the Annapurna Conservation Area, Nepal. *Environmental Development* 8:22–37.

Beniston, M. 2012. Impacts of climatic change on water and associated economic activities in the Swiss Alps. *Journal of Hydrology* 412–413:291–96.

Beniston, M., and M. Stoffel. 2014. Assessing the impacts of climatic change on mountain water resources. *Science of the Total Environment* 493:1129–37.

Bhadwal, S., A. Groot, S. Balakrishnan, S. Nair, S. Ghosh, G. Lingaraj, C. Terwisscha van Scheltinga, A. Bhave, and C. Siderius. 2013. Adaptation to changing water resource availability in northern India with respect to Himalayan glacier retreat and changing monsoons using participatory approaches. *Science of the Total Environment* 468–469:S152–S161.

Black, R., W. N. Adger, N. W. Arnell, S. Dercon, A. Geddes, and D. Thomas. 2011. The effect of environmental change on human migration. *Global Environmental Change* 21 (Suppl. 1): 3–11.

Bliss, A., R. Hock, and V. Radić. 2014. Global response of glacier runoff to twenty-first century climate change. *Journal of Geophysical Research: Earth Surface* 119 (4): 717–30.

Boelens, R. 2014. Cultural politics and the hydrosocial cycle: Water, power and identity in the Andean highlands. *Geoforum* 57:234–47.

Bury, J. 2015. The frozen frontier: The extractives super cycle in a time of glacier recession. In *The high-mountain cryosphere*, ed. C. Huggel, M. Carey, J. Clague, and A. Kääb, 71–89. Cambridge, UK: Cambridge University Press.

Bury, J., B. G. Mark, M. Carey, K. Young, J. McKenzie, M. Baraer, A. French, and M. Polk. 2013. New geographies of water and climate change in Peru: Coupled natural and social transformations in the Santa River watershed. *Annals of the Association of American Geographers* 103 (2): 363–74.

Bury, J., B. G. Mark, J. McKenzie, A. French, M. Baraer, K. In Huh, M. Zapata Luyo, and R. J. Gómez López. 2011. Glacier recession and human vulnerability in the Yanamarey watershed of the Cordillera Blanca, Peru. *Climatic Change* 105 (1–2): 179–206.

Carey, M., M. Baraer, B. G. Mark, A. French, J. Bury, K. R. Young, and J. M. McKenzie. 2014. Toward hydro-social modeling: Merging human variables and the social sciences with climate-glacier runoff models (Santa River, Peru). *Journal of Hydrology* 518:60–70.

Carey, M., A. French, and E. O'Brien. 2012. Unintended effects of technology on climate change adaptation: An historical analysis of water conflicts below Andean glaciers. *Journal of Historical Geography* 38 (2): 181–91.

Carey, M., M Jackson, A. Antonello, and J. Rushing. 2016. Glaciers, gender, and science: A feminist glaciology framework for global environmental change research. *Progress in Human Geography*. Advance online publication. doi: 10.1177/0309132515623368

Chaudhary, P., S. Rai, S. Wangdi, A. Mao, N. Rehman, S. Chettri, and K. S. Bawa. 2011. Consistency of local perceptions of climate change in the Kangchenjunga Himalaya landscape. *Current Science* 101 (4): 504–13.

Drenkhan, F., M. Carey, C. Huggel, J. Seidel, and M. T. Oré. 2015. The changing water cycle: Climatic and socioeconomic drivers of water-related changes in the Andes of Peru. *WIREs Water* 2 (6): 715–33.

Drew, G. 2012. A retreating goddess? Conflicting perceptions of ecological change near the Gangotri-Gaumukh glacier. *Journal for the Study of Religion, Nature and Culture* 6 (3): 344–62.

Einarsson, B., and S. Jónsson. 2010. The effect of climate change on runoff from two watersheds in Iceland. Paper presented at the Conference on Future Climate and Renewable Energy: Impacts, Risks and Adaptation, Oslo, Norway.

Folke, C. 2006. Resilience: The emergence of a perspective for social–ecological systems analyses. *Global Environmental Change* 16:253–67.

French, A., J. Barandiarán, and C. Rampini. 2015. Contextualizing conflict: Vital waters and competing values in glaciated environments. In *The high-mountain cryosphere: Environmental changes and human risks*, ed. C. Huggel, M. Carey, J. Clague, and A. Kääb, 315–36. Cambridge, UK: Cambridge University Press.

Gagné, K., M. B. Rasmussen, and B. Orlove. 2014. Glaciers and society: Attributions, perceptions, and valuations. *WIREs Climate Change* 5:793–808.

Gardarsson, S., and J. Eliasson. 2006. Influence of climate warming on Halslon Reservoir sediment filling. *Nordic Hydrology* 37 (3): 235–41.

Grah, O., and J. Beaulieu. 2013. The effect of climate change on glacier ablation and baseflow support in the Nooksack River basin and implications on Pacific salmonid species protection and recovery. *Climatic Change* 120:657–70.

Hulme, M. 2011. Reducing the future to climate: A story of climate determinism and reductionism. *Osiris* 26:245–66.

Immerzeel, W. W., F. Pellicciotti, and M. F. P. Bierkens. 2013. Rising river flows throughout the twenty-first century in two Himalayan glacierized watersheds. *Nature Geoscience* 6:742–45.

Immerzeel, W. W., L. P. H. van Beek, and M. F. P. Bierkens. 2010. Climate change will affect the Asian water towers. *Science* 328:1382–85.

Intergovernmental Panel on Climate Change. 2014. *Climate change 2014: Impacts, adaptation, and vulnerability. Part A: Global and sectoral aspects. Contribution of Working Group II to the Fifth Assessment Report of the Intergovernmental Panel on Climate Change.* New York: Cambridge University Press.

Ioris, A. A. R. 2012. The geography of multiple scarcities: Urban development and water problems in Lima, Peru. *Geoforum* 43 (3): 612–22.

Jaeger, W. K., A. J. Plantinga, H. Chang, K. Dello, G. Grant, D. Hulse, J. J. McDonnell, et al. 2013. Toward a formal definition of water scarcity in natural–human systems. *Water Resources Research* 49 (7): 4506–17.

Jurt, C., M. D. Burga, L. Vicuña, C. Huggel, and B. Orlove. 2015. Local perceptions in climate change debates: Insights from case studies in the Alps and the Andes. *Climatic Change* 133 (3): 511–23.

Klenk, N., and K. Meehan. 2015. Climate change and transdisciplinary science: Problematizing the integration imperative. *Environmental Science and Policy* 54:160–67.

Konchar, K. M., B. Staver, J. Salick, A. Chapagain, L. Joshi, S. Karki, S. Lo, A. Paudel, P. Subedi, and S. K. Ghimire. 2015. Adapting in the shadow of Annapurna: A climate tipping point. *Journal of Ethnobiology* 35 (3): 449–71.

Lackey, R. 2000. Restoring wild salmon to the Pacific Northwest: Chasing an illusion? In *What we don't know about Pacific Northwest fish runs: An inquiry into decision-making*, ed. P. Koss and M. Katz, 91–143. Portland, OR: Portland State University.

Lampis, A. 2013. Cities and climate change challenges: Institutions, policy style and adaptation capacity in Bogotá. *International Journal of Urban and Regional Research* 37 (6): 1879–1901.

Lave, R. 2015. Exploring the proper relation between phyical and human geography: Early work by John E. Thornes and Ron Johnston. *Progress in Physical Geography* 39 (5): 687–90.

López-i-Gelats, F., J. L. Contreras Paco, R. Huilcas Huayra, O. D. Siguas Robles, E. C. Quispe Peña, and J. Bartolomé Filella. 2015. Adaptation strategies of Andean pastoralist households to both climate and non-climate changes. *Human Ecology* 43 (2): 267–82.

Lynch, B. D. 2012. Vulnerabilities, competition and rights in a context of climate change: Toward equitable water governance in Peru's Rio Santa Valley. *Global Environmental Change* 22 (2): 364–73.

Mantua, N., I. Tohver, and A. Hamlet. 2010. Climate change impacts on streamflow extremes and summertime stream temperature and their possible consequences for freshwater salmon habitat in Washington State. *Climatic Change* 102 (1–2): 187–223.

Margulies, J. D., N. R. Magliocca, M. D. Schmill, and E. C. Ellis. 2016. Ambiguous geographies: Connecting case study knowledge with global change science. *Annals of the Association of American Geographers* 106 (3): 572–96.

Mark, B. G., M. Baraer, A. Fernandez, W. W. Immerzeel, R. D. Moore, and R. Weingartner. 2015. Glaciers as water resources. In *The high-mountain cryosphere: Changes and risks*, ed. C. Huggel, M. Carey, J. J. Clague, and A. Kääb, 184–203. Cambridge, UK: Cambridge University Press.

Mark, B. G., J. Bury, J. McKenzie, A. French, and M. Baraer. 2010. Climate change and tropical Andean glacier recession: Evaluating hydrologic changes and livelihood vulnerability in the Cordillera Blanca, Peru. *Annals of the Association of American Geographers* 100 (4): 794–805.

McDowell, G., J. D. Ford, B. Lehner, L. Berrang-Ford, and A. Sherpa. 2013. Climate-related hydrological change and human vulnerability in remote mountain regions: A case study from Khumbu, Nepal. *Regional Environmental Change* 13 (2): 299–310.

McDowell, J. Z., and J. J. Hess. 2012. Accessing adaptation: Multiple stressors on livelihoods in the Bolivian highlands under a changing climate. *Global Environmental Change* 22 (2): 342–52.

Molden, D. J., R. A. Vaidya, A. B. Shrestha, G. Rasul, and M. S. Shrestha. 2014. Water infrastructure for the Hindu Kush Himalayas. *International Journal of Water Resources Development* 30 (1): 60–77.

National Academy of Sciences. 2012. *Himalayan glaciers: Climate change, water resources, and water security*. Washington, DC: National Academy of Sciences.

Neumann, R. P. 2015. Political ecology of scale. In *The international handbook of political ecology*, ed. R. L. Bryant, 475–86. Northampton, MA: Edward Elgar.

Nolin, A. W., J. Phillippe, A. Jefferson, and S. L. Lewis. 2010. Present-day and future contributions of glacier runoff to summertime flows in a Pacific Northwest watershed: Implications for water resources. *Water Resources Research* 46:W12509.

Nüsser, M., S. Schmidt, and J. Dame. 2012. Irrigation and development in the Upper Indus Basin. *Mountain Research and Development* 32 (1): 51–61.

Nyaupane, G. P., and N. Chhetri. 2009. Vulnerability to climate change of nature-based tourism in the Nepalese Himalayas. *Tourism Geographies* 11 (1): 95–119.

Paerregaard, K. 2013. Bare rocks and fallen angels: Environmental change, climate perceptions and ritual practices in the Peruvian Andes. *Religions* 4 (2): 290–305.

Pelto, M. S. 2011. Skykomish River, Washington: Impact of ongoing glacier retreat on streamflow. *Hydrological Processes* 25 (21): 3356–63.

Petticrew, M., and H. Roberts. 2006. *Systematic reviews in the social sciences: A practical guide*. Malden, MA: Blackwell.

Postigo, J. C., K. R. Young, and K. A. Crews. 2008. Change and continuity in a pastoralist community in the high Peruvian Andes. *Human Ecology* 36 (4): 535–51.

Qureshi, A. S. 2011. Water management in the Indus basin in Pakistan: Challenges and opportunities. *Mountain Research and Development* 31 (3): 252–60.

Raoul, K. 2015. Can glacial retreat lead to migration? A critical discussion of the impact of glacier shrinkage upon population mobility in the Bolivian Andes. *Population and Environment* 36 (4): 480–96.

Rasmussen, M. B. 2015. *Andean waterways: Resource politics in highland Peru*. Seattle: University of Washington Press.

———. 2016. Water futures: Contention in the construction of productive infrastructure in the Peruvian highland. *Anthropologica* 58 (2): 211–26.

Rhoades, R. E., X. Z. Ríos, and J. A. Ochoa. 2008. Mama Cotacachi: History, local perceptions, and social impacts of climate change and glacier retreat in the Ecuadorian Andes. In *Darkening peaks: Glacier retreat, science, and society*, ed. B. Orlove, E. Wiegandt, and B. H. Luckman, 216–25. Berkeley and Los Angeles: University of California Press.

Riedel, J. L., S. Wilson, W. Baccus, M. Larrabee, T. Fudge, and A. Fountain. 2015. Glacier status and contribution to streamflow in the Olympic Mountains, Washington, USA. *Journal of Glaciology* 61 (225): 8–16.

Rodela, R. 2011. Social learning and natural resource management: The emergence of three research perspectives. *Ecology and Society* 16 (4): 30.

Salick, J., A. Byg, and K. Bauer. 2012. Contemporary Tibetan cosmology of climate change. *Journal for the Study of Religion, Nature & Culture* 6 (4): 447–76.

Schaefli, B., B. Hingray, and A. Musy. 2007. Climate change and hydropower production in the Swiss Alps: Quantification of potential impacts and related modelling uncertainties. *Hydrology and Earth System Sciences Discussions* 11 (3): 1191–1205.

Sherpa, P. 2014. Climate change, perceptions, and social heterogeneity in Pharak, Mount Everest region of Nepal. *Human Organization* 73 (2): 153–61.

Sivapalan, M., H. H. G. Savenije, and G. Blöschl. 2012. Socio-hydrology: A new science of people and water. *Hydrological Processes* 26 (8): 1270–76.

Skarbø, K., and K. VanderMolen. 2014. Irrigation access and vulnerability to climate-induced hydrological change in the Ecuadorian Andes. *Culture, Agriculture, Food and Environment* 36 (1): 28–44.

Stone, D., M. Auffhammer, M. Carey, G. Hansen, C. Huggel, W. Cramer, D. Lobell, M. Ulf, A. Solow, L. Tibig, and G. Yohe. 2013. The challenge to detect and attribute effects of climate change on human and natural systems. *Climatic Change* 121 (2): 381–95.

Swyngedouw, E. 2009. The political economy and political ecology of the hydro-social cycle. *Journal of Contemporary Water Research and Education* 142: 56–60.

———. 2010. Apocalypse forever? Post-political populism and the spectre of climate change. *Theory Culture & Society* 27 (2–3): 213–32.

Terrier, S., F. Jordan, A. Schleiss, W. Haeberli, C. Huggel, and M. Künzler. 2011. Optimized and adapted hydropower management considering glacier shrinkage scenarios in the Swiss Alps. In *Dams and reservoirs under changing challenges*, ed. A. Schleiss and R. M. Boes, 497–508. New York: CRC Press, Taylor & Francis Group.

Tiwari, P. C., and B. Joshi. 2012. Environmental changes and sustainable development of water resources in the Himalayan headwaters of India. *Water Resources Management* 26 (4): 883–907.

Vergara, W., A. Deeb, A. M. Valencia, R. S. Bradley, B. Francou, A. Zarzar, A. Grünwaldt, and S. Haeussling. 2007. Economic impacts of rapid glacier retreat in the Andes. *EOS, Transactions American Geophysical Union* 88 (25): 261–68.

Wainwright, J. 2010. Climate change, capitalism, and the challenge of transdisciplinarity. *Annals of the Association of American Geographers* 100 (4): 983–91.

Williams, C., and I. Golovnev. 2015. Pamiri women and the melting glaciers of Tajikistan. In *A political ecology of women, water and global environmental change*, ed. S. Buechler and A.-M. S. Hanson, 206–23. London and New York: Routledge.

Wong, E. 2015. Chinese glacier's retreat signals trouble for Asian water supply. *New York Times* 8 December 2015. http://www.nytimes.com/2015/12/09/world/asia/chinese-glaciers-retreat-signals-trouble-for-asian-water-supply.html?_r=0 (last accessed 2 July 2016).

Wrathall, D. J., J. Bury, M. Carey, B. Mark, J. McKenzie, K. Young, M. Baraer, A. French, and C. Rampini. 2014. Migration amidst climate rigidity traps: Resource politics and social–ecological possibilism in Honduras and Peru. *Annals of the Association of American Geographers* 104 (2): 292–304.

Young, K., and J. Lipton. 2006. Adaptive governance and climate change in the tropical highlands of western South America. *Climatic Change* 78 (1): 63–102.

Agro-environmental Transitions in African Mountains: Shifting Socio-spatial Practices Amid State-Led Commercialization in Rwanda

Nathan Clay

Agricultural commercialization has been slow to take hold in mountain regions throughout the world. It has been particularly limited by challenges of mechanization, transportation access, and governance. Efforts at green-revolution style development have met with persistent failures in highland sub-Saharan Africa, where agricultural systems are often finely tuned to complex and dynamic social–ecological contexts. In Rwanda, a mountainous country in east central Africa, development efforts have long aimed to transition away from largely subsistence-based production that relies on high labor input toward commercial farming systems that are rooted in capital investment for marketable goods. Since 2005, Rwanda's land policy has become increasingly ambitious, aiming to reduce the 85 percent of households involved in agriculture to 50 percent by the year 2020. The country's Crop Intensification Program (CIP) compels farmers to consolidate land and cultivate government-selected crops. Although state assessments have touted the productivity gains created through the CIP, others speculate that households could be losing access to crucial resources. Research from both sides, however, has focused squarely on the CIP's immediate successes and failures without considering how households are responding to the program within the context of the complex and variable mountain environment. Drawing from political ecology and mountain geography, this article describes recent state-led agricultural commercialization in Rwanda as a partial and contested process. By analyzing complex land-use and livelihood changes, it fills an important conceptual and empirical research gap in understanding the environmental and social dynamics of the agrarian transitions of the highlands of Africa.

农业商业化, 缓慢地发生在全球的山区之中。此一过程特别受到机械化, 运输管道和治理等挑战的限制。绿色革命的发展形态之努力, 在亚撒哈拉非洲的高地遭遇了持续的失败, 其中农业系统经常细微地转向複杂且动态的社会生态脉络。在卢旺达这个位于中东非的一个山丘国度中, 发展意图长期以来旨在将仰赖劳力密集投入的大幅生计生产, 转向植基于可交易物品的资本投入之商业耕作系统。自 2005 年以来, 卢旺达的土地政策逐渐展现野心, 旨在 2020 年前, 将百分之八十五参与农业生产的家户数降至百分之五十。该国的耕作集约计画 (CIP) 强迫农人合併土地, 并耕种政府选定的作物。尽管国家评估吹捧透过 CIP 创造的生产力增长, 其他评估则质疑家户可能会损失取得关键资源的管道。但两造研究皆完全聚焦 CIP 的立即性功过, 却未能考量家户如何在複杂和多变的山区环境脉络中回应该计画。本文运用政治生态学和山区地理学, 描绘卢旺达晚近由国家主导的农业商业化, 作为不完全且受到争议的过程。本文透过分析複杂的土地使用和生计变迁, 弥补理解非洲高地农业变迁的环境与社会动态中重要的概念与经验研究之阙如。 *关键词: 农业, 治理, 集约, 山区, 政治生态学, 亚撒哈拉非洲。*

La comercialización agrícola ha sido lenta en afianzarse en las regiones montañosas de todo el mundo. En particular, ha estado muy limitada por retos de mecanización, acceso al transporte y gobernanza. Los esfuerzos emprendidos siguiendo el estilo de desarrollo de la revolución verde han enfrentado fracasos recurrentes en las tierras altas del África subsahariana, donde los sistemas agrícolas a menudo están finamente calibrados con contextos socioecológicos complejos y dinámicos. En Ruanda, un país montañoso del lado oriental del África central, los esfuerzos de desarrollo han buscado desde hace tiempo apartar la agricultura, de una producción basada en gran medida en la subsistencia, dependiente en alto grado del trabajo, hacia sistemas agrarios comerciales que se basen en inversión de capital para una producción orientada al mercado. Desde el 2005, la política de tierras de Ruanda se ha tornado cada vez más ambiciosa, con la pretensión de reducir el 85 por ciento de las familias involucradas en agricultura a solo el 50 por ciento para el año 2020. El Programa de Intensificación de Cultivos del país (CIP) obliga a los cultivadores a consolidar la propiedad rural y a cultivar las cosechas seleccionadas por el gobierno. Aunque las evaluaciones oficiales alaban los supuestos logros del CIP, otros especulan que las unidades familiares del campo podrían estar perdiendo acceso a recursos cruciales. La investigación de los dos lados, sin embargo, se ha concentrado en los éxitos y fracasos inmediatos del CIP sin tomar en cuenta cómo están respondiendo las familias rurales al programa dentro del contexto del complejo y variable entorno montañoso. Con base en la ecología política y la geografía de montañas, este artículo describe la reciente comercialización agrícola en Ruanda manejada por el gobierno como un proceso incompleto y cuestionado. Analizando los cambios complejos del uso del suelo y en la subsistencia, el artículo llena un importante vacío conceptual y empírico para entender las dinámicas sociales y ambientales de la transición agraria en las tierras altas africanas.

The high levels of biophysical and social–political variation in mountain landscapes present distinct and seemingly insurmountable challenges to agricultural development efforts emphasizing mechanization, intensification, and commercialization (Jodha 1997). At the same time, Africa's tropical highlands support some of the highest rural population densities in the world through sophisticated labor and land-use systems that facilitate intensive farming despite scarce resources (Campbell and Riddell 1986; Messerli and Hurni 1990). Paradoxically, development interventions rarely engage with or build from these local strategies, often overlooking and even contradicting complex agroecological realities and practices (Ferguson 1990). An important example of this externally driven approach lies in the government of Rwanda's ambitious policies begun in 2005 to intensify production of six crops deemed economically viable and necessary to the country's development (maize, wheat, bean, cassava, rice, potato). Although government assessments extol the program's success at tripling productivity of these crops (Ministry of Agriculture and Animal Resources [MINAGRI] 2011), numerous scholars challenge that these gains could come at the expense of resource access and livelihood security among the rural poor (Ansoms et al. 2014; Dawson, Martin, and Sikor 2016). Missing from this discussion, however, is empirical evidence about how agricultural intensification intersects with community and household-level power dynamics to shape livelihood and land-use strategies across highly variable biophysical gradients in mountainous Rwanda. I submit that such a perspective will help to provide a more nuanced understanding of how one-size-fits-all intensification solutions can be limited in adoption and success if they are not carefully aligned with spatialized agricultural practices that are common to mountain regions in sub-Saharan Africa (SSA).

In responding to Smethurst's (2000) counsel for mountain geography to investigate linked society–environment processes from a political ecology perspective, this research also contributes much-needed empirical documentation, based on a 2014–2015 study of farm households in southwest Rwanda, of some of the unique challenges inherent in agriculture-led growth for highland Africa. Recent scholarship has speculated that Rwanda's top-down agricultural intensification program has constrained rural residents' access to land and other resources (Huggins 2014; Dawson, Martin, and Sikor 2016). This and other research to date, however, has not considered the diverse ways in which households respond to commercialization efforts with spatial practices of land use and livelihoods across a socially and biophysically dynamic mountain landscape. As I show, some households are well positioned to respond, whereas others are not.

The main theoretical contribution of this research lies in advancing insights into how recent state-led agricultural commercialization is partial and contested amidst mountain biophysical complexes and social–political complexities in the twenty-first century (Scott 1998; Funnel and Price 2003; Peters 2013). Beginning with an overview of geographic approaches (human, physical, and environment–society) to mountain agriculture, I argue that political ecology offers a useful set of tools for understanding commercialization in mountain landscapes. Reviewing literature on agricultural development in African mountains, however, I find that a political ecological perspective is strikingly absent in comparison with research coming from other mountain regions, such as the Andes and Himalayas.

The Rwanda case study demonstrates the value of a political ecological approach to understanding the intersections of topography, seasonality, heterogeneous resource access, labor practices, institutional arrangements, and state-led development initiatives that underlie the observed successes and failures of agricultural commercialization. Through analysis of household and parcel-level data on the spatiotemporal dynamics underlying land use and livelihood practices in the mountain landscape, the article discusses how Rwanda's national vision for agricultural intensification is not well aligned with local intensification practices and thereby risks exacerbating poverty and food insecurity for an important segment of the farm population. The article concludes with reflections on how this approach is broadly suited to dynamic and complex mountain socioenvironments in Africa as well as how the political ecological understanding of commercialization as a partial process can inform future research and policy on agricultural intensification.

Agricultural Intensification in African Mountains

Geographers have long played a vital role in debunking problematic assumptions about the relationship between population, rural livelihoods, and environmental degradation in mountain regions (Blaikie and Brookfield 1987; Ives and Messerli 1989; Li 1999). A notable example is the theory of Himalayan degradation (Eckholm 1976), which invoked Malthusian narratives to claim that growing populations of small-scale upland farmers caused deforestation and erosion that created downstream flooding and silting. Geographers countered with detailed analyses of land-use and livelihood practices, demonstrating that peasant "land managers" make informed decisions within complex political economic as well as cultural ecological contexts (Blaikie 1985; Blaikie and Brookfield 1987; Ives and Messerli 1989). Despite such integrative contributions, the majority of geographical work actively dealing with mountain settings has leaned toward the biophysical, prompting others to call on mountain geographers for more research investigating the intersections of environment and society (Smethurst 2000; Friend 2002; Price et al. 2013).

Whereas human and nature–society geographers rarely engage mountains explicitly (Funnel and Price 2003), political ecologists have made notable and recurrent contributions to mountain geography. Zimmerer (1999, 2011) demonstrated that agroclimatic zone models of Andean cultivation and irrigation overlook how relational topographical aspects shape household land-use decisions. Scott (1985, 2009) depicted the conflictual relationship between the state and highland peoples, who evade government control by swidden cultivation emphasizing tubers. Moore (1998, 1999) explained how the highland Zimbabwe landscape contributes materially and symbolically to gendered resource access struggles. Forsyth (1998, 2011) pointed

to the overlaps and simultaneous validity of local and scientific narratives about soil erosion and other "mountain myths." A common thread in this rich scholarship is that attention to the intersections of environmental governance and land-use practices elucidates how simplified models of agro-environmental change can threaten livelihoods and provoke further environmental degradation in mountain contexts.

In demonstrating that power struggles over resource access have implications for environment and livelihoods, these political ecology insights from mountain settings help frame inquiries into the tenability of large-scale agricultural development programs for highland SSA. Recent literature, including the World Bank's *Agriculture for Development* report (World Bank 2008), asserts that agricultural intensification can usher SSA toward long-term, pro-poor economic growth by facilitating increased production, the emergence of a vibrant rural nonfarm sector, and subsequent livelihood diversification (Diao, Hazell, and Thurlow 2010). Advocates of this agriculture-led growth model often gesture to countrywide increases in food crop production as a measure of success (MINAGRI 2011). Others, however, caution that the vision of private sector-led, longer term macroeconomic enhancement held by promoters such as a Green Revolution for Africa (AGRA) poorly resembles the smallholder focus that enabled short and medium-term poverty reduction in Asia's state-led green revolution of the 1960s and 1970s (Hart 1998; Dorward et al. 2004).

Indeed, research has demonstrated crucial limitations of input-based intensification for small farmers in SSA. For example, fertilizer subsidies can be inaccessible to poorer households that lack access to adequate land and labor to utilize them (Ellis and Maliro 2013), and influxes of hybrid maize seed can displace seeds for local crop varieties seen to perform better under adverse growing conditions (Bezner Kerr 2013). Moreover, there is skepticism that macroeconomic surpluses will trickle down to the rural poor (Haggblade, Hazell, and Reardon 2007), and numerous studies demonstrate that aggregate food crop productivity at the national level does not guarantee food access or even aggregate availability at local levels (Clover 2003). To speak to this debate, local-level research that examines pathways between agricultural growth and poverty reduction in SSA is essential to redressing core deficiencies in understanding agrarian change (Abro, Bamlaku, and Hanjra 2014), namely, how resource access patterns could be embedded in social and political contexts such as gender and class relations (Bezner Kerr 2012).

To these calls for critical engagement with local-level processes of agrarian change, I add that, for highland areas of SSA, it is essential to explicitly address how spatially and temporally variable mountain environments factor into land-use and livelihood practices as well as how households can adjust such practices in their responses to top-down agricultural policy. As I demonstrate later, attention to social–ecological aspects of mountain agriculture enables a fuller picture of commercialization as a process that alternates among strategies of domination, accommodation, and resistance (Zimmerer 1991). Political ecology lays the fundamental groundwork for considering how these trajectories coalesce to produce winners and losers; specifically, how agricultural policies premised on modernization might increase resource scarcity and thereby exacerbate environmental conflict between social groups (Blaikie and Brookfield 1987; Robbins 2012).

Agrarian Change in Rwanda

Located along the Western Rift (or Albertine Rift) mountains, Rwanda has an average elevation of 1,600 m and undulating terrain throughout the country. This geography contributed to settlements dispersed across the mountains and a precolonial feudal political system in which livelihood, land use, and ethnicity were entangled. Pasture for ruling Tutsi pastoralists covered the hills, and small parcels in the western highlands were allotted to Hutu agriculturalists who intensively cultivated sorghum, millet, and beans (Gourou 1953; D. Newbury and Newbury 2000). This unequal land distribution and hierarchical agrarian society persisted through Belgian colonial rule, during which time agriculturalists further intensified production, shortening fallows and replacing grains with calorie-rich tubers (Bart 1993; C. Newbury 1988). Despite repeated attempts to resettle households into villages, dispersed settlements remained the norm well after independence in 1962, and farming systems continued to be characterized as low-input, high-labor smallholder operations in which households countered scarce land and high microlevel variations in soil fertility and household capacity with careful selection of crops (Lewis and Berry 1988; Clay, Reardon, and Kangasniemi 1998). Homes were generally located in middle to upper hillside and surrounded by banana (a valued cash crop), with other important

crops such as sorghum, bean, and sweet potato cultivated nearby. Crop rotation helped to maintain soil fertility and intercropping to mitigate risk of devastating loss due to crop failure (Ford 1990; De Lame 2005).

In the midst of rebuilding following the 1994 genocide and civil war, the Rwandan government initiated an ambitious economic development plan, called Vision 2020. Motivated by both land scarcity concerns and determination to become a middle-income country by the year 2020, a primary directive of this plan is to transition from subsistence-based to commercial agriculture, reducing producer households from 85 percent (in 2004) to 50 percent by 2020. This transformation is premised on a sweeping program of "agricultural modernization," which policy documents define as large size, high-input, monocropped fields of urban and export-oriented crops. Although agriculture is identified as a key driver of medium-term economic growth, this policy has a built-in short-term ultimatum for millions of smallholders: Either intensify and become professional farmers or sell land and diversify livelihood (Government of Rwanda 2004).

Since 2004, Rwanda's agricultural vision has manifested as compulsory land tenure registration to inhibit informal ownership, resettling households from hillsides to villages to open expansive areas for agriculture, and, more recently, the Crop Intensification Program (CIP). This program targets a fourfold yield increase through large-scale agro-engineering (draining marshlands and building terraces), increasing chemical fertilizer use, and obligating producers to consolidate land and cultivate government-selected crops using hybrid seeds. These Western-inspired mechanisms are coordinated through a methodical governmental hierarchy whereby the state has extended its reach into once-isolated rural communities (Ansoms and Rostagno 2012). Many local leaders are appointed and live outside the area they represent and progress is measured regularly through *imihigo* (performance contracts) with higher administrative levels. The main direction of accountability is thus upwards toward the central government, with civil society weakened (Thomson 2010; Ingelaere 2014).

This top-down implementation leaves little possibility for producer households to influence or oppose the government's vision (Ansoms 2008; Ingelaere 2010) despite concerns that the CIP would do little to alter long-standing class-based inequalities and skepticism that the fledgling nonfarm sector would absorb the land-poor following their exit from agriculture (Musahara and Huggins 2005; Pottier 2006). With political economy studies on contract farming and marshlands, authors have demonstrated the increased ease with which the private sector can gain control of land to the detriment of smallholder livelihoods (Ansoms 2009; Huggins 2014) and that the CIP fails to take into account local institutions of resource access (Ansoms 2013; Ansoms et al. 2014). This scholarship speculates that the CIP is adversely positioned to address persistent rural poverty. These conclusions, however, are drawn from unique or pilot locations of the CIP, providing little discussion of agrarian change in more widespread hillside agricultural systems.

Following marshland trials, the CIP has since been extended throughout Rwanda's hillsides (MINAGRI 2011). The few studies on the program in hillside settings rely on data from the first year of its implementation; these considered household perceptions of what potential impacts could be (Pritchard 2013) and of its influence on well-being (Dawson, Martin, and Sikor 2016). Missing from this discussion is empirical material on household livelihood and land-use responses to the CIP, an indispensable element to efforts aimed at potential pro-poor adjustments to agricultural policy. Van Damme, Ansoms, and Baret (2014), with a study of eighteen banana farmers, established the value of considering field-level responses. This article expands on this work, suggesting that explicit engagement with the myriad biophysical and social complexities of the mountain environment and how they intersect with gender and class-laden social–ecological processes (e.g., labor, land, and seed networks) can add constructive nuance to claims of critical shortcomings to agricultural policy.

Methods

The research informing this article was carried out in four *umudugudu* (villages) in Nyamagabe district of southwest Rwanda. Historically one of the country's poorest regions, Nyamagabe is characterized by acidic soils (umbrepts at high elevations; ferrisols on granite at middle elevations), high average elevation, and steep slopes (Olson 1994b). Despite notoriously poor soil, rain-fed agriculture during two growing seasons per year is the primary livelihood in this remote area, albeit with some modifications to cropping patterns. In the 1990s, Nyamagabe produced between 30 and 50 percent less bean and maize than the national average, and sweet potato, which responds well to acidic soil, has taken on increased importance, comprising 38 percent of the diet, compared to 23 percent nationally (Olson 1994a).

The study community—selected following extensive visits throughout the country over several years—is representative of a predominantly subsistence area in which major strides to commercialize agriculture have been taken over the past decade. Active in the community since 2010, the CIP has reshaped the landscape through village-level agro-engineering projects (draining marshlands in valleys and constructing bench terraces on hillsides) as well as subtler institutional transformations that reach into households and livelihoods. The program operates as three overlapping spatial components: drained marshland, terraced hillside, and consolidated but unterraced hillside.

To consider how this governance complexity resonates with and shapes land-use and livelihoods, this article integrates data from qualitative and quantitative methods covering multiple organizational levels: (1) a structured household survey focusing on livelihoods and resource access ($n = 428$) and corresponding parcel-level questionnaire ($n = 3,017$ fields) concentrating on cultivation practices and challenges, both conducted by research assistants with all available households; (2) in-depth semistructured interviews (averaging eighty minutes) conducted by the author and research assistants with thirty-six male and thirty-six female respondents, selected randomly; (3) interviews with local leaders ($n = 38$, purposively sampled) at sector, cell, and village levels and key informants; and (4) participant observation of cultivation practices. Household and field surveys provide detail about livelihoods and land-use practices, to which in-depth interviews and participant observation add further insight about the roles of power structures and institutions in processes of agrarian change. A limitation to this site-specific approach lies in its inability to detect or compare regional differences in agricultural commercialization. On the other hand, an in-depth case study provides invaluable understanding of land-use and livelihood dynamics as they relate to the biophysical and social complexities of the mountain environment.

Reinventing the Socio-spatial Basis of Agriculture

This study finds that hillside CIP activities have, on the whole, served to raise and enable state control over household land-use decisions. Despite vastly divergent agro-ecological contexts, hillside CIP mechanisms closely resemble those of marshland pilot sites. In mid- and upper slope locations, households are obligated to plant a crop selected by cell leaders and a regional agronomist for each village: maize, wheat, peas, potato, or climbing bean. In conjunction with being locked into the selected crop, households are asked not to cultivate four crops that the government has deemed irrelevant to commercial agriculture: sweet potato, sorghum, cassava, and a regionally important grain called *ciraza*. These rules are enforced via direct and indirect means. Each household signs a performance contract pledging to cultivate the selected crop. Biweekly meetings of government-led farmer associations called *Twigire Muhinzi* (self-reliant farmer) serve to disseminate information about the CIP and provide a forum to publicly disgrace households that have cultivated prohibited crops. More directly, local authorities penalize those who plant nonapproved crops via fines, uprooting banned crops, or, more rarely, jail time. In fields located among bench terraces, obligation to cultivate the selected crop (chosen by the agronomist) is stricter still. Households with parcels in terraced areas retain ownership of the land but are required to purchase and plant government-approved seeds. Any terraced field not in use during a growing season is liable for government expropriation.

By limiting household autonomy, the CIP presents an acute constraint to an agrarian system that, respondents emphasized, is premised on flexibility to enable optimizing land use according to variable needs and capacities—specifically, planting the crops households need in locations where they will do well. This flexibility has historically been such an important component of smallholder agriculture in this mountain setting that households have prioritized land in a range of microclimates across the hillside. A crucial oversight of the CIP lies in the preexisting deep unevenness of access among and within households to land in this range of microclimates as well as to agricultural inputs. In addition to debilitating traditional risk management strategies, the CIP risks exacerbating this unevenness by requiring crop cultivation on land poorly suited for those crops and market involvement by households with little means to do so. In the following section I discuss the microlevel processes that shape the degree to which some households might benefit and others not.

Producing Viable Households: Uneven Resource Access

This research demonstrates that households with certain livelihoods and patterns of asset ownership are predisposed to succeed with the CIP vision of commercialization. Using a two-step cluster analysis, four distinct groups were identified, each significantly different from the others in assets (land, livestock, and consumer goods), income, and livelihood: (1) *very poor* households working agricultural labor, with 0.12 ha mean land (near landless), no livestock, and little income; (2) *poor* households predominantly cultivating family farms, with a small ruminant, pig, or a rented bull, average 0.29 ha land, and little income; (3) *relatively wealthy* households with 0.67 ha mean land, who combine cultivation with wage work or business, have some income, and most own one cow; and (4) *wealthy* households averaging 1.57 ha of land, two or more cows, and substantial income from nonfarm work or business (summarized in Table 1).

Despite the CIP's emphasis on agricultural intensification, many respondents (41 percent) reported that agricultural yields had decreased over the past ten years. Whereas 21.0 percent and 43.5 percent of respondents in the two wealthier groups claimed that yields increased, significantly fewer (7.9 and 12.6 percent) poor and very poor households reported an increase. These productivity declines can, in part, be traced to several external events (drought, heavy rain, and pests) as well as constrained access to manure, fertilizer, and seeds. These challenges were unevenly experienced: 57 percent of fields operated by very poor and poor households exhibited one or more of these productivity constraints compared to 45 percent and 33 percent of fields operated by the two wealthier groups.

Productivity challenges appear to have been exacerbated by the compulsory transition to commercial crops. Over three growing seasons, 79.0 percent of fields planted with commercial crops (maize, wheat, climbing beans, peas, and potatoes) were affected by at least one agroclimatic challenge compared to only 43.6 percent of fields with traditional crops (bush beans, sweet potato, ciraza, sorghum, cassava, taro, yam, and banana). For poorer households, the disparity in challenges between commercial crops and traditional crops is particularly acute: More than half of traditional crop fields had no challenges versus only 15 percent of commercial fields.

Interviews helped further clarify factors behind this uneven success with commercial crops. Respondents emphasized the near impossibility of cultivating maize and beans in low-fertility soils, without cash to buy necessary inputs, in the absence of surplus labor, or in less than ideal climatic conditions. Resource access

Table 1. Characteristics of household disaggregation and related findings

Characteristics and findings	Household group			
	Very poor	Poor	Relatively wealthy	Wealthy
Number of households	115	131	121	46
Land[**]	0.12 ha	0.29 ha	0.67 ha	1.57 ha
Livestock[**]	None	Goat, pig, or rented bull	One cow	Two cows
Principal livelihood[**]	Agricultural labor	Cultivate own farm	Diversified farmer	Non-farm/business
Yield increase[*]	8%	13%	21%	44%
Tubura member[*]	21%	42%	48%	59%
Field challenges[*]	57%	56%	45%	33%
Hunger > six months[*]	30%	16%	10%	6%
Valley land[**]	0.013 ha	0.03 ha	0.07 ha	0.21 ha

*Significant at 0.001 level (two-tailed chi-square); percentages reported.
**Significant at 0.001 level (two-tailed analysis of variance); means reported.

constraints and soil variations combine to the extent that poorer households often prefer traditional crops or even fallowing land over commercial crops. As a young woman from a very poor household explained, "We have land here on terraces which is in fallow because they are favorable for sweet potato and cassava and we know that the selected crop [wheat] would not give any harvest."

This response captures the difficulties of grafting a commercial vision of intensive agriculture onto a landscape of variable soil quality and household means. One limit to incorporating crops like wheat and maize is the extent to which these crops increase reliance on purchased inputs. Fields with commercial crops average 157,979 RWF (US$198) per hectare for seed and fertilizer. That contrasts to an average of 16,853 RWF (US$11) for fields with traditional crops, for which households generally use seeds saved from previous harvests or acquired from friends and neighbors. Distribution of seeds and fertilizer for selected crops has been recently privatized, further complicating input access for households with limited cash flow. A U.S. nonprofit company, the One Acre Fund (locally known as *Tubura*), distributes maize seed and fertilizer on credit in Nyamagabe and other regions of Rwanda in a partnership with the government. Although wealthy respondents noted productivity benefits, Tubura's 19 percent interest rate makes it less accessible to poorer households, which account for just 21.0 percent and 41.5 percent of members compared to 58.7 percent relatively wealthy households.

Even when they do manage to purchase seeds and chemical fertilizer, respondents highlighted that manure availability remains a constraint, especially for the 30 percent of households without livestock and the further 21 percent who either rent a bull or own only small livestock. Agricultural labor can be another scarce resource among poorer households that must first work in other households' fields to afford seeds and fertilizer for the obligatory commercial crops. This can mean costly delays in preparing land and sowing seeds that, with brief and unpredictable growing seasons, can be the difference between adequate and mediocre yields. Thirty-three percent of households planted late during the past three seasons, more for female-headed households (42 percent). Formalizing seed distribution might have aggravated this issue, as a woman from a poor household explained: "In past years, that challenge [planting late] was not there because people were saving and even sharing seeds. At that time people were growing in the same week because we had the same seeds. But now with the coming of Tubura, where we planted seeds and got chemical fertilizer some families have the problem of planting late."

Considering the variable capacity to cultivate commercial crops, it is perhaps unsurprising that respondents generally reported decreased food security compared to ten years ago: 87 percent among poorer households run out of own-produced food earlier now compared to 64 percent among the two wealthier groups. Of the poorest households, 29.7 percent reported inadequate food during six or more months of the year, contrasting with just 10 percent and 6 percent of wealthier households. In addition to the factors described earlier, 61 percent of survey respondents directly attributed the banning of key food security crops (sweet potato, cassava, and ciraza) to their declining food access. The following response from a woman in a poor household summarizes the perceived grave consequences of limiting farm-level decision making: "Listen, there are terraces in neighboring villages. Here we are lucky. If the government plans terraces for here, we will follow but we will die."

Although it sounds extreme, this comment illustrates the degree to which food security concerns shape land-use decisions. The ability to harvest various preferred food crops at staggered times throughout the year is often prioritized over discrete harvests of one or two commercial crops. These food security priorities, uncertain income or timely labor, and inadequate postharvest storage result for many households in a preference for traditional crops like sweet potato, which provide reasonable yield with little inputs, for which vines are freely available, and that can remain in the ground for up to twelve months until the crop is needed for either food or sale. In the following section, I discuss how the mountainous terrain alternately facilitates and hinders producers' abilities to respond to decision-making constraints.

Shifting Socio-spatial Practices of Agriculture

Rwanda's CIP is directed at overcoming long-standing limitations to commercializing agriculture that are rooted in the country's mountainous landscape (MINAGRI 2011). In compelling producers to plant crops according to location rather than biophysical or socioeconomic capacities of fields and households, however, the CIP has stymied dynamic and complex agrarian practices that have evolved over generations specifically to counter those limitations. In attempting to make mountain agriculture more legible (Scott 2009), the CIP has, for many households, actually served to make the challenges of mountain agriculture more pronounced than they would otherwise be. As a poignant example, although the mid- to upper slope is generally the most fertile and highest yielding field location (Steiner 1998), the proliferation there of commercial crops appears to have reversed this trend. Midslope areas registered a statistically significant 1.5 times the likelihood of productivity challenges than lower slope areas.

This surprising disparity in challenges experienced at different hill locations owes largely to the fact that in upper and midslope areas cultivation of government-selected crops is heavily enforced, whereas in lower hill and valley areas it is not. Producers have adjusted land-use strategies to take advantage of this institutional discrepancy, planting vital food security crops in what are commonly referred to as "hidden places," meaning areas out of sight of local leaders. A woman from a poor household summarizes this community-wide practice of everyday peasant resistance enabled by the mountain landscape (Scott 1985): "If a household decides it needs to have fresh beans it grows them far away. It is like stealing."

Contrary to much of the research on Rwanda's commercialization program, these findings indicate that household response strategies alternate between accommodation and resistance. These strategies have an important spatial component that is embedded in the topographic reality of land-use management in mountain settings. There is further irony in that, although the CIP endeavors chiefly to streamline farming systems by consolidating land, it has led to more fracturing of farming operations for households striving to continue planting core food security crops. Importantly, this strategy of resistance also has hidden costs: an increased labor burden for women, who are nearly always the household member making the long daily trek to harvest sweet potatoes or beans.

Another effect has been skyrocketing land values and a ballooning rental market for valley land, a statistically significant and exponentially greater proportion of which is controlled by wealthier households (averages of 0.013 ha and 0.03 ha among poorer groups vs. 0.07 ha and 0.21 ha among the two wealthier groups). The ban on sweet potato has, perversely, increased its importance as a commercial crop. More households now have no option but to purchase sweet potato and this heightened demand means that costs at local markets quadrupled at times during the past five years. Only 8 percent of the poorest group of households sell sweet potato, however, compared with 22 percent of relatively wealthy households. This is a particularly troubling shift as sweet potato sales have traditionally been one of the few ways for women and poor households to acquire cash.

Conclusion

A political ecology perspective—one that considers both governance structures and household decision making—enables understanding of agricultural commercialization as a highly spatial process that is closely tied to biophysical and cultural ecologies of the mountain environment. The CIP appears to have contributed to decreased food security for certain producer groups in Nyamagabe district due to its failure to appreciate local social–ecological processes, notably the spatiotemporal intersections of livelihood and land-use practices and high biophysical variability (soil fertility and climate) at the level of individual fields. The reason that these challenges are now so acute is that the state has asked farm households to bear the brunt of the risk involved in adopting its vision for intensified, commercial agriculture, at the same time actively disregarding local agricultural practices that are themselves effective risk management strategies. This has borne disproportionately negative impacts on the poorest groups of households while enabling others to increase productivity. By increasing the importance of scarce valley land as the only place to grow sweet potato, the CIP has further enabled wealthier households to gain control over a principal cash revenue generator for the poorest groups and for women. This reflects concerns that smallholder commercialization could occur at the expense of the poorest households (Akram-Lodhi 2008).

In addition to demonstrating the challenges of compelling households to intensify agriculture along a narrowly prescribed vision and reiterating the need for development planning to recognize the legitimacy of diverse and variable social and agronomic systems (Berry 1993; Scott 1998), these findings illuminate the unique challenges posed by top-down agricultural policies in diverse mountain areas. For Rwanda, the findings reinforce earlier calls to involve producers in on-farm experimentation in search of optimal and sustainable mixes of crops and landraces for their fields (Steiner 1998). For highland agricultural regions of SSA more broadly, these findings expose some of the perils of overlooking complex social–ecological practices and underscore the importance of paying close attention to the intersections of land use and livelihoods. Recognizing that some groups have benefited from Rwanda's ambitious commercialization platform, this article looks to broaden the debate over how the CIP and other similar programs might move beyond simply raising crop productivity among a minority of advantaged households. In approaching commercialization as a complex process, future research on Rwanda and elsewhere in highland SSA will be better positioned to contribute to the dialogue on how poverty reduction, nutrition improvement, and environmental sustainability might be simultaneously pursued and successfully accomplished.

Acknowledgments

Rwanda's Ministry of Education provided permits to conduct this study and I wish to acknowledge the International Center for Tropical Agriculture and the Rwanda Agriculture Board for providing institutional support. Fieldwork would not have been possible without the gracious help of hundreds of people throughout Rwanda. The study took shape with the guidance of Karl Zimmerer and Brian King and benefited from the careful attention of Dan Clay and Kayla Yurco. Feedback from Mark Fonstad and two anonymous reviewers also greatly improved this article. I alone am responsible for any errors in fact or interpretation.

Funding

This study was supported by the U.S. Fulbright Foundation, a Borlaug Fellowship for Global Food Security, and the Geography Department at Penn State University.

References

Abro, Z. A., A. A. Bamlaku, and M. A. Hanjra. 2014. Policies for agricultural productivity growth and poverty reduction in rural Ethiopia. *World Development* 59:461–74.

Akram-Lodhi, A. 2008. (Re)imagining agrarian relations? The World Development Report 2008: Agriculture for development. *Development and Change* 39 (6): 1145–61.

———. 2008. Striving for growth, bypassing the poor? A critical review of Rwanda's rural sector policies. *The Journal of Modern African Studies* 46 (1): 1–32.

———. 2009. Re-engineering rural society: The visions and ambitions of the Rwandan elite. *African Affairs* 108 (431): 289–309.

———. 2013. Large-scale land deals and local livelihoods in Rwanda: The bitter fruit of a new agrarian model. *African Studies Review* 56 (3): 1–23.

Ansoms, A., and D. Rostagno. 2012. Rwanda's Vision 2020 halfway through: What the eye does not see. *Review of African Political Economy* 39 (133): 427–50.

Ansoms, A., I. Wagemakers, M. Walker, and J. Murison. 2014. Land contestation at the micro scale: Struggles for space in the African marshes. *World Development* 54:243–52.

Bart, F. 1993. Montagnes d'Afrique terres paysannes: Le cas de Rwanda [Peasant landscapes of the African Highlands: The case of Rwanda]. Centre d'Etudes de Géographie Tropicale, Collection Espaces Tropicaux No. 7. Bordeaux, France: Presses Universitaires de Bordeaux, Université de Bordeaux III.

Berry, S. 1993. *No condition is permanent: The social dynamics of agrarian change in sub-Saharan Africa.* Madison: University of Wisconsin Press.

Bezner Kerr, R. 2012. Lessons from the old Green Revolution for the new: Social, environmental and nutritional issues for agricultural change in Africa. *Progress in Development Studies* 12 (2–3): 213–29.

———. 2013. Seed struggles and food sovereignty in northern Malawi. *The Journal of Peasant Studies* 40 (5): 867–97.

Blaikie, P. 1985. *The political economy of soil erosion in developing countries.* London and New York: Routledge.

Blaikie, P., and H. Brookfield, eds. 1987. *Land degradation and society.* London: Methuen.

Campbell, D., and J. Riddell. 1986. Agricultural intensification and rural development: The Mandara mountains of north Cameroon. *African Studies Review* 29 (3): 89–106.

Clay, D., T. Reardon, and J. Kangasniemi. 1998. Sustainable intensification in the highland tropics: Rwandan farmers' investments in land conservation and soil fertility. *Economic Development and Cultural Change* 46 (2): 351–77.

Clover, J. 2003. Food security in sub-Saharan Africa. *African Security Review* 12 (1): 5–15.

Dawson, N., A. Martin, and T. Sikor. 2016. Green revolution in sub-Saharan Africa: Implications of imposed innovation for the wellbeing of rural smallholders. *World Development* 78:204–18.

De Lame, D. 2005. *A hill among a thousand: Transformations and ruptures in rural Rwanda.* Madison: University of Wisconsin Press.

Diao, X., P. Hazell, and J. Thurlow. 2010. The role of agriculture in African development. *World Development* 38 (10): 1375–83.

Dorward, A., J. Kydd, J. Morrison, and I. Urey. 2004. A policy agenda for pro-poor agricultural growth. *World Development* 32 (1): 73–89.

Eckholm, E. 1976. Losing ground. *Environment: Science and Policy for Sustainable Development* 18 (3): 6–11.

Ellis, F., and D. Maliro. 2013. Fertiliser subsidies and social cash transfers as complementary or competing instruments for reducing vulnerability to hunger: The case of Malawi. *Development Policy Review* 31 (5): 575–96.

Ferguson, J. 1990. *The anti-politics machine: "Development," depoliticization and bureaucratic power in Lesotho.* Cambridge, UK: Cambridge University Press.

Ford, R. 1990. The dynamics of human–environment interactions in the tropical montane agrosystems of Rwanda: Implications for economic development and environmental sustainability. *Mountain Research and Development* 10 (1): 43–63.

Forsyth, T. 1998. Mountain myths revisited: Integrating natural and social environmental science. *Mountain Research and Development* 18 (2): 107–16.

———. 2011. Politicizing environmental explanations: What can political ecology learn from sociology and philosophy of science. In *Knowing nature: Conversations at the intersection of political ecology and science studies,* ed. M. Goldman, P. Nadasdy, and M. Turner, 31–46. Chicago: University of Chicago Press.

Friend, D. A. 2002. Mountain geography in 2002: The international year of mountains. *Geographical Review* 92 (2): iii–vi.

Funnel, D., and M. Price. 2003. Mountain geography: A review. *The Geographical Journal* 169 (3): 183–90.

Gourou, P. 1953. La densité de la population au Ruanda-Urundi: Esquisse d'une [Population density in Rwanda-Urundi: A case study]. Mem no. 8 XXI-6. Brussels, Belgium: Institut Royale Colonial Belge Etude Géographique.

Government of Rwanda. 2004. *National land policy.* Kigali, Rwanda: Ministry of Lands, Environment, Forestry, Water and Mines.

Haggblade, S., P. B. R. Hazell, and T. Reardon, eds. 2007. *Transforming the rural nonfarm economy: Opportunities and threats in the developing world.* Baltimore: Johns Hopkins University Press.

Hart, G. 1998. Regional linkages in the era of liberalization: A critique of the new agrarian optimism. *Development and Change* 29 (1): 27–54.

Huggins, C. 2014. "Control grabbing" and small-scale agricultural intensification: Emerging patterns of state-facilitated "agricultural investment" in Rwanda. *The Journal of Peasant Studies* 41 (3): 365–84.

Ingelaere, B. 2010. Do we understand life after genocide?: Center and periphery in the construction of knowledge in post-genocide Rwanda. *African Studies Review* 53 (1): 41–59.

———. 2014. What's on a peasant's mind? Experiencing RPF state reach and overreach in post-genocide Rwanda (2000–10). *Journal of Eastern African Studies* 8 (2): 214–30.

Ives, J., and B. Messerli. 1989. *The Himalayan dilemma: Reconciling development and conservation.* London and New York: Routledge.

Jodha, N. S. 1997. Mountain agriculture. In *Mountains of the world: A global priority,* ed. B. Messerli and J. D. Ives, 313–35. New York: Parthenon.

Lewis, L. A., and L. Berry. 1988. *African environments and resources.* London and New York: Routledge.

Li, T. M., ed. 1999. *Transforming the Indonesian uplands: Marginality, power and production.* Amsterdam: Harwood.

Messerli, B., and H. Hurni. 1990. *African mountains and highlands: Problems and perspectives.* Bern, Switzerland: AMA.

Ministry of Agriculture and Animal Resources. 2011. *Strategies for sustainable crop intensification in Rwanda.* Kigali, Rwanda: Ministry of Agriculture and Animal Resources.

Moore, D. 1998. Subaltern struggles and the politics of place: Remapping resistance in Zimbabwe's eastern highlands. *Cultural Anthropology* 13 (3): 344–81.

———. 1999. The crucible of cultural politics: Reworking "development" in Zimbabwe's eastern highlands. *American Ethnologist* 26 (3): 654–89.

Musahara, H., and C. Huggins. 2005. Land reform, land scarcity and post-conflict reconstruction: A case study of Rwanda. In *From the ground up: Land rights, conflict, and peace in sub-Saharan Africa,* ed. C. Huggins and J. Clover, 279–336. South Africa: Institute for Security Studies.

Newbury, C. 1988. *The cohesion of oppression: Clientship and ethnicity in Rwanda, 1860–1960.* New York: Columbia University Press.

Newbury, D., and C. Newbury. 2000. Bringing the peasants back in: Agrarian themes in the construction and corrosion of statist historiography in Rwanda. *The American Historical Review* 105 (3): 832–77.

Olson, J. M. 1994a. Farming systems of Rwanda: Echoes of historic divisions reflected in current land use. Rwanda Society-Environment Project Working Paper 2, Michigan State University, East Lansing.

———. 1994b. Land degradation in Gikongoro, Rwanda: Problems and possibilities in the integration of household survey data and environmental data. Rwanda Society-Environment Project Working Paper 5, Michigan State University, East Lansing.

Peters, P. 2013. Land appropriation, surplus people and a battle over visions of agrarian futures in Africa. *The Journal of Peasant Studies* 40 (3): 537–62.

Pottier, J. 2006. Land reform for peace? Rwanda's 2005 Land Law in context. *Journal of Agrarian Change* 6 (4): 509–37.

Price, M., A. Byers, D. Friend, T. Kohler, and L. Price. 2013. *Mountain geography: Physical and human dimensions.* Berkeley: University of California Press.

Pritchard, M. 2013. Land, power, and peace: Tenure formalization, agricultural reform, and livelihood insecurity in rural Rwanda. *Land Use Policy* 30 (1): 186–96.

Robbins, P. 2012. *B.* 2nd ed. Oxford, UK: Wiley-Blackwell.

Scott, J. C. 1985. *Weapons of the weak: Everyday forms of peasant resistance.* New Haven, CT: Yale University Press.

———. 1998. *Seeing like a state: How certain schemes to improve the human condition have failed.* New Haven, CT: Yale University Press.

———. 2009. *The art of not being governed: An anarchist history of upland Southeast Asia.* New Haven, CT: Yale University Press.

Smethurst, D. 2000. Mountain geography. *The Geographical Review* 90 (1): 35–56.

Steiner, K. 1998. Using farmers' knowledge of soils in making research results more relevant to field practice: Experiences from Rwanda. *Agriculture, Ecosystems and Environment* 69 (3): 191–200.

Thomson, S. 2010. Getting close to Rwandans since the genocide: Studying everyday life in highly politicized research settings. *African Studies Review* 53 (3): 19–34.

Van Damme, J., A. Ansoms, and P. Baret. 2014. Innovation from above and from below: Confrontation and integration on Rwanda's hills. *African Affairs* 113 (450): 108–27.

World Bank. 2008. *World development report 2008: Agriculture for development.* Washington, DC: World Bank.

Zimmerer, K. S. 1991. Wetland production and smallholder persistence: Agricultural change in a highland Peruvian region. *Annals of the Association of American Geographers* 81 (3): 443–63.

———. 1999. Overlapping patchworks of mountain agriculture in Peru and Bolivia: Toward a regional-global landscape model. *Human Ecology* 27 (1): 135–65.

———. 2011. Spatial-geographic models of water scarcity and supply in irrigation engineering and management (Bolivia, 1952–2009). In *Knowing nature: Conversations at the intersection of political ecology and science studies,* ed. M. Goldman, P. Nadasdy, and M. Turner, 167–85. Chicago: University of Chicago Press.

"Water Is Life": Local Perceptions of Páramo Grasslands and Land Management Strategies Associated with Payment for Ecosystem Services

Kathleen A. Farley and Leah L. Bremer

Andean páramo grasslands have long supported human populations that depend on them as forage for livestock and, increasingly, have been recognized as critical water sources with large soil carbon stores and high levels of biodiversity. Recent conservation efforts have used payment for ecosystem services (PES) to incentivize land management that aims to enhance ecosystem services related to water, carbon, and biodiversity, as well as local livelihoods. Data to assess ecological and social outcomes of these programs are limited, however. In particular, a better understanding of how incentivized land management practices affect the local values and uses of páramos is needed. We conducted interviews with PES participants on their perceptions of the value of páramos and of management practices incentivized through PES—afforestation and removal of burning—and linked them with data on ecological outcomes of those practices. We found that local perceptions of páramo values include provisioning, regulating, and cultural ecosystem services, underpinning basic needs, security, health, and social relations. In some cases, local perceptions align with research on ecological outcomes of PES, whereas in others, expectations of PES participants are unlikely to be met. We also found examples of both synergies—where PES land management strengthens an existing páramo value—and trade-offs, in which existing benefits might be diminished. By improving understanding of how people perceive the benefits they obtain from páramos and how participation in PES is likely to affect those uses and values, our findings help connect local perceptions with ecological science to inform policy and management.

安地斯帕拉莫草原，长期以来支持依赖其作為牲口粮草的人口，并且逐渐被认為是具有大量土壤碳储存与高度生物多样性的关键水资源。晚近的保育努力，运用付费生态系统服务 (PES)，以物质刺激来鼓励土地管理，旨在强化与水资源、碳和生态多样性以及地方生计的生态系统管理服务。但取得这些计画的生态与社会结果的数据却相当有限。以物质刺激鼓励的土地管理实践，如何影响地方价值和帕拉莫的使用，则特别需要更进一步的理解。我们访谈 PES 的参与者，理解他们对於帕拉莫的价值以及透过 PES 的物质刺激所鼓励的管理实践——造林和去除焚烧——之感知，并将其连结至这些实践的生态结果之数据。我们发现，对於帕拉莫价值的地方感知，包含物质供给，规范和文化生态系统服务，支撑了基础需求，安全，健康和社会关系。在若干案例中，地方感知与 PES 生态后果的研究相符合，在其他案例中，PES 参与者的期待则难以被满足。我们同时发现协同——其中 PES 土地管理强化既有的帕拉莫价值——以及权衡的案例，其中既有的效益可能会减弱。我们的研究发现透过促进对人们如何感知他们从帕拉莫中获得的益处，以及参与 PES 如何可能影响这些使用与价值的理解，协助将地方感知连结至生态科学，以告知政策及管理。关键词：保育，生态系统服务，环境感知，土地管理，PES。

Durante mucho tiempo, los pastizales paramunos andinos han sostenido poblaciones humanas que dependen de ellos para alimentar sus ganados al tiempo que, también, los páramos han sido reconocidos como cruciales fuentes de agua, con buena capacidad de almacenamiento de carbono y altos niveles de biodiversidad. Los esfuerzos de conservación recientes han usado el pago por servicios ecosistémicos (PES) para incentivar un manejo de tierras que propenda por el fortalecimiento de los servicios ecosistémicos relacionados con agua, carbono y biodiversidad, lo mismo que como medio de sustento local. Sin embargo, los datos que permitan evaluar los resultados ecológicos y sociales de estos programas son limitados. Se requiere, en particular, entender mejor la manera como las prácticas promovidas sobre manejo de la tierra afectan los valores locales y el uso de los páramos. Practicamos entrevistas con participantes de los PES en relación con sus percepciones del valor de los páramos y de las prácticas de manejo que se incentivan a través de los PES—reforestación y eliminación de las quemas—y relacionamos esas entrevistas con los datos de los resultados ecológicos de tales prácticas. Descubrimos que las percepciones locales del valor de los páramos incluyen lo relacionado con abastecimiento, regulación y servicios ecosistémicos culturales, enfatizando necesidades básicas, seguridad, salud y relaciones sociales. En algunos casos, las percepciones locales concurren con la investigación sobre los resultados ecológicos de los PES, mientras en otros se evidencia la escasa probabilidad de que las expectativas de los participantes de los PES se logren. Descubrimos también ejemplos de ambas sinergias—donde el manejo de la tierra con los PES fortalece un valor existente del páramo—y de compensaciones, cuando los beneficios existentes pueden ser disminuidos. Al mejorar la comprensión sobre el modo como la gente percibe los beneficios que pueden derivar de los páramos y sobre cómo quizás afectará esos usos y valores su participación en los PES, nuestros hallazgos ayudan a conectar las percepciones locales con la ciencia ecológica para sustentar políticas y manejo.

Payment for ecosystem services (PES) has become a widely implemented conservation strategy in many parts of the world, in particular in the tropical Andes, where water supply for urban populations, agriculture, and hydropower are regulated by high-altitude páramo grasslands (Farley et al. 2011). Many PES programs target multiple ecological and social objectives; however, analyses of program outcomes are limited and most only assess changes in a single ecosystem service (ES), even though it has been recognized that attempting to maximize one ES might lead to declines in others (Chan et al. 2006; E. M. Bennett, Peterson, and Gordon 2009; Balvanera et al. 2012). Research on ecological outcomes of PES continues to be necessary to understand whether expected benefits are being delivered to the "ecosystem service users" (e.g., downstream or larger scale beneficiaries; Naeem et al. 2015). At the same time, complementary information is needed on benefits and costs to ecosystem service providers (e.g., the communities and landowners who ultimately decide whether plans are implemented). There is a need for more research that addresses social outcomes of PES, including changes in livelihoods, governance, and cultural values, to assess effectiveness in achieving the social outcomes that were included in project objectives and to ensure the sustainability of PES programs (G. Bennett and Carrol 2014; Asbjornsen et al. 2015). In particular, there is a noted gap in understanding PES benefits and costs for different groups of people at different scales, from the global to the community or household scale. Greater understanding of these benefits and costs and how they are distributed among diverse stakeholder groups is critical to planning and implementing effective, equitable, and sustainable PES.

Evaluation of social outcomes of PES requires understanding how incentivized land management practices affect the ways in which local landowners use and value the ecosystems targeted by these programs. Assessing how people perceive the benefits they obtain from ecosystems is key to understanding the ways in which policies and programs are likely to alter those uses and relationships (Asah et al. 2014). Similarly, efforts to change behaviors of landowners require an understanding of the benefits associated with current management and the ways in which modifications in land management will alter those benefits (Asah et al. 2014). This type of information is frequently lacking in PES evaluation, however. As noted by Asah et al. (2014), "We have incomplete understanding of how perception, acquisition, and use of ecosystem services might inform individual and collective behaviors" (181). This understanding is particularly important, since incentivized changes in land management might not be enduring if attitudes and social norms remain unchanged (Pretty and Smith 2004). At the same time, better information on cultural, social, cognitive, and economic factors that influence management decisions can enhance the ability to design PES programs that are best suited to particular contexts and stakeholder groups (Balvanera et al. 2012).

Recent conservation efforts in the Ecuadorian Andes have used PES to compensate landowners for changes in land management that are expected to maintain or improve provision of ES in páramo grasslands (Farley et al. 2013). Limited data, however, have been available to inform land management recommendations as programs are developed or to assess the ecological or social outcomes across diverse stakeholder groups once the programs are in place. In particular, data on the effects of land management changes on the people who rely on these ecosystems have been scarce. These data are particularly needed for the Ecuadorian Andes, where PES programs have rapidly expanded in páramo ecosystems and have provided models for PES development in other regions (Farley et al. 2011; Bremer, Auerbach et al. 2016). Here, our aim is (1) to evaluate the ways in which participants enrolled in PES value páramo grasslands in the Ecuadorian Andes and (2) to understand the likely impacts of changes in land management on the people who rely on the benefits provided by páramo ecosystems. In this research, we address the following questions:

1. How do PES participants perceive the value of the páramo?
2. How do PES participants perceive the value and outcomes of land management strategies related to PES, in particular afforestation and burning?

We then relate these perceptions to measured ecological outcomes of these land management strategies.

Study Area

Páramo grasslands have supported human populations for at least 7,000 to 8,000 years, with activities likely consisting mainly of low-frequency burning and grazing prior to Spanish colonization (Luteyn 1992; Buytaert et al. 2006; Jantz and Behling 2012). Grazing with sheep, cattle, and horses and burning large extensions of páramo began in the 1500s with European colonization (Buytaert et al. 2006; Harden 2006). Currently, the most common land uses are burning coupled with livestock grazing, agriculture, and pine plantations, and páramos remain important as areas of high cultural value (Buytaert et al. 2006; Farley 2007; White 2013). More recently, páramo grasslands have been increasingly recognized as a critical water source, for their large soil carbon stores, and for their high levels of plant diversity and endemism (Mena, Medina, and Hofstede 2001; Sklenar and Ramsay 2001; Farley, Kelly, and Hofstede 2004; Buytaert et al. 2006; Farley et al. 2013). These properties make these ecosystems important from the global perspective as carbon sinks as well as from the regional perspective of water resources, given that they provide water for major Andean cities, hydropower production, and agriculture.

PES programs targeting páramo grasslands have emerged in response to the desire to protect the biodiversity and ecosystem services they provide in an effective and equitable way. These programs, primarily focused on carbon or water, often with biodiversity and livelihood co-objectives, incentivize either a reduction of burning and grazing or afforestation. For example, SocioPáramo was launched in 2009 as a component of the Ecuadorian Ministry of the Environment's SocioBosque program, which sought to conserve the country's remaining privately and communally owned forests and páramo grasslands. As of October 2010, páramo contracts made up 5 percent of total community and 21 percent of individual land entered into the wider SocioBosque program (de Koning et al. 2011). SocioPáramo provides incentives for landowners to eliminate burning and reduce grazing in an effort to conserve carbon stores, biodiversity, and water supplies, while also improving livelihoods. Other programs focus on carbon sequestration through afforestation, and use nonnative pine trees or, more recently, *Polylepis racemosa*, a species native to the Peruvian Andes. Afforestation is not incentivized by SocioParamo, but it is permitted with *Polylepis*. Other PES programs focus on water, including water funds that aim to reduce burning and grazing and promote páramo restoration and reforestation. The most prominent example is FONAG, a widely replicated fund established to protect water sources in páramo grasslands and montane forests (Goldman-Benner et al. 2012; Bremer, Auerbach et al. 2016). In both water and carbon PES programs, biodiversity and local livelihoods are often targeted as cobenefits (Farley et al. 2011).

Information on ecological outcomes of these programs has been limited, but studies were conducted concurrently with this research to evaluate the outcomes of afforestation and burn exclusion on carbon storage, soil hydrology, and plant diversity in two study areas, representing two regions where land is being enrolled in SocioPáramo. In the case of afforestation, pines were found to effectively sequester carbon in their biomass, particularly if they remain unharvested for several decades; however, their effects on soil carbon ranged from large negative to small positive impacts (Farley et al. 2013; Bremer, Farley et al. 2016). Young stands of *Polylepis* stored only modest amounts of carbon above ground and did not alter soil carbon (Farley et al. 2013). Soils under pine plantations were significantly drier than under grasslands, which also regulated the infiltration and subsurface movement of water more uniformly than pine plantations. This supports research done at the catchment scale, indicating a 50 percent decline in water yield with pine afforestation (Buytaert, Iñiguez, and De Bièvre 2007). *Polylepis* did not affect soil moisture but might use more water as tree size increases (Harden et al. 2013; see Table 1). Compared to grasslands, plant species richness in pine plantations was lower in one study area but higher in the other, where native forest was adjacent, but the pine plantation dramatically altered species composition and decreased native species

Table 1. Ecological outcomes of land management strategies associated with payment for ecosystem services

	Afforestation with pine	Afforestation with *Polylepis*	Burn exclusion
Water	Large decrease in soil moisture	No change	Moderate decrease in soil moisture
Carbon	Biomass: large increase	Biomass: small increase	Biomass: increase
	Soil carbon: small increase to large decrease	Soil carbon: no change	Soil carbon: no significant change (trend toward higher soil C)
Biodiversity	Plant species richness: lower to higher	No change	Plant species richness: highest with intermediate burn frequency
	Plant species composition: large change		Plant species composition: shift toward woody species; increase in shrub diversity

Note: Findings are based on results from two study sites in the Ecuadorian Andes published in Farley et al. (2013), Harden et al. (2013), and Bremer, Farley et al. (2016).

abundance. Afforestation with *Polylepis* did not affect species richness (Bremer 2012; see Table 1).

In the case of burn exclusion, above-ground carbon increased and there was a trend toward greater soil carbon storage with burn exclusion (Farley et al. 2013; Bremer, Farley et al. 2016). There was no evidence that burning reduced the water-retention capacities of páramo soils; however, where grass was unburned for forty-five years and woody species were established, soil moisture was intermediate between grass and pine, suggesting that burning promotes moist soil conditions by sustaining the dominance of grasses over woody vegetation (Hartsig 2011; Harden et al. 2013). The highest number of species was found with intermediate levels of burning, and shrub abundance and diversity increased and herbaceous species richness and diversity declined under long periods of burn exclusion (Bremer 2012; see Table 1).

Here, we link these ecological outcomes with the perceptions of páramo landowners who are participants in the SocioPáramo PES program. Our aim is to improve understanding of how people perceive the benefits they obtain from páramo ecosystems and how participation in PES is likely to affect the ways in which communities and individuals use and value them.

Methods

We conducted structured, in-depth interviews with participants who were enrolled in the SocioPáramo program between January and September 2010 and in July 2011. We interviewed all community participants who had joined by May 2011 (nineteen of nineteen) and the majority of individual landowner participants (forty-five of sixty-three) who enrolled by October 2010, for a total of sixty-four participant interviews

(Figure 1). We also interviewed the SocioPáramo program director and four of five extension agents. Interviews included both closed- and open-ended questions focused on participant characteristics, motivations for enrollment, land-use history, land-use and livelihood changes associated with participation, perceived benefits and drawbacks of participation, perceptions of the value of páramos, and perceptions and experiences with afforestation and burning (Bremer, Farley, and Lopez-Carr 2014; Bremer et al. 2014). We collaborated with local assistants who helped conduct interviews, rephrase questions where they were not fully understood, and review the information collected. Interviews lasted one to four hours and were conducted anonymously and with informed consent. Institutional Review Board (IRB) approval was granted through San Diego State University and the

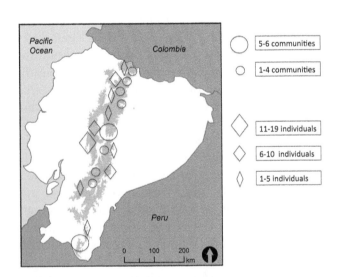

Figure 1. Map of the study area with páramo grasslands in gray and regions in which interviews were conducted indicated by circles (community interviews) and diamonds (individual interviews).

138

University of California, Santa Barbara. All information collected was kept confidential in accordance with IRB protocol and neither individual nor community names were used in communicating or reporting information.

We coded responses from a combination of closed- and open-ended questions to assess perceived benefits described by interviewees. We coded the past, current, and planned practices of grazing, afforestation, and conservation, and perceptions of these practices, to assess how participants value and perceive these land-management options. We also assessed the perceived benefits described by interviewees in the context of the ecosystem services categories used by the Millennium Ecosystem Assessment (MEA; provisioning, regulating, cultural, and supporting ES), which guided one level of coding (MEA 2005; Asah, Blahna, and Ryan 2012). Further, we drew on elaborated responses from interviewees to connect cited ES to MEA's "constituents of well-being": security (including secure resource access and security from disasters), basic material for a good life (including adequate livelihoods and access to goods), health (including access to clean air and water), and good social relations.

Results

Our assessment of values held by communities and individuals participating in SocioPáramo illustrates that those values fall within three of the four categories of ES (provisioning, regulating, and cultural) and relate to all four categories of human well-being (Table 2). Among the provisioning services, fresh water, livestock, and medicine were all cited. Participants emphasized the value of the páramo for water for irrigation and for urban and rural populations, linking it to several constituents of well-being, including providing basic material for life, health in terms of access to clean water, and security in terms of secure resource access. Livestock were perceived as either enhancing or degrading social relations, with some seeing it as a positive cultural tradition and others seeing it as a source of conflict (Table 2).

Regulating ecosystem services were cited by participants in the form of water regulation, erosion regulation, natural hazard regulation, and climate regulation. All four of the regulating ES were linked to security, in terms of secure resource access as well as security from disasters (Table 2). Cultural ecosystem services were referenced in the context of aesthetic values, tourism, and recreation as well as other

Table 2. Ecosystem services valued by community and individual participants in SocioPáramo

	Ecosystem services valued	Related constituents of human well-being
Provisioning	Freshwater	Basic material, health, security
	Livestock	Basic material, security, social relations (positive and negative)
	Medicine	Health
Regulating	Water regulation	Basic material, health, security
	Erosion regulation	Basic material, security
	Natural hazard regulation	Security
	Climate regulation	Health, security
Cultural	Aesthetic	Health
	Recreation and tourism	Basic material, health
	Social cohesion	Social relations
	Sense of place	Social relations
	Cultural heritage and values	Social relations

recognized cultural ES not included in the MEA, such as social relations, sense of place, and cultural heritage and values (Raymond et al. 2009). In some cases, these values were associated with traditional uses such as livestock grazing, whereas in other cases community cohesion and pride were referenced with respect to management practices associated with PES or other recent management strategies (e.g., removal of cattle grazing and transition to alpaca grazing), demonstrating that these values are dynamic. Cultural ES were linked with all categories of human well-being and were connected to a wide range of social relations and security factors such as land tenure security (Table 2). In the following sections, we discuss the specific ways in which these ES values were identified with respect to the páramo and the two land management strategies associated with PES.

Perceptions of the Value of Páramo Grasslands

In response to questions regarding what value the páramo holds for respondents, the most common response related to water supplies (sixteen community respondents and thirty-five individuals), with many stating that water was the primary value of the páramo, and others stating, "Water is life," "The páramo gives water," or that important springs originate there. Many specified that páramos needed to be cared for or conserved to maintain this function and the majority of communities and some individuals had already put all or part of their páramo into reserves prior to enrolling in SocioPáramo. Responses specified the value of the water that originates in their páramos for their own communities—in particular for irrigation—or for nearby cities and counties that rely on it. Two communities and one individual, however, stated that they believe that water comes from montane forests, not páramos, and that grasslands "lose water."

Eight communities and twenty-six individuals also cited values associated with grazing and livestock, including cattle, sheep, and camelids. In two communities, though, only a few families used the páramo for grazing and they were the only ones who obtained this value from it. Another respondent noted that although grazing is an important use, there is "a lot of conflict in the páramo." In many cases, the páramo is used primarily during the dry season when lower elevation pastures are less productive, so it is seen as an important insurance policy. One individual participant specified that he was using light grazing as a form of conservation, and another indicated that some areas of the páramo—in particular, wetlands—should be "untouched." One respondent referred to the cultural importance of raising livestock in the páramo, particularly for indigenous groups, noting that he believed that it was not right to take away grazing from them because that is taking away "their way of life."

Although less prevalent than water and grazing, biodiversity was a value cited by a number of respondents. Five community respondents and five individuals noted values related to biodiversity and more than half of those referred to the value of medicinal plants obtained from the páramo, although one community noted that they used to be more important. One mentioned a specific plant, Chuquiraga jussieui, as having particular value for medicinal purposes. One of the individual respondents referred to the páramo's biodiversity value more broadly—as a place for plants and animals threatened with extinction—rather than for individual use. Finally, one community respondent focused on aesthetic values related to biodiversity, noting that they "love to see the birds and the rabbits" in the community's páramo.

Many community respondents also emphasized that protecting their páramo was an important source of community cohesion, identity, and pride. For example, one community representative expressed pride that her community had protected their páramo long before it was widely believed that the "páramo is like gold." Another community respondent explained how conservation of their páramo and montane forest was intricately related to their fight for land tenure and an important source of community identity. Another community noted that the páramo is the location of the casa comunal, making it a valuable community gathering point.

Perceptions of Afforestation

Interviewees cited a range of values associated with afforestation, including the value of pines for timber production as well as the value of planted pine and Polylepis as a means of demonstrating community land ownership or to prevent the páramo from being used for more intensive agricultural production, constituting a form of protection. One community valued afforestation as a means to help prevent landslides, whereas one individual stated that he planted Polylepis for its aesthetic value and another felt that it would enhance tourism. Another individual who had participated in planting for carbon sequestration stated that planting trees is good for the environment and for wildlife, whether the trees are native or not. He stated that "pine is helping with climate change" and that he believes that pine has made it rain more than before on his property.

Respondents expressed a range of views on the relationship between afforestation and water. Nineteen individual respondents and three community respondents stated that they were opposed to planting pine because it negatively affects the water-related function of the páramo. Many communities explained that they had initially thought that pines would provide benefits, including water and income, but they now realize that "pines suck water" and "the community now does not want pine." Among those opposing pine, one respondent stated that he was also against planting native tree species, noting that even if they are from the páramo they are not from the same páramo. Several respondents, though, saw afforestation as positive for the water function of the páramo, particularly where native species were planted. Four individuals and one community respondent stated that reforestation is the best way to produce water in the páramo or that, whereas pine uses more water than páramo, native species use less water or bring more water. For example, one individual respondent stated that pine was preferable to páramo grasslands for water retention because grasslands are a "sponge" but they dry out, unlike forests where "water is always there." Another individual respondent stated that shrubs and native species "act as a curtain catching water," whereas grasslands lose water.

Perceptions of Burning and Removal of Burning

The majority of interviewees, including sixteen community and nine individual respondents, stated that burning the páramo diminished its value. Many of those who saw burning as negative had already abandoned the practice prior to enrolling in SocioPáramo, including some who have not burned for twenty years and enrolled in the program with the view that the páramo should not be burned. The reasons for this were varied, including views that burning leads to soil erosion, loss of soil fertility, loss of water, air pollution, and loss of biodiversity (primarily wildlife). One of the strongest statements against burning came from an individual respondent who said that burning was a long-standing custom that should be broken, but they lacked someone to inform people that they "should not damage the páramo that produces water." He argued that burning was damaging not only to the water function of the páramo and to wildlife but also to the cattle. Another individual respondent stated that burning was "mala cultura" and that "everyone had learned not to burn," whereas another said that "if no one burned it would be beautiful," and one respondent felt that there should be "a strong punishment" for burning. These perceptions were likely influenced by campaigns promoting burn exclusion as a means to protect the páramo for water, biodiversity, and other ES. The many communities who had already banned burning, however, described their views as stemming from their own observations of reductions in water supplies that they tied to land management practices. The majority of communities (twelve of eighteen) and about 35 percent of individual participants (sixteen of forty-five) had worked with a nongovernmental organization or environmental authority, but we did not find any trends between this and perceptions of burning.

Community participants, in particular, linked protection from burning to improved water supplies. For example, one community representative explained, "We were confused before, thinking that burning would bring rain," but they now believe that burning is negative for the water supply. Another community respondent similarly stated that "cattle trample the sponge," and "without the pajonal we can't capture

water." More specifically, one community respondent stated that the larger the bunchgrasses, the more water, because of increased fog drip and soil protection. Another community respondent stated that grasses and forest are equally good for water production and that removing burning would mean less destruction of water sources. Others were less certain about the association. For example, another community respondent said that, without burning, the grasslands would likely convert to shrubs in the lower altitude areas, but it was unclear what that would mean for water, although he expected that more vegetation is better for capturing water. This view—that an increase in shrubs associated with less burning would be positive for water—was echoed by other interviewees. One individual respondent, though, commented that in his experience of twenty years without burning, grasslands were maintained and shrub cover did not increase; he stated, "If the whole cordillera was conserved, there would be more water."

Many other respondents, however, particularly individuals, did not see burning the páramo as negative, and many saw it as necessary, particularly for anyone owning livestock. Among the reasons for burning, most respondents cited improved regrowth of grasses for livestock, to bring rain, to hunt rabbits, and to bring out bears and other wildlife. Some stated that without burning there would be insufficient regrowth of forage for livestock and that it will become "old and bad" for cattle and the area will dry up. Several individual respondents stated that burning every five to ten years does not damage the páramo and can be beneficial because it prevents more damaging burns from occurring. One individual participant emphasized that the biggest threat to the páramo was expansion of potato agriculture and that burning coupled with alpaca grazing was a much more sustainable alternative. Similarly, some interviewees felt that burning was inevitable, in some cases because neighbors burned even if they did not, and some stated that it was preferable to have a controlled burn. Finally, several interviewees also commented on the cultural value of burning, stating that it is "part of indigenous culture" or that it is a "pre-Hispanic practice" that has been used for 7,000 years and "burning is part of the grassland."

Discussion

Our first goal is to improve understanding of how PES participants perceive the benefits they obtain from páramo ecosystems as well as the benefits obtained or lost through application of PES land management strategies. Second, we aim to connect those perceptions to measured ecological outcomes, contributing to the limited research that links local perceptions to broader measures of environmental values (Raymond et al. 2009). Local people's perceptions of ecosystem services can provide valuable information for developing and adapting policy and management guidelines (Asah et al. 2014; De Oliveira and Berkes 2014; Sandhu and Sandhu 2014). In the case of PES, this information can help improve management strategies through a better understanding of what drives ecosystem management decisions among PES participants (Balvanera et al. 2012). At the same time, understanding how perceived benefits relate to more formal constructions of ecosystem services such as the MEA can help translate those perceptions into a form that can more readily be applied to management (Asah, Blahna, and Ryan 2012). Many PES programs have been developed with limited information available on the links between land management strategies and ecosystem services production (Muradian et al. 2010; Farley et al. 2011); however, Mach, Martone, and Chan (2015) noted that adaptive management can involve initiating management without all of the necessary information but moving toward collecting it over time and changing management accordingly. With the goal of contributing to the collection of this type of information, we assessed the relationship between the ways in which participants value and obtain ecosystem services from páramo grasslands and the outcomes of land management associated with PES.

Afforestation

In the case of pine afforestation, the measured ecological outcomes concur with other existing research (e.g., Buytaert, Iñiguez, and De Bièvre 2007) and with the most broadly held perception among participants that this form of land management will reduce water availability. That view was not universally expressed, however, and for those participants who expect equal or improved water availability or supply with pine afforestation, they are likely to lose a valued ecosystem benefit. For those participants who expect less water with pine plantations, participation in pine afforestation programs involves a known trade-off. Although economic benefits associated with pine can be important for some páramo communities (Farley 2007, 2010), interviewees in this study did not express many economic benefits associated with pine, and many chose to avoid programs that incentivized this form of land use due to that trade-off.

The data on ecological outcomes also point toward effects on biodiversity values held by participants due to shifts in species composition. Consistent with other studies, our ecological data indicated that the largest decline in species richness and abundance was among species most typical of páramo grasslands (Van Wesenbeeck et al. 2003; Bremer and Farley 2010). Given that the main biodiversity value noted by participants related to species used for medicinal purposes, the shift from species typical of páramo grassland to forest species has the potential to affect that value. Other biodiversity values related to wildlife also have the potential to be influenced by a shift in vegetation cover, but further research is needed on these outcomes.

The data on ecological outcomes related to *Polylepis* afforestation do not indicate that this land use increases water availability or production, suggesting that those who adopt it with that expectation likely will not obtain the expected benefits. We did not find a decrease either; however, given the young age of the plantations, continued monitoring will be necessary to understand the medium- and long-term effects. The importance some interviewees placed on value obtained from planting *Polylepis* in terms of demonstrating land ownership supported past research (Farley 2010) and raises the question of whether there might be a trade-off between a regulating ecosystem service (water regulation) and cultural ecosystem services, as well as a trade-off between two forms of security—that associated with secure water resources and that associated with land tenure security. Mach, Martone, and Chan (2015) noted that in many cases there is insufficient information on ES to effectively evaluate trade-offs; in the tropical mountains, which tend to be data-scarce, this is particularly true (Ponette-Gonzalez et al. 2014). This study highlights very specific areas in which additional data are needed. In addition to continued monitoring of water-related ES, both cultural ES and social relations merit further attention. These types of benefits have been found to have "profound influences on human actions including compliance with management and policy" (Asah et al. 2014, 185).

Burning and Removal of Burning

Although most respondents saw burning as damaging to the páramo and removal of burning as a measure to improve water-related ecosystem services, the data on ecological outcomes indicate a moderate decrease in soil moisture from removal of burning. In this case, adopting the PES land management strategy might not produce the benefit expected by participants. Other participants noted that removing burning might increase the density of shrubs or convert grasslands completely to shrublands. The data on ecological outcomes revealed differences between our two study areas—with one showing limited conversion and the other showing strong signs of conversion—so that whether participant expectations are met will vary from place to place. The consequences of shrub conversion for water were seen as neutral or positive by interviewees; however, the ecological data revealed lower soil moisture under shrubs than grasslands, suggesting that there could be some loss of this ecosystem benefit. These outcomes are important to match with participant perceptions because participants might be less willing to continue participation if the management strategies have negative impacts on any of their primary uses (Asah et al. 2014).

Some respondents indicated that intermediate burning was positive for biodiversity, an idea supported by the data on ecological outcomes. Many others, however, held the view that burning has negative consequences for native plants and animals, underscoring that the perception of benefits can differ greatly among landowners even within the same ecosystem (Asah et al. 2014). Our data indicate that although burning can decrease the abundance of shrubs and trees, other species benefit from burning, so the effect of this land management strategy will depend on which species and land covers provide the most benefits.

Conclusion

Marston (2008) noted the "frequent disconnect between mountain science, policymaking, and resource management" (507). In the context of páramo grasslands in the Andes, we have sought to link local perceptions to the ecological science on outcomes to provide a more comprehensive set of data to inform policy and management. At the same time, we contribute to geographic research in the Andes that links local perceptions to biophysical measurements (Mark et al. 2010). It is widely recognized that the knowledge and values of local communities are critical to conservation, and it is increasingly understood that

natural resource management requires understanding social values and perceptions (Pretty and Smith 2004; Bury et al. 2013). Few studies, though, have evaluated perceptions of local communities and individuals regarding the value of ecosystem services (De Oliveira and Berkes 2014). This gap can limit the effectiveness of conservation policies and programs; as noted by Asah et al. (2014), "If managers want to effectively constitute and regulate certain behaviors, to effectively manage ecosystem services, they must first understand what and how people gain or lose (direct or indirect ecosystem benefits) by engaging in those behaviors" (181). A better understanding of local perceptions and values can help in designing more effective policies and programs that are appropriate to the context and stakeholders in the region and can inform adaptation of existing programs (Balvanera et al. 2012; Mach, Martone, and Chan 2015). Our findings contribute to filling this gap in the context of PES in the Ecuadorian Andes.

Acknowledgments

We are extremely grateful to all of the people who took the time to be interviewed and provide their insight. For collaboration and support, we thank Carol Harden, Stuart White, José Alvear, the community of Zuleta, Fundación Cordillera Tropical, SocioPáramo (SocioBosque) staff, José Romero, Karina Paredes, Patricio Padrón, Ecociencia, Naturaleza y Cultura, the park guards of Sangay National Park, Mazar Wildlife Reserve, and Zuleta, Will Anderson, Daisy Cárate, and Sebastian Vasco.

Funding

This material is based on work supported by the National Science Foundation under Grant No. 0851532, a Fulbright Student Grant, and the San Diego State University Grant Program. Any opinions, findings, and conclusions or recommendations expressed in this material are those of the author(s) and do not necessarily reflect the views of these organizations.

Supplemental Material

Supplemental data (interview questions related to the topics of páramo values and perceptions of incentivized land-management practices) for this article are available on the publisher's Web site at http://dx.doi.org/10.1080/24694452.2016.1254020.

References

Asah, S. T., D. J. Blahna, and C. M. Ryan. 2012. Involving forest communities in identifying and constructing ecosystem services: Millennium Assessment and place specificity. *Journal of Forestry* 110 (3): 149–56.

Asah, S. T., A. D. Guerry, D. J. Blahna, and J. J. Lawler. 2014. Perception, acquisition and use of ecosystem services: Human behavior, and ecosystem management and policy implications. *Ecosystem Services* 10:180–86.

Asbjornsen, H., A. S. Mayer, K. W. Jones, T. Selfa, L. Saenz, R. K. Kolka, and K. E. Halvorsen. 2015. Assessing impacts of payments for watershed services on sustainability in coupled human and natural systems. *Bioscience* 65:579–91.

Balvanera, P., M. Uriarte, L. Almeida-Lenero, A. Altesor, F. DeClerck, T. Gardner, J. Hall, et al. 2012. Ecosystem services research in Latin America: The state of the art. *Ecosystem Services* 2:56–70.

Bennett, E. M., G. D. Peterson, and L. J. Gordon. 2009. Understanding relationships among multiple ecosystem services. *Ecology Letters* 12:1–11.

Bennett, G., and N. Carrol. 2014. *Gaining depth: State of watershed investment 2014.* Washington, DC: Forest Trends' Ecosystem Marketplace. www.ecosystemmarketplace.com/reports/sowi2014 (last accessed 7 October 2016).

Bremer, L. L. 2012. Land-use change, ecosystem services, and local livelihoods: Ecological and socioeconomic outcomes of payment for ecosystem services in Ecuadorian páramo grasslands. PhD dissertation, San Diego State University–University of California Santa Barbara, San Diego.

Bremer, L. L., D. A. Auerbach, J. H. Goldstein, A. L. Vogl, D. Shemie, T. Kroeger, J. L. Nelson, et al. 2016. One size does not fit all: Natural infrastructure investments within the Latin American Water Funds Partnership. *Ecosystem Services* 17:217–36.

Bremer, L. L., and K. A. Farley. 2010. Does plantation forestry restore biodiversity or create green deserts? A synthesis of the effects of land-use transitions on plant species richness. *Biodiversity and Conservation* 19 (14): 3893–3915.

Bremer, L. L., K. A. Farley, O. A. Chadwick, and C. P. Harden. 2016. Changes in carbon storage with land management promoted by payment for ecosystem services. *Environmental Conservation* 43 (4): 397–406.

Bremer, L. L., K. A. Farley, and D. Lopez-Carr. 2014. What factors influence participation in payment for ecosystem services programs? An evaluation of Ecuador's SocioPáramo program. *Land Use Policy* 36:122–33.

Bremer, L. L., K. A. Farley, D. Lopez-Carr, and J. Romero. 2014. Conservation and livelihood outcomes of payment for ecosystem services in the Ecuadorian Andes: What is the potential for "win–win"? *Ecosystem Services* 8:148–65.

Bury, J., B. G. Mark, M. Carey, K. R. Young, J. M. McKenzie, M. Baraer, A. French, and M. H. Polk. 2013. New geographies of water and climate change in Peru: Coupled natural and social transformations in the Santa River watershed. *Annals of the Association of American Geographers* 103 (2): 363–74.

Buytaert, W., R. Celleri, B. De Bièvre, F. Cisneros, G. Wyseure, J. Deckers, and R. Hofstede. 2006. Human impact on the hydrology of the Andean páramos. *Earth-Science Review* 79:53–72.

Buytaert, W., V. Iñiguez, and B. De Bièvre. 2007. The effects of *Pinus patula* forestation on water yield in the Andean páramo. *Forest Ecology and Management* 251:22–30.

Chan, K. M. A., M. R. Shaw, D. R. Cameron, E. C. Underwood, and G. C. Daily. 2006. Conservation planning for ecosystem services. *PLoS Biology* 4 (11): 2138–52.

de Koning, F., M. Aguinaga, M. Bravo, M. Chiu, M. Lascano, T. Lozada, and L. Suarez. 2011. Bridging the gap between forest conservation and poverty alleviation: The Ecuadorian SocioBosque program. *Environmental Science & Policy* 14:531–42.

De Oliveira, L. E. C., and F. Berkes. 2014. What value São Pedro's procession? Ecosystem services from local people's perceptions. *Ecological Economics* 107:114–21.

Farley, K. A. 2007. Grasslands to tree plantations: Forest transition in the Andes of Ecuador. *Annals of the Association of American Geographers* 97 (4): 755–71.

———. 2010. Pathways to forest transition: Local case studies from the Ecuadorian Andes. *Journal of Latin American Geography* 9 (2): 7–26.

Farley, K. A., W. G. Anderson, L. L. Bremer, and C. P. Harden. 2011. Compensation for ecosystem services: An evaluation of efforts to achieve conservation and development in Ecuadorian páramo grasslands. *Environmental Conservation* 38 (4): 1–13.

Farley, K. A., L. L. Bremer, C. P. Harden, and J. Hartsig. 2013. Changes in carbon storage under alternative land uses in páramo grasslands: Implications for payment for ecosystem services. *Conservation Letters* 6:21–27.

Farley, K. A., E. F. Kelly, and R. G. M. Hofstede. 2004. Soil organic carbon and water retention following conversion of grasslands to pine plantations in the Ecuadorian Andes. *Ecosystems* 7:729–39.

Goldman-Benner, R. L., S. Benitez, T. Boucher, A. Calvache, G. Daily, P. Kareiva, T. Kroeger, and A. Ramos. 2012. Water funds and payments for ecosystem services: Practice learns from theory and theory can learn from practice. *Oryx* 46 (1): 55–63.

Harden, C. P. 2006. Human impacts on headwater fluvial systems in the northern and central Andes. *Geomorphology* 79:249–63.

Harden, C. P., J. Hartsig, K. A. Farley, J. Lee, and L. L. Bremer. 2013. Effects of land-use change on water in Andean páramo grassland soils. *Annals of the Association of American Geographers* 103 (2): 375–84.

Hartsig, J. 2011. The effects of land-use change on the hydrological properties of Andisols in the Ecuadorian páramo. Master's thesis, Department of Geography, University of Tennessee, Knoxville.

Jantz, N., and H. Behling. 2012. A Holocene environmental record reflecting vegetation, climate, and fire variability at the Páramo of Quimsacocha, southwestern Ecuadorian Andes. *Vegetation History and Archaeobotany* 21:169–85.

Luteyn, J. L. 1992. Páramos: Why study them. In *Páramo: An Andean ecosystem under human influence*, ed. H. Balslev and J. L. Luteyn, 1–15. London: Academic.

Mach, M. E., R. G. Martone, and K. M. A. Chan. 2015. Human impacts and ecosystem services: Insufficient research for trade-off evaluation. *Ecosystem Services* 16:112–20.

Mark, B. G., J. Bury, J. M. McKenzie, A. French, and M. Baraer. 2010. Climate change and tropical Andean glacier recession: Evaluating hydrologic changes and livelihood vulnerability in the Cordillera Blanca, Peru. *Annals of the Association of American Geographers* 100 (4): 794–805.

Marston, R. A. 2008. Land, life, and environmental change in mountains. *Annals of the Association of American Geographers* 98 (3): 507–20.

Mena, P. V., G. Medina, and R. Hofstede. 2001. *Los páramos del Ecuador* [The páramos of Ecuador]. Quito, Ecuador: Abya-Yala.

Millennium Ecosystem Assessment (MEA). 2005. *Ecosystems and human well-being: Synthesis*. Washington, DC: Island.

Muradian, R., E. Corbera, U. Pascual, N. Kosoy, and P. H. May. 2010. Reconciling theory and practice: An alternative conceptual framework for understanding payments for environmental services. *Ecological Economics* 69:1202–08.

Naeem, S., J. C. Ingram, A. Varga, T. Agardy, P. Barten, G. Bennett, E. Bloomgarden, et al. 2015. Get the science right when paying for nature's services. *Science* 347 (6227): 1206–07.

Ponette-González, A. G., K. A. Brauman, E. Marín-Spiotta, K. A. Farley, K. C. Weathers, L. M. Curran, and K. R. Young. 2014. Managing water services in tropical regions: From land cover proxies to hydrologic fluxes. *Ambio* 44 (5): 367–75.

Pretty, J., and D. Smith. 2004. Social capital in biodiversity conservation and management. *Conservation Biology* 18 (3): 631–38.

Raymond, C. M., B. A. Bryan, D. H. MacDonald, A. Cast, S. Strathearn, A. Grandgirard, and T. Kalivas. 2009. Mapping community values for natural capital and ecosystem services. *Ecological Economics* 68:1301–15.

Sandhu, H., and S. Sandhu. 2014. Linking ecosystem services with the constituents of human well-being for poverty alleviation in eastern Himalayas. *Ecological Economics* 107:65–75.

Sklenar, P., and P. M. Ramsay. 2001. Diversity of zonal paramo plant communities in Ecuador. *Diversity and Distributions* 7:113–24.

Van Wesenbeeck, B. K., T. Van Mourik, J. F. Duivenvoorden, and A. M. Cleef. 2003. Strong effects of a plantation with Pinus patula on Andean subparamo vegetation: A case study from Colombia. *Biological Conservation* 114:207–18.

White, S. 2013. Grass páramo as hunter-gatherer landscape. *The Holocene* 23:898–915.

Natural Hazard Management from a Coevolutionary Perspective: Exposure and Policy Response in the European Alps

Sven Fuchs ⓘ, Veronika Röthlisberger ⓘ, Thomas Thaler ⓘ, Andreas Zischg ⓘ, and Margreth Keiler ⓘ

A coevolutionary perspective is adopted to understand the dynamics of exposure to mountain hazards in the European Alps. A spatially explicit, object-based temporal assessment of elements at risk to mountain hazards (river floods, torrential floods, and debris flows) in Austria and Switzerland is presented for the period from 1919 to 2012. The assessment is based on two different data sets: (1) hazard information adhering to legally binding land use planning restrictions and (2) information on building types combined from different national-level spatial data. We discuss these transdisciplinary dynamics and focus on economic, social, and institutional interdependencies and interactions between human and physical systems. Exposure changes in response to multiple drivers, including population growth and land use conflicts. The results show that whereas some regional assets are associated with a strong increase in exposure to hazards, others are characterized by a below-average level of exposure. The spatiotemporal results indicate relatively stable hot spots in the European Alps. These results coincide with the topography of the countries and with the respective range of economic activities and political settings. Furthermore, the differences between management approaches as a result of multiple institutional settings are discussed. A coevolutionary framework widens the explanatory power of multiple drivers to changes in exposure and risk and supports a shift from structural, security-based policies toward an integrated, risk-based natural hazard management system.

本文採用共同演化的观点来理解欧洲阿尔卑斯地区暴露于山区灾害的动态。本文呈现奥地利和瑞士在 1919 年至 2012 年间,对山区灾害 (洪泛、山洪暴发与泥石流) 而言具有风险元素之特定空间且基于对象的时间评估。该评估是根据下列两组不同的数据集: (1) 遵循具法律约束力的土地使用规划限制之灾害信息, 以及 (2) 从不同的国家层级空间数据组合而成的建筑形态信息。我们探讨这些跨领域动态,并聚焦经济、社会与制度间的相互依赖, 以及人类和物理系统的互动。曝险度在回应包括人口成长及土地使用冲突等多重驱力时有所改变。研究结果显示,当若干区域资产与灾害曝险度的显着增加有关时, 其他区域则以低于平均的曝险度为特征。空间与时间的结果, 显示出欧洲阿尔卑斯地区热点的相对稳定性。这些研究与各国家的地志学, 以及各别的经济活动范围与政治环境相符。此外, 本文探讨因多重制度环境所导致的管理方法差异。共同演化架构, 扩张了多重驱力之于曝险度和风险的改变的解释力, 并支持从结构性、以安全为基础的政策转变为整合性的、以风险为基础的自然灾害管理系统。 关键词: 共同演化, 欧洲阿尔卑斯地区, 曝险, 自然灾害管理, 路径依赖。

Se adopta una perspectiva co-evolucionista para entender la dinámica de la exposición a los riesgos de montaña en los Alpes europeos. Se presenta una evaluación temporal espacialmente explícita y basada en objeto de los elementos de riesgo en catástrofes de montaña (inundaciones fluviales, inundaciones torrenciales y flujos de detritos) en Austria y Suiza, para el período de 1919 a 2012. La evaluación descansa en dos conjuntos de datos diferentes: (1) información de riesgos que adhiere a las restricciones de planificación de uso del suelo legalmente obligatorias, y (2) información combinada sobre tipos de construcciones desde diferentes fuentes de datos espaciales a nivel nacional. Discutimos estas dinámicas transdisciplinarias y nos enfocamos en interdependencias e interacciones económicas, sociales e institucionales entre sistemas humanos y físicos. La exposición cambia en respuesta a múltiples controles, incluyendo crecimiento de la población y conflictos por usos del suelo. Los resultados muestran que mientras algunas ventajas regionales están asociadas con un fuerte incremento en exposición a los riesgos, otras están caracterizadas por un nivel de exposición por debajo del promedio. Los resultados espaciotemporales indican puntos calientes relativamente estables en los Alpes europeos. Estos resultados coinciden con la topografía de los países y con el respectivo ámbito de actividades económicas y el contexto político. Adicionalmente, se discuten las diferencias entre los enfoques de administración como resultado de múltiples escenarios institucionales. Un marco co-evolucionario amplía el poder explicativo de múltiples controles a los cambios en exposición y riesgo, y soporta un cambio de políticas estructurales, basadas en seguridad, hacia un sistema integrado de manejo de catástrofes naturales basado en riesgo.

In Europe, approximately 40 percent of the total land area is mountainous and is home to almost 20 percent of the total population (Nordregio 2004). Consequently, mountain regions are characterized by a significant number of settlements and economic and recreational areas. Only about 17 percent of the European Alps is suitable for permanent settlement due to topographic constraints, however (Tappeiner, Borsdorf, and Tasser 2008). As a result, mountain region developments are inherently linked to natural hazard risk, as land development occurs in hazard-prone areas where many settlements are located on alluvial fans and in floodplains. Flood risk management differs remarkably between floodplains along large rivers (e.g., the Rhine in Europe or the Mississippi in the United States) and the floodplains of alpine rivers. Whereas large rivers are predominantly managed with flood retention and levee constructions (Remo, Carlson, and Pinter 2012; Theiling and Burant 2013), mountainous areas are primarily managed by restricting the development of settlements in floodplains. Consequently, spatiotemporal exposure and the vulnerability of elements at risk plays a dominant role in risk management.

The main drivers of natural hazard risk are high reliefs, hydroclimatology, and the effects of climate dynamics on hydrological hazards (Keiler, Knight, and Harrison 2010). Hydrological hazards constitute a major threat to communities and assets, even though they occur episodically (Fuchs et al. 2013), especially if exposure and vulnerability are not properly managed (Zimmermann and Keiler 2015). These two aspects have only received scientific attention relatively recently (Papathoma-Köhle et al. 2011; Totschnig and Fuchs 2013; Fuchs, Keiler, and Zischg 2015; Papathoma-Köhle et al. 2015), whereas the overall concept of risk that combines hazard, exposure, and vulnerability had already been introduced in operational risk management for decades (Keiler et al. 2004; Kienholz et al. 2004).

Despite the considerable efforts to reduce mountain hazard risk, particularly with the implementation of technical means such as levees and retention basins (Holub and Fuchs 2009), the losses due to hydrological hazards in Europe remain significant (Andres, Badoux, and Hegg 2015; Fuchs, Keiler, and Zischg 2015). Although there is some evidence of increasing losses, which can be found in the publications of large reinsurers (Munich Re 2016; Swiss Re 2016), some scholars stated that underlying trends should be carefully interpreted. Mudelsee et al. (2003) analyzed flood magnitudes and concluded that there is no evidence of recent upward trends describing the occurrence of large flood events in central Europe. Similarly, Barredo (2009) reported no clear positive trend in flood losses in Europe once the losses are normalized by socioeconomic development indicators. Furthermore, when flood data in the United States are presented in terms of damage per unit wealth, a slight and statistically insignificant downward trend is observed (Loucks and Stedinger 2007).

Besides hazard dynamics (i.e., changes in the natural frequency and magnitude of events due to climatic change), shifts in hazard losses could result from (1) changing exposure of elements at risk due to overall population migration and associated land development, (2) changing vulnerability due to the presence or absence of technical mitigation measures, and (3) a greater awareness of threats considered in land use planning. In the past, spatially explicit data on elements at risk in Europe were fragmentary; a spatiotemporal assessment of exposure was limited to studies using large-scale, aggregated data (Keiler 2004; Keiler et al. 2006; Fuchs et al. 2013) and neglected any small-scale but supraregional dynamics. Spatially inclusive and comprehensive analyses on national levels were undertaken, for example, on flood risk in The Netherlands (Jongman et al. 2014) and on mountain hazards in Austria (Fuchs, Keiler, and Zischg 2015) when such data became recently available. In the following review, we focus on residential buildings (RBs) exposed to flood hazards in The European Alps, and we show how such data can be used to improve our understanding of hazard exposure and how a coevolutionary framework widens the explanatory power of multiple drivers in exposure dynamics. The coevolutionary framework provides a guideline for analyzing and explaining the linkage between exposure and policy.

Assessing Coevolution in Natural Hazard Management

We attempt to address challenges attributed to institutional changes in natural hazard management by focusing on the exposure of RBs in the European Alps from a coevolutionary perspective. Coevolution includes two or more interdependently evolving systems (Gual and Norgaard 2010). The aim is to analyze and understand the coevolutionary changes within the different interacting systems, where coevolutionary dynamics are path dependent (Kallis 2007). These dynamics include social adaptation to environmental change. A central theme inherent to coevolutionary thinking in social science is the analysis of institutional changes, especially with respect to the development of human behavior. Institutions are defined as a constant (formally legal and informally social) norm over a certain period of time (van den Bergh and Stagl 2003). Institutions are responsible for the organization of structures to optimize for social and economic behaviors (e.g., by minimizing uncertainty). Therefore, institutions have a direct influence on individuals and vice versa. Institutions influence the behavior of individuals (top down); their behavior and habitat are also key drivers for the development of new institutions or institutional changes (bottom up; Hodgson 2006). In summary, institutions define rules or procedures that support decision-making processes.

The aim is to interpret and to holistically explain exposure evolution in the European Alps with respect to policy responses and technological developments. The insights then support the valuation of natural hazard management policies. We identified two evolutionary systems:

- The first system is characterized by population pressures (i.e., demand for increased residences in hazard areas) associated with different behaviors, norms, beliefs, and physical attributes. Over time, the behavior and attributes of the populations in Austria and Switzerland changed. This is exemplified by the increase of single households compared to multihouseholds starting in the 1960s or gradually more numerous requests for secondary residences over the last 100 years (Statistik Austria 2004). Changes were based on socioeconomic developments within the society and external drivers (e.g., influx of homeowners from abroad). Furthermore, societal attributes change (e.g., new designs and uses for RBs, the number of inhabitants from 16.57 people per RB in 1919 to 4.59 people in 2012).

- The second evolutionary system involves changes in natural hazard management policy. For decades since the 1890s, the focus was on the implementation of structural engineering measures (Holub and Fuchs 2009). From the 1970s onward, nonstructural measures (e.g., land use planning) supplemented these engineered structures. Over time, however, key strategies in natural hazard management were incapable of sufficiently addressing the magnitude of associated losses. Institutions and respective policymakers currently rely on a combination of structural and nonstructural measures to reduce natural hazard risk in the European Alps (Fuchs 2009) and beyond (Kubal et al. 2009). There is an evident shift in natural hazard discourse away from exclusively engineered solutions toward broader integrated management strategies. These include land use management and other incentives to discourage developments in high-risk areas (Fuchs 2009). Consequently, this shift has been identified as a key point of contention in policy discussions, especially toward the implementation of nonstructural measures (Wiering and Immink 2006). This was triggered by crises such as the Galtür avalanche event in 1999 (Keiler 2004) and flood events in 2002 and 2005 (Bard, Renard, and Lang 2012). These catastrophic events provide new opportunities for actors from all administrative levels to introduce new management systems. Despite these shifts, natural hazard management still predominantly considers the use of structural measures (Thaler, Priest, and Fuchs 2016). Additionally, the implementation of structural mitigation measures has encouraged increases in the number of buildings in hazard areas.

Assessing Flood Hazard Exposure

Two different data sets were used for this study. Information on flood hazards provided input for the exposure assessment, in addition to data on building inventory in Austria and Switzerland (see Figure 1). Hazards such as river and torrential flooding (i.e., dynamic flooding with sediment transport and debris flows) in mountain rivers were assessed.

Available hazard maps were combined with nation-wide flood modeling results (see supplementary materials) to obtain spatial information on flood hazards. We defined a low- to medium-probability event as a source for the exposure assessment, in accordance with the requirements of the European Union Floods Directive (Commission of the European Communities 2007).

For the building exposure assessment, information on RBs was computed according to Fuchs, Keiler, and Zischg (2015), using specified information related to the entire building inventory. This information is available in a governmental database and contains details about the location and size of each building, the building category, and the year and period of construction (Bundesamt für Statistik 2012; Statistik Austria 2012).

Exposed buildings are defined as built structures that are susceptible to hydrological hazards. The hazard information was overlaid with building inventory data in a geographic information system. Each building was characterized by its main use, which was assessed by the net area of used space allotted for the different purposes of each floor.

Results

Analysis of Exposure Evolution

An overview on the number of RBs is provided in Table 1. A total of 3,574,198 RBs is located in Austria and Switzerland, of which 14.14 percent are exposed to hydrological hazards. The percentage exposed is slightly higher in Switzerland than in Austria. Almost two thirds (62.6 percent) of these buildings are single-family houses (SFHs), and slightly more than one third (37.4 percent) are apartment buildings (ABs). Between 1919 and 2012, the overall share of exposed RBs dropped around 2 percent, whereas the absolute number increased by a factor of 5. Similarly, the overall share of exposed SFHs dropped by around 2 percent, but the overall number of exposed SFHs increased by a factor of 5.6. Finally, the overall share of exposed ABs dropped by around 1.5 percent, but the overall number of exposed ABs increased by a factor of 4.2. In Switzerland, the exposure is generally slightly higher than in Austria.

The temporal development of the total RB stock is shown in Figure 2. Starting with an almost similar number of RBs in 1919 (Austria, 312,962; Switzerland, 307,751), the increase until 2012 was considerably higher in Austria (1,984,475) than in Switzerland (1,589,723). This increase followed a similar shape until 1960; thereafter, the increase was steeper in Austria than in Switzerland. A comparable pattern is observed for SFHs, starting

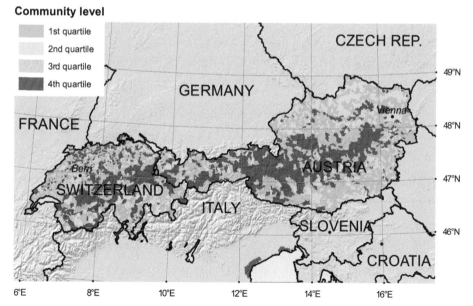

Figure 1. Exposure rate of residential buildings to hydrological hazards in Austria and Switzerland (exposed buildings to all buildings within a local authority, shown in terms of quartiles). (Color figure available online.)

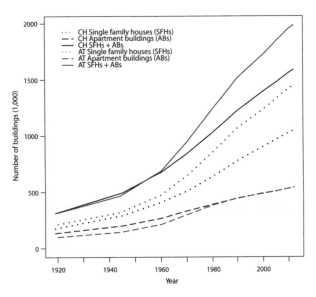

Figure 2. Absolute number of residential buildings in Austria and Switzerland (total number, single-family houses and apartment buildings) between 1919 and 2012. CH = Switzerland; AT = Austria; SFH= single-family house; AB = apartment building. (Color figure available online.)

from a total of 211,586 (Austria) and 173,309 (Switzerland), with a steeper increase in Austria than in Switzerland after 1960 and reaching totals of 1,447,144 (Austria) and 1,048,217 (Switzerland). In contrast, in 1919, there was a higher number of ABs in Switzerland (134,442) than in Austria (101,376). This number increased to almost the same amount for both countries (537,331 in Austria and 541,506 in Switzerland) in 2012.

Starting in 1919, there was a lower number of exposed RBs in Austria (42,219) than in Switzerland (58,446). These numbers increased until 2012, where the total numbers were higher in Austria (267,759) than in Switzerland (237,454). Similarly, the number of exposed SFHs increased from 26,473 (Austria) and

29,371 (Switzerland) to 179,257 (Austria) and 137,129 (Switzerland) between 1919 and 2012. In 1919, the number of exposed ABs started as a moderate amount in both countries (15,746 in Austria and 29,075 in Switzerland), which increased to 88,502 (Austria) and 100,325 (Switzerland) in 2012 (see Table 1).

Spatial analysis of the data reveals that hydrological hazards are an evident threat to municipalities, even if considerable differences between regions exist (Figure 1). In general, the exposure to hydrological hazards is defined as the share of exposed RBs to all existing RBs within a municipality. Exposure is low (first quartile) in communities located in the northern and southern alpine foreland and high (fourth quartile) in municipalities located in the high mountain areas around the main divide. The large river courses (Rhone, Aare, Rhine, Danube, and Mur) coincide with the higher levels of exposure in municipalities situated along these features. Moreover, some regions in the Central Alps are associated with low exposure values, even though there are above-average numbers of hazard events (Fuchs, Keiler, and Zischg 2015). This observation can be partially explained by a rigorous regional spatial planning policy (Thaler 2014; Thaler, Priest, and Fuchs 2016) and is discussed in the following section.

The temporal analysis reveals distinct differences between the Eastern and Western Alps. As shown in Figure 3, the share of exposed SFHs (number and value) compared to the entire number of SFHs decreased from 16.95 percent to 13.08 percent in Switzerland but was more or less constant in Austria (from 12.51 percent in 1919 to 12.39 percent in 2012). Hence, although the absolute number of exposed SFHs is higher in Austria than in Switzerland (Table 1), the relative distribution is reversed. The

Table 1. Overview of residential buildings in Austria and Switzerland

	Total RB N	Total RB exposed		Total SFH N	Total SFH exposed		Total AB N	Total AB exposed	
		N	%		N	%		N	%
CH 1919	307,751	58,446	18.99	173,309	29,371	16.95	134,442	29,075	21.63
CH 2012	1,589,723	237,454	14.94	1,048,217	137,129	13.08	541,506	100,325	18.53
AT 1919	312,962	42,219	13.49	211,586	26,473	12.51	101,376	15,746	15.53
AT 2012	1,984,475	267,759	13.49	1,447,144	179,257	12.39	537,331	88,502	16.47
CH + AT 1919	620,713	100,665	16.22	384,895	55,844	14.51	235,818	44,821	19.01
CH + AT 2012	3,574,198	505,213	14.14	2,495,361	316,386	12.68	1,078,837	188,827	17.50

Note. RB = residential buildings; SFH = single-family house; AB = apartment building; CH = Switzerland; AT = Austria.

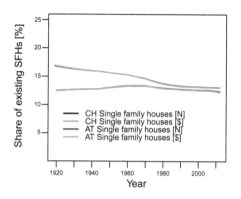
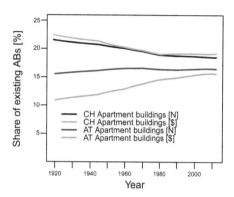

Figure 3. Share of exposed single-family houses (left) and apartment buildings (right) in Austria and Switzerland, relative to the total number of SFHs and ABs per country. The share of existing SFHs (left) is essentially identical in number and value for both Austria and Switzerland. This effect gives the appearance of only two graph lines when in fact there are four. CH = Switzerland; AT = Austria; SFH = single-family house; AB = apartment building. (Color figure available online.)

temporal ABs pattern is comparable to the one for SFHs and shows a slight decrease in the number of exposed ABs from 21.63 percent to 18.53 percent in Switzerland, with a similar progression for the values exposed. The Austrian data, in contrast, show a slightly increasing trend for the relationship between exposed ABs and the total ABs (from 15.53 percent in 1919 to 16.47 percent in 2012; the highest value is 16.62 percent in 1970) and a strong increase in the values. Hence, even if the increase in exposed buildings for the 1919 to 2012 period is lower in Switzerland than in Austria (factor of 4.67 vs. 6.77 for SFHs, 3.45 vs. 5.62 for ABs), the relative share of exposed SFHs and ABs remains higher in Switzerland than in Austria over 1919 to 2012. If the entire population is

considered, the share of RBs (number and value) slightly decreased from 1919 to 2012 (Figure 4), with a higher rate of decrease during the 1970s. If values and numbers are compared, the exposed SFHs and ABs were becoming more expensive since the 1970s.

Analysis of Policy Response in Natural Hazard Management

Strategies to prevent or to reduce the effects of natural hazards in settlement areas can be traced back to medieval times; official authorities were only founded in 1876 (Switzerland) and 1884 (Austria) as a result of legal regulation (Schweizerische Eidgenossenschaft 1876; Österreichisch-Ungarische Monarchie 1884). Since then, efforts to minimize detrimental impacts to civilians and society have been centered on silvicultural measures to prevent erosion and the introduction of engineering structures within the catchment, along channel systems, and in deposition areas. Starting in the 1950s, conventional mitigation concepts, which were aimed at decreasing both the magnitude and frequency of events, were increasingly complemented with technical mitigation measures. The amendment of respective legal regulations marks a turning point in responsibility sharing. Changes were observed in the following examples: the Hydraulic Engineering Assistance Act (Republik Österreich 1848), the Water Act (Republik Österreich 1959), the Disaster Act (Republik Österreich 1966), and the Forest Act (Republik Österreich 1975) in Austria and the Water Act and the Forest Act in Switzerland (Schweizerische Eidgenossenschaft 1991a, 1991b). As a result of these regulations, which were supplemented by multiple

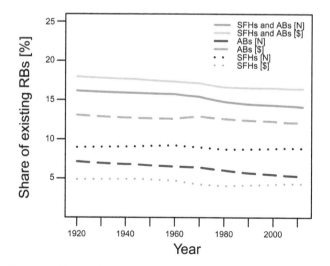

Figure 4. Share of exposed single-family houses and apartment buildings in relation to the total number of residential buildings. SFH = single-family house; AB = apartment building; RB = residential building. (Color figure available online.)

federal directives, protection against natural hazards became a governmental duty. Starting in the 1970s, with the Directive on Hazard Mapping in Austria (Republik Österreich 1976) and the National Spatial Planning Act in Switzerland (Schweizerische Eidgenossenschaft 1979), the use of nonstructural measures for natural hazard protection was implemented. As a result, spatial planning methods such as hazard maps aimed at reducing development activities in hazard-prone areas were introduced (Holub and Fuchs 2009). Multiple directives, such as the Directive on the Assessment of Flood Hazards in Spatial Planning (Bundesamt für Wasserwirtschaft, Bundesamt für Raumplanung, and Bundesamt für Umwelt, Wald und Landschaft 1997) in Switzerland and the Disaster Management Act (Republik Österreich 1996) in Austria supplemented these national laws, and further federal regulations from both countries were set into motion (Kanonier 2006). The European Union Floods Directive (Commission of the European Communities 2007) finally provided the basis for a risk-based management of flood hazards in European countries. In summary, we identified four key periods of natural hazard management in the European Alps, which are attributed to different hazard paradigms:

- In the 1870s and 1880s, a governmental system for natural hazard protection was introduced. The initial legal regulations that were focused on natural hazard management shifted to watershed management, forest-biological, soil bio-engineering measures, and technical measures (construction material: timber and stone masonry) for the first time.
- In the 1950s and 1960s, a shift toward engineering systems was observed. In the European Alps, the mitigation of mountain hazards was predicated on the implementation of structural engineering measures. These targeted the minimization of both the magnitude and frequency of events, which were increasingly complemented by more sophisticated technical mitigation measures.
- In the 1970s and 1980s, the system evolved to include a broader discussion on natural hazard management based on respective national laws. These laws served as responses to various natural hazard events. As a result, nonstructural measures supplemented engineering solutions. In particular, land use planning was introduced. Institutions and policymakers relied on a combination of structural and nonstructural measures to reduce the negative impact of future events.
- Finally, the risk-based approach was introduced in the 1990s. The shift from hazard to risk required a completely different approach to effectively address outstanding management issues. Here, the concept of risk is defined as a function of hazard and consequences. Comprehensive experiences have been documented about the application of the risk concept to mountain hazard management, especially in Switzerland. The risk-based approach was focused on encouraging a discourse on risk within respective societies. By considering different scenarios (including the aspect of residual hazard), a greater focus is placed on stakeholder engagement and bottom-up initiatives, and the implementation of catchment-wide management concepts was observed.

Discussion and Conclusion

The aforementioned results clearly showed that effective exposure reduction has yet to be achieved. In fact, the evolution of new policy instruments in flood risk management has largely been unable to reduce increases in exposure for both countries. We observed that the new flood risk management strategies allowed continuous developments in floodplain areas. In particular, the expectation that engineered measures would protect floodplains had encouraged development instead. This resulted in increases in potential losses (White, Kates, and Burton 2001). In response, the public administration in Austria and Switzerland generated a situation of moral hazard within society (Tarlock 2012), because new buildings in hazard areas were secured by innovative defense strategies. For example, the Austrian housing subsidy system, which changed in 1958, led to an increase in public subsidies that are available for private house owners. As such, the total number of new RBs increased to 1,296,101 between 1960 and 2012, compared to the 375,412 new RBs between 1919 and 1960. Consequently, the availability of housing subsidies contributed to the development of SFHs; during the period of investigation, more than 115,687 were constructed in hazard-prone areas with financial support from the government. This effect is referred to as *perverse subsidies* (van Beers and Van Den Bergh 2001). Furthermore, the Austria Superior Administrative Court decided against enforcing the production of hazard maps as a part of statutory regulation for spatial planning; instead they are only judged as an expert report (Verwaltungsgerichtshof 1995).

Because an absolute decrease in exposed RBs would only be possible if the original buildings are removed from identified hazard zones, we computed the hypothetical development of the buildings based on a scenario where a construction ban is enforced in endangered areas. To demonstrate the effectiveness of such a ban, associated legal regulations were assumed to be effective in the 1970s and the 1990s, respectively. If we hypothetically assume that starting in 1976, which coincides with the amendment to the Directive on Hazard Mapping in Austria, further development in hazard-prone areas would have been stopped, a total of 162,907 buildings would not have been constructed in exposed areas. This number equates to −32 percent of exposure. Similarly, after 1991, which coincides with the amendment of the Water Act and the Forest Act in Switzerland, a total of 102,935 buildings would not have been constructed in hazard-prone zones. This is equivalent to −20 percent exposure. Consideration for these scenarios also demonstrates the importance of time when investigating the effectiveness of nonstructural measures. For instance, it recommends that land use planning policies should be consistently implemented over longer temporal horizons.

The policy system encourages private homeownership, despite associated increases in vulnerability. Another driver that contributed to increased exposure was the interpretation of land use management regulations at local levels. In some of the regions belonging to the fourth quartile of exposure (Figure 1), the regional land use management act allowed the construction of houses outside of defined building zones in land use plans; consequently, around 7,000 out of a total of 12,000 new residential buildings were constructed due to this exemption. Furthermore, governmental organizations interpreted how land was protected by engineering structures differently. In Salzburg, for example, new buildings and settlements were built to create dense urban areas, resulting in an increase in exposure in high-risk areas. Moreover, the public administration seemed to have ignored the problem of exposure, as natural hazard management had little or no impact on the design of local land use plans and strategies. This is explained by the fact that economic growth within administrative boundaries is regularly prioritized above ecological concerns or

protection against hazards (Thaler 2014; Thaler, Priest, and Fuchs 2016).

The observations show that the exposure of RBs has considerably increased over the last ninety years. This rise has been observed despite the introduction of natural hazard management strategies in the European Alps. This development was heavily influenced by the occurrence of disasters, which led to an increased gravitation toward the dependency on technical mitigation measures but did not prevent further unsuitable land use developments. Moreover, acknowledging the levee effect, natural hazard management encouraged further development of hazard-prone areas with the consequence of an increase of exposure dynamics. As such, both systems (exposure dynamics and management paradigms) are profoundly interrelated, where increases in exposure necessitate further mitigation measures. These measures evolve from purely engineered solutions toward risk-based planning approaches. The implementation of key strategies in isolation, however, does not completely eliminate potential losses due to damages over time. Instead, a lock-in situation results, where the reliance on technical mitigation measures that dominate current risk management approaches continues to be more prominent than the perceived impact of land use planning.

To break away from the way exposure has been addressed to date, there is a need to set incentives to ensure responsible natural hazard management. This requires rigorous enforcement of land use planning legislation (e.g., reconsideration of *perverse subsidies* from a political perspective), which would foster the popularization of alternative hazard mitigation measures and promote the implementation of coherent policies. It would also support the development of further incentives to minimize risk. Natural hazard risk management will only be successful if the further development of construction in hazard-prone areas is restricted.

The aforementioned management approaches ensure the availability and accessibility of knowledge on natural hazard risk and how this can be effectively applied to a range of societal conditions. The result would be a paradigm shift in natural hazards management, which would result in decreased vulnerability and increased resilience for the affected population.

Funding

This study received funding from the Austrian Science Fund (FWF): P27400.

Supplemental Material

Supplemental data for this article can be accessed on the publisher's Web site at http://dx.doi.org/10.1080/24694452.2016.1235494.

ORCID

Sven Fuchs (iD) http://orcid.org/0000-0002-0644-2876

Veronika Röthlisberger (iD) http://orcid.org/0000-0003-1911-6268

Thomas Thaler (iD) http://orcid.org/0000-0003-3869-3722

Andreas Zischg (iD) http://orcid.org/0000-0002-4749-7670

Margreth Keiler (iD) http://orcid.org/0000-0001-9168-023X

References

Andres, N., A. Badoux, and C. Hegg. 2015. Unwetterschäden in der Schweiz im Jahre 2014 [Swiss hazard losses in 2014]. *Wasser, Energie, Luft* 107 (1): 47–54.

Bard, A., B. Renard, and M. Lang. 2012. Floods in the alpine areas of Europe. In *Changes in flood risk in Europe*, ed. Z. Kundzewicz, 362–71. Boca Raton, FL: CRC.

Barredo, J. 2009. Normalised flood losses in Europe: 1970–2006. *Natural Hazards and Earth System Sciences* 9 (1): 91–104.

Bundesamt für Statistik. 2012. *Eidgenössisches Gebäude-und Wohnungsregister, Merkmalskatalog* [Swiss federal building register, catalogue of characteristics]. Neuchâtel, Switzerland: Bundesamt für Statistik.

Bundesamt für Wasserwirtschaft, Bundesamt für Raumplanung, and Bundesamt für Umwelt, Wald und Landschaft. 1997. *Berücksichtigung der Hochwassergefahren bei raumwirksamen Tätigkeiten* [Acknowledging flood hazards in spatial planning]. Biel und Bern, Switzerland: BWW, BRP, BUWAL.

Commission of the European Communities. 2007. Directive 2007/60/EC of the European Parliament and of the Council of 23 October 2007 on the assessment and management of flood risks. *Official Journal of the European Union* 288:27–34.

Fuchs, S. 2009. Susceptibility versus resilience to mountain hazards in Austria—Paradigms of vulnerability revisited. *Natural Hazards and Earth System Sciences* 9 (2): 337–52.

Fuchs, S., M. Keiler, S. A. Sokratov, and A. Shnyparkov. 2013. Spatiotemporal dynamics: The need for an innovative approach in mountain hazard risk management. *Natural Hazards* 68 (3): 1217–41.

Fuchs, S., M. Keiler, and A. Zischg. 2015. A spatiotemporal multi-hazard exposure assessment based on property data. *Natural Hazards and Earth System Sciences* 15 (9): 2127–42.

Gual, M. A., and R. B. Norgaard. 2010. Bridging ecological and social systems coevolution: A review and proposal. *Ecological Economics* 69 (4): 707–17.

Hodgson, G. 2006. What are institutions? *Journal of Economic Issues* 40 (1): 1–25.

Holub, M., and S. Fuchs. 2009. Mitigating mountain hazards in Austria—Legislation, risk transfer, and awareness building. *Natural Hazards and Earth System Sciences* 9 (2): 523–37.

Jongman, B., E. E. Koks, T. G. Husby, and P. J. Ward. 2014. Increasing flood exposure in the Netherlands: Implications for risk financing. *Natural Hazards and Earth System Sciences* 14 (5): 1245–55.

Kallis, G. 2007. When is it coevolution? *Ecological Economics* 62 (1): 1–6.

Kanonier, A. 2006. Raumplanungsrechtliche Regelungen als Teil des Naturgefahrenmanagements [Land use planning regulations as part of natural hazard management]. In *Recht im Naturgefahrenmanagement*, ed. S. Fuchs, L. Khakzadeh, and K. Weber, 123–53. Innsbruck, Austria: Studienverlag.

Keiler, M. 2004. Development of the damage potential resulting from avalanche risk in the period 1950–2000, case study Galtür. *Natural Hazards and Earth System Sciences* 4 (2): 249–256.

Keiler, M., S. Fuchs, A. Zischg, and J. Stötter. 2004. The adaptation of technical risk analysis on natural hazards on a regional scale. *Zeitschrift für Geomorphologie* 135 (Suppl.): 95–110.

Keiler, M., J. Knight, and S. Harrison. 2010. Climate change and geomorphological hazards in the eastern European Alps. *Philosophical Transactions of the Royal Society of London A* 368: 2461–79.

Keiler, M., R. Sailer, P. Jörg, C. Weber, S. Fuchs, A. Zischg, and S. Sauermoser. 2006. Avalanche risk assessment—A multi-temporal approach, results from Galtür, Austria. *Natural Hazards and Earth System Sciences* 6 (4): 637–51.

Kienholz, H., B. Krummenacher, A. Kipfer, and S. Perret. 2004. Aspects of integral risk management in practice—Considerations with respect to mountain hazards in Switzerland. *Österreichische Wasser-und Abfallwirtschaft* 56 (3–4): 43–50.

Kubal, C., D. Haase, V. Meyer, and S. Scheuer. 2009. Integrated urban flood risk assessment—Adapting a multicriteria approach originally developed for a river basin to a city. *Natural Hazards and Earth System Sciences* 9 (6): 1881–95.

Loucks, D., and J. Stedinger. 2007. Thoughts on the economics of floodplain development in the U.S. In *Extreme hydrological events: New concepts for security*, ed. O. Vasiliev, P. van Gelder, E. Plate, and M. Bolgov, 3–19. Dordrecht, The Netherlands: Springer.

Mudelsee, M., M. Börngen, G. Tetzlaff, and U. Grünewald. 2003. No upward trends in the occurrence of extreme floods in central Europe. *Nature* 425 (6954): 166–69.

Munich Re. 2016. *Topics geo: Natural catastrophes 2015*. Munich, Germany: Munich Re.

Nordregio. 2004. Mountain areas in Europe: Analysis of mountain areas in EU member states, acceding and other European countries. Final report, Nordregio, Stockholm, Sweden.

Österreichisch-Ungarische Monarchie. 1884. *Gesetz, betreffend Vorkehrungen zur unschädlichen Ableitung von Gebirgswässern* [Law related to the nonhazardous discharge of mountain waters]. Wien, Austria: Kaiserlichkönigliche Hof-und Staatsdruckerei.

Papathoma-Köhle, M., M. Kappes, M. Keiler, and T. Glade. 2011. Physical vulnerability assessment for alpine hazards: State of the art and future needs. *Natural Hazards* 58 (2): 645–80.

Papathoma-Köhle, M., A. Zischg, S. Fuchs, T. Glade, and M. Keiler. 2015. Loss estimation for landslides in mountain areas—An integrated toolbox for vulnerability assessment and damage documentation. *Environmental Modelling and Software* 63:156–69.

Remo, J. W. F., M. Carlson, and N. Pinter. 2012. Hydraulic and flood-loss modeling of levee, floodplain, and river management strategies, Middle Mississippi River, USA. *Natural Hazards* 61 (2): 551–75.

Republik Österreich. 1948. Wasserbautenförderungsgesetz [Federal Hydraulic Engineering Development Act]. BGBl 34/1948.

———. 1959. Wasserrechtsgesetz [Federal Water Act]. BGBl 215/1959.

———. 1966. Bundesgesetz vom 9. September 1966 über den Katastrophenfonds [Federal Act of 9 September 1966 on the Disaster Fund]. BGBl. Nr. 207/1966.

———. 1975. Forstgesetz [Federal Forest Act]. BGBl 440/1975.

———. 1976. Verordnung des Bundesministers für Land- und Forstwirtschaft vom 30. Juli 1976 über die Gefahrenzonenpläne [Decree of the Federal Minister for Agriculture and Forestry of 30 July 1976 related to hazard mapping]. BGBl 436/1976.

———. 1966. Bundesgesetz vom 9. September 1966 über den Katastrophenfonds [Federal Act of 9 September 1966 on the Disaster Fund]. BGBl. Nr. 207/1966.

Schweizerische Eidgenossenschaft. 1876. Bundesgesetz betreffend die eidgenössische Oberaufsicht über die Forstpolizei [Federal law related to the confederate supervision of the Swiss Forest Police]. BS 9 521.

———. 1979. Bundesgesetz über die Raumplanung [Federal law on spatial planning]. SR 700.

———. 1991a. Bundesgesetz über den Wald [Federal Forest Act]. SR 921.0.

———. 1991b. Bundesgesetz über den Wasserbau [Federal Hydrography Act]. SR 721.100.

Statistik Austria. 2004. *Gebäude-und Wohnungszählung 2001—Hauptergebnisse Österreich* [Census of buildings and flats 2001—Main results for Austria]. Wien, Austria: Statistik Austria.

———. 2012. *Adress-GWR Online Handbuch, Teil C, Anhang 2: Merkmalskatalog.* Wien, Austria: Statistik Austria.

Swiss Re. 2016. *Natural catastrophes and man-made disasters in 2015.* Zurich: Swiss Re.

Tappeiner, U., A. Borsdorf, and E. Tasser. 2008. *Alpenatlas* [Atlas of the Alps]. Heidelberg, Germany: Springer.

Tarlock, A. D. 2012. United States flood control policy: The incomplete transition from the illusion of total protection to risk management. *Duke Environmental Law & Policy Forum* 23:151–83.

Thaler, T. 2014. Developing partnership approaches for flood risk management: Implementation of inter-local co-operations in Austria. *Water International* 39 (7): 1018–29.

Thaler, T., S. Priest, and S. Fuchs. 2016. Evolving interregional co-operation in flood risk management: Distances and types of partnership approaches in Austria. *Regional Environmental Change* 16 (3): 841–53.

Theiling, C. H., and J. T. Burant. 2013. Flood inundation mapping for integrated floodplain management: Upper Mississippi River system *River Research and Applications* 29:961–78.

Totschnig, R., and S. Fuchs. 2013. Mountain torrents: Quantifying vulnerability and assessing uncertainties. *Engineering Geology* 155:31–44.

van Beers, C., and J. C. J. M. Van Den Bergh. 2001. Perseverance of perverse subsidies and their impact on trade and environment. *Ecological Economics* 36 (3): 475–86.

van den Bergh, J. C. J. M., and S. Stagl. 2003. Coevolution of economic behaviour and institutions: Towards a theory of institutional change. *Journal of Evolutionary Economics* 13 (3): 289–317.

Verwaltungsgerichtshof. 1995. Erkenntnis 91/10/0090 [Decision 91/10/0090] (27.03.1995) https://www.ris.bka.gv.at/Dokumente/Vwgh/JWT_1991100090_19950327x00/JWT_1991100090_19950327x00.pdf (last accessed 1 December 2015).

White, G., R. Kates, and I. Burton. 2001. Knowing better and losing even more: The use of knowledge in hazards management. *Environmental Hazards* 3 (3–4): 81–92.

Wiering, M., and I. Immink. 2006. When water management meets spatial planning: A policy-arrangements perspective. *Environment and Planning C: Government and Policy* 24 (3): 423–38.

Zimmermann, M., and M. Keiler. 2015. International frameworks for disaster risk reduction: Useful guidance for sustainable mountain development? *Mountain Research and Development* 35 (2): 195–202.

Bringing the Hydrosocial Cycle into Climate Change Adaptation Planning: Lessons from Two Andean Mountain Water Towers

Megan Mills-Novoa, Sophia L. Borgias, Arica Crootof, Bhuwan Thapa, Rafael de Grenade, and Christopher A. Scott

Glaciers, snowpack, rivers, lakes, and wetlands in mountain regions provide freshwater for much of the world's population. These systems, however, are acutely sensitive to climate change. In Andean water towers, which supply freshwater to more than 100 million people, climate change adaptation planning is critical. Adaptation plans, however, are more than just documents; they inform and are informed by sociopolitical processes with major implications for hydrosocial relations in mountain water towers. Noting the inadequate scholarly attention to climate change in relation to the hydrosocial cycle, we draw on the hydrosocial literature to examine and compare climate change adaptation plans from mountain water tower regions of Piura, Peru, and the Santiago metropolitan region in Chile. Through a hydrosocial lens, we find that these plans reinforce hydrosocial relations such as upstream–downstream disparities that tend to exclude those who access water informally, have differing ontologies of water, or have livelihoods outside of dominant economic sectors. Our analysis suggests that the Andean plans reinforce current water access patterns, missing a key opportunity to reenvision more inclusive hydrosocial relationships in the context of a changing climate. This study encourages further engagement between the climate change adaptation and hydrosocial literature within and beyond mountain water tower regions. Critical hydrosocial analysis of adaptation plans reveals gaps that must be addressed in future planning and implementation efforts if adaptation is going to provide meaningful pathways for change.

山区的冰川, 积雪场, 河流, 湖泊以及溼地, 为全球为数众多的人口提供乾淨的水源。但这些系统却对气候变迁极其敏感。在提供超过一亿人口乾淨水源的安第斯水塔中, 气候变迁调适规划十分关键。但调适计画绝非仅只是纸上作业; 它们提供信息给社会政治过程并受其告知, 对于山区水塔的水文社会关系具有重要的意涵。我们指出学术界对气候变迁之于水文社会循环的关注不足, 并引用水文社会学文献, 检视并比较秘鲁皮乌拉山区水塔地区和智利圣地牙哥大都会区的气候变迁调适计画。我们透过水文社会学的视角, 发现这些计画增强了诸如倾向排除以非正式管道获取水资源, 具有不同的水本体论, 或是在支配性的经济部门之外生活的人的上下游不均水文社会关系。我们的分析指出, 安第斯计画强化了当前的水资源取得模式, 错失了在气候变迁的脉络中重新想像更具包容性的水文社会关系的关键契机。本研究鼓励对山区水塔区域之中和之外的气候变迁调适和水文社会学进一步进行研究。对调适计画的批判水文社会学分析, 揭露了若期待调适能够提供有意的改变途径的话, 未来的规划和执行所需应对的落差。 关键词: 调适规划, 气候变迁, 水文社会循环, 山岳, 水资源。

Los glaciares, el manto de nieve, los ríos, lagos y humedales de las regiones montañosas suministran agua fresca para una gran parte de la población del mundo. Sin embargo, estos sistemas son extremadamente sensibles al cambio climático. En las torres de agua andinas, que suministran agua potable a más de 100 millones de personas, planear la adaptación al cambio climático es una cuestión crítica. Los planes de adaptación, sin embargo, son algo más que simples documentos; ellos informan y son informados por procesos sociopolíticos con serias implicaciones para las relaciones hidrosociales en las torres de agua montañosas. Notando la inadecuada atención de los eruditos al cambio climático en relación con el ciclo hidrosocial, nos apoyamos en la literatura hidrosocial para examinar y comparar los planes de adaptación al cambio climático en las regiones de torres de agua montañosas de Piura, Perú, y los de la región metropolitana de Santiago, en Chile. A través de una lente hidrosocial, hallamos que estos planes refuerzan relaciones hidrosociales tales como las disparidades observadas río arriba o río abajo que tienden a excluir a quienes acceden al agua de manera informal, tienen diferentes ontologías sobre el agua, o que tienen medios de sustento por fuera de los sectores económicos dominantes. Nuestro análisis sugiere que los planes andinos refuerzas los actuales patrones de acceso al agua, dejando pasar una excelente oportunidad para reformular relaciones hidrosociales más incluyentes dentro de un contexto de cambio climático. Este estudio promueve compromisos de mayor vuelo entre la adaptación al cambio climático y la literatura hidrosocial dentro y más allá de las regiones de torres de agua montañosas. El análisis crítico hidrosocial de los planes de adaptación revela vacíos que deben tenerse en cuenta en la planificación futura y en los esfuerzos de implementación si se quiere que la adaptación provea rutas significativas de cambio.

Mountains are home to glaciers, snowpack, rivers, lakes, and wetlands that provide freshwater for much of the world's population. Because mountains provide a disproportionate amount of runoff to adjoining lowland areas, they are symbolically referred to as "mountain water towers" (Viviroli et al. 2007). These mountain water towers are experiencing faster than average warming (Pepin 2015), making them unique harbingers of climate change. Glacial retreat, decreasing snow cover, and variable precipitation affect the timing and volume of water resources (Beniston 2003), and these changes are rapidly unfolding in many of the world's mountain regions (McDowell, Stephenson, and Ford 2014).

Mountains, however, are more than headwater regions and canaries for climate change; they are home to more than 1.2 billion people worldwide (Körner and Ohsawa 2005). Highland communities, especially those with land-based livelihoods, are highly vulnerable to drought, flooding, and extreme events (Valdivia et al. 2010). Vulnerability, however, is context specific and shaped by interrelated sociocultural, political, and economic dynamics (Ribot 2011). The vulnerability of mountain communities is not solely determined by environmental change but can become exacerbated by it (Valdivia et al. 2010).

Dependence on mountain water resources by highland and lowland communities makes these mountain regions critical interfaces for adaptation—sites where upstream and downstream actions are intimately interconnected. Upstream and downstream users, however, have different needs and might exercise different levels of power over decision making (Budds 2009; Rodríguez-de-Francisco and Boelens 2016).

Climate change adaptation planning informs and is informed by hydrosocial relationships. These plans have major implications for the future distribution of water resources in mountain water towers. We consider water access a key analytic for understanding how climate change adaptation planning maintains or transforms patterns of water distribution. We define access as the ability to benefit from water resources (Ribot and Peluso 2003). We also consider how affected communities access adaptation planning.

In this research, we analyze adaptation plans for two mountain water tower regions in Peru and Chile (Figure 1). We begin by introducing the hydrosocial cycle and bring it into conversation with scholarship on adaptation planning assessment. We describe our comparative approach, discuss key themes arising from our analysis, and conclude by reflecting on the implications of our findings for new directions in climate change adaptation planning.

The Hydrosocial Cycle

Understanding the upstream–downstream dynamics of climate change adaptation in mountain water towers requires moving beyond hydrologic connections to hydrosocial relationships. The hydrologic cycle describes water's eternal physical journey from precipitation to infiltration to storage and evaporation. The hydrosocial cycle integrates the physical properties and processes of water with the social actors and institutions shaping these flows (Swyngedouw 2004; Linton and Budds 2014). Through the hydrosocial cycle, the sociopolitical and physical factors influencing the flow of water become inseparable. Water is not just shaped by society but is also embedded in and, in turn, shapes society (Swyngedouw 2004, 2009; Budds 2009; Boelens et al. 2016). The hydrosocial cycle is in a constant state of reconfiguration, prompted by social, physical, and technical interventions such as new water regulations, changes in water availability, and technological innovations (Linton and Budds 2014). Every intervention within the hydrosocial cycle shifts power relations and social structures (Swyngedouw 2009).

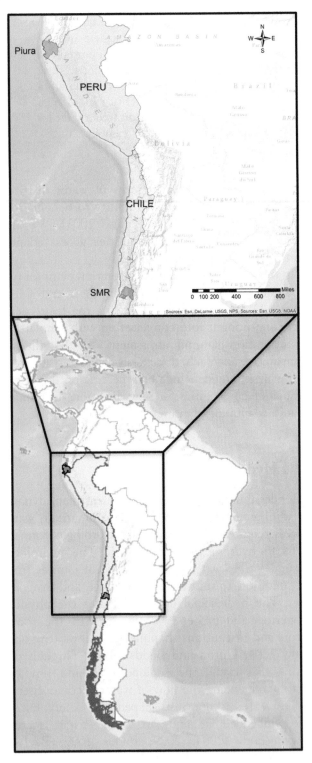

Figure 1. Location of water tower regions: Piura, Peru, and Santiago Metropolitan Region (SMR). (Color figure available online.)

Climate change and subsequent adaptation pose major interventions in the hydrosocial cycle that can fundamentally alter hydrosocial relationships. Yet there have been few articles that consider climate change in relation to the hydrosocial cycle (cf. Swyngedouw 2013). Whereas hydrosocial literature continues to evolve (cf. Linton and Budds [2014] on waterscapes and Boelens et al. [2016] on hydrosocial territories), climate change assemblages of physical and sociopolitical change remain largely unexplored.

We employ a hydrosocial lens to assess climate change adaptation plans. Adaptation planning has the potential to address issues of uneven resource access, but without critical analysis, it can reinforce or exacerbate existing social inequities (Eriksen, Nightingale, and Eakin 2015). The hydrosocial cycle provides a productive conceptual framework for this analysis, elucidating how power relations shape and are shaped by climate change adaptation planning.

Assessing Adaptation Planning

Climate change adaptation planning assessment literature has primarily focused on developed countries, particularly Australia, European nations, and the United States, where planning processes are structured and sustained (Biesbroek et al. 2010; Preston, Westaway, and Yuen 2011; Bizikova et al. 2014). This scholarship provides qualitative and quantitative assessment criteria but few that elucidate sociopolitical issues in planning processes, plan content, and implementation. Preston, Westaway, and Yuen (2011) compiled nineteen criteria from an extensive review of adaptation planning assessment studies. These included assessment of human and social capital, vulnerability and risk, and stakeholder engagement. They found that adaptation plans were overly "climate-centric" and often neglected institutional challenges to adaptation. We seek to enrich the criteria proposed in this literature by drawing on the hydrosocial cycle to better understand how these plans consider and address uneven resource access in mountain water towers.

Site Selection

This study analyzes and compares climate change adaptation plans from the Santiago metropolitan region in Chile (Clima Adaptación Santiago [CAS] 2012) and Piura, Peru (Gobierno Regional Piura [GRP] 2011; Table 1). These regions were selected because they (1) share similar demands on mountain-sourced water resources, (2) face escalating water conflicts, (3) are experiencing intensifying lowland drought under climate change, (4) are pilot regions for

Table 1. Comparative characteristics of the regions

	Santiago Metropolitan Region, Chile	Piura, Peru
Area (km^2)	15,403	29,853
Population	6.69 million (2010)	1.84 million (2013)
Elevation range (m)	343–6,000	0–3,644

national water governance strategies, and (5) have finalized regional climate change adaptation plans.

The Santiago metropolitan region (SMR) is home to more than 40 percent of Chile's population, largely concentrated in the capital city of Santiago. The Maipo River supplies the SMR with 90 percent of water for urban and residential use and 70 percent of the water for irrigation (Dirección de General de Aguas [DGA] 2015). In the headwater region, the river supports mountain communities reliant on smallholder agriculture and ecotourism that have come into conflict with hydropower and mining development (DGA 2015). Extended drought since 2010 has raised concerns about climate change (Centro de Ciencia del Clima y la Resiliencia 2015), and the SMR has hosted some of the first regional-level adaptation planning processes to prepare for increasing climatic variability (Barton, Krellenberg, and Harris 2015). The SMR Climate Change Adaptation Plan was developed from 2009 to 2012 and remains the only completed plan for the region.

Piura is the second most populous region in Peru. Within the Piura region, nearly 95 percent of all surface water, sourced from the Piura and Chira rivers, is used for irrigation (Instituto Nacional de Estadística e Informática [INEI] 2012b). Piura has been a frontier for agricultural expansion, with 1.9 million hectares of cultivated land up from 1.1 million in 1994 (INEI 1994, 2012a). Both the Chira River and the Piura River are fed by high-elevation neotropical wetlands called *páramos*, which have sociocultural importance to highland communities. These communities have fervently opposed mining projects that threaten their water access (Bebbington and Williams 2008). Climate change is projected to increase the water deficit in the region due to elevated temperatures and less frequent but more intense rain events (Autoridad Nacional de Agua [ANA] 2013). Piura was a pilot region for the implementation of the 2009 Water Resources Law and was prioritized for climate change adaptation

planning due to its vulnerability to El Niño (GRP 2011). The Piura plan was finalized in 2011.

Located on the western coast of the Andes, both Piura and the SMR are semiarid regions that rely on Andean water towers to meet growing lowland water demands. The SMR and Piura regions face similar climate change challenges and are models of adaptation planning in their respective countries, with rich potential for comparative analysis.

Methods

We employed qualitative content analysis (Julien 2008) to evaluate hydrosocial themes in the water sections of the SMR and Piura adaptation plans. Drawing on hydrosocial literature, we selected seven hydrosocial themes (Table 2) for coding (Ayres 2008). We conducted thematic coding of adaptation plans using NVivo qualitative analysis software (Version 10, QSR International Pty Ltd., Melbourne, Australia), examining and comparing the compiled passages for each theme across plans. Results were triangulated with contextual knowledge gained during fieldwork within the study regions in 2015 and 2016. Interviews with adaptation planning facilitators and participants captured additional details about the stakeholder engagement process, implementation, and lessons learned. Participant observation of water management practices and mountain livelihoods provided further understanding of how adaptation planning interfaces with broader hydrosocial relations.

Results and Discussion

A summary of the content analysis results can be found in Table 3. This analysis revealed several key findings.

Socioenvironmental and Cultural Values of Water Neglected

Water plays significant social, cultural, and environmental roles in the SMR and Piura regions, yet water is described primarily in the language of supply, demand, and economic efficiency. In the mountainous Maipo headwaters, in-stream uses of the river for ecotourism, conservation, subsistence, and pastoralism lack formal water rights and are often overlooked. These nontitled uses regularly collide with titled water use for mining and hydropower development (DGA 2015). Although

Table 2. Hydrosocial themes for climate change adaptation plan assessment

Themes	Questions	References
Value of water	How is the value of water discussed?	Linton and Budds (2014)
	Are socioenvironmental and cultural values acknowledged?	
Water access	Are gaps in water supply acknowledged?	Swyngedouw (2004)
	Do proposed measures address these gaps?	
Upstream measures	What is proposed? Who benefits?	Rodríguez-de-Francisco and Boelens (2016)
Downstream measures		
Participation	What stakeholders are included? When and in what capacity?	Preston, Westaway, and Yuen (2011); Bizikova et al. (2014)
	How does participation influence the outcomes?	
Upstream–downstream relationships	Are upstream and downstream connections identified?	Rodríguez-de-Francisco and Boelens (2016)
	Are certain areas privileged?	
Vulnerability	How is vulnerability identified and addressed?	Ribot (2011)

the SMR plan attends to the vulnerability of hydropower production to reduced stream flow and prioritizes highland areas for reforestation, it falls short of addressing the socioenvironmental value of in-stream water uses central to mountain livelihoods.

The Piura highlands are home to *comunidades campesinas*, who view the páramos that feed the Chira and Piura rivers as sacred sites for healing and purification. The shamanic ceremonies centered on these lakes reflect the spiritual link between water and highland communities (Recharte 2006). *Comunidades campesinas* have fiercely rejected two copper mining projects that threaten their access to these waters for cultural and livelihood practices (Bebbington and Williams 2008). The Piura adaptation plan omits the sociocultural values held by highland communities, depicting a hydrosocial cycle that does not engage with indigenous ontologies of water and portrays the headwaters as uninhabited natural spaces in need of reforestation.

Although both plans include upstream measures (i.e., riparian reforestation) aimed at preserving downstream water supply, these measures do not account for the socioenvironmental and cultural values of highland communities. This omission reinforces an existing hydrosocial order that does not acknowledge informal water users and people with differing ontologies of water.

Marginalized Hydrosocial Actors Absent from Participatory Processes

Both plans emphasize participatory processes, integrating diverse stakeholders through advisory councils and workshops. They primarily engaged government agency representatives; key players from the urban, agricultural, and industrial sectors; and a few high-profile civil society groups. Meanwhile, local community representatives (i.e., informal water users and *comunidades campesinas*) were not included in these processes.

The SMR plan was created out of a participatory process sustained from 2010 to 2012. This process was contracted by the German Ministry of Environment as a collaboration among German research institutes, Chilean universities, and regional government agencies. Ten roundtable discussions facilitated by university researchers aimed to build on existing institutional relationships. Participants (see Table 3) were offered the opportunity to provide feedback on several iterations of the plan. Highland representatives and community groups responsible for ensuring rural water access were largely absent, however, and their concerns were left unaddressed.

The creation of the Piura climate change adaptation strategy was sparked by a federal mandate issued by the Ministry of the Environment (MINAM) in 2009. MINAM mandated that all regions in Peru produce climate change strategies for 2011 to 2021, mirroring national efforts. The formulation of the Piura adaptation plan took place in 2010 and 2011. In total, twenty-three organizations participated in eight provincial meetings with listening sessions related to the observed impacts of climate variability and climate change. As in the SMR process, these meetings preceded the drafting of the plan. Unlike the SMR, though, there was no opportunity for iterative feedback. Whereas environmental and agricultural users of water were included in workshops, none of the leaders of Piura's 122 *comunidades campesinas* were included.

We find that, although planning involved robust participatory processes, the exclusion or inclusion of stakeholders implicitly defines who is considered an actor in the hydrosocial cycle. Without the presence of actors representing differing social–environmental and cultural values, certain hydrosocial dimensions remain absent from the discussion.

Table 3. Content analysis results by hydrosocial theme

Hydrosocial theme	Santiago Metropolitan Region	Piura
Value of water	• Valued for urban drinking water, ecosystem services, and economic production: irrigation, hydropower, industry	• Valued for ecosystem services and economic production: irrigation, industry
Participation	• 10 themed roundtables: energy, water, land use, vulnerability • 15–20 government, university, private sector, water user organization, and NGO representatives • Participatory process informed iterative drafting	• 8 province-specific workshops on climate change impacts and disaster risk mitigation • 10–30 participants in each session, primarily government and NGO representatives • Participatory process informed predrafting stage
Upstream–downstream relationships	• Recognizes downstream dependence on mountain-sourced water • Upstream–downstream connections absent in measures	• Recognizes importance of upstream forests and wetlands for downstream water supply • Emphasis on upstream as headwaters and downstream as water users
Downstream measures	• Reduce water demand and increase efficiency • Flood control • Gray water reuse • Low-flow installations	• Increase irrigation efficiency • Construction and restoration of reservoirs and water harvesting systems • Upgrade water tariff system
Upstream measures	• Restoration and conservation of precordillera forests and riparian areas	• Restoration and conservation of headwater ecosystems
General measures	• Education and awareness programs • Diversify regional energy matrix • Reform water laws • Create basin organizations to coordinate water management	• Awareness building • Evaluate water quality and availability • Increase user coordination • Wastewater treatment • Hydrometeorological monitoring and alert systems
Vulnerability	• Notes high vulnerability in areas of urban expansion • No mention of mountain or rural communities	• Calls for identification of vulnerable populations • No mention of mountain or rural communities
Water access	• Emphasizes sustaining urban and agricultural water supply • No mention of communities without water access • Urban water supplier included but not rural water committees	• Mentions asymmetrical water access between large versus small irrigators • Goal of 10 percent reduction in water poverty via water tariffs adjustment and headwater conservation
Implementation	• Includes detailed implementation manual • Has not been implemented	• Includes adaptation goals for government implementation • No timeline or budget • Fragmented implementation

Note: NGO = nongovernmental organization.

Regional Economic Development Driving Proposed Adaptation Measures

In both plans, we see adaptation measures targeted to sustain economic development trajectories despite climate change impacts. Under current hydrosocial configurations, most of the water flows to prominent agricultural and urban water sectors, and adaptation plans seem to reinforce, not reconfigure, these relationships.

In the Chilean case, the SMR adaptation plan emphasizes minor technological interventions, such as low-flow installations, gray water reuse systems, and canal improvements, aimed at increasing the efficiency of current usage patterns. Centered on ensuring urban and agricultural water supplies, the plan does not address long-standing water scarcity in rural margins of the region, such as the Maipo River headwaters (Gobierno Regional Metropolitano de Santiago [GORE] 2013). Meanwhile, in Piura, Peru, water sector adaptation measures focus on sustaining the lowland agro-export industry by recuperating highland ecosystems, increasing irrigation efficiency, repairing and constructing irrigation storage infrastructure, and improving water pricing mechanisms to "reflect the value of conservation and the risk of water scarcity" (GRP 2011, 54).

We acknowledge that sustaining regional economic development is an important aspect of adaptation, but the SMR and Piura adaptation plans seek to sustain current flows of water to urban centers and high-value agricultural fields, proposing no interventions to redirect or expand the distribution systems to communities currently excluded from access. These adaptation planning processes thus miss the opportunity to reconfigure hydrosocial relationships and address long-standing water scarcity in mountain communities.

Missed Opportunity to Address Key Water Sector Vulnerabilities

Vulnerability is an important guiding concept in climate change adaptation plans and holds the potential to open dialog among stakeholders and planners about access and equity. Although both the SMR and Piura plans identify vulnerable areas, neither targets measures to rigorously address issues of socioeconomically and spatially differentiated vulnerability. Specifically, our hydrosocial analysis raises questions about why enduring disparities in potable water access were not considered. The plans emphasize emerging climate-related risks and ignore preexisting challenges.

Roundtable discussions convened by SMR plan facilitators mentioned the vulnerability of the urban periphery to flooding and extreme heat but omitted any mention of vulnerability in the rural or mountainous areas. Uneven access to potable water in the mountainous Maipo River canyon is not addressed in the plan. Whereas billions of dollars have been invested to secure the infrastructure needed to supply Santiago with drinking water, efforts to improve rural water supply have been stalled by lack of funding and political will (DGA 2015). Many isolated communities in the headwater region rely on informal water supply systems that are sensitive to drought, as explained by a General Water Directorate representative:

> It is a critical situation because there are many people that depend on small drainages that are disappearing since it hasn't rained. . . . They don't have water rights. (Interview 24 June 2015)

The informality of potable water supply in the headwater region likely contributes to its exclusion from climate change adaptation planning, as it lies outside the legal framework of the institutions expected to implement the plan.

Within the Piura plan, the first "strategic objective" is to have "regional actors identify climate change vulnerabilities in Piura and propose implementation measures for adaptation" (GRP 2011, 48). The document, however, does not put forth a concrete plan for identifying vulnerable communities, nor do the proposed measures target these communities. For example, the highland provinces of Ayabaca and

Huancabamba have poverty rates over 75 percent, compared to the regional average of 40.5 percent (INEI 2007), and yet these provinces are entirely excluded from the proposed measures. The Piura plan aims to reduce water poverty by 20 percent but does not propose measures that address economic and sociopolitical issues behind enduring potable water scarcity, which affects 65 percent of the rural population and 21 percent of the urban population (ANA 2013). The two proposed measures for reducing water poverty are to upgrade the water tariff system and increase upstream reforestation, neither of which will improve potable water access.

The SMR and Piura plans are designed for implementation by regional governments and, therefore, people and issues that fall outside state, legal, and institutional frameworks are not included. Informal water supply systems of mountain communities and their vulnerability to increasing water scarcity are key issues that are thus avoided, missing a critical opportunity to explore avenues toward effective and equitable adaptation.

Challenges to Implementation and New Directions

As climate change impacts escalate and adaptation planning advances globally, planners recognize a need to move from "premise to practice" (Medema, McIntosh, and Jeffrey 2008). Yet, the step from adaptation planning to implementation requires funding, institutional capacity, cooperation, and time, which are often in short supply.

The SMR plan was carefully designed to facilitate implementation, yet no actions have been taken:

> They have done nothing with the plan. . . . To implement the measures in the plan, first you need money. . . . Second, for a lot of them—they have to be implemented at a local scale, by municipalities, who have even fewer resources. And they were not involved. (Workshop coordinator, interview 10 August 2015)

The challenges of implementing the SMR plan have inspired new adaptation planning initiatives that build on its participatory approach, incorporating new actors at the municipal and basin scales (Vicuña et al. 2014). Yet, resource conflicts and uneven access remain unaddressed by these initiatives. The director of the basin-scale initiative described the challenge of maintaining collaboration: "This is a fragile system . . . if we open up certain issues to debate, the conversation is over, so we have to be careful" (Interview 2 June 2015). Thus, planning centers on broad mutual goals, avoiding topics like communities without water, which lead to heated and disruptive debates. Instead, adaptation measures remain focused on sustaining current water usage under climate change projections. Although this is important, it does not aid communities currently excluded from the supply and demand equation and leaves them little hope of gaining water access in the future.

Although the Piura plan is intended to inform the operations of regional government entities, implementation of specific measures depends on funding from bilateral donors or nongovernmental organizations (NGOs). Peru's rise to the status of a lower middle-income country has led to a contraction of aid money (AidData 2016), which presents challenges to the plan's implementation. Within the water sector, however, a water user organization and NGO have created a fund using lowland irrigator fees to pay for highland reforestation. This project holds potential to build sustained upstream–downstream dialogue but raises concerns about how upstream communities are perceived:

> The person [upstream] doesn't receive money, they receive trees. Here is your payment: 10,000 trees. Now plant them. Why? They [downstream irrigators] have a distorted idea about compensation. What they need to do is invest in those people. (NGO director, interview 19 January 2016)

This program reflects enduring assumptions about headwater regions, which are perceived as sites for reforestation without consideration of the livelihood needs of highland communities. Implementation of the Piura adaptation plan will likely require innovative funding models, but it must also examine the hydrosocial relationships that it reinforces.

Neither the SMR nor Piura adaptation plans have been fully implemented, but these plans set an important precedent for ongoing adaptation efforts. Both cases suffer from a lack of funding and institutional capacity for implementation. Nonetheless, these plans are snapshots of hydrosocial relationships perceived by those privileged in the climate change adaptation planning process. As we have discussed, the plans reveal critical gaps in how the hydrosocial cycle is conceptualized. As adaptation planning and implementation proceeds in new directions, critical analysis of these past plans can inform more comprehensive measures to address overlooked aspects of the hydrosocial cycle, such as long-standing disparities in water access.

Conclusion

Climate change adaptation plans are more than just documents; they inform and are informed by sociopolitical processes. By viewing them through a hydrosocial lens, we explore how climate change adaptation plans reflect certain understandings of hydrosocial relations that tend to exclude those with informal water access, differing water ontologies, or livelihoods outside of dominant economic sectors.

Adaptation planning processes are advancing globally and should be considered within the hydrosocial literature. We have taken a hydrosocial approach to adaptation planning assessment, with an emphasis on its implications for mountain water towers. This, however, is just one way to bring hydrosocial literature into conversation with climate change adaptation. Other avenues include hydrosocial studies of implementation processes, ethnographic research on how autonomous adaptation reconfigures hydrosocial relationships, and analysis of how site-based adaptive action reshapes upstream–downstream linkages. Mountain water towers are critical sites for these studies, as key interfaces between upstream and downstream water users with intimately connected hydrosocial relationships.

Comparative content analysis of climate change adaptation plans in two Andean water tower regions reveals several key findings. Both the SMR and Piura adaptation plans privilege the economic value of water, omitting socioenvironmental and cultural values. Although both plans were developed through robust participatory processes, they did not include marginalized hydrosocial actors from rural and mountainous areas. Adaptation measures were primarily targeted to sustain dominant economic sectors, reinforcing rather than reconfiguring water access patterns. The plans do not acknowledge socioeconomically and spatially differentiated vulnerability within the water sector, missing a key opportunity to redress long-standing disparities. Although neither plan has been implemented, these documents set an important precedent for future efforts and therefore merit critical analysis.

The December 2015 Paris Agreement sparked climate change adaptation aid pledges by developed nations and renewed emphasis on adaptation planning and implementation. Meanwhile, the rise in climate justice movements underlines the need for social and environmental justice to be at the core of climate change adaptation. Adaptation planning provides a critical moment to reorient this process toward addressing uneven resource access. This key issue is particularly salient for mountain communities, which are highly vulnerable to climate change and face resource challenges often overlooked in adaptation planning. The two Andean plans reinforce current water access patterns, missing a key opportunity to envision more inclusive hydrosocial relationships under a changing climate. Critical hydrosocial analysis of these plans reveals gaps that must be addressed in future planning and implementation efforts if adaptation is going to provide meaningful pathways for change.

Acknowledgments

The authors would like to thank Dr. Ben Preston, Sarah Kelly-Richards, Noah Silber-Coats, and our two anonymous reviewers for their invaluable insights.

Funding

This work would not have been possible without the generous support of the Inter-American Institute for Global Change Research project CRN3056 (which is supported by U.S. National Science Foundation [NSF] Grant GEO-1128040); U.S. Agency for International Development and U.S. National Academies of Sciences Project PEER II 2-359 (linked to NSF Grant DEB-101049); the International Water Security Network funded by Lloyd's Register Foundation (a charitable foundation in the United Kingdom helping to protect life and property by supporting engineering-related education, public engagement, and the application of research); the CGIAR Research Program on Water, Land and Ecosystems; and the International Centre for Integrated Mountain Development under its Himalayan Adaptation, Water and Resilience program.

References

AidData. 2016. AidDataCore_ResearchRelease_Level1_v3.0 Research Releases data set. Williamsburg, VA: AidData. http://aiddata.org/research-datasets (last accessed 12 June 2016).

Autoridad Nacional de Agua (ANA). 2013. *Plan de Gestión de los Recursos Hídricos de la Cuenca Chira-Piura: Informe Final* [Water resources management plan for the Chira-Piura River Basin: Final report]. Piura, Peru: ANA.

Ayres, L. 2008. Thematic coding and analysis. In *The Sage encyclopedia of qualitative research methods*, ed. L. M. Given, 868–69. Thousand Oaks, CA: Sage.

Barton, J. R., K. Krellenberg, and J. M. Harris. 2015. Collaborative governance and the challenges of participatory climate change adaptation planning in Santiago de Chile. *Climate and Development* 7 (2): 175–84.

Bebbington, A., and M. Williams. 2008. Water and mining conflicts in Peru. *Mountain Research and Development* 28 (3–4): 190–95.

Beniston, M. 2003. Climatic change in mountain regions: A review of possible impacts. *Climatic Change* 59:5–31.

Biesbroek, G. R., R. J. Swart, T. R. Carter, C. Cowan, T. Henrichs, H. Mela, M. D. Morecroft, and D. Rey. 2010. Europe adapts to climate change: Comparing national adaptation strategies. *Global Environmental Change* 20:440–50.

Bizikova, L., E. Crawford, M. Nijnik, and R. Swart. 2014. Climate change adaptation planning in agriculture: Processes, experiences and lessons learned from early adapters. *Mitigation and Adaptation Strategies for Global Change* 19:411–30.

Boelens, R., J. Hoogesteger, E. Swyngedouw, J. Vos, and P. Wester. 2016. Hydrosocial territories: A political ecology perspective. *Water International* 41 (1): 1–14.

Budds, J. 2009. Contested H2O: Science, policy and politics in water resources management in Chile. *Geoforum* 40 (3): 418–30.

Centro de Ciencia del Clima y la Resiliencia (CR2). 2015. *La megasequía 2010–2015: Una lección para el futuro* [The mega-drought 2010–2015: A lesson for the future]. Santiago de Chile: Informe a la Nación.

Clima Adaptación Santiago (CAS). 2012. *Plan de Adaptación al Cambio Climático para la Región Metropolitana de Santiago de Chile* [Climate change adaptation plan for the Santiago metropolitan region]. https://www.ufz.de/export/data/403/46050_PlanAdaptacion_121126.pdf (last accessed 20 January 2016).

Dirección de General de Aguas (DGA). 2015. *Diagnóstico Plan Maestro de Recursos Hídricos Región Metropolitana de Santiago: Informe Final Vols. 1 & 2* [Water resources master diagnostic plan for the metropolitan region: Final report volume 1 & 2]. Santiago de Chile: Ministerio de Obras Publicas, Chile.

Eriksen, S. H., A. J. Nightingale, and H. Eakin. 2015. Reframing adaptation: The political nature of climate change adaptation. *Global Environmental Change* 35:523–33.

Gobierno Regional Metropolitano de Santiago (GORE). 2013. *Política Regional para el Desarrollo de Localidades Aisladas* [Regional policy for the development of isolated communities]. Santiago de Chile: Gobierno Regional Metropolitano de Santiago.

Gobierno Regional Piura (GRP). 2011. Estrategia Regional de Cambio Climático [Regional climate change strategy]. http://siar.regionpiura.gob.pe/admDocumento.php?accion=bajar&docadjunto=854 (last accessed 27 January 2016).

Instituto Nacional de Estadística e Informática (INEI). 1994. *Censo Agropecuario* [Agricultural census]. Lima, Peru: INEI.

———. 2007. *Censos Nacional 2007: XI de Población y VI de Vivienda* [National census 2007: 11th population and 6th housing]. Lima, Peru: INEI.

———. 2012a. *Censo Agropecuario* [Agricultural census]. Lima, Peru: INEI.

———. 2012b. *Peru en Cifra* [Peru deciphered]. Lima, Peru: INEI.

Julien, H. 2008. Content analysis. In *The Sage encyclopedia of qualitative research methods*, ed. L. M. Given, 121–23. Thousand Oaks, CA: Sage.

Körner, C., and M. Ohsawa. 2005. Mountain systems. In *Millennium ecosystem assessment, ecosystems and human well-being: Current state and trends*, ed. T. Schaaf and C. Lee, Vol. 1, 681–716. Washington, DC: Island.

Linton, J., and J. Budds. 2014. The hydrosocial cycle: Defining and mobilizing a relational-dialectial approach to water. *Geoforum* 57:170–80.

McDowell, G., E. Stephenson, and J. Ford. 2014. Adaptation to climate change in glaciated mountain regions. *Climatic Change* 126:77–91.

Medema, W., B. S. McIntosh, and P. J. Jeffrey. 2008. From premise to practice: A critical assessment of integrated water resources management and adaptive management approaches in the water sector. *Ecology and Society* 13 (2): 29.

Pepin, N. 2015. Elevation-dependent warming in mountain regions of the world. *Nature Climate Change* 5 (5): 424–30.

Preston, B. L., R. M. Westaway, and E. J. Yuen. 2011. Climate adaptation planning in practice: An evaluation of adaptation plans from three developed nations. *Mitigation and Adaptation Strategies for Global Change* 16:407–38.

Recharte, J. 2006. Sacred lakes and springs in the northern Andes and the Huascarán World Heritage Site and Biosphere Reserve, Peru. In *Proceedings from UNESCO-MAB: Tokyo Proceedings*, 112–17. Paris: UNESCO.

Ribot, J. 2011. Vulnerability before adaptation: Toward transformative climate action. *Global Environmental Change* 21:1160–62.

Ribot, J. C., and N. L. Peluso. 2003. A theory of access. *Rural Sociology* 68 (2): 153–81.

Rodríguez-de-Francisco, J. C., and R. Boelens. 2016. PES hydrosocial territories: De-territorialization and re-patterning of water control arenas in the Andean highlands. *Water International* 41 (1): 140–56.

Swyngedouw, E. 2004. *Social power and the urbanization of water: Flows of power*. Oxford, UK: Oxford University Press.

———. 2009. The political economy and political ecology of the hydrosocial cycle. *Universities Council on Water Resources Journal of Contemporary Water Research and Education* 142:56–60.

———. 2013. Into the sea: Desalination as hydro-social fix in Spain. *Annals of the Association of American Geographers* 103 (2): 261–70.

Valdivia, C., A. Seth, J. L. Gilles, and M. García. 2010. Adapting to climate change in Andean ecosystems: Landscapes, capitals, and perceptions shaping rural livelihood strategies and linking knowledge systems. *Annals of the Association of American Geographers* 100 (4): 818–34.

Vicuña, S., S. Bonelli, E. Bustos, and T. Uson. 2014. Beyond city limits: Using a basin perspective to assess urban adaptation to climate change: The case of the city of Santiago in Chile. In *Proceedings of the Resilient Cities 2014 Congress*, ed. F. Schreiber and L. Kavanaugh, 1–23. Bonn Germany: ICLEI.

Viviroli, D., H. H. Durr, B. Messerli, M. Meybeck, and R. Weingartner. 2007. Mountains of the world, water towers for humanity: Typology, mapping, and global significance. *Water Resources Research* 43:1–13.

Nanga Parbat Revisited: Evolution and Dynamics of Sociohydrological Interactions in the Northwestern Himalaya

Marcus Nüsser ⓘ and Susanne Schmidt ⓘ

Regular availability of glacier and snow meltwater is essential for irrigated crop cultivation in the northwestern Himalaya. Based on a case study from the Nanga Parbat region in Gilgit-Baltistan, Pakistan, general patterns and site-specific particularities of irrigation networks in semiarid high mountain regions are conceptualized as continuously evolving sociohydrological interactions. These interactions are shaped by an interplay of glacio-fluvial runoff, water distribution, socioeconomic setting, institutional arrangements, external development interventions, and historical trajectories. Building on the paradigm of sociohydrology that changes in water availability coevolve with socioeconomic and land use transitions, this article explores glacier fluctuations and associated developments in meltwater-dependent crop cultivation in the Rupal Valley. The evolution of irrigation networks is analyzed using multitemporal high-resolution satellite imagery, repeat photography, and primary socioeconomic data collected in successive field surveys. Changes are historically contextualized with the help of archival material such as colonial reports and cadastral maps. This integrative study discovered the extension of cultivated areas, an increase in individual field numbers, and a reduction in average field size against the background of population increase and glacier retreat. Despite socioeconomic and environmental changes, the strong coupling of the human–water system remains intact, demonstrating a high degree of persistence of sociohydrological features over time. Adaptive strategies, however, often fail in the face of unpredictable natural processes.

取得冰川和融雪水的稳定管道,是喜马拉雅西北部灌溉作物耕作的关键。本文根据巴基斯坦吉尔吉特—巴尔蒂斯坦的南迦帕尔巴特地区的案例研究,将半乾旱高山地区灌溉网络的一般模式与特定场所的特殊性,概念化为持续演化的社会水文互动。这些互动,由冰川沉积径流、水资源分佈、社会经济环境、制度安排、外部发展介入,以及历史轨迹的交互作用所形塑。本文根据水资源可及性的改变随着社会经济及土地使用变迁共同演化的社会水文范式,探讨鲁帕尔谷的冰川变动以及依赖融雪水的作物耕种之相关发展。本文运用多重时间、高辨识率的卫星影像、重复摄像以及在连续田野调查中取得的初步社会经济数据,分析灌溉网络的演化。本文透过诸如殖民报告和地籍图的档案材料之协助,将变迁历史进行脉络化。此一整合性的研究,发现在人口增加与冰川倒退的背景下,种植面积有所扩张,个别田地数量增加,且平均田地面积减少。尽管面对社会经济与环境的变迁,人类—水资源系统的强健联结仍然完整无缺,证实了社会水文特徵随着时间仍然高度续存。但面对无法预测的自然过程,调适策略却经常失败。 关键词: 冰河变化, 喜马拉雅, 山区农业, 南迦帕尔巴特峰, 社会水文学。

La disponibilidad regular de agua de deshielo proveniente de glaciares y nieve es esencial para el cultivo de cosechas de riego en los Himalayas del noroeste. Con base en un estudio de caso de la región de Nanga Parbat en Gilgit-Baltistan, Pakistán, los patrones generales y particularidades específicas del sitio de las redes de irrigación en regiones semiáridas de alta montaña se conceptualizan como interacciones sociohidrológicas de evolución continua. Estas interacciones están configuradas por unas relaciones mutuas de la escorrentía glacio-fluvial, la distribución del agua, el escenario socioeconómico, el aparato institucional, las intervenciones de desarrollo externas y las trayectorias históricas. Construyendo sobre el paradigma de la sociohidrología de que los cambios en la disponibilidad hídrica co-evolucionan con las transiciones socioeconómicas y de usos del suelo, este artículo explora las fluctuaciones glaciarias y desarrollos asociados en el cultivo de cosechas dependientes de agua de deshielo del Valle Rupal. L evolución de las redes de irrigación se analizó mediante el uso de imágenes satelitales multitemporales de alta resolución, fotografía repetida y datos socioeconómicos primarios recogidos en sucesivos estudios de campo. Los cambios se contextualizan históricamente con la ayuda de materiales de archivo tales como informes coloniales y mapas catastrales. Este estudio integrativo descubrió la extensión de las áreas cultivadas, un incremento del número de campos individuales, y una reducción del tamaño promedio del campo, contra un trasfondo de incremento de la población y retroceso del glaciar. A pesar de los cambios socioeconómicos y ambientales, el fuerte acoplamiento del sistema hombre–agua permanece intacto, demostrando un alto grado de persistencia de los rasgos sociohidrológicos a través del tiempo. Sin embargo, las estrategias de adaptación a menudo fallan frente a procesos naturales impredecibles.

Against the background of the prominent Himalayan glacier debate of the past decade, global concerns were raised about the severe consequences of detected and expected changes in the South Asian cryosphere (Cogley 2011; Bolch et al. 2012; Hewitt 2014; Bajracharya et al. 2015). The sensitive water tower function of the Hindu Kush, Karakoram, and northwest Himalayan ranges, whose meltwaters feed the upper Indus Basin and secure water availability in the adjacent large-scale irrigation network of the Punjab, gained scientific interest and media coverage as a critical issue in climate change scenarios (Viviroli and Weingartner 2004; Archer et al. 2010; Immerzeel, van Beek, and Bierkens 2010; Kaser, Großhauser, and Marzeion 2010; Bocchiola et al. 2011; Nüsser and Baghel 2014). Such discussions, however, mainly focus on larger scales, neglecting the adaptation strategies of mountain communities that are in close proximity to these glaciers and use ice and snow meltwater for crop cultivation and basic needs (Kreutzmann 2011). To understand local practices within their immediate social and physical environments and their historical evolution across multiple scales and feedback loops, the concept of sociohydrology is particularly useful.

With its focus on studying the evolution of coupled human–water systems, sociohydrology has acquired increasing currency. Although it emerged as a rather technical approach based on hydrological engineering, it highlighted the importance of including social and economic dynamics in human–water interactions (Sivapalan, Svanije, and Blöschl 2012). In this article, general patterns and particularities of irrigation systems in arid and semiarid mountain regions are framed as sociohydrological interactions shaped by an interplay of glacio-fluvial runoff, water distribution, socioeconomic setting, institutional arrangements, external development interventions, and historical trajectories. Human–water interactions have been analyzed in different parts of the upper Indus Basin in northern Pakistan (Kreutzmann 2000, 2012; Nüsser 2001; Parveen et al. 2015) and in Ladakh, northern India

(Labbal 2000; Dame and Mankelow 2010; Dame and Nüsser 2011; Nüsser, Schmidt, and Dame 2012). Besides certain differences in water abstraction and distribution systems, these studies identified commonalities such as sophisticated designs and technologies adapted for sustainable resource utilization. In Gilgit-Baltistan (known until 2009 as Northern Areas of Pakistan), water channels fed by snow and glaciers comprise 65.8 percent of the total irrigated area (Government of Gilgit-Baltistan 2013).

Building on the paradigm of sociohydrology that changes in water availability coevolve with socioeconomic and land use changes, this article explores glacier fluctuations and associated developments in meltwater-dependent crop cultivation in the Nanga Parbat region at the northwestern end of the Himalayan range. To analyze the coupling of glacio-hydrological and socioeconomic dynamics influencing water abstraction and distribution networks, we adopt a combined approach based on land use mapping, multiscale satellite imagery, repeat photography, and primary socioeconomic data collected by the authors through field surveys conducted between 1992 and 2010. The analysis is further complemented by archival material, such as unpublished cadastral maps, reports prepared during colonial times, and material collected by members of the German Himalaya expeditions in 1934 and 1937 (Finsterwalder, Raechl and Misch 1935; Finsterwalder 1938; Troll 1939). These expeditions produced meticulous maps and detailed data on glaciers, hydrology, and vegetation. Their research suffered, however, due to strict disciplinary separation, an emphasis on cataloguing geophysical data, a lack of curiosity about socioeconomic issues, and unfamiliarity with the local language (Shina) and customs. The resulting gaps in our knowledge of historical sociohydrology can be filled through colonial administrative records, which primarily served the purpose of revenue assessment and built on a greater familiarity with customary law and local socioeconomic conditions. Although each of these sources has specific issues and biases, together they tend to complement each other and cancel out some biases. Together, they offer a rare long-term perspective for a better

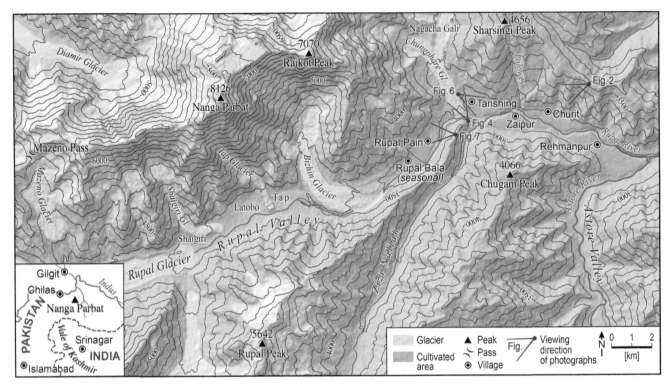

Figure 1. Map of Rupal Valley showing the location of villages and the viewing direction of photographs. (Color figure available online.)

understanding of sociohydrological interactions. The field research for this article conducted between 1992 and 1997 focused primarily on pastoralism and crop cultivation (Nüsser and Clemens 1996; Nüsser 2000), whereas later expeditions between 2006 and 2010 focused on glacier changes and updating previously collected socioeconomic data (Schmidt and Nüsser 2009). Due to the availability of these diverse data sets, the Rupal Valley allows for an integrated and historically informed analysis of land use patterns at a plot level against the background of regional development trajectories and changing human–water interactions.

Hydrological and Socioeconomic Setting

The Rupal Valley, located to the south of the main ridge of Nanga Parbat, releases its waters into the Astore River, which is a tributary of the Indus. Based on hydrometeorological data, Farhan et al. (2015) concluded that more than 75 percent of the total runoff between June and August in the Astore Basin depends on glaciers and seasonal snowmelt. In the Rupal Valley, the permanent settlements of Rehmanpur (called Rampur until 1947), Churit, Tarishing, and Zaipur are located at altitudes between approximately 2,630 and 2,910 m, where fluvial dissection of

the basal moraine has resulted in isolated plateaus up to 50 m above the river (Figure 1). Due to the favorable topographical setting, most cultivated areas are located here, with smaller strips of additional fields on the adjoining south-facing slopes. The size of landholdings in the valley is relatively homogenous and varies between 0.4 and 1.2 ha per household (Clemens 2001). It is a single-cropping area with dominant cultivation of wheat (locally *goom*) and barley (locally *yo*), up to 3,350 m, and to a lesser extent also maize (up to 2,750 m), potatoes, vegetables, and fodder plants (Nüsser and Clemens 1996). The production of potatoes, largely as cash crops for sale in the lowlands, has shown the largest increase over the past twenty years. This is mostly driven by the initiative of the Aga Khan Rural Support Programme (AKRSP), which plays a significant role in rural development in Gilgit-Baltistan. Depending on altitude, land preparation in the Rupal Valley generally takes place in April and May. Fields are irrigated in intervals of seven to twelve days until the crops are harvested and threshed in September. Together with seasonal pastoral migrations, these practices shape the spatial and temporal organization of labor requirements of combined mountain agriculture (Clemens and Nüsser 2000).

Institutional arrangements between households, neighbors, and villages have evolved in response to agropastoral resource scarcity and for coping with labor shortages. These shortages exist due to alternative income sources and the requirements of multilocational agriculture with widely distributed landholdings and grazing grounds. Irrigation and labor-intensive construction, maintenance, and restoration of gravity-dependent water channels (locally *kuhls*) are generally conducted by community institutions (*jirgas*). When an increase in village population leads to the construction of new irrigation channels, the *jirga* recalculates and allots water rights or turns of access (locally *weygon*) to households, based on their share of land (Bilal, Haque, and Moore 2003). This demonstrates the adaptive and flexible response of community institutions to demographic and hydrological changes.

Recent land use change in the Rupal Valley is characterized by the expansion of villages and irrigated area, massive reduction of sparsely distributed juniper bushes (*Juniperus semiglobosa*) on south-facing slopes and moist coniferous forests on north-facing slopes, and a slight increase in hygrophilous poplars and willows within the cultivated area (Figure 2). Based on satellite data and ground truthing, the number of buildings in Tarishing increased from 320 to 460 and the total irrigated area increased from 0.38 km^2 to 0.44 km^2 between 1965 and 2003. The highest permanent village is Rupal Pain (3,100 m), a former summer settlement of Tarishing (Troll 1939; Nüsser 1998), which can only be reached by crossing the debris-covered tongue of Chungphare Glacier. Although this village remains cut off from the lower valley and is completely reliant on animal transport, the number of buildings increased from 140 to 260 and the total irrigated area more than doubled from 0.17 km^2 to 0.37 km^2 between 1965 and 2003. These increases parallel changes in population size typical for the whole of Astore District, with a continuous growth rate of 3.1 percent (Government of Gilgit-Baltistan 2013).

Historically, the limited natural resource base has never been able to meet subsistence needs and supplementary food provisions had to be purchased in the lowlands of Pakistan; this continued until the implementation of subsidized government programs in the 1970s (Nüsser and Clemens 1996). Despite an increase in off-farm income through government jobs, mountain tourism, and remittances from migrating household members as complementary components of livelihood security, irrigated agriculture and livestock remain important sources of food security, with 98 percent of households in Astore District relying on them to some extent (Government of Gilgit-Baltistan 2013).

The almost complete decline of international trekking tourism in northern Pakistan since 11 September 2001 significantly reduced alternative income opportunities of mountain farmers. The 2013 terrorist attack that killed eleven mountaineers in the Diamir Base Camp of Nanga Parbat further reduced international climbing, trekking, and research activities in the region.

Recent land use and water management patterns have evolved under changing historical conditions, receiving a major impetus through colonial interventions and population transfer, when the region was transformed into an imperial landscape during the nineteenth century.

Sociohydrological Interventions under Colonial Rule

The Astore Valley and its tributaries, including Rupal Valley, came under the rule of the Maharaja of Jammu and Kashmir in 1846 and remained as a princely state under colonial administration for a century, until the partition of British India in 1947. The valley gained strategic importance as a mountainous transit corridor between the capital of Srinagar in the Vale of Kashmir and the military outposts in Gilgit (General Staff India 1928). In the early colonial period, the administration forced the local population to deliver agricultural products, mainly wheat, barley, and fodder, for the supplies of the troops (*hukmi kharid*) and to provide required labor for transport purposes (*begar*). The sensitive geopolitical setting during colonial rule put severe pressure on local agriculture to produce surplus crops and fodder for military use (Nüsser 1998). Until the 1850s, regular raids by the people of Chilas, who live on the western side of Nanga Parbat, resulted in abandoned fields and deserted villages in the Rupal Valley. After British military interventions, the slave and animal raids from the people of Chilas decreased in number and villagers were encouraged to settle back. According to colonial sources and oral history, additional recruitment of farmers among the population of Baltistan was initiated to stimulate the reoccupation of villages (Drew 1875; Neve 1913; Singh 1917). The fact that even today the inhabitants of Tarishing and Rupal Pain predominantly claim their origin from Baltistan results from this demographic intervention. This particular history of settlement is also reflected in the territorial distribution of overlapping utilization rights of pastures and common land between both villages (Nüsser 1998).

During colonial rule and shortly after the first population census (Gazetteer 1890), various assessments by

19 July 1934, Photo: Richard Finsterwalder

24 August 2006, Photo: Marcus Nüsser

Figure 2. The lower Rupal Valley in 1934 and 2006: The most significant visible changes are the expansion of villages and irrigated area, a reduction of natural vegetation, and an increase in willow and poplar plantations within the cultivated area. (Color figure available online.)

cadasters were carried out in Astore in the years 1893, 1903–1904, and 1916 to collect land revenues from the local population (Lawrence 1908; Singh 1917). For this, detailed land use assessments and cadastral maps were produced for individual villages, together with information on land use rights, which were mainly based on customary law (*rawaj*).[1] All lands were categorized according to their quality and suitability for irrigated agriculture and estimated crop yields. This classification distinguished between the categories cultivated land, wasteland (*banjirkadim*), stony area (*sanglakh*), and habitation (*abadi*; Figure 3).

Although colonial travelers described Rupal Valley as a fertile valley (Neve 1913), agrarian production in the single-cropping areas was usually insufficient to meet the demands of the growing population, and food security could only be ensured through regular supplies from Kashmir (Singh 1917). The colonial administration prioritized the increase of agricultural production by initiating and improving various irrigation schemes and development programs to cultivate topographically suitable areas. In

the Rupal Valley, avalanche protection of channels and the construction of an aqueduct to reduce water scarcity of the village Zaipur received special attention (Singh 1917). These examples demonstrate the ways in which the British–Kashmiri revenue system transformed early sociohydrological interactions and shaped the evolution of the coupled human–water system at the regional scale. Just as cadasters have provided insights into land use change, accounts of explorers are useful in determining glacio-fluvial dynamics and adaptive responses taken to maintain irrigation networks.

Glacier Dynamics and Adaptive Irrigation Practices

The first glacier exploration in the Rupal Valley was carried out by Adolph Schlagintweit in 1856, around the glacial extent of the Little Ice Age. Besides landscape paintings, sketch maps, and descriptions, he produced the first regional glacier inventory with a differentiation of debris-covered and clean ice glaciers. His material, partly published by his brother Hermann, indicates higher glacier surfaces and glacial damming of a meltwater lake at the base of Chungphare Glacier, the glacier closest to Tarishing (Schlagintweit 1872). These findings were confirmed by Drew (1875), who provided a detailed report of this glacier, mentioning a less crevassed ice and debris surface, which had lost some 30 m in height after the outburst of the glacier-dammed lake. Based on field visits in 1887 and 1906, Neve (1907) described the rapid advance of this glacier by about 90 m and its subsequent breaching by the main river.

A comparison of the sketch map by Schlagintweit and the topographic map produced by the German Himalaya Expedition shows that the Chungphare Glacier retreated by about 600 m between 1856 and 1934. Since then, it has further retreated by approximately 600 m between 1934 and 2013, as can be derived by juxtaposing the historical topographic map and contemporary satellite imagery. A significant thinning of the glacier is confirmed by terrestrial photogrammetry data from 1934, 1958, and 1987 (Kick 1994; Figure 4). In 2013 the Chungphare Glacier was 11.7 km long, covered a total area of 24.6 km^2, and terminated at an altitude of 2,920 m, as derived from remote sensing data.

Direct water abstraction from the Chungphare Glacier and irrigation channels across lateral moraines are described for Tarishing, when its surface

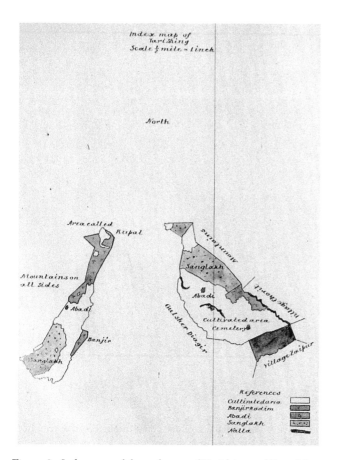

Figure 3. Index map of the cadasters of Tarishing and Rupal from colonial land records. (Color figure available online.)

22 June 1934, Photo: Richard Finsterwalder

16 August 2010, Photo: Marcus Nüsser

Figure 4. Chungphare Glacier in 1934 and 2010: Significant changes are the retreat of the glacier terminus and the incomplete road from Tarishing to Rupal Pain, partly washed away by torrential rains in August 2010. (Color figure available online.)

was level with the upper lateral moraine in 1906 (Neve 1913). Up until at least the mid-1930s, some water was diverted from the glacier to the fields as is shown in topographic maps (Finsterwalder 1938; Troll 1939). Two former intakes on the lateral moraine and fragments of desiccated diversion structures are detectable in satellite imagery (Figure 5). Due to retreat and thinning, however, meltwater from this large valley glacier does not contribute to irrigation anymore, as the debris-covered surface is located too far below the lateral moraines to allow water inflow from this source. As the cultivated area of Tarishing is located above the glacier snout, its irrigation system is disconnected from the

Chungphare Glacier. Instead, the channel intakes are mostly constructed at tributaries of different sizes along the south-facing slopes of the mountain range. This is similar to the case of Gojal in the Hunza Karakoram, where glacier downwasting led to desiccation of moraine-crossing channels and considerable adaptations to irrigation networks had to be made (Parveen et al. 2015).

Today, Tarishing mainly receives irrigation water from a torrent to the east of Chungphare Glacier, which is fed by meltwater from rock glaciers and avalanche accumulations below Sharsingi Peak (Figure 1). From the analyses of satellite imagery, the number of irrigated fields in Tarishing has more

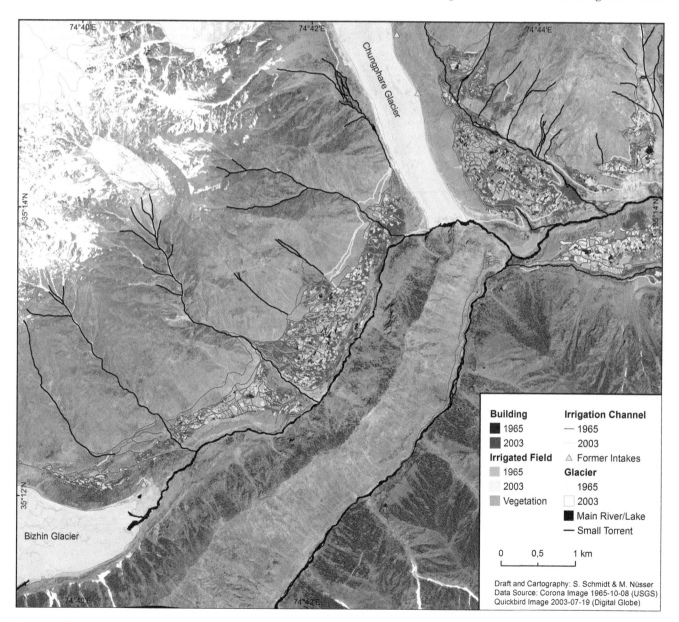

Figure 5. Changes in irrigation patterns in Rupal Valley between 1965 and 2003. (Color figure available online.)

12 June 1937, Photo: Carl Troll

14 August 2010, Photo: Marcus Nüsser

Figure 6. Tarishing in 1937 and 2010: A significant increase in the number of houses and hygrophilous trees within the cultivated area. (Color figure available online.)

21 October 1992, Photo: Marcus Nüsser

5 September 2010, Photo: Marcus Nüsser

Figure 7. Rupal Pain in 1992 and 2010: A significant increase in the number of houses and fields together with a reduction in field size. (Color figure available online.)

than doubled (from 700 to 1,760), although their average size decreased from 438 m^2 to 251 m^2 between 1965 and 2003. According to interviewees, the steady decrease in average field size mainly results from the principal practice of partible land inheritance. In the case of Rupal Pain, water is diverted from small torrents, the largest one fed by a cirque glacier above an altitude of 4,250 m and the eastern one supplied by a rock glacier, located at 4,200 m (Figure 5). New channels have been constructed from the orographic right ablation valley of Chungphare Glacier to irrigate former wastelands and grasslands on the pleistocene till terrace. In Rupal Pain, the former summer settlement of Tarishing, the number of irrigated fields has increased almost four times (from 420 to 1,610), although their average size has decreased from 393 m^2 to 227 m^2. In both villages, not only cropland but also meadows are irrigated to increase fodder production, and various water channels have been constructed on small dams and hardened grassy areas. In Rupal Bala, the summer settlement of Churit village, the number of irrigated fields has increased almost six times (from 130 to 800), although their average size has decreased to less than half their former size (from 824 m^2 to 302 m^2). An effort of the inhabitants of Rupal Bala in the 1990s to create an irrigation intake at the Bizhin Glacier (Figure 5) failed due to changes in the water course. External development interventions such as the effort to increase water availability by installing flexible water pipes over the Chungphare Glacier were funded by the AKRSP in the 1990s. The aim of this farm–forestry project was to irrigate deciduous tree plantations along the lateral moraine above Tarishing, but it failed because pipes and channels were destroyed by glacier movements and rockfalls.

Whereas water supply in Tarishing (Figure 6) and Rupal Pain (Figure 7) is sufficient to meet basic irrigation demands, interviewees in the villages of Zaipur and Churit reported regularly experiencing severe water scarcity. Zaipur only receives water through two channels from Zizi Nallah, a southern tributary of the Rupal River. An aqueduct first built during colonial times transferred water until the early 1990s. Nowadays, water is diverted through a government-funded pipe construction that fords the small side valley before reaching the fields a few hundred meters downstream (Figure 5). Due to the limited availability of irrigation water, some tracts of arable land are left uncultivated. With 722 m^2 in 2003, however, the

average field size in Zaipur is larger than that in any other village of Rupal Valley. The inhabitants of Churit face similar problems, as a 4- to 5-km-long water transfer scheme from the Lolowey (Figure 1), located east of Tarishing, constructed in the 1980s is still not operational because of regular landslides. Despite changes in water availability and the cumulative effects of labor shortage and subsidized food programs, the irrigated area shows highly persistent structures. In general, comparative interpretation of satellite imagery (Figure 5) and repeat photography (Figures 6 and 7) reveals the extension of cultivated areas, an increase in individual field numbers, and a reduction in average field size.

Conclusion

This article demonstrates the relevance of repeated field visits and historical sources to reveal underlying drivers of change, beyond deterministic simplifications. The Nanga Parbat region is a well-suited study area because of the availability of a large multitemporal database that can be used for a better understanding of the complex sociohydrological interactions and dynamics in the mountains of northern Pakistan and beyond. The necessity to integrate multiple data sources of varying reliability, however, requires a clear awareness of biases, missing links, and fragmentation of data. It is also important to take account of the sensitive geopolitical setting of the broader northwestern Himalayan region with its contested boundaries and overlapping territorial claims (Baghel and Nüsser 2015; Kreutzmann 2015) in the choice of study sites.

A mapping of irrigation channels enables a visualization of temporal and spatial variations in sociohydrological interactions. An improved understanding of the functioning of irrigation networks helps to detect adaptation strategies to changing glacio-fluvial conditions. Current human–water relations need to be contextualized against socioeconomic and historical trajectories. In this case this can be seen in the way the British–Kashmiri revenue system shaped the early form of sociohydrological interactions through population transfer, land classification, and surplus extraction of agricultural produce and labor. As sociohydrological interactions evolve through adaptive responses to environmental change, customary law and institutional arrangements are critical factors in the mobilization of community members for water and labor sharing. Although a steady increase in population in a peripheral mountain region would imply a lower availability of land and water resources per capita, the ongoing expansion of cultivated area allows a continued maintenance of subsistence production. This shows how measures taken to adapt irrigation networks to changing glacio-fluvial conditions in the Rupal Valley are key to continued food security. Innovative adaptation strategies are not always successful and often fail in the face of natural processes. This situation requires site-specific and sophisticated responses, something often lacking in external development interventions. Until today, the strong coupling of the human–water system remains intact and shows a high degree of persistence of sociohydrological features over time. Extrapolating these findings to the Himalayan scale, it is of the utmost importance to examine human adaptations to ongoing and anticipated cryosphere changes within a broader frame of sociohydrological interactions.

Acknowledgments

The authors thank the anonymous reviewers for their insightful and constructive comments. Special thanks to Ravi Baghel (Heidelberg) for proofreading and for refining the arguments in this article. We also thank Nils Harm (Heidelberg) for his cartographic contribution. We further appreciate the help of Akhtar Hussain (Tarishing, Gilgit) for his continuous support during various fieldwork campaigns between 1992 and 2010.

ORCID

Marcus Nüsser ⓘ http://orcid.org/0000-0002-8626-8336

Susanne Schmidt ⓘ http://orcid.org/0000-0003-4649-2445

Note

1. Colonial cadasters and information on resource utilization rights were consulted and photographed at the revenue office in Astore Town, the regional administrative center, in May 1994. Although all land revenues were abolished in 1972, these files are still used to settle conflicts about land possession and village-wide land use rights.

References

Archer, D. R., N. Forsythe, H. J. Fowler, and S. M. Shah. 2010. Sustainability of water resources management in the Indus Basin under changing climatic and socio-economic conditions. *Hydrology and Earth System Sciences* 14 (8): 1669–80.

Baghel, R., and M. Nüsser. 2015. Securing the heights: The vertical dimension of the Siachen conflict between India and Pakistan in the eastern Karakoram. *Political Geography* 48:24–36.

Bajracharya, S. R., S. B. Maharjan, F. Shrestha, W. Guo, S. Liu, W. Immerzeel, and B. Shrestha. 2015. The glaciers of the Hindu Kush Himalayas: Current status and observed changes from the 1980s to 2010. *International Journal of Water Resources Development* 31 (2): 161–73.

Bilal, A., H. Haque, and P. Moore. 2003. *Customary laws: Governing natural resource management in the Northern Areas*. Karachi, Pakistan: International Union for Conservation of Nature.

Bocchiola, D., G. Diolaiuti, A. Soncini, C. Mihalcea, C. D'Agata, C. Mayer, and C. Smiraglia. 2011. Prediction of future hydrological regimes in poorly gauged high altitude basins: The case study of the upper Indus, Pakistan. *Hydrology and Earth System Sciences* 15 (7): 2059–75.

Bolch, T., A. Kulkarni, A. Kääb, C. Huggel, F. Paul, J. G. Cogley, H. Frey, et al. 2012. The state and fate of Himalayan glaciers. *Science* 336 (6079): 310–14.

Clemens, J. 2001. *Ländliche Energieversorgung in Astor: Aspekte des nachhaltigen Ressourcenmanagements im nordpakistanischen Hochgebirge* [Rural energy supply in Astore: Aspects of sustainable resource management in the high mountains of northern Pakistan]. Sankt Augustin, Germany: Asgard.

Clemens, J., and M. Nüsser. 2000. Pastoral management strategies in transition: Indications from the Nanga Parbat Region (NW-Himalaya). In *High mountain pastoralism in northern Pakistan*, ed. E. Ehlers and H. Kreutzmann, 151–87. Stuttgart, Germany: Steiner.

Cogley, J. G. 2011. Present and future states of Himalaya and Karakoram glaciers. *Annals of Glaciology* 52 (59): 69–73.

Dame, J., and J. S. Mankelow. 2010. Stongde revisited: Land use change in central Zangskar. *Erdkunde* 64 (4): 355–70.

Dame, J., and M. Nüsser. 2011. Food security in high mountain regions: Agricultural production and the impact of food subsidies in Ladakh, northern India. *Food Security* 3 (2): 179–94.

Drew, F. 1875. *The Jummoo and Kashmir territories: A geographical account*. London: E. Stanford.

Farhan, S. B., Y. Zhang, Y. Ma, Y. Guo, and N. Ma. 2015. Hydrological regimes under the conjunction of westerly and monsoon climates: A case investigation in the Astore Basin, northwestern Himalaya. *Climate Dynamics* 44 (11): 3015–32.

Finsterwalder, R. 1938. *Die geodätischen, gletscherkundlichen und geographischen Ergebnisse der deutschen Himalaja Expedition 1934 zum Nanga Parbat* [The geodetic, glaciological and geographical results of the German Himalaya expedition to Nanga Parbat]. Berlin: K. Siegismund.

Finsterwalder, R., W. Raechl, and P. Misch. 1935. The scientific work of the German Himalayan Expedition to Nanga Parbat 1934. *The Himalayan Journal* 7: 44–52.

Gazetter. 1890. *Gazetteer of Kashmir and Ladák. Together with routes in the territories of the Maharája of Jamú and Kashmír*. Calcutta, India: The Superintendent of Government Printing.

General Staff India. 1928. *Military report and gazetteer of the Gilgit Agency and the independent territories of Tangir and Darel*. Simla, India: Government of India Press.

Government of Gilgit-Baltistan. 2013. *Gilgit-Baltistan at a glance*. Gilgit, Pakistan: Planning and Development Department.

Hewitt, K. 2014. *Glaciers of the Karakoram Himalaya: Glacial environments, processes, hazards and resources*. Dordrecht, The Netherlands: Springer.

Immerzeel, W., L. P. H. van Beek, and M. F. P. Bierkens. 2010. Climate change will affect the Asian water towers. *Science* 328 (5984): 1382–85.

Kaser, G., M. Großhauser, and B. Marzeion. 2010. Contribution potential of glaciers to water availability in different climate regimes. *Proceedings of the National Academy of Sciences of the United States of America* 107 (47): 20223–27.

Kick, W. 1994. *Gletscherforschung am Nanga Parbat 1856–1990* [Glacier research in the Nanga Parbat region]. Munich, Germany: Deutscher Alpenverein.

Kreutzmann, H. 2000. Water management in mountain oases of the Karakoram. In *Sharing water: Irrigation and water management in the Hindukush-Karakoram-Himalaya*, ed. H. Kreutzmann, 90–115. Karachi, Pakistan: Oxford University Press.

———. 2011. Scarcity within opulence: Water management in the Karakoram Mountains revisited. *Journal of Mountain Science* 8 (4): 525–34.

———. 2012. After the flood: Mobility as an adaptation strategy in high mountain oases. The case of Pasu in Gojal, Hunza Valley, Karakoram. *Die Erde* 143 (1–2): 49–73.

———. 2015. Boundaries and space in Gilgit-Baltistan. *Contemporary South Asia* 23 (3): 276–91.

Labbal, V. 2000. Traditional oases of Ladakh: A case study of equity in water management. In *Sharing water: Irrigation and water management in the Hindukush–Karakoram–Himalaya*, ed. H. Kreutzmann, 163–83. Karachi, Pakistan: Oxford University Press.

Lawrence, W. 1908. *The imperial gazetteer of India: Kashmir and Jammu*. Calcutta, India: The Superintendent of Government Printing.

Neve, A. 1907. Rapid glacial advance in the Hindu Kush. *Alpine Journal* 23:400–401.

———. 1913. *Thirty years in Kashmir*. London: E. Arnold.

Nüsser, M. 1998. *Nanga Parbat (NW-Himalaya): Naturräumliche Ressourcenausstattung und humanökologische Gefügemuster der Landnutzung* [Nanga Parbat (NW Himalaya): Natural resources and human-ecological land use patterns]. Bonn, Germany: F. Dümmler.

———. 2000. Change and persistence: Contemporary landscape transformation in the Nanga Parbat area, northern Pakistan. *Mountain Research and Development* 20 (4): 348–55.

———. 2001. Understanding cultural landscape transformation: A re-photographic survey in Chitral, eastern Hindukush, Pakistan. *Landscape and Urban Planning* 57 (3–4): 241–55.

Nüsser, M., and R. Baghel. 2014. The emergence of the cryoscape: Contested narratives of Himalayan glacier dynamics and climate change. In *Environmental and climate change in South and Southeast Asia: How are local cultures coping?*, ed. B. Schuler, 138–56. Leiden, The Netherlands: Brill.

Nüsser, M., and J. Clemens. 1996. Impacts on mixed mountain agriculture in the Rupal Valley, Nanga Parbat, northern Pakistan. *Mountain Research and Development* 16 (2): 117–33.

Nüsser, M., S. Schmidt, and J. Dame. 2012. Irrigation and development in the upper Indus Basin: Characteristics and recent changes of a socio-hydrological system in central Ladakh, India. *Mountain Research and Development* 32 (1): 51–61.

Parveen, S., M. Winiger, S. Schmidt, and M. Nüsser. 2015. Irrigation in upper Hunza: Evolution of socio-hydrological interactions in the Karakoram, northern Pakistan. *Erdkunde* 69 (1): 69–85.

Schlagintweit, H. 1872. *Reisen in Indien und Hochasien. Dritter Band*. [Travels in India and High Asia. Volume 3]. Jena, Germany: Hermann Costenoble.

Schmidt, S., and M. Nüsser. 2009. Fluctuations of Raikot Glacier during the last 70 years: A case study from the Nanga Parbat Massif, northern Pakistan. *Journal of Glaciology* 55 (194): 949–59.

Singh, T. 1917. *Assessment report of the Gilgit Tahsil*. Lahore, Pakistan: Khosla Bros.

Sivapalan, M., H. H. G. Svanije, and G. Blöschl. 2012. Socio-hydrology: A new science of people and water. *Hydrological Processes* 26 (8): 1270–76.

Troll, C. 1939. Das Pflanzenkleid des Nanga Parbat. Begleitworte zur Vegetationskarte der Nanga Parbat-Gruppe (Nordwest-Himalaja) 1:50.000 [The plant cover of Nanga Parbat: Commentary to the vegetation map of the Nanga Parbat group (NW Himalaya)]. *Wissenschaftliche Veröffentlichungen des Deutschen Museums für Länderkunde zu Leipzig* N.F. 7:149–93.

Viviroli, D., and R. Weingartner. 2004. The hydrological significance of mountains—From regional to global scale. *Hydrology and Earth System Sciences* 8 (6): 1016–29.

Applied Montology Using Critical Biogeography in the Andes

Fausto O. Sarmiento, J. Tomás Ibarra, Antonia Barreau, J. Cristóbal Pizarro, Ricardo Rozzi, Juan A. González, and Larry M. Frolich

More than most other landforms, mountains have been at the vanguard of geographical inquiry. Whether promontories, cultural works on slopes, or even metaphorical/spiritual heights, mountain research informs current narratives of global environmental change. We review how montology shifts geographic paradigms via the novel approach of critical biogeography in the Andes. We use it to bridge nature and society through indigenous heritage, local biodiversity conservation narratives, and vernacular nature–culture hybrids of biocultural landscapes (BCLs), focusing on how socioecological systems (SES) enlighten scientific query in the Andes. In our Andean study cases, integrated critical frameworks guide the understanding of BCLs as the product of long-term human–environment interactions. With situated exemplars from place naming, wild edible plants, medicinal plants, sacred trees, foodstuffs, ritualistic plants, and floral and faunal causation, we convey the need for cognition of mountains as BCLs in the Anthropocene. We conclude that applied montology allows for a multimethod approach with the four Cs of critical biogeography, a model that engages forward-looking geographers and interdisciplinary Andeanists in assessments for sustainable development of fragile BCLs in the Andes.

山岳较其他诸多土地形式而言，更处于地理学探问的前沿。无论是海角，斜度的文化工作，甚或是隐喻|精神性的高度，山岳研究告知了全球环境变迁的当代叙事。我们透过安第斯山的批判生物地理学之崭新方法，回顾山岳本体论如何改变地理范式。我们运用此一方法，透过原着民族袭产，地方生物多样性保存论述，以及生物文化地景 (BCLs) 的风土自然—文化混合，连结自然与社会，并聚焦社会生态系统 (SES) 如何启发安第斯山的科学探问。在我们的安第斯山研究案例中，整合性的批判架构，指引着对BCLs的理解，作为长期人类自然互动的产物。透过地方命名，野生可食用植物，医疗植物，神圣树木，粮食，仪式性植物，以及动植物的因果关系之脉络性范例，我们传达认可山岳在人类世中作为 BCLs 的必要性。我们于结论中主张，应用的山岳本体论，使得具有四大批判生物地理学的多重方法——一个让具前瞻性的地理学者和跨领域的安第斯山研究者参与至安第斯山中脆弱的 BCLs 之可持续发展评估的模式。 关键词： 安第斯山, 生物文化地景, 批判生物地理学, 民族志生物学, 非物质袭产, 山岳本体论, 帕拉莫。

Más que cualquier otro tipo de geoformas, las montañas han estado a la vanguardia de la investigación geográfica. Así sea en promontorios, trabajos culturales en las laderas, o incluso en alturas metafórico/espirituales, la investigación de montañas nutre las narrativas actuales del cambio ambiental global. Hacemos una revisión sobre el alcance de la *montología*, o estudio de las montañas, en la transformación de los paradigmas geográficos en los Andes al aplicar el novedoso enfoque de la biogeografía crítica. Lo usamos para tender un puente entre naturaleza y sociedad por medio de la heredad indígena, las narrativas locales sobre conservación de la biodiversidad y los paisajes bioculturales de híbridos vernáculos de naturaleza–cultura (BCLs), enfocándonos en la manera como los sistemas socioecológicos (SES) iluminan la indagación científica en los Andes. En nuestros estudios de casos andinos, el entendimiento de los BCLs como producto de interacciones humano–ambientales a largo plazo es guiado por esquemas críticos integrados. Por medio de ejemplos destacados a partir del proceso de nomenclatura de lugares, plantas silvestres comestibles, plantas medicinales, árboles sagrados, comidas, plantas rituales y la causalidad florística y faunística, hacemos notar la necesidad del conocimiento de las montañas como BCLs en el Antropoceno. Concluimos que una montología aplicada permite llegar a un enfoque multi-metodológico con los cuatro de la biogeografía crítica, un modelo que involucre a los geógrafos de miras abiertas y a especialistas interdisciplinarios sobre los Andes en evaluaciones del desarrollo sustentable de los frágiles BCLs de aquellas montañas.

Mountainscapes, or the appropriated interpretation of topographies and lifestyles, have driven epistemologies through elucidations of highland–lowland dynamics of many locales. These summit–abyss allusions drive environmental cognition of mountains while stimulating new ecoregional classifications worldwide (Lewis and Wigen 1997; Vallega 1999; Fouberg and Moseley 2015). Mountainous regions foster comparative studies incorporating not only the physicalities of the prominence but also verticality and accessibility (Allan 1986; Price et al. 2013) with significant human impacts; these human–environment relations define the heights in space–time in the public imaginary (Funnell and Price 2003; Welberry 2005; Macfarlane 2009). There is a Mountain *Problématique* that requires a "Mountain Agenda" (Messerli and Ives 1997, 455) with seven prerequisites: (1) perspective, (2) reciprocity, (3) devastation, (4) hazards, (5) awareness, (6) knowledge and research, and (7) policy, all of them pointing to the creation of *montology* (Neustadtl 1977). Montology is not only "the interdisciplinary study of the physical, chemical, geological, and biological aspects of mountain regions" but also is "the study of lifestyles and economic concerns of people living in these regions" (*Oxford English Dictionary* 2002).

Mountain landscapes' scientific disciplines, traditional cultures, and artistic creations suggest that these landforms are best understood as more than simply material entities of scientific curiosity; mountains are historically and socially constructed, and these constructions shape broader knowledge systems about society, place, and ecology (Debarbieux and Rudaz 2015). This fact explains the constructivist view using montology (Haslett 1998; Sarmiento 2000; Rhoades 2007), where cultural landscapes are central to define their identity, with place naming and biota distribution reflecting their sociopolitical and historical context (cf. critical biogeography) of Andean countries. Currently, epistemologies equate biodiversity with the physical setting and conservation policies favor tangible mountain biotas migrating upward with global warming (Borsdorf and Stadel 2015). Applying montology, we have increased understanding of the Andes as a socioecological system (SES) including the intangible heritage of the human driver of change.

Unlike chasing a chimera in the past (Messerli and Ives 1997), the cognate fields of biogeography, geoecology, and ethnobiology intertwine today, giving montology the wherewithal to define mountains holistically. Mountain research continues to elicit bridging the epistemological crevasse between the biophysical and the human models of "natural" (Gade 1996, 2016; Castree 2014) insofar as the biota distribution in the biocultural landscape (BCL; Cocks 2006; Hong, Bogaert, and Min 2014). Further, BCL requires longtime human manipulations. They are "complexes of biotic and cultural elements interconnected by historical and ecological evolutionary feedback, making them holistic assemblages yet dynamic and emergent social constructs with rich ancestral cultural practices" (Pungetti 2013, 56). Thus, mountains situate resource use with political ecology as sources of mineral and other environmental services and sinks of governance, marginalization, poverty, food (in)security, and globalized (in)equality within the hegemony of empire and indigenous affairs, historicity, and ethics (Rozzi 2015). Moreover, BCLs are made of

majestic mountains, sacred forests, indigenous seeds, revered rivers which give life, renewal, inspiration and spiritual satisfaction. "The Source" is much more than just an awesome physical feature, it also comprises those mystical elements in a BCL that are less tangible, particularly with English language. It is the sacred essence of a natural spring that make it part of a creation story and not just a watery hole in the ground. It is the vast genetic universe inside of a single locally-adapted seed, or the connection you feel when you hike a special mountain and something just feels right, like you belong. A powerful natural energy emanates within a thriving BCL. To outsiders it might be overwhelming, or indiscernible, but to the stewards of that BCL it is as essential as the air, water and soil. (Christensen Fund 2016)

Most heights are prone for awe-generating sources whose intangibles comprise the "mountain heritage." For the traditionalist, the physical driver is pivotal: Descriptions of highland people and nature made by Humboldt in the 1800s are still as valid today as when biogeography was born (Wulf 2015). Mountains, however, are not just lowlands at higher elevations.

Montology emphasizes disciplinary hybrids to understand mountains holistically by challenging long-held beliefs. For instance, even basic premises for measuring the vertical dimension give montology a niche. Depending on the convention utilized, the "tallest mountain" on Earth might be either

Sagarmantha (Asia), Mauna Kea (Hawai'i, Polynesia), Chimborazo (South America), Denali (North America), Kosciuszko (Australia), Kilimanjaro (Africa), Sierra Nevada de Santa Marta (South America), or Mt. Lamlam (Guam, Micronesia). Depending on the choice of criteria, montologists might employ either (1) elevation above sea level, (2) continuous vector slope, (3) planetary radius toward the troposphere, (4) edifice prominence, (5) Z proportion of X length, (6) Z proportion of Y width, (7) shore/summit direct line of sight, or (8) trench/summit ratio (Sarmiento 2016a). This physical disaccord has a cultural counterpart in the disputed names of some peaks, most notably *Denali* (Mt. McKinley), *Sagarmantha* (Mt. Everest), and *Tayrona* (Sierra Nevada de Santa Marta; see Figure 1). Some groups deem demonic buttes as the highest mountains, such as Devil's Tower for the *Arapaho* in the United States or *Auyán-Tepuy* for the *Pemón* in Venezuela. Other groups consider menacing volcanoes as the highest mountains, such as *Reventador* for the *Cofán* in Ecuadorian Amazonia or *Popocatepetl* for the *Mixtec* of the central plateau of Mexico. Hitherto, the Western predicament of conventionalism favored scientific over vernacular descriptors. This practice is currently contested in the Global South, however (Gudynas 2013).

Montology as a New Paradigm

Social construction of mountains grapples with paradigms that are undermined by vapid interpretations of desultory phenomena; the resulting poorly scrutinized rhetoric misguided conservation, inscribing mountains as protected areas only if they were pristine or assessing highland communities as if they were peacefully bonded (Berkes, Folke, and Colding 2000; Arpin and Cosson 2015). Their mythical Shangri-La, Xanadu, Meru, or even Zomia are imagined paradises for the tired, the lost, the pure, or the anarchist on mountainscape territories. Mountain imaginaries thus vary according to geographic, scientific fashions without theoretical grounding in complex BCLs (Gould 1979; Bradshaw and Bekoff 2001; Koutsopoulos 2011; Fu and Jones 2013; Rozzi et al. 2015; Convery and Davis 2016).

The waves of paradigmatic change in mountain geography provide alternative ways of knowing, especially from traditional ecological knowledge (TEK) handed down through generations, often through songs, stories, and beliefs (Berkes, Folke,

and Colding 2000). TEK is kept by indigenous, metis, mestizos, and other locals, creating sustainable lifescapes with time-tested practices that exemplify mountains as BCLs (Allan, Knapp, and Stadel 1988). Mountain people are developmental subjects, as required by the new spiritual dimension, the so-called sacred ecology transition, while societies become affluent. This transition reflects an inverse Kuznets environmental curve for Earth stewardship, as opposed to demographic or forestry transition curves of environmental degradation (Figure 2). By looking through a critical lens at sacred ecology (Berkes 2012; Rozzi et al. 2012), the financially richest postindustrial economies with disposable income can better attune with a comprehension of BCLs (Hong, Bogaert, and Min 2014) by increased spiritual awakening.

Shifting Paradigms and Mountain Methodologies

The topographic, geodesic, and chorological maps of mountains no longer suffice. New tools in the montologist's arsenal include telemetry, remotely sensed data, relational databases with geographic information systems, 3D plotting, modeling, geovisualization, flight-in software, ground-penetrating radar, and cloud stripping networks (Sarmiento, Box, and Usery 2004). These revolutionary research instruments, informatics, and analytics, although offering a new, faster mode of mountain cognition, also benefit from anecdotal, ground truthing, and direct observational data gathering and local knowing (Graham and Shelton 2013). The role of TEK itself has undergone its own paradigmatic shift, going from the vertical "dendritic" approach of top-down, reactive fixes to problems into a horizontal "rhizomic" approach of bottom-up, proactive planning for development with participatory communal benefits (Guattari 1995). In postcontemporary discourse, the conservation toward sustainability paradigm emphasizing biocultural heritage replaces the prior paradigm of conservation toward nature pristine that emphasized wilderness (Estevez et al. 2010; Rotherham 2015; see Table 1).

Two methodological aspects of the paradigmatic shift in the Andes bring currency to BCLs: (1) demystifying of hinterlands and (2) reaffirmation of mountain identity. Throughout the tropical Andes, a plethora of *pueblos originarios* struggle to maintain

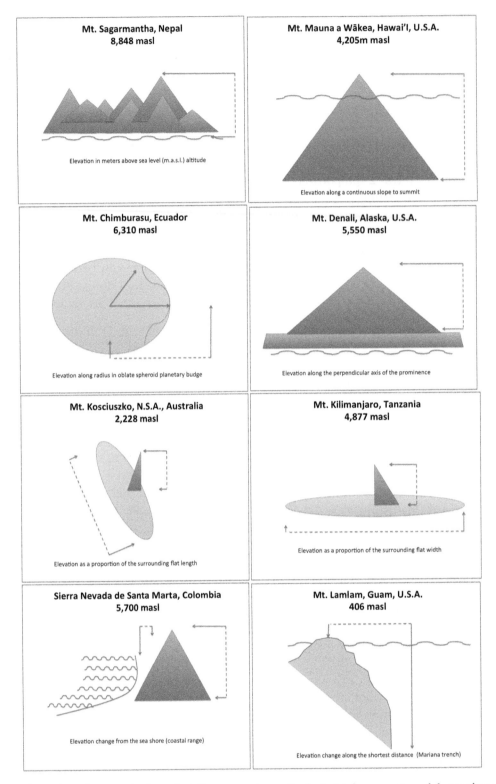

Figure 1. Conceptual models used to consider the eight tallest mountains on Earth. A simple convention of choice plays a major role in the new understanding of mountain socioecological systems. (Color figure available online.)

their identity and sovereignty amidst globalized acculturation; they recently gained political clout and social agency at the national and local levels, as in the community conserved areas and *resguardos indígenas* (e.g., Rozzi 2012; Sarmiento and Hitchner forthcoming). Because mountains harbor high

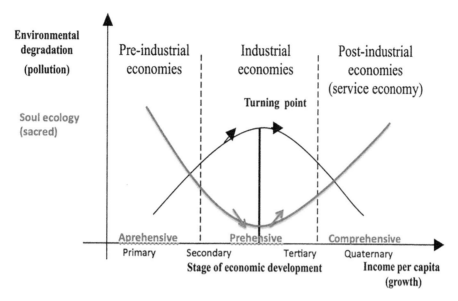

Figure 2. A quadratic equation representing the inverse Kuznets environmental degradation curve exhibited by the changing soul ecology dimension of the sacred, as societal forces move from apprehended cultures of primitive, preindustrial stages to prehensive, modern industrial cultures and to comprehensive, postindustrial stages. *Source*: Panayotou (1993). (Color figure available online.)

biodiversity values and provide the majority of landscape services (Bastian et al. 2014), we posit the notion of highland identity to incorporate insights from folk literature and eco-critical narratives of regional saliency for climate change scenarios, particularly those dealing with Andean biota affected by global warming and rural-to-urban migration.

Table 1. Paradigms of biogeography, including main exponents and the timeline period of major shifts in the different scientific trends observed

Shifting paradigm	Period	Scientific trend	Main exponent
Geodesic shape	1736	Planetocentric	Charles Marie de La Condamine
Binomial systematics	1738	Divine taxonomic order	Carl Linnaeus
Natural history	1777	Phenomenology	Georg Adam Forster
Altitude/latitude	1802	Romanticism	Alexander von Humboldt
Plant taxonomy	1817	Revisionism	Agustin-Pyramus de Candolle
Natural selection	1838	Pragmatism	Charles Darwin
Natural selection	1858	Spiritualism	Alfred Russel Wallace
Field biology	1863	Mimetism	Henry Bates
Plant ecology	1890	Reductionism	Carl G. Oscar Drude
Phytogeography	1896	Continentality	Heinrich G. Adolf Engler
Zoogeography	1899	Dispersal	Robert F. Scharff
Chorology	1907	Utilitarianism	Alfred Hettner
Landscape ecology	1939	Positivism	Carl Troll
Historical geography	1952	Culturalism	Carl O. Sauer
Island biogeography	1967	Insularity	Robert MacArthur
Evolutionary biology	1970	Phylogeneticism	Ernest Mayr
Ecosystem ecology	1971	Holism	Eugene P. Odum
Sociobiology	1975	Altruism	Edward O. Wilson
Bioenergetics	1995	Emergism	Odum, Odum, & Odum
Biodiversity	1998	Conservationism	Norman Myers
Human impact	2005	Integrationism	Jared Diamond
Montology	2007	Postmodernism	Robert Rhoades

Criticality of Mountain Biogeography Theory

Biogeography falls within physical geography. Conventional mountain biogeography describes why organisms occupy the vertical gradient (Resler and Sarmiento 2016), with inferences on distribution and migrations, such as when finding the equilibrium between colonization and extinction on insular peaks (MacArthur and Wilson 1967). Linking vegetation with climate and elevation was the driver of physiographic studies of mountain lands. Island biogeography theory, however, did not employ critical approaches in science and society (Slaughter and Rhoades 2005) needed to (re)present the interweaved BCL fabric matrix (Sarmiento 2012). We agree with Lave et al. (2014) when affirming "Critical Physical Geography (CPG) combines critical attention to power relations with deep knowledge of biophysical science or technology in the service of social and environmental transformation" (2). In montology, it is clear that critical biogeography necessitates the political ecology angle that must be emphasized in the training of geographers (Castree 2000). Separating nature and culture is impracticable in SES: "Socio-biophysical landscapes [cf. BCLs] are as much the product of unequal power relations, histories of colonialism, and racial and gender disparities as they are of hydrology, ecology, and climate change" (Lave et al. 2014, 6).

Andean biogeography played a key role in our understanding of mountain systems. Humboldtian views included not only field measurements and experimentation in geoecology but also poetry and nature paintings (Wulf 2015). Tropandean BCLs henceforth became the "birthplace of ecology" and remain a powerful inspiration for montological research (Figure 3). The name *Andes* itself comes from terracing, the impressive manufactured feature that conquistadors encountered exploring the cordillera or *Ritisuyu*. Critical biogeography explains the essence of BCLs in the Andean Holocene (White 2013).

Regional Saliency of Montological Research

Describing the Equatorial Andes, Humboldt wrote of palm trees in the inter-Andean valleys,

Figure 3. The biocultural landscape is exemplified by Mt. Chimburasu (in Spanish, Chimborazo) as the tallest mountain on Earth and as a telluric presence of this *Apu* in mountain communities. The *páramo* and the *llamakuna* represent the inseparable nature–culture coupling of biocultural heritage, similar to what Humboldt found in 1802. *Photo*: Fausto O. Sarmiento, 1998. (Color figure available online.)

without considering that three centuries ago the *Inka*'s northward expansion had introduced the royal palm (*Parahubaea cocoides* Burret) as a marker of *Inka* nobles settled toward the Empire's confines. Today those palms cannot be found wild in Ecuador but still decorate the entrance to homesteads and central plazas of mountain villages as the *coco chileno* palm. Unlike the Columbian Exchange of 1492 (Crosby 2003), the pre-Columbian exchange of antiquity awaits scientific scrutiny. Moreover, Helferich (2011) wrote that Humboldt noted extensive *páramo* grasslands above the treeline, but he did

Table 2. Examples of how local viewpoints influence current understanding of the Andes mountainscape.

Field of knowledge and topical area	Actual term *Latinized origin*	Popular meaning *Hegemony*	Vernacular meaning *Indigenous revival*
Place naming: *how the name influences common understanding*			
Onomastics	Andes cordillera *Antisuyu*	A tribe said to live east of Cuzco	*Ritisuyu*: the zone of high mountains with snow in the upper reaches
Political ecology	Río Santa Rosa *Roman sanctorum*	Honoring a Catholic saint from Lima, with no ecological value	*Chakapata*: the description of the actual ecological character of the river
Etymology	Páramo *Alpine para-moor*	Highland grasslands	*Paramuna*: the meteorological condition of cold drizzle
Medicinal plants: *how usage guides common practice*			
Placebo effect	Cedrón *Aloysia citrodora*	Lemon verbena tea	*Shunguyaku*: infusion to calm your nerves
Energy boost	Coca *Erythroxylon coca*	Illegal plant for high alkaloid content	*Kuka*: sacred leaf to offer limitless energy and reduce hunger and thirst
Disinfectant	Sangre de drago *Croton lechleri*	Sap with antibacterial properties	*Draku*: sap with curative antiseptic properties
Sacred trees: *how myth explains the existence in place*			
Origin	Lechero *Euphorbia laurifolia*	Fence post and browsing source	*Pinllu*: sacred tree of the *Atawallu runakuna* of Ecuador
Oracle	Araucaria *Araucaria araucana*	Monkey-puzzle tree	*Pewen*: *Mapuche* sacred food (*ngülliw*) or drink (*chavid*)
Destiny	Arbol de dios *Buddleja incana*	Stunted tree of the highlands	*Kiswar*: sacred tree for the *Inka*, a source of awe, inspiration, and timber
Foodstuff: *how utilization favors traits in place*			
Nutrition	Quinoa *Chenopodium quinoa*	The Andean cereal, now popular in organic food stores	*Kinwa*: sacred food of the *Inka* with domesticated hybrids
Wholesome	Mashua *Tropaeolum tuberosum*	The root from the altiplano, aphrodisiac	*Maswara*: sacred food of the *Aymara*
Security	Chuño *Solanum tuberosum andinum*	Freeze-dried potato of the Andes	*Chuñu*: potatoes pressed and dehydrated in the cold mountain air
Rituals: *how observance and tradition inform practice*			
Psychotropic	Ayahuashca *Banisteriopsis caapi*	Native psychoactive brew	*Yagué*: Spiritual medicine for the Pan Amazonian tribes of the verdant
Respect	Alubillo *Toxicodendrum striatum*	Manzanillo, the poison ivy of the Andes	*Ninakaspi*: a guardian who can hurt with allergic rashes
Cleansing	Ortiga *Caiophora superba*	Ortiga, the nettle of the Andes	*Itapalla*: rituals for cleansing and traditional medicine of *yachag*
Flora causation: *how plants are distributed in the landscape*			
Planting	Paja *Calamagrostis*	Páramo pajonal, a wet grassland system	*Ichumanta*: Planted where grazing pressure is felt after burning for clearings
Selection	Aliso *Alnus jorulensis*	Alder growing in disturbed coves	*Jatun Kaspi*: Agri-forestal system grew on the terraces along with other crops
Externality	Cacaotillo *Miconia robinsoniana*	Shrub-dominated area in the mountains of the Galapagos Islands	*Colquita*: Its appearance in the fossil record is recent, just after introduced cattle grazing
Fauna causation: *how animals are distributed in the landscape*			
Domestication	Llama *Lama glama*	Beast of burden	*Llama*: Provides everything but milk to Andean societies
Selection	Cobayo *Cavia porcellus*	Pets for laboratory experiments	*Kuy*: Clean the homes and provide meat for Andean households
Externality	Oso de anteojos *Tremarctos ornatus*	Indicator species for the Páramo ecosystem	*Ucumari*: an inhabitant of the cloud forest belt, uses the grasslands to scout for terrestrial bromeliads

not mention that throughout colonial times, the straw from *paja* (*Calamagrostis* spp., *Festuca* spp., *Stipa* spp.) was a most valuable commodity for construction, roofing, textile, handcrafts, transportation, insulation, and fodder to sheep, cattle, and goats introduced with the "mediterranization" of the Andean farmscape. This made *pajonales* an attractive economic alternative in the highlands, leading to an anthropogenic, planted, and burned grassland and heathland or moorland composed of pyrophitic species. Following the views of pastoral thinkers from Europe at the time, Humboldt recorded extensive grazing by some 40,000 heads of sheep in the *páramos* of Antisana (Bunkse 1981) as his utopic portrait of the Andes, maximizing the catholic bucolic ideals, yet forgetting the devastation due to overgrazing, ignoring the manufactured BCL evidence, and affixing the segmented view of tropical mountains.

Some species remind us of intricate coupled nature–culture interactions (Gade 1999). Fauna–flora interactions better exemplify what is now generally accepted: The human impact of change during the Holocene is responsible for what we see today in Andean mountainscapes (Orlove 1985; Sarmiento 2002; Tellkamp 2014). Inter-Andean

valleys are now lacking natural vegetation and are composed of a BCL matrix of crops and introduced grasses, hedges, and patches of planted woodlots of pine and eucalyptus, all interacting with cows, sheep, goats, chickens, and the native *kuy* or guinea pig (*Cavia porcellus* Linnaeus). The misnomer of *kuy* as "African swain" reflects a critical hegemonic discourse on how and why species have been named without regard to the vernacular. Critical biogeography, thus, using the 4 Cs multimethod, explains how fortuitous distributions can only partially account for spatial interpretations and that novel ways should provide not only *content* (e.g., the cattle egret, *Bubulcus ibis* Linnaeus, as a disperser) but also *continent* (its distributional range along herds of livestock on wet American grasslands after their arrival in North America in 1941 from Africa), *contestation* (e.g., why all cattle egrets are now gone from Andean wetlands, reflecting cattle ranching woes and climate change), and *conveyance* (e.g., their inclusion on the bird list as emblematic of the grasslands, although in reality cattle egrets are ephemeral), thus making it human-dependent as driving change (see Table 2).

Ibarra et al. (2012) used the iconic Andean condor (*Vultur gryphus* Linnaeus) to translate the

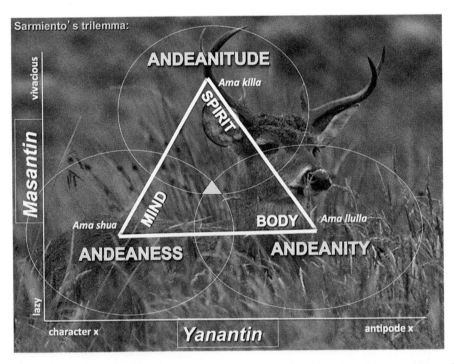

Figure 4. An example of the white-tail deer as an iconic biocultural landscape marker that aids in the construction of the Andean identity. Note the polarity vector of *Yanantin* being influenced by the energizing vector of *Masantin* affecting the trifecta of Andeanity, Andeaness, and Andeanitude to understand the vortex of being Andean (*Lo Andino*). *Source*: Adapted from Sarmiento (2015). (Color figure available online.)

cosmological vision onto the BCL, which is essential for Andean stewardship (Rozzi 2004). In the same vein, the Andean Lapwing (*Vanellus resplendens* Tschudi) is a proxy for farmscape transformation, abandoning the nature–culture dichotomy in favor of a BCL perspective that reciprocally explains the nature–culture continuum (Sarmiento 2016b).

Taxonomic groups of species of wide distribution and diverse cultural meanings along, across, and within the Andes, like the *caracaras* (e.g., *Milvago chimango* Vieillot, *Caracara plancus* Miller, and *Phalcoboenus megalopterus* Meyen), reflect the human–environmental interaction from the lowlands toward the highlands, including Amazonian (McMichael et al. 2012) and marine-coastal environments (Pizarro, Anderson, and Rozzi 2012). Other species such as llamas (*Lama glama* Linnaeus; May 2015) and white-tailed deer (*Odocoileus virginianus* Zimmermann; Sarmiento 2015) exemplify the bridging of the social and biological sciences from the critical biogeography perspective of Andean identity by integrating mysticism and the spiritual dimension (Figure 4). Barreau et al. (2016) showed

Figure 5. Gathering *piñones* (seeds of the monkey-puzzle tree *Araucaria araucana* Molina) and other wild edible plants. Restricted access to Andean temperate forests, due to land grabbing and biodiversity preservation initiatives, is endangering the continuity of this practice for the *Mapuche* people. *Photo*: J. Tomás Ibarra, 2014. (Color figure available online.)

how land grabbing, lack of access to mountain forests, and formal Chilean school regime have eroded plant knowledge transmission to children, thereby limiting local food sovereignty (Figure 5). In northwestern Argentina, González et al. (2014) recognize quinoa (*Chenopodium quinoa* Willd) as a "superfood" highlighted by the International Year of Quinoa in 2013. This United Nations

(A)

ALEXANDER VON HUMBOLDT–23 JUNE 1802

IN MEMORY OF HIS CONTRIBUTIONS TO MOUNTAIN GEOECOLOGY

The Andean Mountains, especially Chimborazo, stirred the imagination and scientific labor of this great man. In addition to his many other contributions, it was in this tropandean landscape, beneath the eternal snows of our majestic volcano, where he laid the foundations of "mountain geoecology," or "montology," that continues to mold world society. The Rio de Janeiro Earth Summit of 1992 ensured international recognition of the importance of our mountains, in part from United Nations University research, and created an awareness that is finally transcending into action. This advance culminated in November 1998, when the General Assembly of the United Nations declared AD 2002 as the International Year of the Mountains

"FOR A BETTER BALANCE BETWEEN MOUNTAIN ENVIRONMENT, DEVELOPMENT OF RESOURCES, AND THE WELL-BEING OF MOUNTAIN PEOPLES"

Chimborazo, the birthplace of Mountain Geoecology
December 15, 1998

Indigenous Committees of Chimborazo
Jack D. Ives (IMS/UNU) International Mountain Society
Fausto O. Sarmiento (AMA) Andean Mountains Association
Lawrence S. Hamilton (WCPA-IUCN) World Conservation
Union, Commission on Protected Areas: Mountains

Bruno Messerli (IGU) International Geographical Union
Juan Hidalgo (CEPEIGE) Pan American Center for
Geographical Studies and Research
Patricio Hermida (INEFAN) Chimborazo Reserve Manager

(B)

Figure 6. (A) Montology plaque at the mountaineering refuge at the snowline of Mt. Chimburasu, Ecuador. Pictured among local indigenous leaders and governmental officials are mountain scientists who promote montological research. (B) Text of the English version of the montology plaque on the cairn of Mt. Chimburasu. The Spanish and *Kichwa* versions were included to relate to regional, national, and local audiences.

recognition is due to its high-quality protein content and its plasticity for climate change adaptation on marginal and saline lands of the Altiplano.

Conclusion

Mountain investigation begs the inclusion of conventional hard and modern soft science, including indigenous and TEK, to create the new transdisciplinary science of montology. Just as environmental geography is envisioned to capture the interaction of physical and human dimensions in the essence of place, so is montology intended to capture the meaning of mountains and their lifescape. With examples of Andean plants and animals, the coupling of human–environment interactions has shown integration of physical, psychological, and spiritual dimensions to understand mountain territoriality on biota distribution. In addition, the transdisciplinary montological approach includes little-known facts from indigenous cosmologies and habitual usages that help better protect charismatic, even totemic, species from endangerment. Critical biogeography, hence, must drive research in mountain scenarios of climate change and biota distribution (Figure 6).

Funding

Funds for Fausto O. Sarmiento came from the University of Georgia Latin American and Caribbean Studies Institute's NRC Title VI (ED/NRC/P015A140046) to the Neotropical Montology Initiative and the University of Georgia Willson Center to the Research Cluster on Indigenous Foods and Fabrics. J. Tomás Ibarra was supported by the Center for Intercultural and Indigenous Research (CONICYT/FONDAP/15110006).

References

Allan, N. 1986. Accessibility and altitudinal zonation models of mountains. *Mountain Research and Development* 6 (3): 185–94.

Allan, N., G. Knapp, and C. Stadel. 1988. *Human impact on mountains.* Lanham, MD: Rowman & Littlefield.

Arpin, I., and A. Cosson. 2015. The category of mountain as source of legitimacy for national parks. *Environmental Science & Policy* 49:57–65.

Barreau, A., J. Ibarra, F. Wyndham, A. Rojas, and R. Kozak. 2016. Generational change in Mapuche knowledge of wild edible plants in Andean temperate ecosystems of Chile. *Journal of Ethnobiology* 36:412–32.

Bastian, O., K. Grunewald, R.-U. Syrbe, U. Walz, and W. Wende. 2014. Landscape services: The concept and its practical relevance. *Landscape Ecology* 29:1463–79.

Berkes, F. 2012. *Sacred ecology.* London and New York: Routledge.

Berkes, F., C. Folke, and J. Colding. 2000. *Linking social and ecological systems: Management practices and social mechanisms for building resilience.* Cambridge, UK: Cambridge University Press.

Borsdorf, A., and C. Stadel. 2015. *The Andes: A geographical portrait.* New York: Springer.

Bradshaw, G., and M. Bekoff. 2001. Ecology and social responsibility: The re-embodiment of science. *Trends in Ecology & Evolution* 16:460–65.

Bunkse, E. 1981. Humboldt and an aesthetic tradition in geography. *The Geographical Review* 71 (2): 127–46.

Castree, N. 2000. Professionalisation, activism, and the university: Whither "critical geography"? *Environment and Planning A* 32 (6): 955–70.

———. 2014. *Making sense of nature: Representation, politics, and democracy.* London and New York: Routledge.

Christensen Fund. 2016. *Backing the stewards of cultural and biological diversity.* https://www.christensenfund.org (last accessed 6 October 2016).

Cocks, M. 2006. Biocultural diversity: Moving beyond the realm of indigenous and local people. *Human Ecology* 34 (2): 185–200.

Convery, I., and P. Davis, eds. 2016. *Changing perceptions of nature.* Martlesham, UK: Boydell.

Crosby, A. 2003. *The Columbian exchange: Biological and cultural consequences of 1492.* Westport, CT: Greenwood.

Debarbieux, B., and G. Rudaz. 2015. *The mountain: A political history from the Enlightenment to the present.* Chicago: University of Chicago Press.

Estevez, A., D. Sotomayor, A. Poole, and C. Pizarro. 2010. Creating a new cadre of academics capable of integrating socio-ecological approach to conservation biology. *Revista Chilena de Historia Natural* 83:17–25.

Fouberg, E., and W. Moseley. 2015. *Understanding world regional geography.* Hoboken, NJ: Wiley Blackwell.

Funnell, D., and M. Price. 2003. Mountain geography: A review. *The Geographical Journal* 169 (3): 183–90.

Fu, B., and K. Jones, eds. 2013. *Landscape ecology for sustainable environment and culture.* New York: Springer.

Gade, D. 1996. Carl Troll on nature and culture in the Andes. *Erdkunde* 50:301–16.

———. 1999. *Nature and culture in the Andes.* Madison: University of Wisconsin Press.

———. 2016. *Spell of the Urubamba.* New York: Springer.

González, J., S. Eisa, S. Hussin, and F. Prado. 2014. Quinoa: An Incan crop to face global changes in agriculture. In *Quinoa: Improvement and sustainable production,* ed. K. S. Murphy and J. Matanguihan, 1–18. Hoboken, NJ: Wiley-Blackwell.

Gould, P. 1979. Geography 1957–1977: The Augean period. *Annals of the Association of American Geographers* 69 (1): 139–51.

Graham, M., and T. Shelton. 2013. Geography and the future of big data, big data and the future of geography. *Dialogues in Human Geography* 3 (3): 255–61.

Guattari, F. 1995. *Chaosophy*, ed. S. Lotringer. New York: Semiotext(e).

Gudynas, E. 2013. Debates on development and its alternatives in Latin America: A brief heterodox guide. In *Beyond development*, ed. M. Lang and D, Mokrani, 15–39. Amsterdam: Transnational Institute.

Haslett, J. 1998. A new science: Montology. *Global Ecology and Biogeography Letters* 7 (3): 228–29.

Helferich, G. 2011. *Humboldt's cosmos: Alexander von Humboldt and the Latin American journey that changed the way we see the world*. Old Saybrook, CT: Tantor eBooks.

Hong, S.-K., J. Bogaert, and Q. Min, eds. 2014. *Biocultural landscapes: Diversity, functions and values*. New York: Springer.

Ibarra, J., A. Barreau, F. Massardo, and R. Rozzi. 2012. The Andean condor: A biocultural keystone species of the South American landscape. *Boletín Chileno de Ornitología* 18 (1–2): 1–22.

Koutsopoulos, K. 2011. Changing paradigms of geography. *European Journal of Geography* 1:54–75.

Lave, R., W. Wilson, E. Barron, C. Biermann, M. Carey, C. Duvall, L. Johnson, K. Lane, N. McClintock, D. Munroe, and R. Pain. 2014. Intervention: Critical physical geography. *The Canadian Geographer* 58 (1): 1–10.

Lewis, M., and K. Wigen. 1997. *The myth of continents: A critique of metageography*. Berkeley: University of California Press.

MacArthur, R., and E. Wilson. 1967. *Theory of island biogeography*. Princeton, NJ: Princeton University Press.

Macfarlane, R. 2009. *Mountains of the mind*. London: Granta.

May, R. H., Jr. 2015. Andean llamas and Earth stewardship. In *Earth stewardship*, ed. R. Rozzi, F. S. Chapin III, J. B. Callicott, S. T. A. Pickett, M. E. Power, J. J. Armesto, and R. H. May, Jr., 77–86. New York: Springer.

McMichael, C., D. Piperno, M. Bush, M. Silman, A. Zimmerman, M. Raczka, and L. Lobato. 2012. Sparse pre-Columbian human habitation in western Amazonia. *Science* 336 (6087): 1429–31.

Messerli, B., and J. Ives. 1997. *Mountains of the world: A global priority*. New York and Carnforth: Parthenon.

Montology. 2002. *Abridged Oxford English dictionary*. https://en.oxforddictionaries.com/definition/montology (last accessed 13 December 2016).

Neustadtl, S. 1977. Montology: The ecology of mountains. *Technology Review* 79 (8): 64–66.

Orlove, B. 1985. The history of the Andes: A brief overview. *Mountain Research and Development* 5:45–60.

Panayotou, T. 1993. *Empirical tests and policy analysis of environmental degradation at different stages of economic development*. Geneva, Switzerland: International Labour Office.

Pizarro, J., C. Anderson, and R. Rozzi. 2012. Birds as marine–terrestrial linkages in sub-polar archipelagic systems: Avian community composition, function and seasonal dynamics in the Cape Horn Biosphere Reserve (54–55°S), Chile. *Polar Biology* 35:39–51.

Price, M., A. Byers, D. Friend, T. Kohler, and L. Price, eds. 2013. *Mountain geography: Physical and human dimensions*. Berkeley: University of California Press.

Pungetti, G. 2013. Biocultural diversity for sustainable ecological, cultural and sacred landscapes: The biocultural landscape approach. In *Landscape ecology for sustainable environment and culture*, ed. B. Fu and K. B. Jones, 55–76. New York: Springer.

Resler, L., and F. Sarmiento. 2016. Mountain geographies. In *Oxford bibliographies in geography*, ed. B. Warf. New York: Oxford University Press.

Rhoades, R. 2007. *Listening to the mountains*. Dubuque, IA: Kendall.

Rotherham, I. 2015. Bio-cultural heritage and biodiversity: Emerging paradigms in conservation and planning. *Biodiversity & Conservation* 24 (13): 1–25.

Rozzi, R. 2004. Implicaciones éticas de narrativas yaganes y mapuches sobre las aves de los bosques templados de Sudamérica austral [Ethical implications of Yagan and Mapuchenarratives on birds of the temperate forests of southern South America]. *Ornitología Neotropical* 15:435–44.

———. 2012. South American environmental philosophy: Ancestral Amerindian roots and emergent academic branches. *Environmental Ethics* 34:343–65.

———. 2015. Earth stewardship and the biocultural ethic: Latin American perspectives. In *Earth stewardship*, ed. R. Rozzi, F. S. Chapin, III, J. B. Callicott, S. T. A. Pickett, M. E. Power, J. J. Armesto, and R. H. May, Jr., 87–112. New York: Springer.

Rozzi, R., J. Armesto, J. Gutiérrez, F. Massardo, G. Likens, C. Anderson, and T. Arroyo. 2012. Integrating ecology and environmental ethics: Earth stewardship in the southern end of the Americas. *BioScience* 62 (3): 226–36.

Rozzi, R., F. Chapin, III, J. Callicott, S. Pickett, M. Power, J. Armesto, and R. May, Jr. 2015. Introduction: Linking ecology and ethics for an interregional and intercultural Earth stewardship. In *Earth stewardship*, ed. R. Rozzi, F. S. Chapin III J. B. Callicott, S. T. A. Pickett, M. E. Power, J. J. Armesto, and R. H. May, Jr., 1–14. New York: Springer.

Sarmiento, F. 2000. Human impacts in man-aged tropandean landscapes: Breaking mountain paradigms. *Ambio* 29 (7): 423–31.

———. 2002. Anthropogenic landscape change in highland Ecuador. *The Geographical Review* 92 (2): 213–34.

———. 2012. *Contesting Páramo: Critical biogeography of the northern Andean highlands*. Charlotte, NC: Kona.

———. 2015. The antlers of a trilemma: Rediscovering Andean sacred sites. In *Earth stewardship*, ed. R. Rozzi, F. S. Chapin III, J. B. Callicott, S. T. A. Pickett, M. E. Power, J. J. Armesto, and R. H. May, Jr., 49–64. New York: Springer.

———. 2016a. GEOG 3290: Mountain geography. University of Georgia course syllabus. http://bulletin.uga.edu/link.aspx?cid=GEOG3290 (last accessed 12 October 2016).

———. 2016b. Identity, imaginaries and ideality: Understanding the cultural landscape of the Andes through the iconic Andean lapwing (*Vanellus resplendens*). *Revista Chilena de Ornitología* 22 (1): 38–50.

Sarmiento, F., E. Box, and L. Usery. 2004. GIScience and tropical mountains: A challenge for geoecological research. In *Geographic information science and mountain geomorphology*, ed. M. Bishop and J. Shroder, Jr., 289–307. New York: Springer.

Sarmiento, F., and S. Hitchner, eds. Forthcoming. *Indigeneity and the sacred: Indigenous revival and the conservation of sacred natural sites in the Americas.* New York: Berghahn.

Slaughter, S., and G. Rhoades. 2005. From "endless frontier" to "basic science for use": Social contracts between science and society. *Science, Technology & Human Values* 30 (4): 536–72.

Tellkamp, M. 2014. Habitat change and trade explain the bird assemblage from the La Chimba archaeological site in the northeastern Andes of Ecuador. *Ibis* 156 (4): 812–25.

Vallega, A. 1999. Ocean geography vis-a-vis global change and sustainable development. *The Professional Geographer* 51 (3): 400–14.

Welberry, K. 2005. Wild horses and wild mountains in the Australian cultural imaginary. *PAN: Philosophy Activism Nature* 3:23–32.

White, S. 2013. Grass páramo as hunter-gatherer landscape. *The Holocene* 23 (6): 898–915.

Wulf, A. 2015. *The invention of nature: Alexander von Humboldt's new world.* New York: Knopf.

Snowlines and Treelines in the Tropical Andes

Kenneth R. Young, Alexandra G. Ponette-González, Molly H. Polk, and Jennifer K. Lipton

Examination of the dynamism of snowlines and treelines could provide insights into environmental change processes affecting land cover in the tropical Andes Mountains. Further, land cover at these ecotones represents a powerful lens through which to monitor and understand ecological processes across biophysical gradients while acknowledging their socioenvironmental dimensions. To illustrate this approach, we draw on recent research from two sites in the high tropical Andes where, at the regional scale, land cover assessments document retreating glaciers and changing amounts of forest cover, even though steep topographic gradients impose spatial shifts at much finer scales. Our results show that heterogeneous patterns of glacier recession open up new ecological spaces for plant colonization, potentially forming new grasslands, shrublands, and wetlands. In addition, treeline shifts are tied to changes in woody plant dominance, which can vary in rate and pattern as a result of aspect, past land use, and current livelihoods. We suggest that the telecoupling of regional and global biophysical and socioeconomic drivers of land use and land cover change to specific landscape combinations of elevation, aspect, and slope position might explain much of the spatial heterogeneity that characterizes landscape stasis and flux in mountains.

检视雪线和树线的动态, 能够为影响热带安第斯山脉的土地覆盖之环境变迁过程提供洞见。此外, 这些交错群落的土地覆盖, 呈现一个有力的视角, 藉此监控并理解生物物理梯度中的生态过程, 同时认识其社会环境之面向。为了阐述此一方法, 我们运用在高山热带安第斯山中的两地进行的晚近研究, 在区域尺度上, 这两个地方中的土地覆盖评估, 记录了后退的冰川以及森林覆盖的改变量, 尽管陡峭的地形梯度是在更细微的尺度上加诸空间变迁。我们的研究显示, 冰川倒退的异质模式, 开展了植物定殖的崭新生态空间, 并有可能构成新的草原、灌丛带和湿地。此外, 树线变迁与木本植物优势的改变有关, 并可能因坡向、过往的土地使用和当前的生计, 而在速率和模式上有所不同。我们主张, 驱动海拔、坡向和坡位的特定地景组合的土地使用和土地覆盖变迁之区域及全球生物物理及社会经济驱力的远程耦合, 或能大幅解释以山区地景静止和流动为特征的空间异质性。关键词: 安地斯山脉, 气候变迁, 生态演替, 冰川倒退, 土地使用/土地覆盖变迁。

El examen de la dinámica en las líneas de nieve y las líneas arbóreas podría suministrarnos una mayor comprensión de los procesos de cambio ambiental que afectan la cubierta de la tierra en las montañas andinas tropicales. Más todavía, esa cobertura en estos ecotones representa una lente poderosa a través de la cual monitorear y entender los procesos ecológicos a través de gradientes biofísicos, al tiempo que se reconocen sus dimensiones socioambientales. Para ilustrar este enfoque, nos apoyamos en investigación reciente conducida en dos sitios de los altos Andes tropicales donde, a escala regional, las evaluaciones de la cubierta de la tierra documentan el retroceso de los glaciares y las cantidades cambiantes de cobertura forestal, incluso si los fuertes gradientes topográficos imponen cambios espaciales a escalas mucho más finas. Nuestros resultados muestran que los patrones heterogéneos de la recesión glaciaria abren nuevos espacios ecológicos de colonización por las plantas, formando potencialmente nuevos pastizales, matorrales y humedales. Además, los cambios de la línea arbórea están ligados a cambios en el dominio de plantas leñosas, las cuales pueden variar en rata y patrón como resultado del aspecto, uso del suelo anterior y actuales medios de subsistencia. Sugerimos que el acoplamiento a distancia de los controles biofísicos y socioeconómicos regionales y globales del uso del suelo y del cambio de la cubierta de la tierra, por combinaciones específicas de paisajes de elevación, aspecto y posición de las laderas, podrían explicar gran parte de la heterogeneidad espacial que caracterizan la movilidad y el flujo del paisaje en las montañas.

Shifting climate regimes and socioeconomic processes are altering perceptions and decisions by people in the Andes Mountains, thus affecting their land uses, while concurrent changes in ecological processes alter land cover types, ecosystem responses and functions, and environmental services used by people (Young 2009; Ponette-González et al. 2014). Such global-to-local connectivity has been referred to as *telecoupling* (Liu et al. 2013). For example, glacier recession in the Andes is driven by climate change

(Rabatel et al. 2013), with many implications for livelihoods (Postigo, Young, and Crews 2008) and watersheds (Bradley et al. 2006; Carey et al. 2014), whereas other studies report increased woody plant cover, especially in drier areas (Aide et al. 2013). Demographic shifts to urban areas (Álvarez-Berríos, Parés-Ramos, and Aide 2013) and massive investments in irrigation and energy infrastructure have pulled people and their environmental influences to lower elevations (e.g., Bury et al. 2013).

Evaluation of shifting ecological transition zones, or ecotones, might provide an important means of monitoring these global as well as localized influences on land cover dynamics in the Andes and in similar places worldwide. Snowlines separate the cryosphere, including places with permanent or seasonal snow or ice cover, from other landscapes, and treelines distinguish places along biophysical gradients where land cover switches from dominance by woody plants to herbaceous vegetation or nonvegetated cover types. At a small cartographic scale, such lines can be used to indicate locations of important biophysical changes that control, for example, the mass balance of glaciers and the altitudinal limit of closed forest (Gerrard 1990). These ecotones also might be highly sensitive to global- and local-scale changes.

Tropical glacier snowlines are located near the 0° isotherm, so any shift in temperature or precipitation can trigger shifts in mass balance. Thus, tropical glaciers fluctuate over yearly and decadal time periods (Rodbell 1993; Thompson et al. 2006; Stansell et al. 2013). During the Little Ice Age, tropical Andean glaciers reached maximum extensions, but then later—beginning in 1880—they experienced significant and accelerated recession that continues today (Vuille et al. 2008). Additionally, steep elevation gradients and tropical seasonality influence glacier fluctuations. Drivers of contemporary tropical glacier retreat are thought to include combinations of shifts in air temperature, humidity, precipitation, cloudiness, incoming shortwave radiation, and related anthropogenic climate change (Kaser 1999; Vuille et al. 2003; Thompson et al. 2006; Vuille et al. 2008).

Tropical Andean treelines are affected by multiple ecological processes (e.g., seed dispersal, seedling growth) in a grassland–forest ecotone that can experience cool (6–10°C) and humid (500–2,000 mm per year) conditions, frequent fogs, and episodic frost events (Young and León 2007). Since 1939, surface temperatures in the tropical Andes have increased ∼0.10°C per decade (Vuille et al. 2008). Sediment

records show that treeline is responsive to such climatic fluctuations on millennial timescales, shifting up- and downslope in response to warm–wet and cool–dry periods, respectively (Bush et al. 2005). Human-set fires and grazing, however, play a decisive role in shaping treeline structure and distribution as well (White 2013). Where grasslands are grazed and burned, altitudinal treeline is often depressed by several hundred meters (Young 1993a); woodlands might be absent or switch abruptly into grasslands (Ponette-González et al. 2016). Future dynamics of tropical forest–grassland ecotones are thus likely to be far more complex than the upslope shift predicted with simple climate models.

In this research, we examined the landscape-scale consequences of three decades of change in both glacier extent and representation of woody and nonwoody vegetation in Peru. A case study in Huascarán National Park, in the Cordillera Blanca, documents and interprets recent land cover change using satellite imagery. Another case study evaluates treeline response to past land use in Río Abiseo National Park, on the Eastern Andean Cordillera. We posit that such examinations reveal additional complexity in spatial patterns, which in turn might lead to insights into processes of change associated with the cryosphere, the biosphere, and interactions with human land use. Regional trends of receding glaciers and shifting vegetation mosaics, although indicative of coupling to global processes, do not necessarily provide useful parameters for predicting future change in particular montane landscapes, given high place-to-place variation. This article attempts to elucidate some of the implications of spatial heterogeneity, both for improving knowledge of mountain environments and for better assessing results of land use/land cover change studies.

Glacier Recession in the Cordillera Blanca of Peru

Study Site and Methods

The effects of glacier recession on downslope ecosystems were studied in the Cordillera Blanca range in northern Peru (8.5° to 10° S), where glacier retreat is widespread (Kaser and Osmaston 2002; Mark and Seltzer 2005; Burns and Nolin 2014). According to the Peruvian government's first detailed countrywide glacier inventory in 1970, glacier area in the Cordillera

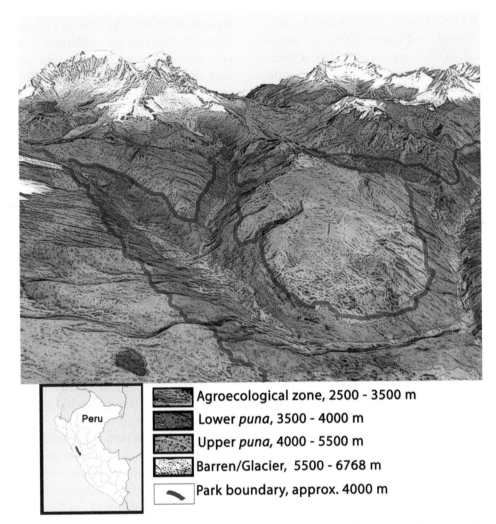

Figure 1. Ecological zonation near and inside Huascarán National Park. The agroecological zone (green) has a mixed tenure regime of private land, sectorial fallow, and park buffer zone; the lower puna (orange) has private and communal land, sectorial fallow, and park buffer zone; the upper puna (yellow) includes communal land and the core area of the park; and the barren rock and glacial ice area (gray/white) is under the tenure of Huascarán National Park. (Color figure available online.)

Blanca was 723 km². By 2003, glacier area had decreased about 30 percent to 527 km². The remaining glaciers have also become more fragmented: In 1970 there were 722 glaciers compared to 755 in 2003 (Autoridad Nacional del Agua, Unidad de Glaciologia 2013). With increased surface-to-area (or mass) ratios, these small glaciers are likely to disappear. Yet, relatively little is known about effects of recession on lower elevation ecosystems, including grasslands, shrublands, forests, and wetlands (Young 2015).

The Cordillera Blanca is oriented on a northwest–southeast axis and is composed of a series of parallel valleys separated by high ridges. Much of the Cordillera Blanca is inside Huascarán National Park (3,400 km²). The precipitation regime is defined by a dry season from June to September and a wet season from October to May (Kaser and Osmaston 2002). Land

cover in the park is heterogeneous, with patches of wetlands, shrubs, *Polylepis* forests, and rock outcroppings embedded in a grassland matrix (i.e., "*puna*" ecological zone); scree, bare rock, and snow and ice cover dominate landscapes above 5,000 m. In addition to rising temperatures (Vuille et al. 2003), topography is an important variable affecting biophysical gradients and land use (Figure 1).

Roughly 270,000 people inhabit the Santa River basin (Bury et al. 2011). Livelihoods are predominately based on subsistence agriculture, livestock production, mining, and tourism (Lipton 2014). Semistructured interviews were conducted with 117 informants (89 men, 28 women); interviews were conducted in Spanish and Quechua with the aid of field assistants in the buffer zone and core of Huascarán National Park to identify land use and tenure arrangements and to make land cover observations. Multiple elevation and ecological gradients are simultaneously used for agropastoral and community land use systems (Figure 1) and operate under distinct tenure regimes (private lands, sectorial fallow lands, communal lands), modified by national park regulations since 1975. Households typically maintain croplands from 2,500 to 3,500 m, whereas fields in middle elevations (3,500–4,000 m) are seasonally planted for household or community use. Throughout the region, the upper *puna* (4,000–5,500 m) is used for transhumance of cattle, sheep, and other stock. Cattle and horses typically graze freely in upland valleys, often on wetland or riparian sites. Land use intensity has lessened in recent years given increased migration to the coast; availability of jobs in other sectors including mining, tourism, and corporate farms; and decreased returns on crops.

Land use/land cover in the Cordillera Blanca was mapped by classifying Landsat TM images acquired on 15 May 1987, 20 August 1999, and 18 August 2010 (downloaded from http://glovis.usgs.gov). For each date, two Level 1T images (path-rows 8–66, 8–67) from the dry season were mosaicked together. Level 1T processing provides systematic, radiometric, and geometric accuracy by using a digital elevation model;

image coregistration is consistent and spatial errors are less than one-half pixel (Hansen and Loveland 2012). A hybrid supervised–unsupervised classification was performed (using ERDAS Imagine 2014), a technique that uses all seven bands in 30-m spatial resolution (Walsh et al. 2003). Band 6, the Landsat TM thermal band, was resampled from 120-m to 30-m spatial resolution. This technique was selected due to previous good performance (Kintz, Young, and Crews-Meyer 2006). First, an unsupervised classification technique (ISODATA) clustered the spectral information into 255 signatures. Signatures were then evaluated using the transform divergence method, a test that measures the statistical distance between signatures on a scale of 0 to 2,000. We removed signatures with poor spectral separability (defined as <1,950). Next, the remaining signatures were classified using a supervised classification, resulting in thirty-five to forty-five classes. Using expert knowledge of the study area, 360 ground control points, and photographs, these classes were individually identified as one of seven land cover classes: barren, puna (nonwetland tropical alpine vegetation), wetland, snow and ice, lake, shadow, and cloud. Classes were selected based on previous remote sensing analyses in the study area (Lipton 2008; Silverio and Jaquet 2009). Following Ozesmi and Bauer (2002), we

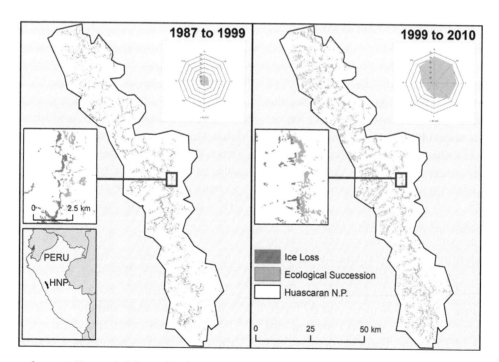

Figure 2. Land cover change in Huascarán National Park, northern Peru, for 1987 to 1999 (left) and 1999 to 2010 (right). Glacier recession (brown) could be followed by primary succession on barren spaces exposed by glacier retreat and by woody plant establishment on scree slopes and other expansions of shrublands and grasslands (shown in green). Inset shows aspect differences in ecological succession for the two time periods. HNP = Huascarán National Park. (Color figure available online.)

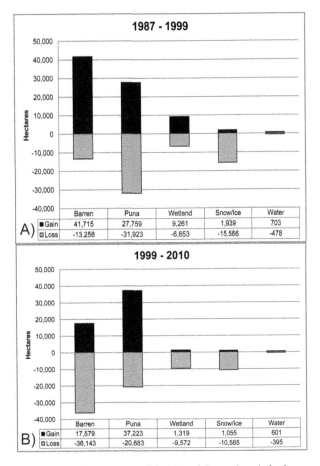

Figure 3. Land cover gains (black) and losses (gray) for barren, puna, wetland, snow and ice, and water in Huascarán National Park, northern Peru, from 1987 to 1999 (top) and 1999 to 2010 (bottom).

added a near infrared–red band ratio (Landsat TM Band 4/Band 3) to the classified image to improve the accuracy of difficult-to-classify wetlands. Lake colors vary widely due to suspended glacial flour in the study area, giving poor classifications. Therefore, lakes were digitized manually using the GLIMS data set and added to the classified image (http://www.glims.org/; Kargel et al. 2014).

The final product was a categorical map for each date. Accuracy assessments for image classification were within standard norms of >85 percent with the exception of 2010. For 1987, overall accuracy was 89.4 percent and overall kappa was 87.5 percent; for 1999, the respective values were 91.5 percent and 89.8 percent. For 2010, overall accuracy was 78.5 percent and overall kappa was 74.2 percent; lower accuracy values in 2010 were due to confusion between shadow and puna classes. In the final step, a geographic information system (GIS; Land Change Modeler 2.0, Clark Labs, Clark University, Worcester, MA, USA) was

used to evaluate the extent of changes from snow and ice to barren (interpreted as glacier recession) and from barren to puna (interpreted as ecological succession, including primary succession) for 1987 to 1999 and 1999 to 2010 (Figure 2). Aspect is an important spatial organizing feature (see Figure 1) and for this analysis was derived from an ASTER GDEM v.2 product and categorized into eight bins representing the cardinal and intercardinal directions. Using a GIS (ArcGIS, Version 10.2, Esri, Redlands, CA, USA), the area of change from barren to puna was calculated for each of the bins.

Principal Findings: Glacial Recession Affects Downslope Ecosystems

Land cover assessments showed that glacier loss resulted in opportunities for ecological succession (Figure 3), as barren lands were exposed and plants colonized open areas. The area subject to plant colonization was, however, much larger in extent than the area affected directly by recession (Figure 2). Field observations suggest that much of this increase was due to forbs and graminoids appearing on newly exposed substrates, in addition to increasing shrub presence on rocky slopes. From 1987 to 1999, there were no differences in ecological succession by aspect, whereas from 1999 to 2010 barren-to-puna change was concentrated on north- and northeast-facing slopes (Figure 2, insets).

The land change analysis revealed additional complexities, including cover changes that were not consistent over time (Figure 3) and from–to class changes that were interconnected with other class changes: (1) from 1987 to 1999, there was a slight net loss of puna, suggesting that some tropical alpine vegetation areas contracted to expose additional barren sites; (2) from 1999 to 2010, there was a large increase in puna due to barren lands converted as succession occurred and grasses and shrubs increased in extent, mostly on north- and northeast-facing slopes previously lacking dense vegetation (Figure 2; we note that confusion in classifying shadows makes some dynamics associated with the puna class ambiguous); and (3) from 1987 to 1999, wetlands experienced a net increase in area but then decreased in area from 1999 to 2010 (Figure 3). The effects of glacier recession (and other concurrent changes) thus extend downslope, directly affecting ecosystems that expand with new sites available for colonization and those affected by changing amounts of glacier meltwater. Gains and losses in land cover varied by aspect, however, and the direction and type of change were not consistent over time.

Woody Plant Encroachment at Treeline in the Eastern Andean Cordillera, Peru

Study Site and Methods

Spatial patterns of woody plant encroachment into grasslands were examined at treeline in the Eastern Andean Cordillera after nearly three decades of grazing reduction and fire exclusion. In the glacially sculpted landscapes of northern Peru (Rodbell 1993), treeline is found in a relatively narrow transition zone (3,200–3,600 m) that extends from the upper limit of closed-canopy montane forest (i.e., timberline) to humid alpine grassland (i.e., puna or páramo). Above the closed forest limit, woody plants encroach into grasslands, decreasing in stature, area, and contiguity with increasing elevation (Kintz, Young, and Crews-Meyer 2006). These high-elevation woodlands or shrublands are frequently located in topographically sheltered sites, such as ravines or atop small boulder fields, and are considered a characteristic and enigmatic feature of high Andean sites (Kessler 2002). The elevation, patterning, and abruptness of this Andean forest–grassland ecotone vary spatially as a result of differences in climate, land use, and terrain (Young 1993b, 1993c). Timberline and woodlands are found higher on valley sides than on parts of valley bottoms, where cold air drainage, waterlogged soils, and occasional fires prevent tree establishment, creating an inverted treeline (Young 1993b).

This study was conducted in a U-shaped valley above the closed forest limit in Río Abiseo National Park (7°56.88′ S, 77°21.35′ W). The site is characterized by a wet (ca. 1,000–2,000 mm per year) and foggy climate, with extreme topographic complexity (Fry, Ponette-González, and Young 2015). Prior to the park's creation in 1983, herders maintained grasslands with fire, resulting in the creation of a mosaic landscape composed of closed forest, different-aged (and different-sized) woodlands, and grassland. Small-statured C_4 grasses exist at the study site (e.g., Paspalum, Muhlenbergia, and Bothriochloa) alongside the dominant grass genera Calamagrostis and Festuca. After 1983, fire was prohibited and the number of cattle in the park's alpine zone decreased by about 70 percent. Although limited grazing by local people is still permitted, in 2007 an estimated dozen cattle remained in the valley, a stocking rate of 0.017 cattle ha^{-1} (Ponette-González et al. 2016).

Vegetation and soil sampling were conducted across the valley in 2010 and 2011. All forested woodlands larger than 250 m^2 were mapped and stratified by valley position (low, mid, upper) and aspect (east, west). Seventeen (3,400–3,900 m) were randomly selected from these strata. To assess patterns of woody plant encroachment into alpine grasslands, fifty transects were laid perpendicular to forest-grassland boundaries on the upper, lower, windward, and leeward sides of the woodlands, except where steep topography prevented access. Along these transects, three 2×1 m^2 plots were established in each habitat: at the forest–grassland edge (0 m) and 5 m on either side of the edge (edge habitat); between 10 and 40 m into forest (forest habitat); and between 10 and 40 m into grassland (grassland habitat). Within these plots, all seedlings and ramets (<1 m tall) were counted; those counts were pooled for analysis and are hereafter referred to as regeneration. Soil sampling was conducted along seventeen of these transects parallel to slope contours and perpendicular to north-facing woodland edges. Composite soil samples were collected in the regeneration plots ($n = 3$ plots per habitat). In each plot, three soil cores were collected to 20 cm depth, combined, and air-dried in the field. Bulk soil samples were homogenized and passed

Table 1. Stable carbon isotopic composition of soils and fine roots, proportion of plots with regeneration, observations of shrubs and cattle dung, and sodium concentrations in soils on east- and west-facing slopes in a U-shaped valley at treeline in northern Peru

	East	West
Forest fine root $\delta 13C$ (‰)	−28.3	−28.0
Forest soil $\delta 13C$ (‰)	−26.1	−26.2
Edge fine root $\delta 13C$ (‰)	−26.6	−26.5
Edge soil $\delta 13C$ (‰)	−24.1	−25.1[*]
Grassland fine root $\delta 13C$ (‰)	−24.6	−25.7
Grassland soil $\delta 13C$ (‰)	−23.7	−24.6[*]
Regeneration in edge habitat (% of plots)	84	91
Regeneration in grassland habitat (% of plots)	59	74[*]
Shrub presence in grassland (no. of transects)[a]	3/9	8/8[*]
Cattle dung presence in grassland (no. of transects)[a]	7/9[*]	0/8
Na concentration (ppm ± SE)[a]	110 ± 9.7[*]	68 ± 5.9

Note: [a]Data from Ponette-González et al. (2016).

*Significant difference between east- and west-facing slopes ($p < 0.1$).

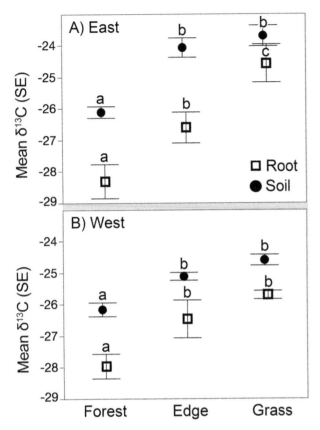

Figure 4. Mean δ13C (SE) of soils and fine roots for forest, edge, and grassland habitat (n = 17 transects) on east- and west-facing slopes sampled at treeline in northern Peru. Different letters represent significant differences among habitats (one-way analysis of variance, $p < 0.1$).

through a 2-mm sieve, and live fine roots (<2 mm) were separated. A subsample of soil and fine root material was dried, ground to a fine powder, and analyzed for total carbon and δ13C on a mass spectrometer. Stable carbon isotope natural abundance is expressed as $\delta(‰) = \left(R_{sample}/R_{standard} - 1\right) \times 1,000$, where R is the ratio of 13C/12C. Further details are in Ponette-González et al. (2016).

The proportion of plots with regeneration in forest, edge, and grassland habitat was calculated for the study valley; Fisher's exact test was used to test for aspect-related differences in the proportion of plots with regeneration for each habitat. Natural abundance of δ13C in soil and fine roots was used to detect vegetation shifts along the forest–grassland ecotone. Given the presence of C4 grasses or species-specific differences in the δ13C of Andean C3 trees and grasses (Szpak et al. 2013), we hypothesized that edge and grassland soils would be enriched in δ13C compared to forest soils. Further, we hypothesized that soil and fine root δ13C would have similar values unless vegetation change had occurred. Differences in the stable isotopic

composition of soil and fine roots among habitat types were assessed for the entire valley, and by aspect, using one-way analysis of variance (ANOVA) with Tukey's honestly significant difference post hoc comparisons, and Kruskal–Wallis with Steel Dwass pairwise comparisons where data were nonnormally distributed. Significance was set at $p < 0.1$.

Principal Findings: Woody Plant Encroachment Varies by Aspect

Previous analysis of soil and fine root δ13C at this site revealed that the mean δ13C values of edge and grassland soils at treeline were similar but enriched compared to forest soils (Ponette-González et al. 2016). Yet, the δ13C of roots displayed a different pattern, with edge fine roots intermediate between forest and grassland. These results suggest that forest roots are extending into the grasslands and that this treeline ecotone might be advancing.

The new analyses presented here indicate that patterns of woody plant encroachment additionally differ by aspect (Table 1). The stable carbon isotopic composition of soil and fine roots was nearly identical in forest habitat irrespective of aspect. This was not the case at forest–grassland edges, where mean soil and fine root δ13C differed more on east-facing compared to west-facing slopes (Figure 4). The smaller difference between soil and fine root δ13C observed at west-facing edges was due to the depletion of soil δ13C (Table 1). Compared to east-facing edges, soils at west-facing edges were depleted in δ13C ($p = 0.0096$) and thus isotopically more similar to fine roots. Grassland soils were also more depleted in δ13C on west-facing than on east-facing grasslands ($p = 0.032$). Seedling and ramet counts showed that regeneration was present in 93 percent of forest, 87 percent of edge, and 66 percent of grassland plots across the valley. Aspect differences for the forest and edge plots were not statistically significant (for forest, east = 91 percent, west = 96 percent; for edge, east = 84 percent, west = 91 percent). There was, however, a significantly greater proportion of plots with regeneration on west-facing (74 percent) than on east-facing (59 percent) grasslands ($p = 0.056$). Taken together, the soil and fine root δ13C data and the large proportion of plots in edge and grassland habitat with regeneration indicate that woody plants are encroaching into grasslands above the closed forest limit. Yet, our findings also highlight the influence of aspect on patterns of woody plant encroachment and, by extension, treeline migration.

Discussion

Land cover at snowline and treeline showed much dynamism in our tropical Andean sites. From 1999 to 2010, about a quarter of the area inside Huascarán National Park underwent some form of change (i.e., gains or losses) in land cover (Figure 3). We documented glacier loss in the Cordillera Blanca as well as gains in shrublands and grasslands in both national parks. These changes conform to observations of glacier retreat across the Andes (e.g., Burns and Nolin 2014) and some indications of treeline migration on tropical mountains (Kintz, Young, and Crews-Meyer 2006; Morueta-Holme et al. 2015). For example, our landscape-scale evidence of woody plant encroachment at treeline in Río Abiseo National Park supports the findings of a park-wide remote sensing analysis by Kintz, Young, and Crews-Meyer (2006). That study detected an upward shift in the treeline ecotone between 1987 and 2001, revealing a significant increase in shrubland area (34 percent) as well as in the number, size, and connectivity of forest patches. Our case studies nevertheless show that although there might be a tendency for ecological zones to shift upward, there are also many exceptions and time lags obscured by regional-scale assessments.

Exceptions and Time Lags

In the high mountains of the Cordillera Blanca, ecological succession is occurring on surfaces exposed by glacier recession. Plant colonization is heterogeneous, though, under different biophysical constraints than glacier loss. Spatial patterns of plant colonization did not always mirror those of glacier loss as we had expected: We found that ecological succession was not necessarily restricted by aspect (1987–1999) or was concentrated on north- and northeast-facing slopes (1999–2010; Figure 2, inset). Succession is perhaps also affected by distance to sources of colonizing plants from nearby vegetation types, in addition to aspect, which would affect duration of solar radiation and photosynthetic capacity.

Wetlands occupy a relatively small part of the landscapes studied (Figure 3), but their change during glacier recession was also complex. Similar to findings from this case study, Polk (2016) found that wetland area showed an overall decrease from 1987 to 2010 (losing approximately 5,581 ha or 45 percent), but that loss was not consistent, with some five-year intervals showing increases. The effects of reduced discharge from melting glaciers presumably could be observable in the spatial distributions of wetlands and other water-controlled ecosystems (Young 2015; Polk 2016). These fluctuations are expected to be associated with hydrologic connectivity, including groundwater interconnections. It is possible that area changes approximate the humpback curve showed by stream discharge as controlled by glacier retreat (Baraer et al. 2012). With decreased future hydrological input into wetlands, desiccation is likely to occur and these ecosystems might transition to increased dominance by seasonal wetlands and precipitation-dependent bogs.

Along the Eastern Andean Cordillera, forest edges at treeline are expanding in some areas, but there is considerable heterogeneity associated both with aspect and with current and past land use. We found that woody plant regeneration was more abundant on west-facing slopes, where differences in soil and fine root $\delta 13C$ were also less pronounced, possibly suggesting a longer history of plant establishment. Working in the same study valley, Ponette-González et al. (2016) observed that recent cattle dung in alpine grasslands was concentrated on east-facing slopes, where soil concentrations of sodium, an animal-limiting nutrient, were also twofold higher compared to west-facing slopes (Table 1). They attributed the lower apparent grazing on west-facing slopes to the lower proportion of grazable area (Fry, Ponette-González, and Young 2015).

Recent studies at tropical Andean treelines, including ours, show considerable variation in the direction, rate, and pattern of treeline migration (Kintz, Young, and Crews-Meyer 2006; Harsch et al. 2009; Lutz, Powell, and Silman 2013). For example, Morueta-Holme et al. (2015) redid observations first carried out by Alexander von Humboldt in Ecuador and showed that native plant distributions have shifted upward, accompanied by both upslope and downslope expansions of humanized landscapes affected by agriculture and burning. That study, however, was restricted to the same southeast-facing slope studied originally by von Humboldt, so there might be additional aspect-related changes not considered in their research. Other studies document little to no change in treeline position (Lutz, Powell, and Silman 2013), suggesting that Andean treelines might exhibit a lagged response to climate change as a result of multiple and often interacting factors (Rehm and Feeley 2015): biophysical controls (e.g., solar radiation; Bader, van Geloof, and Rietkerk 2007); barriers to reproduction (Rehm and Feeley 2013); topography (Coblentz and Keating 2008); and land use. Our findings indicate that the effects of herbivory and topography, and their interactions, on tropical treeline ecotones warrant further study. More broadly, both case studies suggest that accounting for topographic controls on plant colonization and growth at snowlines, tree-lines, and other transition zones could help improve predictability of future landscape change.

Socioenvironmental Dimensions

Protected areas, such as the national parks studied here, have a mandate to allow natural processes to regulate land cover but in reality often must serve additional land use needs for local people (Zimmerer 2011; Lipton 2014). In addition, they are exposed to global biophysical changes and are under increasing demands to provide ecosystem services downstream; for example, for water used for irrigation, domestic needs, and hydroelectric facilities. As species shift their locations with future climate change, protected area systems could provide needed habitat connectivity to permit movement (e.g., Dullinger et al. 2012). For example, Figure 1 suggests that wild plant species moving upslope would find ~3-km high habitat corridors within Huascarán National Park. Such habitat connectivity might not exist outside that park or among the different protected areas in the region, requiring conservation corridors (Young and Lipton 2006) or assisted migration (Richardson et al. 2009). Rates of soil development or slope instability might limit upward expansion in cases of primary succession, and we would also expect interactions and feedbacks with land use to affect ecological succession, as livestock and other herbivores can potentially graze near the highest peaks.

What people do with and to a particular landscape might be explainable by reference to their land uses, environmental governance, tenure rights, and economic goals (Huber et al. 2013; Liu et al. 2013). Predicting such landscape outcomes through time, however, might also require, in addition to downscaling from global climate models, information on the social telecoupling that provides additional or new information through the media or Internet that could alter the decision making of local land users. Considering such dynamics in the context of the landscapes of the tropical Andes suggests that aspect and other topographically controlled features such as soil depth and type could be important local controls. They would direct vegetation dynamics and constrain human land use in ways that might provide a useful framework in which to assess the resulting landscape change.

In conclusion, at the regional scale in the tropical Andes, our land cover assessments, and those of other researchers, continue to elucidate the downslope processes associated with retreating glaciers (Aubry-Wake et al. 2015) and changing amounts of forest cover (Holtmeier 2009; Toivonen et al. 2011; Balthazar et al. 2015). Steep spatial gradients impose land cover shifts at fine scales, though. As seen in our data from northern Peru (Figures 1–4, Table 1), places a few kilometers apart might have distinct climates, natural disturbance regimes, responses to glacier recession, soil and vegetation types, land use patterns, and land tenure. Topography is specific to a particular place given its history of bedrock formation, orogeny, and interactions with other Earth surface processes. The discipline of geomorphology evaluates specific landscape controls but attempts also to make generalizations that permit its development as a science (e.g., Murray et al. 2009). Equally, ecology tries to create generalizations that provide predictions useful for species and climate other than those studied directly (e.g., Billick and Price 2010). Nevertheless, these investigative approaches could tend to downgrade the significance of understanding the place-specific consequences of interacting geospatial, ecological, and social contingencies (cf. Crews and Young 2013). Global changes could be expressed locally in variable ways as seen in effects on changing species distributions (García et al. 2014; Pardikes et al. 2015), altered vegetation structure (Malanson et al. 2011; Higgins and Scheiter 2012), and the effects of fire–climate–human interactions (Butsic, Kelly, and Moritz 2015; Tepley and Veblen 2015). Here, we explored those topics in relationship to land cover in the Andes, which telecouples to distant biophysical and social processes but is expressed in landscapes through local combinations of soil, slope, and exposure. We suggest that change (or stasis) in land cover could offer insights into relevant processes and potentially provide a general tool for evaluating the multiple contingencies acting to create place-to-place uniqueness.

Acknowledgments

SERNANP, APECO, and the Museo de Historia Natural (UNMSM) facilitated our research in Peru.

Funding

We are grateful to the National Science Foundation for funding (CNH #1010381, SES #8713237, DEB #1146446, BCS #0117806, SBE #0905699, BCS #1333141), in addition to a National Geographic

Society Committee for Research and Exploration grant to Alexandra G. Ponette-González (#8877–11).

References

Aide, T. M., M. L. Clark, H. R. Grau, D. López-Carr, M. A. Levy, D. Redo, M. Bonilla-Moheno, G. Riner, M. J. Andrade-Núñez, and M. Muñiz. 2013. Deforestation and reforestation of Latin America and the Caribbean (2001–2010). *Biotropica* 45:262–71.

Álvarez-Berríos, N. L., I. K. Parés-Ramos, and T. M. Aide. 2013. Contrasting patterns of urban expansion in Colombia, Ecuador, Peru, and Bolivia between 1992 and 2009. *Ambio* 42:29–40.

Aubry-Wake, C., M. Baraer, J. M. McKenzie, B. G. Mark, O. Wigmore, R. A. Hellström, L. Lautz, and L. Somers. 2015. Measuring glacier surface temperatures with ground-based thermal infrared imaging. *Geophysical Research Letters* 42 (20): 8489–97.

Autoridad Nacional del Agua, Unidad de Glaciologia. 2013. *Inventario Nacional de Glaciares y Lagunas* [National inventory of glaciers and lakes]. Lima, Peru: Autoridad Nacional del Agua.

Bader, M. Y., I. van Geloof, and M. Rietkerk. 2007. High solar radiation hinders tree regeneration above the alpine treeline in northern Ecuador. *Plant Ecology* 191:33–45.

Balthazar, V., V. Vanacker, A. Molina, and E. F. Lambin. 2015. Impacts of forest cover change on ecosystem services in high Andean mountains. *Ecological Indicators* 48:63–75.

Baraer, M., B. G. Mark, J. M. McKenzie, T. Condom, J. Bury, K. I. Huh, C. Portocarrero, J. Gomez, and S. Rathay. 2012. Glacier recession and water resources in Peru's Cordillera Blanca. *Journal of Glaciology* 58:134–50.

Billick, I., and M. V. Price, eds. 2010. *The ecology of place: Contributions of place-based research to ecological understanding.* Chicago: University of Chicago Press.

Bradley, R. S., M. Vuille, H. F. Diaz, and W. Vergara. 2006. Threats to water supplies in the tropical Andes. *Science* 312:1755–56.

Burns, P., and A. Nolin. 2014. Using atmospherically-corrected Landsat imagery to measure glacier area change in the Cordillera Blanca, Peru from 1987 to 2010. *Remote Sensing of Environment* 140:165–78.

Bury, J., B. G. Mark, M. Carey, K. R. Young, J. McKenzie, M. Baraer, A. French, and M. H. Polk. 2013. New geographies of water and climate change in Peru: Coupled natural and social transformations in the Santa River watershed. *Annals of the Association of American Geographers* 103:363–74.

Bury, J., B. G. Mark, J. M. McKenzie, A. French, M. Baraer, K. I. Huh, M. Zapata Luyo, and R. Gómez López. 2011. Glacier recession and human vulnerability in the Yanamarey watershed of the Cordillera Blanca, Peru. *Climatic Change* 105:179–206.

Bush, M. B., B. C. S. Hansen, D. T. Rodbell, G. O. Seltzer, K. R. Young, B. León, M. B. Abbott, M. R. Silman, and W. D. Gosling. 2005. A 17,000-year history of Andean climate and vegetation change from Laguna de Chochos, Peru. *Journal of Quaternary Science* 20:703–14.

Butsic, V., M. Kelly, and M. A. Moritz. 2015. Land use and wildfire: A review of local interactions and teleconnections. *Land* 4:140–56.

Carey, M., M. Baraer, B. G. Mark, A. French, J. Bury, K. R. Young, and J. M. McKenzie. 2014. Toward hydro-social modeling: Merging human variables and the social sciences with climate-glacier runoff models (Santa River, Peru). *Journal of Hydrology* 518:60–70.

Coblentz, D., and P. L. Keating. 2008. Topographic controls on the distribution of tree islands in the high Andes of south-western Ecuador. *Journal of Biogeography* 35:2026–38.

Crews, K. A., and K. R. Young. 2013. Forefronting the socio-ecological in savanna landscapes through their spatial and temporal contingencies. *Land* 2:452–71.

Dullinger, S., A. Gattringer, W. Thuiller, D. Moser, N. E. Zimmermann, A. Guisain, W. Willner, et al. 2012. Extinction debt of high-mountain plants under twenty-first-century climate change. *Nature Climate Change* 2:619–22.

Fry, M., A. G. Ponette-González, and K. R. Young. 2015. A low-cost GPS-based protocol to create high-resolution digital elevation models for remote mountain areas. *Mountain Research and Development* 35:39–48.

García, R. A., M. Cabeza, C. Rahbek, and M. B. Araújo. 2014. Multiple dimensions of climate change and their implications for biodiversity. *Science* 344:1247579.

Gerrard, A. J. 1990. *Mountain environments: An examination of the physical geography of mountains.* Cambridge, MA: MIT Press.

Hansen, M. C., and T. R. Loveland. 2012. A review of large area monitoring of land cover change using Landsat data. *Remote Sensing of Environment* 122:66–74.

Harsch, M. A., P. E. Hulme, M. S. McGlone, and R. P. Duncan, R. P. 2009. Are treelines advancing? A global meta-analysis of treeline response to climate warming. *Ecology Letters* 12:1040–49.

Higgins, S. I., and S. Scheiter. 2012. Atmospheric CO_2 forces abrupt vegetation shifts locally, but not globally. *Nature* 488:209–13.

Holtmeier, F.-K. 2009. *Mountain timberlines: Ecology, patchiness, and dynamics.* Dordrecht, The Netherlands: Springer.

Huber, R., A. Rigling, P. Bebi, F. S. Brand, S. Briner, A. Buttler, C. Elkin, et al. 2013. Sustainable land use in mountain regions under global change: Synthesis across scales and disciplines. *Ecology and Society* 18 (3): 36.

Kargel, J. S., G. J. Leonard, M. P. Bishop, A. Kääb, and B. H. Raup, eds. 2014. *Global land ice measurements from space 2014 edition.* New York: Springer.

Kaser, G. 1999. A review of the modern fluctuations of tropical glaciers. *Global and Planetary Change* 22:93–103.

Kaser, G., and H. Osmaston. 2002. *Tropical glaciers.* Cambridge, UK: Cambridge University Press.

Kessler, M. 2002. The "*Polylepis* problem": Where do we stand? *Ecotropica* 8:97–110.

Kintz, D. B., K. R. Young, and K. A. Crews-Meyer. 2006. Implications of land use/land cover change in the buffer zone of a national park in the tropical Andes. *Environmental Management* 38:238–52.

Lipton, J. K. 2008. Human dimensions of conservation, land use, and climate change in Huascaran National Park, Peru. PhD dissertation, The University of Texas at Austin.

———. 2014. Lasting legacies: Conservation and communities at Huascaran National Park, Peru. *Society & Natural Resources* 27:820–33.

Liu, J., V. Hull, M. Batistella, R. DeFries, T. Dietz, F. Fu, T. W. Hertel, et al. 2013. Framing sustainability in a telecoupled world. *Ecology and Society* 18 (2): 26.

Lutz, D. A., R. L. Powell, and M. R. Silman. 2013. Four decades of Andean timberline migration and implications for biodiversity loss with climate change. *PLoS ONE* 8: e74496.

Malanson, G. P., L. M. Resler, M. Y. Bader, F.-K. Holtmeier, D. R. Butler, D. J. Weiss, L. D. Daniels, and D. B. Fagre. 2011. Mountain treelines: A roadmap for research orientation. *Arctic, Antarctic, and Alpine Research* 43:167–77.

Mark, B. G., and G. O. Seltzer. 2005. Evaluation of recent glacier recession in the Cordillera Blanca, Peru (AD 1962–1999): Spatial distribution of mass loss and climatic forcing. *Quaternary Science Reviews* 24:2265–80.

Morueta-Holme, N., K. Engemann, P. Sandoval-Acuña, J. D. Jonas, R. M. Segnitz, and J.-C. Svenning. 2015. Strong upslope shifts in Chimborazo's vegetation over two centuries since Humboldt. *Proceedings of the National Academy of Sciences* 112:12741–45.

Murray, A. B., E. Lazarus, A. Ashton, A. Baas, G. Coco, T. Coulthard, M. Fonstad, et al. 2009. Geomorphology, complexity, and the emerging science of the Earth's surface. *Geomorphology* 103:496–505.

Ozesmi, S. L., and M. E. Bauer. 2002. Satellite remote sensing of wetlands. *Wetlands Ecology and Management* 10:381–402.

Pardikes, N. A., A. M. Shapiro, L. A. Dyer, and M. L. Forister. 2015. Global weather and local butterflies: Variable responses to a large-scale climate pattern along an elevational gradient. *Ecology* 96:2891–2901.

Polk, M. H. 2016. "They are drying out": Social–ecological consequences of glacier recession on mountain peatlands in Huascaran National Park, Peru. PhD dissertation, The University of Texas at Austin.

Ponette-González, A. G., H. A. Ewing, M. Fry, and K. R. Young. 2016. Soil and fine root chemistry at a tropical Andean timberline. *Catena* 137:350–59.

Ponette-González, A. G., E. Marín-Spiotta, K. A. Brauman, K. A. Farley, K. C. Weathers, and K. R. Young. 2014. Hydrologic connectivity in the high-elevation tropics: Heterogeneous responses to land change. *BioScience* 64:92–104.

Postigo, J. C., K. R. Young, and K. A. Crews. 2008. Change and continuity in a pastoralist community in the high Peruvian Andes. *Human Ecology* 36:535–51.

Rabatel, A., B. Francou, A. Soruco, J. Gomez, B. Cáceres, J. L. Ceballos, R. Basantes, et al. 2013. Current state of glaciers in the tropical Andes: A multi-century perspective on glacier evolution and climate change. *The Cryosphere* 7:81–102.

Rehm, E. M., and K. J. Feeley. 2013. Forest patches and the upward migration of timberline in the southern Peruvian Andes. *Forest Ecology and Management* 305:204–11.

———. 2015. The inability of tropical cloud forest species to invade grasslands above treeline during climate change: Potential explanations and consequences. *Ecography* 38:1167–75.

Richardson, D. M., J. J. Hellmann, J. S. McLachlan, D. F. Sax, M. W. Schwartz, P. Gonzalez, E. J. Brennan, et al. 2009. Multidimensional evaluation of managed relocation. *Proceedings of the National Academy of Sciences* 106:9721–24.

Rodbell, D. 1993. The timing of the last deglaciation in Cordillera Oriental, northern Peru, based on glacial geology and lake sedimentology. *Geological Society of America Bulletin* 105:923–34.

Silverio, W., and J. M. Jaquet. 2009. Prototype land-cover mapping of the Huascarán Biosphere Reserve (Peru) using a digital elevation model, and the NDSI and NDVI indices. *Journal of Applied Remote Sensing* 3:33516.

Stansell, N. D., D. T. Rodbell, M. B. Abbott, and B. G. Mark. 2013. Proglacial lake sediment records of Holocene climate change in the western Cordillera of Peru. *Quaternary Science Reviews* 70:1–14.

Szpak, P., C. D. White, F. J. Longstaffe, J. F. Millaire, and V. F. V. Sánchez. 2013. Carbon and nitrogen isotopic survey of northern Peruvian plants: Baselines for paleodietary and paleoecological studies. *PLoS ONE* 8:e53763.

Tepley, A. J., and T. T. Veblen. 2015. Spatiotemporal fire dynamics in mixed-conifer and aspen forests in the San Juan Mountains of southwestern Colorado, USA. *Ecological Monographs* 85:583–603.

Thompson, L. G., E. Mosley-Thompson, H. Brecher, M. Davis, B. León, D. Les, P. Lin, T. Mashiotta, and K. Mountain. 2006. Abrupt tropical climate change: Past and present. *Proceedings of the National Academy of Sciences of the United States of America* 103:10536–43.

Toivonen, J. M., M. Kessler, K. Ruokolainen, and D. Hertel. 2011. Accessibility predicts structural variation of Andean *Polylepis* forests. *Biodiversity and Conservation* 20:1789–1802.

Vuille, M., R. S. Bradley, M. Werner, and F. Keimig. 2003. 20th century climate change in the tropical Andes: Observations and model results. *Climatic Change* 59:75–99.

Vuille, M., B. Francou, P. Wagnon, I. Juen, G. Kaser, B. G. Mark, and R. S. Bradley. 2008. Climate change and tropical Andean glaciers: Past, present and future. *Earth-Science Reviews* 89:79–96.

Walsh, S. J., R. E. Bilsborrow, S. J. McGregor, B. G. Frizzelle, J. P. Messina, W. K. Pan, K. A. Crews-Meyer, G. N. Taff, and F. Baquero. 2003. Integration of longitudinal surveys, remote sensing time series, and spatial analyses. In *People and the environment: Approaches for linking household and community surveys to remote sensing and GIS*, ed. J. Fox, R. Rindfuss, S. Walsh, and V. Mishra, 91–130. Boston: Kluwer.

White, S. 2013. Grass páramo as hunter-gatherer landscape. *The Holocene* 23:898–915.

Young, K. R. 1993a. National park protection in relation to the ecological zonation of a neighboring human community: An example from northern Peru. *Mountain Research and Development* 13:267–80.

———. 1993b. Tropical timberlines: Changes in forest structure and regeneration between two Peruvian timberline margins. *Arctic and Alpine Research* 25:167–74.

———. 1993c. Woody and scandent plants on the edges of an Andean timberline. *Bulletin of the Torrey Botanical Club* 120:1–18.

———. 2009. Andean land use and biodiversity: Humanized landscapes in a time of change. *Annals of the Missouri Botanical Garden* 96:492–507.

———. 2015. Ecosystem change in high tropical mountains. In *The high-mountain cryosphere: Environmental changes and human risks*, ed. C. Huggel, M. Carey, J. Clague, and A. Kääb, 227–46. Cambridge, UK: Cambridge University Press.

Young, K. R., and B. León. 2007. Tree-line changes along the Andes: Implications of spatial patterns and dynamics. *Philosophical Transactions of the Royal Society B: Biological Sciences* 362:263–72.

Young, K. R., and J. K. Lipton. 2006. Adaptive governance and climate change in the tropical highlands of western South America. *Climatic Change* 78:63–102.

Zimmerer, K. S. 2011. "Conservation booms" with agricultural growth? Sustainability and shifting environmental governance in Latin America, 1985–2008 (Mexico, Costa Rica, Brazil, Peru, Bolivia). *Latin American Research Review* 46:82–114.

Mountain Ecology, Remoteness, and the Rise of Agrobiodiversity: Tracing the Geographic Spaces of Human–Environment Knowledge

Karl S. Zimmerer, Hildegardo Córdova-Aguilar, Rafael Mata Olmo, Yolanda Jiménez Olivencia, and Steven J. Vanek

We use an original geographic framework and insights from science, technology, and society studies and the geohumanities to investigate the development of global environmental knowledge in tropical mountains. Our analysis demonstrates the significant relationship between current agrobiodiversity and the elevation of mountain agroecosystems across multiple countries. We use the results of this general statistical model to support our focus on mountain agrobiodiversity. Regimes of the agrobiodiversity knowledge of scientists, government officials, travelers, and indigenous peoples, among others, interacting in mountain landscapes have varied significantly in denoting geographic remoteness. Knowledge representing pre-European mountain geography and diverse food plants in the tropical Andes highlighted their centrality to the Inca Empire (circa 1400–1532). The notion of semiremoteness, geographic valley–upland differentiation, and the similitude-and-difference knowledge mode characterized early Spanish imperial rule (1532–1770). Early modern accounts (1770–1900) amplified the remoteness of the Andes as they advanced global ecological sciences, knowledge standardization, and racial representations of indigenous people as degraded, with scant attention to Andean agriculture and food. Global agrobiodiversity knowledge increasingly drew on corresponding representations of mountain remoteness. Our integration of the biogeophysical–social sciences with the geohumanities reveals distinctive geographies of agrobiodiversity knowledge. Assumed remoteness of mountain agrobiodiversity is not inherent but rather is actively formed in relation to global societies and knowledge systems and is thus relational. Connectivity and claims to territorial and indigenous autonomy distinguish newly emergent characteristics of agrobiodiversity. The multifunctionality and political geography of agrobiodiversity are integral to current mountain environments, societies, and sustainability.

我们运用科学、科技与社会研究和地理人文学科中原有的地理架构与洞见,探讨热带山区中全球环境知识的发展。我们的分析证实,当前农业生态多样性和各国间的山区农业生态系统的高度之间有着显着关联性。我们运用这个一般统计模型的结果来支持我们对山区农业生态多样性的关注。科学家、政府官员、旅行者和原住民与其他人在山区地景互动的农业生态多样性的知识体系,在指称地理的偏远性上有显着的差异。呈现热带安第斯山在欧洲殖民前的山区地理和粮食作物多样性的知识,强调其之于印加帝国 (大约在公元 1400 年至 1532 年间) 的重要性。半偏远的概念、地理的谷地—高地差异,以及相似—差异的知识模型,是西班牙帝国统治初期 (1532 年至 1770 年) 的特徵。早期的现代性解释 (1770 年志 1900 年),则在推进全球生态科学,知识标准化并将原住民族的种族再现视为退化之中,将安第斯的偏远性放大,且鲜少关注安第斯的农业与粮食。全球农业生态多样性的知识,逐渐引用呼应山区偏远性的再现。我们对生物自然地理—社会科学和地理人文学科进行的整合,揭露出农业生态多样性知识的特殊地理。山区农业生物多样性所预设的偏远性并非是内在固有的,而是在与全球社会和知识系统的关系中积极建构而成。连结性与领土及原住民自主性的宣称,辨别了农业生态多样性崭新浮现的特徵。农业生态多样性的多重功能性和政治地理,是当前山区环境、社会及可持续性的一部分。 关键词: 农业生态多样性, 地理人文学科和偏远性, 地理与环境科学的历史, 山区生态学, 可持续性与粮食作物。

Para investigar el desarrollo global del conocimiento ambiental sobre las montañas tropicales, usamos un marco geográfico original y conocimientos generados en estudios de ciencia, tecnología y sociedad, y en las geohumanidades. Nuestro análisis demuestra la relación significativa existente entre la agrobiodiversidad actual y la elevación en agroecosistemas de montaña, a través de múltiples países. Utilizamos los resultados de este modelo estadístico general en apoyo de nuestro enfoque sobre la agrobiodiversidad de montaña. El conocimiento de los regímenes de la agrobiodiversidad de científicos, agentes del gobierno, viajeros y pueblos indígenas, entre otros, en interacción en los paisajes montañosos, varían de manera significativa en lo que concierne a denotar lejanía geográfica. Para el imperio incaico (*circa* 1400–1532) era crucial el conocimiento que representaba a la geografía de montaña pre-europea y a las diversas plantas comestibles de los Andes tropicales. La noción de semi-lejanía, la diferenciación geográfica valle–tierras altas y el modo de conocimiento de similitud-y-diferencia caracterizan el gobierno español inicial (1532–1770). Los recuentos del período moderno temprano (1770–1900) amplificaron lo remoto de los Andes a medida que se avanzaba en las ciencias ecológicas globales, la estandarización del conocimiento y las representaciones raciales de los pueblos indígenas, caracterizados como degradados, dedicándole mínima atención a la agricultura y a los alimentos andinos. El conocimiento de la agrobiodiversidad global crecientemente se apoyó en las correspondientes representaciones del carácter remoto de las montañas. La integración que hacemos de las ciencias biogeofísicas-sociales con las geohumanidades revela distintas geografías del conocimiento de la agrobiodiversidad. El presumido carácter remoto de la agrobiodiversidad no es inherente sino, mejor, formado activamente en la relación con sociedades globales y sistemas de conocimiento, por lo que es relacional. La conectividad y los reclamos de autonomía territorial e indígena distinguen características de la agrobiodiversidad de reciente aparición. La multifuncionalidad y la geografía política de la agrobiodiversidad son parte integral de los actuales entornos y sociedades de montaña, y de la sustentabilidad.

Tropical and subtropical mountains are recognized for the dynamics of cultural and biological diversity entwined with tenuous remoteness, fraught political geographies, and increased integration through global socioeconomic and environmental connections. Examples include snow leopards and tribal groups amid major global conservation initiatives in the Himalaya of Pakistan (Hussain 2015); diverse resource management and sporadic cash crop economies in montane Papua New Guinea (Brookfield 2001); and complex emergent agri-food systems amid the mountain landscapes, international migration, and part-time land use of indigenous smallholders in the Andes of western South America (Zimmerer 2014).

The world's tropical and subtropical mountains are increasingly subject to global environmental changes (e.g., biodiversity, water, climate) that deeply entangle sociocultural and political issues (e.g., indigeneity, poverty, food security, migration, political marginalization, geopolitical importance, and extractive industries; Bebbington 2000; Marston 2008; Young 2009; Price et al. 2013; Radcliffe 2015). Through the multiscale interactions of biogeophysical changes with the cultural landscapes and politics of complex societies, these mountain sites have been globally important to development of the resource, ecological, and human–environment sciences (Sarmiento 2000; Harden 2012), as well as nature–society frameworks (Blaikie and Brookfield 1987).

Agrobiodiversity is integral to tropical mountain environments and societies. Defined as the biological diversity of plants and animals for food and livelihoods, its meaning also incorporates production and consumption networks, cultural knowledge and resource systems, social relations, and institutions. Agrobiodiversity is an expanded focus in geography and interdisciplinary fields (Lewis and Chambers 2010; Kerr 2014). It potentially enhances capacities of land use and food systems for social–ecological resilience in the context of global environmental transitions and transformation related to climate change (Zimmerer 2010). Agrobiodiversity can also address sustainability issues of food equity and justice, security and quality, and resources in development (Perreault 2005). Other facets are political issues of governance and social access (Graddy 2013), consumption and markets, health and human well-being (Johns et al. 2013), and agroecology, ecosystem services, and geogenetics (de Haan et al. 2010).

Agrobiodiversity growth and use have been correlated at the global scale with continued concentrated distribution in tropical and subtropical mountains (Vavilov 1992; Brush 2004; Nabhan 2008). Table 1 sketches the global overview.[1] More comprehensive information on global mountain agrobiodiversity is provided in the Supplemental Material. Notwithstanding significant depletion in the Green Revolution (Zimmerer 1996), the resilience capacity of agrobiodiversity-incorporating land and food systems in global mountain systems has been recognized since the early 1990s (Zimmerer 1992). These insights have effectively overturned earlier scientific narratives of the immanent collapse of mountain agrobiodiversity and the presumed catastrophic declines of landscape functions and local food systems.

Table 1. Brief overview of high-agrobiodiversity food complexes in major mountains and associated watersheds (see also Supplemental Material)

Mountain systems and ranges	Examples of high-agrobiodiversity food complexes[a]	Major watersheds
Himalayan (Hindu Kush, Karakoram, and ranges of south China), southeast and central Asia	Upland rice, millets, beans. sesame, apple, apricot, peach, nectarine, walnut	Ganges–Hooghly–Padma, Indus, Brahmaputra–Tsangpo, Mekong, Yangtze, Ayeyarwady
Zagros, Elburz, Taurus, Caucasus; Atlas, Sierra Nevada, and mountains of Mediterranean	Wheats, barley, rye, oat, pea, chickpea, lentil, lupine, fava bean, grass pea, cherry, almond	Aras, Tigris–Euphrates, Ebro, Po, Guadalquivir
East African highlands	Sorghum, millets, tef, ensete, Ethiopian oats and barley, coffee, yam, pigeon pea, cowpea	Upper Nile–Kagera, Upper Congo–Chambeshi–Lualaba
Tropical Andes and foothills, Nevada de Santa Marta, Venezuela south	Potato, quinoa, Andean maize, Andean bean, amaranth, tubers, squash, pepper, peanut, lima, lupine	Upper Amazon, Upper Orinoco, Pilcomayo–Paraná, Magdalena, western South America rivers
Sierras of Mexico and Central America to southwestern United States	Maize, Mesoamerican common bean, pepper, squash, grain amaranth, tomato, jícama	Rio Grande, coastal and interior river valleys of Mexico
East Asia, Southeast Asia, and Pacific	Upland rice, Job's tears, winged beans, rattan (fruits), pit-pit, highland breadfruit, pandanus, taros, eddoes, citrus	Coastal and interior river valleys

[a]Sources include Vavilov (1992). Other main sources are listed in the Supplemental Material.

Sustainability contributions of partially persistent agrobiodiversity that continues to emerge in mountain agrifood systems include human dietary diversity, nutrition, and health, as well as links to soil and aboveground biotic diversity and sustainability-promoting practices such as seed-keeping and exchange, gardening, and soil and water management. The issues of agrobiodiversity access, intellectual property, and conservation are domains of vibrant current policy and politics (additional text and sources detailed in the Supplemental Material). Further contributions include cultural heritage (Sarmiento 2008; Graddy 2013).

The development of past and present agrobiodiversity knowledge is rooted in the wide-ranging accounts of observers, travelers, scientists, officials, and others interacting with diverse local social actors amid varied landscapes and livelihoods. Our study examines the rise of such knowledge in relation to the geographic contexts of subtropical and tropical mountains. In this case study we formulate an original conceptual framework integrating the analysis of current agrobiodiversity and the historical development of related knowledge that has occurred in the human–environment and nature–society dynamics of particular geographic spaces, both material (biogeophysical, social) and symbolic (meanings, ideas).

We focus on the representations and functions of the notion of mountain remoteness vis-à-vis the accounts of agrobiodiversity knowledge. Our work defines remoteness through the overall low degree of connectedness to powerful national and global territories and sites. Amid high levels of poverty and other socioeconomic and environmental challenges, the remoteness of tropical and subtropical mountains is assigned either credit or blame in causal accounts of environmental sustainability or collapse and development (Debarbieux and Rudaz 2015). This influence of remoteness in mountain environmental scientific narratives motivates our research.

Our approach views mountain remoteness as a complex entanglement of biophysical and environmental conditions, geopolitical and social power, and the influence of geographic ideas. It is guided by the geohumanities and, in particular, studies of science, technology, and society in geography and other fields (Goldman, Nadasdy, and Turner 2011; Lave et al. 2014). Drawing on the geohumanities we conceptualize remote areas and the idea or notion of remoteness as relational and potentially integral to modernity, rather than these areas existing entirely outside modern ways. Building on new insight into the environmental knowledges of diverse social actors, institutions, and political powers, our definition of remote areas rests on meanings of connectivity (limited but emphasized), multifactor distancing (physical, social, rhetorical), and edginess (examples include periurban areas of large cities in the Global South; Harms et al. 2014; Hussain 2015).

The idea or notion of remoteness in our framework is thus distinct yet related to political ecology's

Figure 1. The tropical Andes Mountains and the adjoining foothills and lowlands of the Amazon and coasts (present-day Bolivia, Peru, Ecuador, Colombia, and Venezuela).

concept of marginality and also dialogues with the conceptual spatial and social framing of margins, frontiers, borderlands, and contact zones that Pratt (2007) and other authors have developed using literary and ethnographic theory and grounded cultural–historical analysis in anthropology (see Supplemental Material). Finally, our focus on the representation of mountain remoteness enables fresh insights on the spatial historical dynamics of enduring global human–environment

narratives that include the so-called pristine myth of nature, colonial landscape representation as "emptying the land," and mountain gloom and glory (Nicholson 1997; Debarbieux and Rudaz 2015).

We ask four questions using the case study of the tropical Andes Mountains (present-day Bolivia to Venezuela; Figure 1). Can we demonstrate, as general context, the present-day association of increased agrobiodiversity with the elevation of mountain food-growing environments? What ideas of mountain remoteness have arisen across the principal historical geographic periods? How has agrobiodiversity knowledge been related to the representations of remoteness? (*Agrobiodiversity knowledge* is defined to include the evidentiary accounts of broad-based descriptions as well as scientific investigations.) Finally, how does this research enhance the social and environmental sustainability of agrobiodiversity as food growing and use in potential political alternatives?

Methods

Twelve published accounts of approximately 3,350 pages comprised the principal textual sources (Table 2, last column; details in the Supplemental Material) for analysis covering a triad of major historical geographic periods from the pre-Inca to the present, as detailed further later.[2] Analysis was applied also to thirty-one publications belonging to the 1900 to 1970 and 1970 to present periods.

The principal texts, in chronological order, consist of self-designated chroniclers and accounts of the pre-European period by indigenous colonial elites (Guaman Poma de Ayala, de la Vega); a soldier scribe and imperial chronicler (*cronista*) of the nascent colonial period (Cieza de León); a cleric representing environmental knowledge of the seventeenth century (Cobo); a Spanish naval officer, early botanical scientist, and colonial administrator (Ulloa); non-Spanish scientists and natural historians including major historical figures in contemporary geography (Humboldt, Bonpland) and a regional specialist (Haenke); and various early modern travelers including a utopian feminist (Tristán), a proto-archaeological observer (Wiener), notable polymaths (Markham, Raimondi), and a scientific mountaineer (Whymper).

These sources are widely regarded as principal works on the tropical Andes as global sites of pre-European-to-modern environmental, landscape, sociocultural, and political transformations (Salomon 1985; Denevan 1992; Gade 1999, 2015; Knapp 2007; Borsdorf and Stadel 2015; Zimmerer and Bell 2015). Close reading was undertaken to enable literary discourse analysis, contextual situating involving historical and cultural interpretation to ensure a nonpresentist approach, cross-source comparisons, and the consideration of vibrant secondary source literatures (see Supplemental Material).

Statistical analysis was used to examine the present-day relationship of agrobiodiversity to mountain environments. This statistical model was designed to provide a background overview by using recent data on high-agrobiodiversity crops (Andean potatoes, beans, and maize) in Colombia (2011), Ecuador (2013), and Perú–Sistema de Consulta de Cuadros Estadísticos (2013). It consists of information at the subnational level on elevation and the percentage of farms using only farmer varieties as an estimate of the agrobiodiversity level.[3]

Overview: A Preliminary General Model of Current Agrobiodiversity in Mountain Environments

Regression results show significant statistical relationships between the current level of agrobiodiversity and the elevation of growing areas in the tropical Andes Mountains of Colombia, Ecuador, and Peru (Figure 2). High-level statistical significance ($p < 0.0001$) was demonstrated (Figure 2; Supplemental Material). Results reveal greater levels of agrobiodiversity with the increased elevations of tropical mountain food-growing environments across multiple countries. They suggest potential global-scale analogues and underscore the need for additional modeling research currently underway.

Pre-European and the Early and Middle Colonial Periods (5–10,000 BP–1770 CE): Agrobiodiversity in the Andes Mountains

The tropical Andes and the roles of agriculture, land use, and food were geographically central to representations of the pre-European period both in its longer reach and in the epoch of the Inca Empire (circa 1400–1532). By 5,000 to 10,000 years ago, domestication and incipient agriculture had occurred in the main mountain ranges and adjoining foothills and lowlands. Its early timing and extensive scope were globally significant (Moseley 1997; Piperno 2012), followed by the diversification of crop repertoires and foodways and the intensification of production. Political, economic, and cultural developments, together with environmental change, culminated in Inca imperial control (circa 1400–1532).

The geographic centrality of the tropical Andes under Inca rule was framed by Garcilaso de la Vega and Felipe Guaman Poma de Ayala (Table 2; see also Supplemental Material). De la Vega, who referred to

Table 2. Historical geographic overview of the Andes Mountains and the accounts of agrobiodiversity

Time period	Spatial representations[a]	Estimated populations[b]	Modes of geographic information	Connectivity levels and technology	Representation of people of the Andes	Agrobiodiversity knowledge presentations	Authors of major textual sources[c]
Inca (1400–1532)	Center or "navel" of world	15.7 million (tropical Andes); 1.5–2.5 million (Andes Peru)	Distance (units of delivery time represented in khipus)	High (imperial roads-and-runner system)	Inca subjects	Khipus; Imperial and commoner organization; extensive Quechua vocabulary	Guaman Poma; Garcilaso de la Vega
Early and middle colonial (1532–1770)	Andean valleys as agricultural paradise; Andean uplands as remote areas	500,000–1 million (Andes Peru)	Distance (units of leagues contained in textual information)	Medium (roads)	Savages or barbarians	Spanish common names based on similarity and difference	Cieza de León; Garcilaso de la Vega
Late colonial and early republics (1770–1900)	Tropical Andes as remote areas; focus on commerce and exports	300,000–500,000 (Andes Peru) in late 1700s; 1.0–1.8 million by 1900	Elevation (often as diagrams) and cartography	Low (accounts focus on roads, postal system, river transport)	Degraded Indigenous populations; archaeological framing	Nominal	Ulloa; Humboldt and Bonpland; Haenke; Tristán; Wiener; Whymper; Markham; Raimondi
1900–1970	Remote and connected	23 million (Andes, all countries); 6 million (Peru, 1972)	Cartographic and textual	Medium (roads, air travel)	Biological framing; Indigenous, agrarian (later in period)	Sciences of genetics and taxonomy; agronomy; Indigenous lexicography	Vavilov; Cook; Bukasov; León; Ochoa
1970–present	Hybrid	20 million (Andes, all countries); 5.3 million (Peru, 2012)	Contextual and diverse visualizations	High	Indigenous territorial movements, rural–urban networks	Geography; anthropology; human and political ecology	Brush; Gade; Graddy; Nazarea; Rhoades; Sarmiento; Zimmerer

[a]These representations are discussed in the text.

[b]Sources include Denevan (1992). Additional sources are listed in the Supplemental Material.

[c]Sources correspond to the time period of the each author's experience in the Andes. References to some sources are listed in the Supplemental Material to save space.

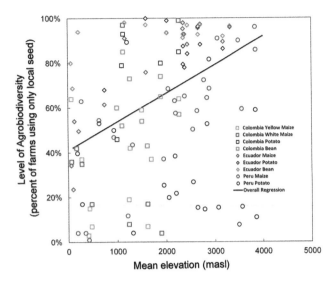

Figure 2. Regression of agrobiodiversity level (percentage local seed use only) versus mean elevation (119 subnational administrative units in Colombia, Ecuador, and Peru). *Note.* Regression outputs for linear regression on the main effects of elevation and country-crop data sets, and their interaction. See Supplemental Material for full regression analysis outputs. (Color figure available online.)

himself as "El Inca," introduced to European audiences the influential translation of Cuzco, the Inca capital, as the world's "center" or "navel" (de la Vega [1609, 1616–1617] 1943). Guaman Poma's Renaissance-style mappamundi, contained in his 1,300-page letter to Spain's King Phillip III, located the tropical Andes and its string of urban–rural spaces as the central cartographic axis of world geography (Guaman Poma de Ayala [1613] 1980).

Their writings actively contested the geographic marginalization of core Andean territories following Spanish conquest in 1532. Both texts also presented Andean food plants as active elements of nonremote Andean bodies, landscapes, and livelihoods. Geographic, political, agricultural, and other types of information were integrated through the *khipu* recording systems of Inca rulers (Table 2).

Spanish colonialism partly marginalized the tropical Andes by displacing the nodes of political and economic power to Pacific coastal cities such as Lima, Guayaquil, and Cartagena. Yet this marginalization was neither geographically complete nor undifferentiated. Colonial control also remained anchored in the sizable urban–rural networks centered in the Andean valleys (Zimmerer and Bell 2015). Spanish imperial rulers generated a new spatial model of colonial geography hinged on the relational contrasts evidenced in their concentrated control of the Andean valleys

(agricultural plenitude, inhabitability) versus the meanings attributed to surrounding uplands (barrenness, inhospitable)—although the latter were partly inhabited.

Imperial power geometries and environmental factors reinforced the colonial contrasts and separateness of well-controlled Andean valleys versus wild upland spaces through the combination of forced urbanization and climate change of the Little Ice Age (Zimmerer and Bell 2015). The demographic collapse of indigenous populations became catastrophic, but it was not a driver per se of this colonial geographic model since mortality rates lessened at the higher elevations (Table 2). In sum, the new two-part colonial geography of the Andes depicted the uplands as remote areas that were marginally under Spanish control and beyond the reach of state knowledge systems (see also Craib 2004; Hussain 2015).

The representation of barrens, and therefore remoteness, was encapsulated in the application of "deserts" as a descriptor of upland spaces (Ulloa ([1772] 1944; Cieza de León [1553] 1984). The dichotomous knowledge–power configuration contrasting Andean valleys and uplands was complex and relational in its geographic dynamics. It differed significantly from the spatially undifferentiated assumption applied elsewhere and often accepted as a global model of colonial landscape change based on "emptying the land" (Pratt 2007).[4]

In contrast to the one-dimensional spatial transformation suggested in the now influential landscape trope of "emptying the land," the development of colonial Andean geography in the 1500s and 1600s evoked a two-part relational design and new colonial knowledge–power geometries that ensured the control of populated areas and indigenous labor. It both resembled and reworked the far-reaching dualism of upland herders and valley-based cultivators that was deeply rooted in Inca and pre-Inca political systems, ideology, and experience (Urton 2012).

The contrasts embedded in the valley–upland model of colonial Andean geography reflected the knowledge mode of similitude-and-difference applied to nature and peoples in the 1500s and 1600s. Resembling the "successive approximation" of the early modern categorization of plants and animals in the New World (Gerbi [1975] 2010), it conjoined the overlapping episteme of the European Renaissance (similitude) and the classical period (differences; Foucault 2002). Andean valleys were evoked as welcoming landscapes well-suited to colonial occupation versus the forbidding and unfamiliar uplands of Andean *puna* and *páramo* (Cieza de León [1553] 1984; see Zimmerer and Bell 2015).

The new mode of colonial landscape knowledge was so influential that the chronicler Cieza de León deployed it as a convincing template to describe the area of Cajamarca without having visited (Table 3). Such textual usages underscored the discursive power of this colonial geographic model and the relationality of the enhanced visibility of Andean valleys together with the obscured spaces of upland landscapes (cf. Scott 1998). The general knowledge mode of similitude and difference was widely applied also to the South American food plants. One early description delineating the distinctiveness of Andean potatoes drew general similarities to the truffle (underground growth), chestnut (texture), and poppy (flower; Cieza de León [1553] 1984). Stylistically similar, the early description of quinoa marshaled comparisons to rice and Moorish chard in the connotation of similitude and difference.

These deployments of similitude and difference—used also in several other early colonial accounts—fostered trust with the reading audience through the rhetoric of firsthand testimony. The latter, also referred to as the autoptic imagination, relied on references to object- and place-based knowledge (Bourguet, Licoppe, and Sibum 2002). It conjured the general impression of the food plants and landscapes of the tropical Andes as distinct yet familiar. This acknowledgment of partial familiarity in the early colonial representation of new geographic spaces and nature, including food plants, bridged the tone of deep-seated knowledge in the works of the Inca interlocutors Guaman Poma and de la Vega with the predominant representation of remoteness that was to follow.

Long Nineteenth Century (1770–1900): Modernity and the Remoteness of Mountain People and Ecology

Alexander von Humboldt sojourned in the tropical Andes between Venezuela and northern Peru between March 1801 and January 1803. Humboldt interacted extensively with scientists in the Andes and completed field excursions with his traveling companion Aimé Bonpland that led to the copious plant collections and environmental measurements on Chimborazo (Sarmiento 2000; Cañizares-Esguerra 2006; Zimmerer 2006). The interest in crops of Humboldt and Bonpland was confined mostly to "economic productions" (e.g., cacao, coffee, sugar, wheat, barley) that anchored their now well-known model of tropical ecological zonation in tiered bands (Humboldt and Bonpland 1821–1829, Vol. 6).

Table 3. From agricultural paradise (1500s) to remote area (1800s): Accounts of the Cajamarca Valley and surrounding uplands (present-day Peru) in the Tropical Andes

Period	Excerpt of account	Source: Author and publication	Interpretation
Early colonial	"Cajamarca is fertile in the extreme, for wheat does as well here as in Sicily, and cattle thrive, and there is an abundance of edible roots." (Cieza de León did not personally visit Cajamarca, which underscores this depiction as template-like; see text.)	Cieza de León, Pedro de ([1553] 1984). *Crónica del Perú.*	Andean Valleys as highly habitable (suggestion of agricultural paradise); Andean uplands as remote areas
Late colonial	Upland: "The natural characteristics of the wildernesses of these Cordilleras are heightened . . . from its being in these very regions that we still see admirable remains of the gigantic work, the artificial road of the Inca." Valley: "The travelers looks down with increased pleasure on the fertile valley of Caxamarca. The soil is extremely fertile, and the plain full of cultivated fields and gardens. . . . Wheat yields [well except for frost]."	Humboldt, A. von. (1849). *Aspects of nature, in different lands and different climates; with scientific elucidations* (415 [upland], 427 [valley]).	Andean uplands as remote areas and Andean valleys as productive; also archaeological framing
Early republic	Valley: "The valley of Cajamarca . . . we saw open before us the immense and wild valley . . . ancient residence of the last Inca."	Wiener, Charles. ([1880] 1993). *Perú y Bolivia: Relato de viaje* (129).	Tropical Andes as remote areas; sublime; distant in space and time

This standardization of environmental knowledge in the Andes and other mountain regions worldwide presaged the contemporary global change and planetary sciences. Humboldt's select choice of crop suite similarly corresponded to the demands of expanding international commerce of the modern world economy as filtered through Spain's Bourbon reforms that were initiated in the mid-1700s. Such focus is characteristic also of the landmark accounts of the Spanish scientist Ulloa ([1772] 1944) and the Bohemian Haenke ([1799] 1974) who settled in present-day Bolivia (see Table 2 and Supplemental Material).

The diverse and plentiful crops and foods of the tropical Andes were mostly insignificant in the texts of Humboldt and Bonpland ([1807] 2009) save the mention of quinoa and Andean potatoes that they categorized among the widely diffused "social plants." This near absence contrasted their extensive ecological descriptions, mapping, and collections of more than 300 species of uncultivated plants, many distributed on Andean slopes. It was also starkly different than their ample accounts of the widespread human resource use and humanized landscapes of the Amazon lowlands. Their omission was not a consequence of lacking scientific knowledge. Indeed, the botanical naming of such Andean food plants as *oca*, *ulluco*, and *mashua*—all staple foods grown from Colombia and Venezuela to Bolivia—was being conducted at this time and during preceding decades by well-known scientists such as Caldas, Molina, Ruíz, Pavón, and Dombey.

Nor did the lack of attention to the diversity of mountain food plants owe to the demographically driven disintensification of agriculture, for in general the indigenous Andean people experienced somewhat less catastrophic depopulation (Table 2). In short, this near invisibility of the Andean peoples and their food plants seemed to mirror their characterization as "experts at concealment" (Humboldt and Bonpland 1821–1829).

Accounts of the early modern period increasingly portrayed the tropical Andes as remote, wild, majestic, and gigantic (Table 3 and Supplemental Material). This shift generally viewed mountains as landscapes of glory rather than gloom. The ideal of the sublime, and the derivative myth of pristine nature (a term coined in present-day critiques), emanated from nineteenth-century Romanticism. Its arc of meaning therefore might have differed from the context of white settler colonization such as in North America. In the tropical Andes the experience of the sublime was cast only in part as nature primeval, symbolized in the "Heart of the Andes" painting of Frederick Edwin Church.

Equally important, the notion of remoteness was a first-order signifier of similar meaning (Gould 2002).

The evolving view of the tropical Andes as remote can be illustrated in a series of historical descriptions of Cajamarca in the Andes of present-day northern Peru (Table 3). This study's focus is on the accounts of travelers and other observers during the mid- and late-1800s also illustrate the geopolitics of the early postcolonial state as an agent of both the incorporation and differentiation of the Andean regions within new national political geographies (Radcliffe 2015; see also Subramanian 2003).

During the 1800s the writings of modern travelers became a significant genre in the tropical Andes that included information on environments and livelihoods. Some but not all of these travel writings evoked the Andes as remote areas. Utopian feminist Flora Tristán ([1838] 2003) found the Andean peaks of Arequipa "isolated" and "desolate" even as she created remarkable descriptions of everyday foods. Markham ([1912] 1990), who traveled in 1852 and 1853, juxtaposed the modernizing technology and investments of Peru's coastal agro-industry with the meager cash cropping of Andean hacienda estates and the generic category of "Indian crops." During a protracted sojourn in Peru (1850–1890), Raimondi described a wide range of food plants including indigenous crops, although little on their uses. These copious observations stemmed from Raimondi's extensive travel and integration into the intellectual and political life of Peru (Raimondi 1874–1913).

Slightly later accounts included those by Charles Wiener, an Austrian-French traveler influenced by the sublime whose account features the extensive mapping of ruins sites (Wiener [1880] 1993). Wiener considered the tropical Andes for the most part as a monolithically remote hinterland, exemplified in the wildness and ancient ruins of the Cajamarca valley (Table 3). To Wiener, the excellence of indigenous Andean agriculturalists was a past phenomenon. Other early archaeology of the nineteenth and twentieth centuries was more varied. Mountaineering in combination with scientific activities added to this tendency. Whymper ([1892] 1942), for example, climbed eleven Andean summits while he focused on human physiological effects and such Andean biota as lichens.

Meanwhile, the demographic decline of Andean populations dropped to a nadir between the mid-1700s and mid-1800s (Table 1), although it was unlikely to have resulted in a preponderance of remote and pristine tracts (Knapp 2007). In sum, numerous early modern accounts (1780–1900) envisioned the remoteness of the tropical Andes, and they also contributed to major scientific advances on ecological and biogeophysical systems as well as early archaeology. By contrast, the food plants, uses, and associated landscapes of agrobiodiversity were not a systematic focus and were often omitted in these accounts.

Discussion: Agrobiodiversity of Mountain Societies Amid Remoteness and Connectedness (1900–1970, 1970–present)

The notion of the tropical Andes as remote set the stage in the subsequent development of modern agrobiodiversity knowledge and science in the early twentieth century (Table 2). The latter's preponderant focus was genetic resources and biological taxonomy and, secondarily, biogeography. Characterized biologically and geographically as "an amazing accumulation of cultivated plants and endemic animals" (Vavilov 1992, 349), the Andes became integral in the itineraries of genetic resource missions of the United States, Soviet Union, England, Canada, and European countries along with those of international scientific institutions. This strand of agrobiodiversity science reinforced the idea of remoteness. It also has raised geographic issues of unacknowledged spatial bias in the biogeography of germplasm collections (Hijmans et al. 2000) and the politics of conservation, including genetic resource access, intellectual property rights, and biopiracy.

Perspectives of cultural and human ecology in particular guided agrobiodiversity studies in both the 1900 to 1970 phase and in recent decades, and people–plant interactions and ethnobotany (including traditional ecological knowledge [TEK]) were also a focus of studies. Ideas of remoteness were important and even anchored these accounts, although countercurrents were notable. For example, one impressive description of existing Andean food plant complexes in the early twentieth century was situated in research that otherwise focused on the archaeology of ancient civilizations (Cook 1925). Meanwhile, the idea of remote areas was initially used as a means of categorizing Andean indigenous groups as tribes (cultural ecology) and closed corporate communities (human ecology).

Socioeconomic and environmental connectivity, rather than remoteness, are now widely recognized as distinguishing the agrobiodiversity and geographic dynamics of places in the Andes and the world's other mountain complexes. It is driven through migration,

urbanization, and urban–rural continua, national and international policy, and transformative factors such as resource extraction industries, water resource management amid climate change impacts, and tourism (Bebbington and Batterbury 2001; Wrathall et al. 2014).

During recent decades this connectedness has reshaped and partially reduced agrobiodiversity practices (Zimmerer 1992). Conversely, certain people, places, and their agrobiodiversity have also shown emergent properties of resilience amid changes and transformation (Zimmerer 2013). Resilience-enabling factors, such as multifaceted cultural values and viable seed systems, can be compatible with increased migration and extralocal flows under certain conditions, although this compatibility requires adequate resource access and the innovation, spatial and otherwise, of knowledge systems, land use, and foodways.

Earlier spatial models of land use and agrobiodiversity in tropical mountains had assumed elevation-based tiers or belts adapted through local cultural and biogeophysical interactions within these community-scale systems. This model assumed local elevation-based crop adaptation. In recent decades, however, mountain land use is recognized as more spatially heterogeneous and prone to patch dynamics ("overlapping patchworks"), rather than conforming to strict tiers. Owing to insights from the 1970s and 1980s, the Convention on Biological Diversity in 1991 recognized that mountain agrifood systems are the result of nonlocal influences. In the context of extralocal and often global factors, research highlights the decline of agrobiodiversity growth and use in various places although emergent viability is potentially characteristic of certain agrifood spaces, such as particular places in periurban areas and national-level food movements and cuisine trends (Zimmerer 2016).

Our findings offer various fresh insights for current studies of the so-called pristine myth of nature, colonial and postcolonial environments, and the metahistorical idea of mountain gloom and glory. The narratives associated with these areas of study have largely overlooked the formative stage of European global empires, conquest, and early colonialism that spanned the 1500s, 1600s, and early 1700s in the tropical Andes. Historical geography of early colonialism in the tropical Andes did not promulgate the pristine myth per se or the parallel ideas of mountain gloom and glory and the colonial state's "emptying the land" border to border. Instead, our analysis indicates the predominance of early imperial discourses and knowledge systems that geographically differentiated mountain spaces as either utilizable colonial land-scapes or mostly uninhabitable indigenous barrens. These geographic concepts resembled the widespread use of the similitude-and-difference mode of knowl-edge applied to biota and other realms encountered by the early European empires.

Today's multidimensional connectivity of landscapes and livelihoods in the Andes and other tropical moun-tains is tending to eclipse the previous portrayal of remoteness (Sevilla-Callejo and Mata-Olmo 2007), even as it is made more complex by the current territo-rial movements of indigenous and peasant communities. These social movements embrace ideas of living well in their multifaceted efforts to enhance territorial auton-omy in conjunction with food security and sovereignty.

Agrobiodiversity-based food growing and use are integral to these current territorial movements along with their political claims and initiatives for sustain-ability, justice, and human health and well-being (Zim-merer 2012). Although connectivity has expanded, current movements seek the control of space to create territories of partial autonomy. In effect, their territo-rial politics and aspirations seek to define the types and degree of connectivity, potentially inventing new and strategic kinds of selective remoteness.

Tropical and subtropical mountains of other world regions share several of the principal trajectories just indicated. For example, studies focused on wildlife conservation and science in the Himalaya suggest a similar arc of spatial ideas (including remoteness) as well as modern scientific developments and changing attitudes toward mountains (Hussain 2015). East Africa also shows the interaction of ideas of mountain spaces—including remoteness—and the development of present-day prominence of agrobiodiversity knowl-edge. Initiatives mixing mountain agrobiodiversity with indigenous territorial movements, sustainability issues, local food access, ethnobotany, and the factors of urbanization, periurban spaces, and migration are common across the previously mentioned world regions and are essentially globalizing in character.

Conclusions: Remoteness, Resource Sciences, and Geographies of Sustainability

The geographic depiction of remote areas has func-tioned in powerful, varied, and contested ways across time in the tropical Andes. Colonial geography enlarged the scope of remoteness between the 1500s and 1800s as a function of prevalent ideas forged amid imperial rule and indigenous resistance interacting with environmental, technological, and various socio-spatial transformations. Late colonial and early modern accounts (1770–1900) amplified the perception of mountain remoteness through influences that ranged from global shifts of predominant political economies to scientific and cultural knowledge systems.

Our study presents the results of a grounded case study showing mountain remoteness as a product of modernity, rather than its omission. Theoretically our framework offers an alternative to the tendency to cast remoteness as an intrinsic characteristic of mountains and, in particular, the world's tropical and subtropical mountains. This study offers an alternative analysis and interpretation. It highlights the powerful role of moun-tain spaces in the historical rise of the environmental sciences and generates insights of present relevance and potential future importance. Specifically, the strategies of current indigenous movements suggest the aspiration to promote territorial autonomy and control connect-edness by aspiring to a kind of selective remoteness. These recent and new territorial strategies contest the prevalent politics and meanings of remoteness deployed among colonial and postcolonial state powers, eco-nomic brokers, and many environmental institutions.

Distinct ideas of tropical mountains as remote areas have been entangled with the rise of diverse forms of agrobiodiversity knowledge. The rise of modern agro-biodiversity science depicted the tropical Andes and other major mountain systems as remote global centers of agrobiodiversity. International political economy and modern industrialization, ranging from genetic resource collections and biotechnology to the history of increasingly efficient rail transportation familiar to cosmopolitan travelers and scientists, reinforced these depictions. The suite of such factors attributed remote-ness and unique value to mountain-grown and utilized agrobiodiversity at the global scale.

By contrast, the preceding phase of agrobiodiversity knowledge (1770–1900) occurred in the context of aspirations to export agriculture and world commerce and only partially adopted the notion of mountain remoteness. The knowledge systems of the previously mentioned periods (1770–1900, 1900–1970) differed significantly from both earlier and later phases. Agro-biodiversity knowledge in the Inca period (1400–1532) and the early and middle colonial periods (1532–1770) was coded to symbolize, respectively, Inca imperial power and mountain-centered political control and subsequently the geographically differentiated poten-tial of colonial utility to Spanish imperialism.

The use of agrobiodiversity knowledge continues to expand and diversify in the tropical Andes as in other mountain environments globally. Agrobiodiversity is being recognized as part of emergent social–ecological systems taking shape amid global social and environmental changes involving such factors as climate, urbanization, and migration (Zimmerer 2013). This new perspective on agrobiodiversity also recognizes the importance of land use planning and the territorial strategies among indigenous, peasant, and smallholder groups allied to broader social movements and diverse institutions including nongovernmental organizations (NGOs), national agencies, and international organizations.

These new social–ecological and political agendas stand to benefit from reflexive understandings of past and present agrobiodiversity knowledge systems. Our research expands significantly the scope of the current agrobiodiversity knowledge. It is intended, moreover, to enable and support reflexive agrobiodiversity science and studies that are rigorous and policy-relevant as well as linked to broader sociocultural concerns, politics, and goals. Focusing on both the past and present of the geographic dynamics and spaces of agrobiodiversity knowledge is vital to these endeavors.

We have designed and implemented a framework guided by studies of science, technology, and society, as advanced recently in nature–society geography as well as cognate fields (Goldman, Nadasdy, and Turner 2011; Lave et al. 2014; Subramanian 2015) to integrate new theorization and grounded empirical research of mountain environments, resources, sociocultural complexity, and knowledge systems. Mountains are among the research sites most distinguished by multistrand engagements with global sustainability goals. Fusing the biogeophysical and social sciences with the powerful role of the geohumanities demonstrates how the geographic transformations of mountains have been integral to past and present understandings of agrobiodiversity and potential pathways contributing to the sustainability of food growing and use.

Acknowledgments

Much of the research and writing of this article was supported through a pair of research fellowships, first in its initial stage at the Agrarian Studies Program of Yale University (2004–2005) and then at the David Rockefeller Center for Latin American Studies of Harvard University (2016) where productive conversations, meetings, and colloquia occurred with numerous colleagues. Sabbatical funding for the lead author was received through the University of Wisconsin–Madison and Pennsylvania State University, respectively. Research on earth system models of agrobiodiversity was funded in part through the GeoSyntheSES Lab. Conversations important to this article also occurred in the contexts of 2014 meeting of the American Anthropological Association, Fulbright research in Peru in 2015, and ongoing research collaborations in South America, Africa, South Asia, and Europe (2015–2019). Community and social movement cooperation and NGO, agency, and university affiliations are gratefully acknowledged. Martha Bell produced the map in Figure 1. The lead author was able to present earlier versions as pieces of keynote presentations to the Melamid Award Colloquium of the American Geographical Society (December 2013); the Altman Symposium on "Rethinking the Anthropocene" at the Humanities Center of Miami University (April 2015); and universities in Lima, Arequipa, and Háunuco in Peru (May 2015, July 2016).

Supplemental Material

Supplemental data for this article can be accessed on the publisher's Web site at http://dx.doi.org/10.1080/24694452.2016.1235482

Notes

1. See also the more comprehensive global data in the Supplemental Material. The data compiled in these tables do not suggest that these mountains are the only globally important sites of concentrated agrobiodiversity.
2. Six additional texts were examined that are not contained in the main article due to space constraints (see Supplemental Material).
3. The seeds circulated through local seed systems include farmer varieties (FVs), which have significantly higher levels of agrobiodiversity than modern varieties (MVs). High-agrobiodiversity varieties known as landraces are a common form of FVs. We ran ordinary least squares regression with continuous and categorical predictors in JMP statistical software Version 11 (SAS Institute, Cary, NC).
4. On this critique see Zimmerer and Bell (2015).

References

Bebbington, A. 2000. Reencountering development: Livelihood transitions and place transformations in the Andes. *Annals of the Association of American Geographers* 90 (3): 495–520.

Bebbington, A. J., and S. P. Batterbury. 2001. Transnational livelihoods and landscapes: Political ecologies of globalization. *Cultural Geographies* 8 (4): 369–80.

Blaikie, P., and H. Brookfield. 1987. *Land degradation and society*. New York: Methuen.

Borsdorf, A., and C. Stadel. 2015. *The Andes: A geographical portrait*. Heidelberg, Germany: Springer.

Bourguet, M.-N., C. Licoppe, and H. O. Sibum. 2002. *Instruments, travel and science: Itineraries of precision from the seventeenth to the twentieth century*. London and New York: Routledge.

Brookfield, H. C. 2001. *Exploring agrodiversity*. New York: Columbia.

Brush, S. B. 2004. *Farmers' bounty: Locating crop diversity in the contemporary world*. New Haven, CT: Yale.

Cañizares-Esguerra, J. 2006. *Nature, empire, and nation: Explorations of the history of science in the Iberian world*. Palo Alto, CA: Stanford.

Cieza de León, P. de. [1553] 1984. *Crónica del Perú* [The chronicle of Peru]. Lima, Peru: Pontificia Universidad Católica del Perú.

Colombia—Producción Estadística. 2011. *Resultados Encuesta Nacional Agropecuaria ENA* [Colombia—2011 national statistical information]. Departamento Dirección de Metodología Administrativo Nacional de Estadística (DANE), Bogotá, Colombia. http://www.dane.gov.co/files/investigaciones/agropecuario/ (last accessed 10 September 2015).

Cook, O. F. 1925. Peru as a center of domestication: Tracing the origin of civilization through the domesticated plants. *Journal of Heredity* 16 (3): 95–110.

Craib, R. B. 2004. *Cartographic Mexico: A history of state fixations and fugitive landscapes*. Durham, NC: Duke.

Debarbieux, B., and G. Rudaz. 2015. *The mountain: A political history from the Enlightenment to the present*. Chicago: University of Chicago.

de Haan, S., J. Núñez, M. Bonierbale, and M. Ghislain. 2010. Multilevel agrobiodiversity and conservation of Andean potatoes in central Peru: Species, morphological, genetic, and spatial diversity. *Mountain Research and Development* 30 (3): 222–31.

Denevan, W. M. 1992. The pristine myth: The landscape of the Americas in 1492. *Annals of the Association of American Geographers* 82 (3): 369–85.

Ecuador—Instituto Nacional de Estadísticas y Censos. 2013. Encuesta de superficie y producción agropecuaria continúa [Ecuador—2013 report of the National Institute for Statistics and Census]. http://www.ecuadorencifras.gob.ec/encuesta-de-superficie-y-produccion-agropecuaria-continua-bbd/ (last accessed 10 September 2015).

Foucault, M. 2002. *The order of things: An archaeology of the human sciences*. London and New York: Routledge.

Gade, D. W. 1999. *Nature and culture in the Andes*. Madison: University of Wisconsin.

———. 2015. *Spell of the Urubamba: Anthropogeographical essays on an Andean Valley in space and time*. Heidelberg, Germany: Springer.

Gerbi, A. [1975] 2010. *Nature in the New World: From Christopher Columbus to Gonzalo Fernandez de Oviedo*, trans. J. Moyle. Pittsburgh: University of Pittsburgh Press.

Goldman, M. J., P. Nadasdy, and M. D. Turner, eds. 2011. *Knowing nature: Conversations at the intersection of political ecology and science studies*. Chicago: University of Chicago.

Gould, S. J. 2002. Art meets science in The Heart of the Andes: Church paints, Humboldt dies, Darwin writes and nature blinks in the fateful year of 1859. In *I have landed: The end of a beginning in natural history*, 90–112. New York: Harmony Books.

Graddy, T. G. 2013. Regarding biocultural heritage: *In situ* political ecology of agricultural biodiversity in the Peruvian Andes. *Agriculture and Human Values* 30 (4): 587–604.

Guamán Poma de Ayala, F. [1613] 1980. *El primer nueva corónica y buen gobierno* [First new chronicle and good government]. Mexico City, México: Siglo Veintiuno.

Haenke, T. [1799] 1974. Introducción a la Historia Natural de la Provincia de Cochabamba y circunvencias. In *Tadeo Haenke, su obra en los Andes y la selva, boliviana* [Tadeo Haenke, his work in the Andes and lowlands of Bolivia], 15–116. La Paz, Bolivia: Editorial Los Amigos del Libro.

Harden, C. P. 2012. Framing and reframing questions of human–environment interactions. *Annals of the Association of American Geographers* 102 (4): 737–47.

Harms, E., S. Hussain, S. Newell, C. Piot, L. Schein, S. Shneiderman, and J. Zhang. 2014. Remote and edgy: New takes on old anthropological themes. *HAU: Journal of Ethnographic Theory* 4 (1): 361–81.

Hijmans, R. J., K. A. Garrett, Z. Huaman, D. P. Zhang, M. Schreuder, and M. Bonierbale. 2000. Assessing the geographic representativeness of genebank collections: The case of Bolivian wild potatoes. *Conservation Biology* 14 (6): 1755–65.

Humboldt, A. von. 1849. *Aspects of nature, in different lands and different climates; with scientific elucidations*. London: Longman, Brown, Green, and Longmans.

Humboldt, A. von, and A. Bonpland. [1807] 2009. Essay on the geography of plants with a physical tableau of the equinoctial regions. In *Essay on the geography of plants*, ed. S. T. Jackson, trans. S. Romanowski, 49–144. Chicago: University of Chicago.

———. 1821–1829. *Personal narrative of travels to the equinoctial regions of the new continent during the years 1799–1804*, vols. 1–7. London: Longman, Hurst, Rees, Orme, Brown, and Green.

Hussain, S. 2015. *Remoteness and modernity: Transformation and continuity in northern Pakistan*. New Haven, CT: Yale.

Johns, T., B. Powell, P. Maundu, and P. B. Eyzaguirre. 2013. Agricultural biodiversity as a link between traditional food systems and contemporary development, social integrity and ecological health. *Journal of the Science of Food and Agriculture* 93 (14): 3433–42.

Kerr, R. B. 2014. Lost and found crops: Agrobiodiversity, indigenous knowledge, and a feminist political ecology of sorghum and finger millet in Northern Malawi. *Annals of the Association of American Geographers* 104 (3): 577–93.

Knapp, G. 2007. The legacy of European colonialism. In *Physical geography of South America*, ed. T. T. Veblen, K. R. Young, and A. R. Orme, 289–304. Oxford, UK: Oxford.

Lave, R., M. W. Wilson, E. S. Barron, C. Biermann, M. A. Carey, C. S. Duvall, and C. Van Dyke. 2014. Intervention: Critical physical geography. *The Canadian Geographer* 58 (1): 1–10.

Lewis, L. R., and K. J. Chambers. 2010. Introduction: Geographic contributions to agrobiodiversity conservation. *The Professional Geographer* 62 (3): 303–04.

Markham, C. R. [1912]1990. *Markham in Peru: The travels of Clement R. Markham, 1852–1853.* Austin: University of Texas.

Marston, R. A. 2008. Land, life, and environmental change in mountains. *Annals of the Association of American Geographers* 98 (3): 507–20.

Moseley, M. E. 1997. *The Incas and their ancestors: The archaeology of Peru.* London: Thames and Hudson.

Nabhan, G. P. 2008. *Where our food comes from: Retracing Nikolay Vavilov's quest to end famine.* Washington, DC: Island.

Nicolson, M. H. 1997. *Mountain gloom and mountain glory: The development of the aesthetics of the infinite.* Seattle: University of Washington.

Perreault, T. 2005. Why *chacras* (swidden gardens) persist: Agrobiodiversity, food security, and cultural identity in the Ecuadorian Amazon. *Human Organization* 64 (4): 327–39.

Perú—Sistema de Consulta de Cuadros Estadísticos. 2013. IV Censo Nacional Agropecuario 2012. Instituto Nacional de Estadística e Informática [Peru—System for the consultation of 2013 national statistical tables]. http://censos.inei. gob.pe/cenagro/tabulados (last accessed 30 May 2015).

Piperno, D. R. 2012. New archaeological information on early cultivation and plant domestication involving microplant (phytolith and starch grain) remains. In *Biodiversity in agriculture: Domestication, evolution, and sustainability,* ed. P. Gepts, T. R. Famula, R. L. Bettinger, S. B. Brush, A. B. Damania, P. E. McGuire, and C. O. Qualset, 136–59. Cambridgeshire, UK: Cambridge University Press.

Pratt, M. L. 2007. *Imperial eyes: Travel writing and transculturation.* London and New York: Routledge.

Price, M. F., A. C. Byers, D. A. Friend, T. Kohler, and L. W. Price, eds. 2013. *Mountain geography: Physical and human dimensions.* Berkeley and Los Angeles: University of California.

Radcliffe, S. A. 2015. *Dilemmas of difference: Indigenous women and the limits of postcolonial development policy.* Durham, NC: Duke.

Raimondi, A. 1874–1913. *El Perú.* 6 vol. Lima, Peru: Impresa del estado.

Salomon, F. 1985. The historical development of Andean ethnology. *Mountain Research and Development* 5 (1): 79–98.

Sarmiento, F. O. 2000. Breaking mountain paradigms: Ecological effects on human impacts in man-aged Trop-Andean landscapes. *AMBIO: A Journal of the Human Environment* 29 (7): 423–31.

———. 2008. Agrobiodiversity in the farmscapes of the Quijos River in the tropical Andes, Ecuador. In *Protected landscapes and agrobiodiversity values,* ed. T. Amend, J. Brown, A. Kothari, A. Phillips, and S. Stolton, 22–30. Gland, Switzerland: International Union for Conservation of Nature and Deutsche Gesellschaft für Technische Zusammenarbeit.

Scott, J. C. 1998. *Seeing like a state.* New Haven, CT: Yale.

Sevilla-Callejo, M., and R. Mata-Olmo. 2007. Introducción a las dinámicas territoriales en el área oriental del Parque Nacional y ANMI Cotapata (Depto. de La Paz, Bolivia) [Introduction to territorial dynamics in the eastern part of Cotapata, national park and natural area under integral management]. *Ecología en Bolivia* 42 (1): 34–47.

Subramanian, A. 2003. Modernity from below: Local citizenship on the south Indian coast. *International Social Science Journal* 55 (175): 135–44.

Tristán, F. [1838] 2003. *Peregrinaciones de una paria, 1833–1834* [Peregrinations of a pariah, 1833–1834]. Arequipa, Peru: Lector.

Ulloa, A. de. [1772] 1944. *Noticias americanas, entretenimiento físico-histórico sobre la América Meridional y la Septentrional oriental* [News of the Americas and physical-historical account about South America and Eastern North America]. Buenos Aires: Editorial Nova.

Urton, G. 2012. The herder–cultivator relationship as a paradigm for archaeological origins, linguistic dispersals, and the evolution of record-keeping in the Andes. *Proceedings of the British Academy* 173:321–43.

Vavilov, N. I. 1992. *Origin and geography of cultivated plants.* Cambridgeshire, UK: Cambridge University Press.

Vega, G. de la [1609, 1616–1617] 1943. *Comentarios reales de los Incas* [Royal commentaries of the Inca]. Buenos Aires: Emecé.

Whymper, E. [1892] 1942. *Travels amongst the great Andes of the equator.* New York: Scribner's.

Wiener, C. [1880]1993. *Perú y Bolivia: Relato de viaje: Seguido de estudios arqueológicos y etnográficos y de notas sobre la escritura y los idiomas de las poblaciones indígenas* [Peru and Bolivia: Account of travel: Following the route of archeological and ethnographic studies and notes about the writing and language of indigenous populations], trans. E. Rivera Martíez. Lima, Peru: Instituto Francés de Estudios Andinos and Universidad Nacional Mayor de San Marcos.

Wrathall, D. J., J. Bury, M. Carey, B. Mark, J. McKenzie, K. Young, M. Baraer, A. French, and C. Rampini. 2014. Migration amidst climate rigidity traps: Resource politics and social-ecological possibilism in Honduras and Peru. *Annals of the Association of American Geographers* 104 (2): 292–304.

Young, K. R. 2009. Andean land use and biodiversity: Humanized landscapes in a time of change. *Annals of the Missouri Botanical Garden* 96 (3): 492–507.

Zimmerer, K. S. 1992. The loss and maintenance of native crops in mountain agriculture among indigenous people. *GeoJournal* 27 (1): 61–72.

———. 1996. *Changing fortunes: Biodiversity and peasant livelihoods in the Peruvian Andes.* Los Angeles: University of California.

———. 2006. Humboldt's nodes and modes of interdisciplinary environmental science in the Andean world. *Geographical Review* 96 (3): 335–60.

———. 2010. Biological diversity in agriculture and global change. *Annual Review of Environment and Resources* 35:137–66.

———. 2012. The indigenous Andean concept of *kawsay,* the politics of knowledge and development, and the borderlands of environmental sustainability in Latin America. *Publications of the Modern Language Association* (PMLA) 127 (3): 600–606.

———. 2013. The compatibility of agricultural intensification in a global hotspot of smallholder agrobiodiversity (Bolivia). *Proceedings of the National Academy of Sciences* 110 (8): 2769–74.

———. 2014. Conserving agrobiodiversity amid global change, migration, and nontraditional livelihood networks: The dynamic uses of cultural landscape knowledge. *Ecology and Society* 19 (2): 1.

———. 2016. Agrobiodiversity as jazz improvisation: Foodscape change and continuity. *ReVista: Harvard Review of Latin America* 16 (1): 15–23.

Zimmerer, K. S., and M. G. Bell. 2015. Time for change: The legacy of a colonial Euro-Andean model of landscape versus the need for landscape connectivity. *Landscape and Urban Planning* 139:104–16.

Heritage as Weapon: Contested Geographies of Conservation and Culture in the Great Himalayan National Park Conservation Area, India

Ashwini Chhatre, Shikha Lakhanpal, and Satya Prasanna

Mountains are one of the last refuges of biodiversity worldwide. As the global discourse on nature conservation becomes prominent within sustainability debates and local populations continue to be blamed for environmental destruction, projected territorial expansion of protected areas will likely lead to high levels of conflict and contestation around mountains of the world. At the same time, deeper penetration of transnational advocacy networks and wider connections of civil society will bring new tools of resistance to bear on this conflict. We propose that democracy plays an increasingly critical role in assisting local opposition to thwart new restrictions on access to natural areas prioritized for conservation. We illustrate this larger argument through the case of the Great Himalayan National Park Conservation Area (GHNPCA), recently designated as a UNESCO World Heritage Site in the Indian Himalayas. In their opposition to exclusion, local communities have employed heritage as a weapon, successfully marshaling the representation of the region as the "Valley of the Gods" and putting their cultural heritage to work against global conservation agendas. Tourism posters depicting the sacred geography of numerous local deities allow local communities to justify opposition to the conservation status that restricts access to their gods, while channeling their demands through elected representatives. The state navigates this complex territory between global and local heritage uneasily, primarily through a series of compromises at the local level. This article focuses on the ways in which mountain heritage—local and global, cultural and natural—is negotiated in the crucible of democracy.

山区是全世界仅存的生物多样性庇护场所之一。当自然保育的全球论述, 在可持续性的辩论中受到突显, 而在地人口持续因环境破坏而被怪罪时, 保育地区预计的领域扩张, 将有可能导致全世界山区的高度冲突和竞夺。于此同时, 跨国倡议网络更深刻的渗透, 以及公民社会更为广泛的连结, 将会为反抗带来新的工具, 对此一冲突施加影响。我们提出, 民主在协助地方反抗加诸以保育为优先的自然地区的管道的新限制中, 逐渐扮演关键的角色。我们透过大喜马拉雅国家公园保育地区 (GHNPCA) 的案例描绘此一更大的主张, 该地区在印度喜马拉雅的部分, 最近被指定为联合国教科文组织的世界遗产地点。在地方社群反对被迫排除的运动中, 他们运用了遗产作为武器, 成功地将该地的再现编织为"神的谷地", 并运用其文化遗产来对抗全球保育议程。描绘出汇聚众多地方神明的神圣地理之观光广告, 让在地社群能够正当化反抗限制其接近他们的神的保育主张, 同时将其需求透过选举代表进行传达。国家主要透过在地层级的一系列妥协, 在此一介乎全球和在地遗产的复杂领域中困难地航行。本文聚焦山地遗产——地方与全球, 文化与自然——在民主的严刻考验中进行协商的方式。关键词: 保育, 发展, 遗产, 山区, 联合国教科文组织。

A escala mundial, las montañas son uno de los últimos refugios de la biodiversidad. En cuanto el discurso global sobre la conservación de la naturaleza se hace prominente dentro de los debates de la sustentabilidad, y las poblaciones locales continúan siendo acusadas de destrucción ambiental, la expansión territorial proyectada de áreas protegidas llevará probablemente a altos niveles de conflicto y confrontación sobre las montañas del mundo. Al mismo tiempo, una más profunda penetración de las redes transnacionales de defensa y conexiones más amplias de la sociedad civil atraerá nuevas herramientas de resistencia a intervenir en este conflicto. Nuestra proposición es que la democracia juegue un papel cada vez más crítico para ayudar a la oposición local a boicotear nuevas restricciones sobre el acceso a áreas naturales priorizadas para conservación. Ilustramos este argumento mayor por medio del caso del Área de Conservación del Parque Nacional del Gran Himalaya (GHNPCA, acrónimo en inglés), designado recientemente como Patrimonio de la Humanidad de la UNESCO, en los Himalayas de la India. En su oposición a la exclusión, las comunidades locales han utilizado la heredad como arma, ordenando con éxito la representación de la región como "Valle de los dioses" y poniendo a trabajar su heredad cultural contra las agendas de conservación global. Los avisos turísticos que aluden a la geografía sagrada de numerosas deidades locales permiten a las comunidades locales justificar su oposición a las condiciones de conservación que restringen el acceso a sus dioses, al tiempo que canalizan sus demandas a través de representantes elegidos. El estado navega este complejo territorio entre heredades locales y globales con preocupación, primariamente a través de una serie de compromisos a nivel local. Este artículo se enfoca en las maneras como la heredad de la montaña—local y global, cultural y natural—es negociada en el crisol de la democracia.

There is a bird, *tutru*, which toils in the forest to build a nest for its young. But when the time comes, another bird, *juraun*, forces *tutru* out and takes over the nest that *tutru* has built with such effort and skill. The *sarkar* [government] is doing the same to us. We have raised these forests. We have nurtured the birds and animals. Now the *sarkar* comes and throws us out of our forests.
—Jai Ram, local medicine man, village Majgraon, Raila

Is it not our duty, as a civilization, that we leave some area, just a small part, for nature, for future generations, for our own sanity?
—Vinay Tandon, (former) Principal Chief Conservator of Forests, Himachal Pradesh

Biomass-based rural livelihoods persist rather uneasily in conservation landscapes throughout the world, with the tension resulting primarily from an apparent contradiction between local and global interests. In one of its ongoing enactments, Indian conservationists have pitted the globally endangered western tragopan, a brilliantly colored pheasant endemic to the western Himalaya and the associated biodiversity, against the grazing and biomass collection activities of local populations in the Great Himalayan National Park Conservation Area (GHNPCA) in Kullu Valley, Himachal Pradesh. This trend is symptomatic of the progression of protected areas around the world: As biodiversity consolidates its status as a common heritage of all humankind and natural areas are destroyed in the densely populated plains and valleys, the spaces for its conservation are increasingly to be found in the mountains (Holmes 2013). Despite lip service to the involvement of communities in biodiversity conservation, the emphasis remains on a command-and-control approach of protection with fences and guns (Adams and Hutton 2007).

Efforts at conservation continue to draw on the assumption (and the often forgone conclusion) that local populations are not only incapable of conserving biodiversity as trustees for all humankind but that the biodiversity in question, in fact, needs to be protected from local use (Neumann 1998; West and Brockington 2006; Dowie 2011). Following this script, executed virtually identically across the world over the last century, local communities have been progressively excluded from the GHNPCA. In May 1999, fifteen years after the first formal demarcation, the Himachal Pradesh government issued the final notification for the park, potentially leading to severe restrictions on use of resources by local populations. Despite continued tensions and opposition, the GHNPCA received World Heritage status from UNESCO in June 2014 under the criteria of "outstanding significance for biodiversity conservation" and was added to the list of more than 1,000 such sites globally (UNESCO 2014).

This article traces the journey of GNHPCA over a thirty-year period from first demarcation in 1984 to World Heritage status in 2014. In particular, we document the efforts of local communities and their representatives to challenge their exclusion from the park. The inscription of the GHNPCA as a site to protect global heritage at the expense of local livelihoods has been fiercely contested at the local level, with limited success. In this struggle, local communities have employed their cultural heritage as a weapon in their opposition to encroachments on local livelihoods. Democracy provides the context within which local opposition has successfully marshaled the Tourism Department's ubiquitous representation of the region as the "Valley of the Gods" and put their religious heritage to work against conservation agendas. Tourism posters celebrating the sacred geography of numerous local deities allow local communities to justify opposition to the global conservation heritage that restricts access to their gods. In doing so, this article joins the work of other scholars in illuminating the struggles and tactics through which local people contest, negotiate, and force a reimagination of global heritage (Smith 2006). Regional and national governments often deploy heritage to exclude local people through essentialized meanings that map poorly onto the lived experiences of local communities. Scholarship in heritage studies and anthropology, on the other hand, has focused on how natural heritage is privileged over cultural heritage (Meskell 2012). Meskell (2012), in particular, traced this privilege to the salience of biodiversity conservation within global and national debates. We argue that the state navigates this complex territory between global and local interpretations of heritage on the one hand and between conservation and development on the other, primarily through a series of compromises. This article focuses on the ways in which mountain heritage—local and global, cultural and natural—is negotiated in the crucible of democracy through the analysis of these negotiations and compromises in the GHNPCA.

The narrative content of this article draws on continuing engagement of the authors with issues and communities surrounding the GHNPCA, cumulatively adding to over forty years since 1995. At least one of the coauthors was witness to each of the events described in the rest of the article between 1995 and 2014. In the following section, we provide insights into the lived experience of heritage, with a focus on the interpellation of cultural and natural aspects. We then describe the inscription of global biodiversity conservation priorities into this landscape and its contradictions with local practices. We illustrate the ways in which local opposition to exclusion incorporates specific interpretations of local heritage in negotiating access to resources, leading to a series of compromises to the professed ideal of global heritage. We then conclude with a discussion of the role of mountain geography in understanding the use of heritage in contests over access to nature and natural resources and the role of elected representatives in mediating the conflicts and effecting local compromises.

Sacred Geography

Kullu, tourism department pamphlets proclaim, is the valley of the gods. It is certainly difficult to contest that claim in light of the profusion of local deities that abound in Kullu. With overlapping jurisdiction as well as a hierarchical arrangement, local deities govern the social life of Kullu far more than laws of the Indian Union govern the civic life of the nation. Many villages have their own deities; a single side valley might have up to a dozen or more. These deities are significant to the state administration as well. In older times, the Kullu *Raja* (ruler) used to invite each of these deities to the annual *Dushehra* (the tenth and final day of the Hindu festival of Navaratri, usually held between September and October) festivities in the autumn. Today, the district collector—the highest ranked civil servant in the district—sends a personal invitation to each deity to attend this annual megaevent, which draws tourists from around the world. On the eve of *Dushehra*, each of these deities is carried by palanquin to Kullu town, the drums and pipes of each procession echoing through the narrow Kullu Valley.

This sacred geography is best reflected in the myriad rites and ceremonies of passage as part of local life. In the GHNPCA, one often comes across majestic *deodar* trees (Himalayan cedar), standing alone near settlements, adorned with sickles and other iron ornaments. These are *banshiras*, living foundation stones of habitation. When an adult son wishes to colonize a fresh area by clearing forest and breaking new land, he will choose a *deodar* tree in the new area and nominate it as the *banshira*. Literally, the *banshira* is charged with the protection of the newly settled family and homestead from evil spirits and demons. Carrying a few iron adornments from the *banshira* of the parent village and putting these on the new one solemnizes a new *banshira*. Thus a tree becomes the guardian angel of the new settlement. The villages and settlements in the Sainj and Jiwa valleys in the GHNPCA can be traced back to eight parent *banshiras*, providing vital clues to the colonization of the valleys through succeeding generations of *banshiras*.

Local deities (*devtas*) govern almost every aspect of rural life in the GHNPCA. All community celebrations and assemblages are presided over by the local *devta*. Permission of the relevant deity is routinely sought before every step in the cycles of agriculture, pastoralism, or other livelihood activities. In April, herders seek divine permission to assemble their herds from the numerous households to whom the animals belong. Then they seek permission to leave the village and cross the boundary into *gahar* or forest pasture. In July, they return to the village, sans the animals, to seek permission to enter the *nigahar* or the alpine meadows above the tree line. This sequence of permissions continues until October, when the herds are disbanded for the winter, each small subherd of fifteen to twenty animals returned to its owner. All along the routes are scattered numerous less important but potentially pernicious deities, who demand appeasement in exchange for safe transit. In several cases, particular deities control hunting concessions, as the one at the head of Tirthan Valley, residing in village Pekhri. By custom, hunters from all neighboring villages seek permission to hunt, which has been traditionally restricted to one animal per person at any time, with strict sanctions against the hunting of females and young ones. In the virtual reality of this sacred geography, deities are the signposts that mark boundaries, which are as much ecological as they are social.

The morality of this social order is not fixed in space and time but serves to delimit boundaries for local negotiation. In June 2000, a year after the disruption caused by the final notification of the Park in May 1999, a group of herders from village Grahan located in the adjoining Kanawar Wildlife Sanctuary came to visit and propitiate the *devta* of Upraila at the head of Jiwa Valley on the edge of GHNPCA. They came to seek his permission to graze in his pastures, now within the newly enclosed National Park. The moral authority of the *devta* being far in excess of those of the park authorities, the herders found it prudent to seek his permission and then take on the might of the law with his blessings. In a few days, they had their permission and returned home to organize their summer migration, due to start in a couple of weeks. Draped in ritual and concealed by ceremony, this institutional complex of negotiation is the bedrock on which rests the ecological sustainability of pastures and economic viability of the herding occupation. The omnipresent *devta* provides the moral terrain for this negotiation as well the peg for temporal continuity.

Local Threats to Global Heritage

The preservation of biodiversity by excluding human pressure is the dominant paradigm in conservation (Adams and Hutton 2007) and runs counter to local livelihoods that are heavily dependent on using the same resources. Science provides the overriding justification for the exclusionary paradigm that dominates biodiversity conservation, providing "evidence" to support the exclusion of local communities (Lewis 2003). Yet this science is contested. It has been contested first and foremost by the people who use the park, but there is also evidence from other parts of the Himalaya, and from elsewhere, that contests key assumptions made by conservation scientists about human influences on biological diversity (Holmes 2013). There is an emerging body of theoretical and empirical work that suggests that disturbance is an integral factor in shaping the diversity of many ecosystems (Willis, Gillson, and Brncic 2004). Many researchers have used this work to argue for human coexistence with diverse biotic communities (Holmes 2013).

The GHNPCA lies in a relatively isolated part of the Kullu Valley, in Himachal Pradesh. It was established in 1984, following a survey conducted by an international team of scientists. It was the judgment of this survey team that this area was the most promising for the location of the national park being planned for the state, based on its low population density and exceptional condition of the forests (Himachal Wildlife Project 1981). The park is noted for harboring one of only two protected populations of the western tragopan in the world (thought to number 1,600 animals in the wild), among four other pheasant species; sizable, contiguous populations of Himalayan tahr and blue sheep; and an endangered population of musk deer (Wildlife Institute of India 1999).

GHNPCA is also used by local communities for a variety of resources. Approximately 20,000 people live in a 5-km-wide belt, ringing the western side of the conservation area. All families own and cultivate land outside the National Park. For the most part these are small parcels of land that provide subsistence for some portion of the year. The bulk of the population depends on a variety of additional resources within the GHNPCA to meet their annual income requirements, including the grazing of sheep and goats, the extraction of medicinal herbs to be sold to a burgeoning pharmaceutical and cosmetics industry, and the collection and sale of morel mushrooms (*guchhi*), consumed in restaurants around the world (Chhatre and Saberwal 2006).

Biologists and officials of the Forest Department have considered these various activities to pose a serious threat to the biological diversity of the region (Himachal Wildlife Project 1981). The presence of herders with their sheep is considered responsible for overgrazing the meadows, with consequences for plant diversity as well as soil erosion. Their movement through the forests while on the spring migration up to the alpine meadows is seen as disturbing the western tragopan when it is nesting, with potentially serious consequences for chick survival. This is also the time of year when large numbers of people comb the forest floor looking for *guchhi*, again seen as disturbing the nesting birds. Researchers report that dogs accompanying *guchhi* collectors and shepherds chase western tragopans and animals such as musk deer (Wildlife Institute of India 1999). Medicinal herb extraction is seen to have escalated to a point where some species are, reportedly, on the decline, far less visible, and smaller in size than just a few years ago.

The point for most villagers around GHNPCA, contrary to the prevailing conservation logic, is that the conservation value placed on the park today cannot be disassociated from its socioecological history. It could be argued that the villagers deserve credit for having taken good care of the park's resources, which is why we find it in such good condition. Villagers argue that it is not despite their presence in the park but because of their seasonal activities that the animal, bird, and plant populations have flourished. Despite a significant amount of scientific research over the last forty years, there is little evidence beyond anecdotes to support the notion that grazing and herb collection—the two main forms of villager dependence on park resources—are detrimental to biological diversity in GHNPCA (Chhatre and Saberwal 2006). Because of poor design, the research is unable to rule out a series of alternative explanations regarding the negative impacts of human activity and resource use on conservation outcomes or even establish that the outcomes are indeed negative. Yet science provides the overriding justification for the exclusionary paradigm that dominates biodiversity conservation in the park (Wildlife Institute of India 1999).

Heritage as Weapon

Villagers, who have had access to park resources for as long as they and the elders in their village can remember, were suddenly informed that they could no longer extract resources from the park to protect the biological resources within the area. By not going into the park, people were told, they were supporting a greater common good. Local residents drew the only conclusion that can be inferred: Their actions are seen as responsible for destroying these same biological resources. More important, though, it is their actions in the future that will lead to this destruction of wildlife. Local communities feel that they must have been doing something right, as the region is being valued for its biological diversity today, given the fact that they grazed their sheep and harvested herbs over the past few decades. How else could the biodiversity have survived? Given the apparent contradiction, as well as the perception that outsiders are trying to appropriate a resource they have used and managed on their own for many centuries, there has been steady opposition and often downright hostility toward any person or agency attempting to enforce an exclusion of people from the area. In their pursuit of warding off this unwanted threat to their access, the opposition to exclusion has used elements of local cultural heritage with creativity and to great effect.

We present three such incidents where local activists weave narratives of their cultural heritage to resist the incursion of global natural heritage and biodiversity on their everyday lives. The three incidents also illustrate the role of democracy as a necessary

condition for the local activism to take shape and successfully use their cultural heritage to their advantage.

Caught between the *Devta* and the World Bank

As social costs of protected areas mounted, the global response to criticism during the 1980s took the form of integrated conservation and development projects designed to reconcile the contradiction between exclusionary conservation and local livelihoods. In India, this took the form of ecodevelopment, a suite of investments undertaken to reduce the dependence of local populations on natural resources inside protected areas through income-generating activities at the household and community levels. Villages around GHNPCA were included as one of the two pilot sites for the ecodevelopment experiment in India, with support from the World Bank in 1994 (Baviskar 2003). It turned out to be far from easy, however.

In September 1992, in a drama being played out in the distant Narmada Valley in central India, the World Bank had asked the government of India to furnish details of people affected by the Sardar Sarovar Dam on the Narmada River within six months and ensure voluntary resettlement of such people, as a condition for continued support to the dam. In what was a period of horrendous repression in the Narmada Valley, the state cracked down heavily on the antidam movement. The repression culminated in January 1993 with police firing on protesting *adivasis*. Amidst a global outcry against the blatant abuse of human rights and repression of indigenous peoples, the World Bank was forced to withdraw support for the project in March 1993 (Baviskar 1995). The GHNPCA ecodevelopment pilot was being negotiated at around the same time that the World Bank was withdrawing from the Narmada Valley (World Bank 1996). Although all was clear on other fronts, GHNPCA had a few villages inside its boundaries that would have to be relocated under existing law. The World Bank would tolerate no infringement of its Operational Directives on Indigenous Peoples and on Involuntary Displacement, thus putting the ecodevelopment funding in jeopardy.

In April 1994, Sat Prakash Thakur, the local elected representative to the State Legislative Assembly and Minister for Horticulture, took a retinue of officials to two villages inside the park—Shakti and Maraur—and held meetings to convince the residents to agree to relocate outside the park. This would have satisfied the World Bank because it did not violate its directive on involuntary displacement. The headman of Shakti, a shaman himself, as well as an elected representative in local government, invited the minister and his party to consult the local *devta* in the presence of a large assortment of residents from nearby villages.

The *devta* refused to permit Shakti villagers to relocate. After repeated attempts and several hours of efforts to placate the *devta* failed, the minister had to return empty-handed. Without permission from the *devta*, residents declined the generous resettlement package on offer and decided to remain inside the park. The precedent of the Narmada Valley and the presence of the World Bank ensured that the two villages could not simply be "thrown out." With time running out for the project to be finalized, the government turned to the last resort—a legal sleight of hand. In June 1994, in deference to the wishes of the World Bank of not displacing people involuntarily and of the residents unwilling to move, the state government renotified the National Park with new boundaries and carved out an area of 90 km^2, including the two recalcitrant villages, to be constituted as the Sainj Wildlife Sanctuary. With the law allowing villages to remain inside sanctuaries but not national parks, the thorny problem of involuntary displacement was thus avoided. The residents of several villages, including those outside the park boundary, could now access resources inside what was now a slightly downgraded protected area.

A Hybrid Moral Economy

During the summer and monsoon of 1999, several villagers were threatened with prosecution and some graziers were fined as the overzealous park staff intimidated local people with threats of prosecution. Villagers responded by enlisting the support of their politicians, namely, Maheshwar Singh, Member of Parliament, and Karan Singh, Member of the State Legislative Assembly, mounting pressure on the Forest Department to turn a blind eye. It is interesting to examine the nature of this engagement between the political leadership and the people, the dimensions of the interaction being determined by a history of political relationships and a peculiar moral economy that was part traditional and part modern, both democratic and imperial. The crucial link here is that Maheshwar Singh was both a democratically elected representative and the erstwhile *raja* of Kullu.

The *raja* of Kullu presides over the congregation of all of the local deities from several valleys that takes place during the annual Dushehra festival in autumn. All of the important deities of the Sainj and Tirthan

valleys along with villagers who accompanied them attended the Dushehra of 1998. The *raja*, bestowed with the honor of presiding over the assembly, is also accorded certain duties, the prime of which is the appeasement of all deities. It is considered a bad omen if important deities do not attend in a particular year and reflects poorly on the *raja*. Therefore, the *raja* takes particular care that no deity feels neglected and, consequently, insulted. In October 1998, during a ritual assembly, the priest of the *devta* of Raila village brought up the issue of the GHNPCA before the *raja*. Other priests and villagers supported the complaint that the new park director was threatening access to the park. The *raja*, subdued by the public threat of these deities to boycott the Dushehra assembly, promised to look into the matter immediately.

In May 1999, after the compensation award for GHNPCA had been announced, there was great consternation in Shainshar and Raila villages, as none of their villagers received any compensation. The logic of the authorities that these villagers did not have recorded rights and were therefore excluded from receiving compensation was hotly contested and refuted by the villagers themselves. They were clear that they had codified rights, no matter where they were written. Many consultations later, it was determined that the British had set aside a large area to be administered by the deposed *raja* of Kullu during the recording of rights in 1897 (Chhatre 2003). This area, called *Waziri Rupi*, was precisely the area where Raila and Shainshar villagers exercised their rights of medicinal plant collection as well as grazing. In this area, colonial officers had codified grazing rights but had left the rest for the use of the *raja* and, at his discretion, to his people (Chhatre 2003). Whereas other villages falling in British-administered territories, such as Shangarh and Nohanda, were mentioned in the Settlement Report, the *Rupi* villages were bypassed by the codification process and, understandably, their names did not figure in the list culled out in 1999.

When this became clear to the Shainshar villagers, they petitioned the *raja*, Maheshwar Singh, for succour. The *raja* accepted his moral responsibility in looking after his subjects and roundly and publicly condemned the award for not including the *Waziri Rupi* villagers. Interestingly, the moral economy of the lord–subject relationship was tempered in this case by liberal democracy. With the news of his censure of the award, other aggrieved subjects followed suit. These were graziers who had been denied entry into the park. Their claims on moral capital were based on being his voters. Having once sided with Shainshar villagers, Maheshwar Singh was involuntarily drawn into supporting all local communities against the park authorities, a web that was as moral as it was political. Eventually, Maheshwar Singh compelled the park authorities to turn a blind eye toward the use of park resources by local communities.

Heritage as Weapon: Local versus Global, Cultural versus Natural

In 2009, the government of Himachal Pradesh submitted a proposal to UNESCO to accord World Heritage status to the GHNPCA under its Natural Heritage category. The state government demanded the World Heritage tag for the outstanding biodiversity of the protected area, obscuring the community-based forest management practices. After a few rounds of discussions during between 2010 and 2012, an expert team from UNESCO visited the GHNPCA in October 2012 and held meetings with local residents in adjacent villages. Ostensibly, the purpose of these meetings was to consult local stakeholders and garner feedback on the proposed World Heritage status. The agency in charge of organizing the visit—the Forest Department—deliberately neglected to advertise the team's visit and purpose and organized a few closed-door meetings with interested individuals in selected villages. Local representatives and activists discovered this subterfuge, but it was too late. The UNESCO team had already left the GHNPCA.

In response, representatives of local communities, aided by nongovernmental organizations (NGOs) and activists, formulated a challenge to the state government's application for World Heritage status. This challenge had a twist, though. Using the different categories in UNESCO's charter, the communities submitted a petition to demand the same status for GHNPCA but for its cultural heritage. A detailed memorandum, signed by elected representatives of local government, a former member of the State Legislative Assembly, and several NGOs, was submitted to UNESCO in November 2012. The document outlined the sacred geography of GHNPCA, the many spiritual sites within it that are regularly used by a large number of communities, and the way of life that is organically linked to the *devta* system. In pointing to sites that were held sacred by a large population, access for whom would be restricted or prohibited after the World Heritage status solely for outstanding biodiversity, the memorandum attempted to invert UNESCO's logic to the same end but through different means.

UNESCO approved World Heritage status for GHNPCA in June 2014 (UNESCO 2014) for its natural heritage, almost exactly fifteen years after the final notification of the National Park status and thirty years after the first notification. Local activists continued to protest, however, and redoubled their efforts to minimize the impact of such a move on their traditional livelihoods. In particular, there was concern that UNESCO would now condone, as required under the covenant, the restriction of human activities through merging Sainj and Tirthan Wildlife Sanctuaries with the National Park. Activists intensified their lobbying with state and national government officials and corresponded with UNESCO officials to present compelling evidence of involuntary displacement and denial of access to local heritage sites. The presented documents listed traditional cultural practices that serve to secure biodiversity in the region and pointed to the lack of support from public agencies in preserving these practices. They also highlighted the stealth with which the Forest Department bypassed the local communities in pursuing the World Heritage status for GHNPCA solely on the basis of natural heritage.

At the formal announcement ceremony in Doha, Qatar, on 12 June 2014, UNESCO directed the state of Himachal Pradesh to "expedite, in accordance with legislated processes, the resolution of community rights based issues with respect to local communities and indigenous peoples in the GHNPCA" (UNESCO 2014). It still stated, however, that the Sainj and Tirthan Wildlife Sanctuaries be accorded a higher protection status and merged with the National Park. A few months later, Forest Department officials met local community members to discuss the UNESCO World Heritage status and its implications. As the meeting commenced, a leader from Shakti village, located inside the Sainj Wildlife Sanctuary, spoke on behalf of the *devta* and threatened the officials against such a merger. Representatives from other villages also presented their concerns regarding World Heritage status and its impact on their livelihoods. The officials finally declared that Sainj and Tirthan Wildlife Sanctuaries would not be merged with the National Park and that access of local people to cultural heritage sites inside GHNPCA would not be curtailed.

Conclusion

Politics is central to the outcomes of conservation contests, especially in the case of protection of biodiversity, with the state as the primary interlocutor and arbiter of conflicting claims (Neumann 1998; Adams and Hutton 2007). Many of these studies, however, document a harsh state, bent on the exploitation of nature and labor. Yet, the notion of the omnipotent state, capable of exerting its will over disparate, fragmented communities, has been challenged (Sivaramakrishnan 1999; Peluso and Vandergeest 2001). An emerging literature has provided more nuanced understandings of both community and state and the means by which access to resources is negotiated or contested locally (Agrawal and Gibson 2001; Subramanian 2003).

In this article, we argue that the state traverses the complex terrain of global and local heritage through a series of compromises at the local level. The downgrading of the portion inside GHNPCA to the Sainj and Tirthan Wildlife Sanctuaries is a clear indication of how the Indian state exploited the flexibility in wildlife laws to appease both the World Bank and local communities simultaneously. The sacred ritual of seeking permission from the *devtas* is the medium through which the local communities seek compromises from the state. By locating the sacred ritual in the representation of the area as "Valley of the Gods," communities are able to embed their claims within state rhetoric. This representation, although used to draw tourists to the valley, also serves as ammunition for the local activists to stave off predatory encroachment by global interests. For the most part, the compromises appeal to local desires only partially and pander largely to global imperatives but also serve to cement local claims to GHNPCA as cultural heritage.

Further, we contend that procedural elements of democracy, formal and broadly defined, are a crucial element of these local compromises and provide the context within which local and global heritage is contested. The pressure of electoral consequences ensures that local politicians are receptive to the demands of local communities and find ways to reach these compromises. The local communities were able to represent their concerns to UNESCO through active lobbying with state government officials and with support from both current and former elected representatives. Finally, state officials use ambiguity in statutory provisions to create the possibility of compromises, such as during the ecodevelopment pilots in 1994. Even now, although UNESCO recognizes the need to involve communities in forest management, it still wants to increase legal protection to GHNPCA by upgrading the two wildlife sanctuaries to National Park status. The two objectives are clearly incompatible under

Indian law, and it is precisely this kind of ambiguity that is exploited by the state as it traverses the murky terrain between local and global heritage.

This article contributes to global debates around the social and political processes that produce conservation landscapes. We present a new dimension to understand the role of democracy in shaping geographies of biodiversity conservation. The insertion of global processes, like UNESCO World Heritage status, into local contests over access to natural resources is usually seen as detrimental to local communities. We show that democracy provides the crucible where such insertions are effectively challenged and compromised, producing new landscapes of conservation. Simultaneously, they open up new possibilities that facilitate different ways to establish control over and legitimize access to natural resources. Because mountain geographies represent the frontier for the territorial expansion of biodiversity conservation, our research suggests potential avenues for continuing research on the politics of heritage, both cultural and natural, as well as local and global.

Acknowledgments

The authors wish to thank Dilaram Shabab, ex-member of legislative assembly and Guman Singh, leader of Himalayan Niti Abhiyan, for their valuable inputs.

References

Adams, W. M., and J. Hutton. 2007. People, parks and poverty: Political ecology and biodiversity conservation. *Conservation and Society* 5 (2): 147–83.

Agrawal, A., and C. Gibson. 2001. *Communities and the environment: Ethnicity, gender and the state in community-based conservation.* New Brunswick, NJ: Rutgers University Press.

Baviskar, A. 1995. *In the belly of the river: Tribal conflicts over development in the Narmada Valley.* New Delhi, India: Oxford University Press.

———. 2003. States, communities and conservation: The practice of ecodevelopment in the Great Himalayan National Park. In *Battles over nature: Science and politics of conservation,* ed. V. Saberwal and M. Rangarajan, 267–99. New Delhi, India: Permanent Black.

Chhatre, A. 2003. The mirage of permanent boundaries: Politics of forest reservation in the Western Himalayas, 1875–97. *Conservation and Society* 1 (1): 137–46.

Chhatre, A., and V. Saberwal. 2006. *Democratizing nature: Politics, conservation and development in India.* New Delhi, India: Oxford University Press.

Dowie, M. 2011. *Conservation refugees: The hundred-year conflict between global conservation and native peoples.* Cambridge, MA: MIT Press.

Himachal Wildlife Project. 1981. *The wildlife of Himachal Pradesh, Western Himalayas.* Orono, ME: Himachal Wildlife Project.

Holmes, G. 2013. Exploring the relationship between local support and the success of protected areas. *Conservation and Society* 11 (1): 72–82.

Lewis, M. 2003. *Inventing global ecology: Tracking the biodiversity ideal in India 1945–1997.* New Delhi, India: Orient Longman.

Meskell, L. 2012. *The nature of heritage: The new South Africa.* Oxford, UK: Blackwell.

Neumann, R. 1998. *Imposing wilderness: Struggles over livelihoods and nature preservation in Africa.* Berkeley: University of California Berkeley.

Peluso, N. L., and P. Vandergeest. 2001. Genealogies of the political forest and customary rights in Indonesia, Malaysia and Thailand. *The Journal of Asian Studies* 60 (3): 761–812.

Sivaramakrishnan, K. 1999. *Modern forests: State-making and environmental change in India.* Stanford, CA: Stanford University Press.

Smith, L. 2006. *Uses of heritage.* London and New York: Routledge.

Subramanian, A. 2003. Community, class and conservation: Development politics on the Kanyakumari Coast. *Conservation and Society* 1:177–208.

UNESCO. 2014. *Decisions adopted by the World Heritage Committee at its 38th session.* Doha, Qatar: UNESCO.

West, P., and D. Brockington. 2006. An anthropological perspective on some unexpected consequences of protected areas. *Conservation Biology* 20 (3): 609–16.

Wildlife Institute of India. 1999. *Project report: An ecological study of the conservation of biodiversity and biotic pressures in the Great Himalayan National Park Conservation Area—An ecodevelopment approach.* Dehradun, India: Wildlife Institute of India.

Willis, K., L. Gillson, and T. Brncic. 2004. How "virgin" is virgin rainforest? *Science* 304:402–03.

World Bank. 1996. *India ecodevelopment project: Staff appraisal report.* Washington, DC: International Bank for Reconstruction and Development.

Perestroika to Parkland: The Evolution of Land Protection in the Pamir Mountains of Tajikistan

Stephen F. Cunha

This article traces the evolution of land protection in the Pamir Mountains of Tajikistan. The Pamirs form the "Roof of the World," where the Hindu Kush, Karakoram, Tian Shan, and Kunlun Shan ranges converge. Field and archival research identified (1) the origin and diffusion of parks and protected areas across the globe, (2) the biophysical properties of the Pamir Mountains that inspired the conservation efforts, (3) the sequence of land protection from national park to supranational World Heritage recognition, and (4) the characteristics of the Pamir Mountains that justify UNESCO Biosphere Reserve status. Stalin forcefully depopulated these highlands in the 1930s. Tense Soviet–Sino relations in the 1960s and the prolonged Soviet–Afghan war further restricted human movements. When Gorbachev's *perestroika* allowed return migration in the mid-1980s, Tajik farmers and Kirghiz pastoralists resettled a landscape of thriving plants and wildlife. Concurrently, a nascent coalition of citizen scientists and government officials began advocating for a park. In 1992 the government established the Tajik National Park to protect environmental and sacred sites, promote traditional economic activity, and develop tourism. The antecedent Soviet collapse, civil war, economic upheaval, and renewed conflict in Afghanistan, however, complicated land protection. In 2013, UNESCO designated the Tajik National Park as a World Heritage Site. Establishing a Biosphere Reserve is the next step to promoting transboundary conservation with the adjacent protected areas in China, Pakistan, and Afghanistan. The potential reserve size, terrain, and demographic trajectory are consistent with the Man and the Biosphere model.

本文追溯塔吉克斯坦帕米尔山地中的土地保护之演化。兴都库山, 喀喇崑崙山, 天山与崑崙山脉汇聚的帕米尔山岳, 组成了 "世界的屋嵴"。田野与文献研究指出 (1) 全球公园与保护地区的起源及扩散, (2) 鼓舞了保育努力的帕米尔山岳的生物物理特徵, (3) 从国家公园到超国家世界文化遗产认定的一系列土地保护, 以及 (4) 证明 UNESCO 生物圈保育情况的帕米尔山岳特徵。在 1993 年间, 史达林强迫迁这些住在高原的人口。1960 年代, 紧张的中苏关系, 以及持续的苏联—阿富汗战争, 进一步限制了人类的移动。1980 年代中期, 戈尔巴乔夫的改革计画准许返乡移民, 塔吉克的农民和吉尔吉斯的放牧者于是重新定居于成长茁壮的植物和荒野地景之中。于此同时, 晚近公民科学家和政府官员的结盟, 开始倡议兴建一座公园。1992 年, 政府成立了塔吉克国家公园, 用来保护环境和圣地, 提倡传统经济活动, 以及发展观光。但苏联解体的先驱, 内战, 经济动盪, 以及阿富汗重燃的冲突, 却复杂化了土地保护。 2013 年, UNESCO 将塔吉克国家公园指定为世界文化遗产, 而建立 "生物圈保护区", 则是下一步提倡中国, 巴基斯坦和阿富汗的邻近保育地区的跨疆界保育。可能的保育区面积, 范围和人口轨迹, 与 "人类及生物圈模型" 相符合。 关键词: 关键词: 中亚, 帕米尔山岳, 公园与受保护地区, 塔吉克斯坦。

Este artículo sigue el paso del proceso de protección de la tierra en las Montañas del Pamir de Tayikistán. Los Pamir constituyen el "Techo del Mundo", donde convergen las cordilleras del Hindu Kush, el Karakorum, el Tian Shan y el Kunlun Shan. Se identificaron mediante investigaciones de campo y de archivo (1) el origen y difusión de parques y áreas protegidas a través del globo, (2) las propiedades biofísicas de las Montañas del Pamir que inspiraron los esfuerzos conservacionistas, (3) la secuencia de la protección de las tierras, desde el reconocimiento del parque nacional hasta la Heredad Mundial supranacional, y (4) las características de las Montañas del Pamir que justifican su estatus de Reserva Biosférica de la UNESCO. En los años 1930 Stalin despobló estas montañas a la fuerza. Unas tensas relaciones soviético–chinas en los 1960 y la prolongada guerra soviético–afgana contribuyeron a restringir aun más los movimientos humanos. Cuando la perestroika de Gorbachov permitió la migración de retorno a mediados de los años 1980, agricultores tayik y pastores kirguises repoblaron un paisaje boyante de plantas y vida silvestre. Al propio tiempo, una naciente coalición de ciudadanos científicos y agentes del gobierno empezó a propugnar por un parque. En 1992 el gobierno estableció el Parque Nacional Tayik para proteger sitios ambientales y sagrados, promover la actividad económica tradicional y desarrollar el turismo. Sin embargo, el colapso soviético antecedente, la guerra civil, la agitación económica y un nuevo conflicto en Afganistán, complicaron la protección de la tierra. En 2013 la UNESCO designó al Parque Nacional Tayik como lugar Patrimonio de la Humanidad. El establecimiento de una Reserva Biosférica es el siguiente paso para promover la conservación transfronteriza en conjunto con las áreas de protección adyacentes de China, Pakistán y Afganistán. El tamaño potencial de la reserva, el terreno y la trayectoria demográfica son consistentes con el modelo del Hombre y la Biósfera.

This article traces evolving land protection in the Pamir Mountains of Tajikistan. The first tangible plans to establish a large mountain protected area in this region emerged in the late 1980s, when Tajikistan was the poorest, smallest, and most peripheral Soviet republic. Although distant from Moscow, the influence of Soviet leader Mikhail Gorbachev's *perestroika* (restructuring) inspired fresh thinking about and new possibilities for nature protection among a small cadre of citizen scientists and government officials. From their effort, the Pamir Mountains evolved from a largely depopulated military zone where plant and animal life flourished in the absence of human presence to the recent designation of the Tajik National Park as a UNESCO World Heritage site. The time frame from 1933 to 2016 entwines with the Soviet collapse, Tajik Civil War, the rise of dual professionalism, and eventual Tajik cooperation with supranational environmental organizations.

We begin with a brief introduction to the origin and diffusion of the parks and protected areas concept across the globe. The relevant geographic characteristics of the Pamir Mountains are then presented to support the legislative justification for park status. The sequence of land protection from national park to supranational World Heritage recognition brings us to the present. The final section presents the biophysical and human characteristics of the Pamir Mountains that justify inclusion in the UNESCO Biosphere Reserve System—an important final element to creating a large central Asian transfrontier reserve. This study focuses on the administrative history of land protection in the Pamir Mountains. The detailed scientific and scenic rationale that moved individuals and governments to protect the Pamir is found in other supporting documents (Cunha 2004; Middleton 2010; UNESCO 2010, 2013; Committee for Environment Protection under the Government of the Republic of Tajikistan and National Center for Mountain Regions Development in Kyrgyzstan 2011; Hauser 2011; International Union for the Conservation of Nature [IUCN] 2013; Kreutzmann and Watanabe 2016).

The Origin and Diffusion of Parks and Protected Areas

The concept of a park—nature enclosed for human delight—originated west of the Pamir with Persian kings in the sixth century BC. Their name for parks is the etymological origin of *paradise*. During the rule of Cyrus the Great, elaborate Persian gardens symbolized Eden and the Zoroastrian elements of sky, earth, water, and plants. East of the Pamir, Chinggis Khan bestowed protection on landscapes from northern Mongolia south to Lake Baikal. During the twelfth and thirteeth centuries, the khan of Khereid Aimag forbid logging and hunting on Bogd Kahn Mountain above Ulaanbaatar. Mongolian officials in 1709 invoked *Khalkh Juram* (rules) to protect other mountains and lakes as sacred sites. In 1778, the Manchu government that then ruled Mongolia also conferred state protection on the Bogd Khan, Khan Khentii, and Otgontenger Mountains. Some of these sites now are integrated into contemporary Mongolian national parks, arguably the oldest protected lands on Earth.

A Medieval Latin word of Germanic origin, *parricus* is the progenitor of the English *park*. As places for royalty to chase stags and boar, these protected lands were closed to commoners. Despite Old World origins, egalitarian codification of parks began in the New World in 1864 when the U.S. Congress set aside Yosemite Valley and a giant sequoia grove to be "held inalienable for future generations as a park and pleasuring ground" (Johnson 2008, 18). In 1872, members of the Hayden Survey proposed Yellowstone as a national park because it fell within a territory and not a state. Congress agreed and designated Yellowstone as the world's first national park. From this modest foundation, more than 200 countries now share the Yosemite idea of protecting land for all citizens to enjoy.

The initial spread of government reserves resulted from visits to new U.S. national parks (Australia in 1879, Canada in 1885, New Zealand in 1887). By 1930, Russia, Japan, Korea, and Iceland, among others, had established parks on their own enterprise without visiting the United States. The Russian *zapovedniks* (strict nature reserves) protected areas analogous to designated wilderness in the United States. In the 1950s, supranational efforts of the World Conservation Union (now IUCN) and United Nations Development Program (UNDP) led to new parks in developing Asia, South America, and Africa. Protected area classification and legal codification diversified as the concept spread. Many reserves are superimposed on existing cultural landscapes, some

Table 1. IUCN Protected Area Management categories and equivalent in Tajikistan

IUCN category	Tajik equivalent	No.
1a. Strict nature reserve	Zapovednik	4
1b. Wilderness area		
2. National park	National park	2
3. Natural monument		
4. Habitat-species management	Zakaznik	14
5. Protected landscape-seascape	Zakaznik	4
6. Protected area with sustainable natural resource use	Many variations	

Note: IUCN = International Union for the Conservation of Nature.
Sources: Biodiversity Conservation Center (2003); Committee for Environment Protection Under the Government of the Republic of Tajikistan and National Center for Mountain Regions Development in Kyrgyzstan (2011); and Stolton, Shadie, and Dudley (2013).

densely settled and modified by human agency. The IUCN Protected Area Management categories scheme is a *lingua franca* that reflects diverse reserve purpose and design (Table 1). Category assignment follows the primary management objective that applies to at least 75 percent of the protected area (Stolton, Shadie, and Dudley 2013). Not all protected areas classify as parks. The Tajiks retain some taxonomy from Soviet rule. In addition to zapovedniks, there are *zakazniks*, which are analogous to IUCN Categories IV and V. In one form or another, more than 200,000 protected areas now cover 14.6 percent of the world's land and 2.8 percent of the oceans, including 5,000 national parks (Marton-Lefève, Badman, and Kormos 2014).

Two new supranational designations emerged during the 1970s. Biosphere Reserves (1971) benchmark the environmental status of specific biomes. Potential reserves must first earn home-country protection on some level of the IUCN scheme. World Heritage Sites (1972) recognize areas of global significance belonging to all humans, irrespective of territorial location. These include natural reserves, cultural sites, individual structures, regional cuisine, and art. Many mountain protected areas extend across international borders where state perimeters follow thinly settled ridgelines. These transfrontier areas promote regional conservation, tourism, and bilateral relations. Although Poland and Czechoslovakia agreed to this concept in 1924, the United States and Canada designated the first International Peace Park in 1932. In the multistate Pamir region, cross-border conservation is a difficult but important element to securing supranational designation. As of this writing, the Tajik National Park is a federally designated national park and World Heritage Site.

Geographic Setting

The Pamir Mountains form the complex orographic node where the Hindu Kush, Karakoram, Alai, Tian Shan, and Kunlun Shan ranges converge (Figure 1). The range stretches across Tajikistan, Afghanistan, China, Kyrgyzstan, and Pakistan, with the core occupying Gorno Badakshan Autonomous Oblast (GBAO) in eastern Tajikistan. Collectively referred to as the "Pamir Knot" or "Roof of the World," the rectangular orographic magnitude is unequaled on Earth. The echelon grouping of ranges extends 135,000 km^2 and exceeds 7,000 m on three summits, the world's third highest mountain chain. Extreme vertical relief blocks air masses from every cardinal direction. The intercepted moisture sustains the Amu Darya, the principal tributary to the Aral Sea. Other geographic traits include intense seismic events, giant landslides, deep gorges, and the largest glacier between Alaska and Antarctica. Key wildlife species include snow leopard (*Panthera uncia*), Asian black bear (*Ursus thibetanus*), Ibex (*Capra sibirica*), and Marco Polo sheep (*Ovis poli*). Isolated riparian birch and mesophyllous juniper stands provide vital resources and habitat for numerous other species.

The Pamir-Za Alai and associated ranges comprise 93 percent of Tajikistan's 143,100 km^2. This cultural crossroads of central Asia shaped the centerpiece of the nineteenth-century Great Game. The only level terrain is to the southwest, where the foothill ranges merge with the Tajik Depression near the capital city of Dushanbe. The east–west axis stretches for 700 km; the north–south distance is 300 km. Almost half of the southern and eastern 3,000 km frontier is shared with China (430 km) and Afghanistan (1,030 km). The narrow (15- to 65-km-wide) Afghan Wakhan Corridor separates southeastern Tajikistan from Pakistan. Eastward, the Sarikol Range towers over the Pamir Plateau and Kara Kul Lake, forming a contested frontier with China. To the north, Kyrgyzstan and Uzbekistan complete a landlocked multistate circle that isolates Tajikistan from the outside world.

The Pamirs dissect Tajikistan into subregions of distinct climate, landforms, resources, and population. These include the Western Pamir Foothills, High Pamir (*Academia Nauk*), Eastern Pamir and Pianj River Corridor, and Pamir Plateau (Figures 2–5). Terrain and winter snowpack insulates ethnic groups and encumbers economic development. The Tajik National Park and World Heritage Site includes part

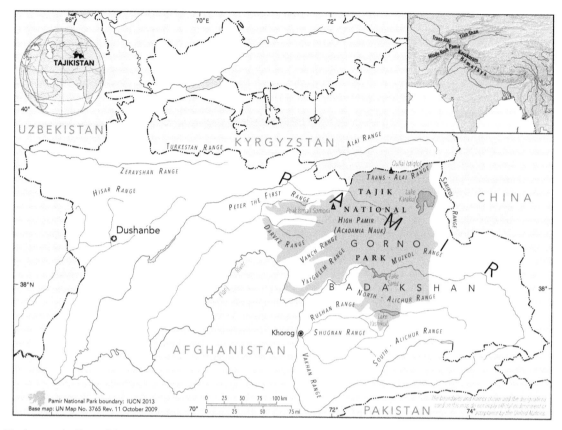

Figure 1. Tajikistan, the Pamir Mountains, and the Tajik National Park, with insets of Tajikistan (left) and the mountains of central Asia. *Source:* United Nations Environment Programme World Conservation Monitoring Centre (2009), International Union for the Conservation of Nature (2013).

of each subregion. The biophysical and cultural diversity enriches the case for national and supranational protection but greatly complicates managing this large and corrugated landscape.

Origin of the Tajik National Park

Members of the 1933 Russian and German Tajikistan-Pamir expedition first proposed a nature reserve in the High Pamir. Scientific personnel discussed the idea while in the field and again during a 1934 summer expedition (State Committee for Nature Protection of Republic of Tajikistan [SCNP] 1992a). Prior to the Bolshevik Revolution, successive czars decreed forest reserves scattered throughout Russia. Just one year before the 1917 Revolution, the Russians established Barguzin Zapovednik on the shores of Lake Baikal. Although this created bureaucratic momentum for a Pamir reserve, political and research constraints stemming from two world wars, the Afghan War (1979–1989), and the Cold War (1946–1990) stifled further collaboration between Russian and Western European

scientists. While studying the Fedchenko Glacier during the 1970s, V. M. Kotlyakov proposed that both the glacier and surrounding Academy Range receive protected status, an idea he promoted throughout his long career with the Soviet Academy of Sciences (SCNP 1992a). Both the Soviet and Tajik Academy of Science journals contain numerous references to Kotlyakov's idea. One enthusiastic Russian botanist suggested "the whole of Tajikistan should be decreed a zapovednik" (Anonymous 1982). On various occasions dating from 1987, Jack Ives (University of California, Davis), Bruno Messerli (University of Bern), Yuri Badenkov (Russian Academy of Sciences), and S. Honglie (Chinese Academy of Sciences) considered the Pamirs an integral component of a potential transfrontier peace park (Thorsell and Hamilton 2002). Attendees of the 1986 Mohonk Himalayan-Ganges Conference regarded the Chinese Pamir and adjacent Taxkorgan Wildlife Refuge as of particular importance (Ives and Ives 1987).

The Karakoram Highway opening in 1985 ignited tourism in adjacent northern Pakistan and western China. This drew the interest of Tajik entrepreneurs.

Figure 2. Western Pamir foothills. *Source*: Author photo. (Color figure available online.)

In 1989, Tajik and Russian government officials, along with private businessmen, formed the Social–Ecological Union of the Republic of Tajikistan (SEURT) to promote a national park in the High Pamir (Buzurukov 1990, 1993; Cunha 1993). SEURT organized to (1) identify lands within Tajikistan for inclusion into the Soviet Zapovednik system, (2) guide proposed new parks through republic (Tajik) and federal (Soviet) government approvals, (3) promote and cooperate in joint research, and (4) establish and regulate tourism within the parks. Financial gain from tourism similar to Nepal and Pakistan motivated SEURT members. In late 1990 this potential new revenue stream helped SEURT earn the support of various *rayon* (county)

Figure 3. The High Pamir above the Moskveen Glacier. *Source*: Author photo. (Color figure available online.)

Figure 4. The Pianj River separates Tajikistan and Afghanistan (distant). *Source:* Author photo. (Color figure available online.)

leaders looking to broaden their local economy beyond agriculture and mining.

In 1990 the United Nations University, University of California, Davis, and the Russian and Tajik Academy of Sciences agreed to provide research support for the proposed Pamir National Park as part of a larger investigation of mountain environments under a Tajikistan 21 proposal (Badenkov 1991). Representatives

signed the pact during a joint Tajik–Russian–U.S. Pamir field survey during the summer of 1990. The SCNP (later renamed State Agency of Natural Protected Areas [SANPA]), the government arm responsible for nature reserves in Tajikistan and a branch of *Goskomproda* (Tajik Ministry of Environment), organized the expedition that included key SUERT members. Although logistical problems and inclement

Figure 5. A Kirghiz yurt between Lake Karakul and the Sarikol Range, Gorno-Badakshan Autonomous Oblast. *Source:* Author photo. (Color figure available online.)

weather plagued the field team, they completed a resource inventory near Sangvor in the Western Pamir foothills and a helicopter reconnaissance to identify potential boundaries. The Tajik Academy of Sciences then collated the scientific justification for the proposed park.

In January the SCNP released the *Scientific Basis of Preliminary Proposals of the Pamir National Park* (SCNP 1992, 1992b). Chief Pamir Park Project Specialist Anvar Buzurukov coordinated this report, coauthored by leading scientists from the Tajik Academy of Sciences, Goskomproda officials, private businessmen, university faculty, and other government leaders. All vigorously supported a Pamir National Park and expected rapid endorsement. In the months following their report, however, coincidental ethnic riots and political crisis diverted parliamentary attention away from the park issue. As the legislature reconvened in May 1992, a minority of SEURT members stated that the Pamir idea was too unknown and important for rapid judgment. They demanded more consultation with several rayon and Gorno-Badakshan Autonomous Oblast (GBOA) officials whose lands comprised much of the proposed park.

On 20 June 1992, eleven months after independence, the Cabinet of Ministers of the Republic of Tajikistan approved Resolution Number N267, recommending to Parliament the creation of a provisional Tajik National Park for

> The purpose of the intensification of the protection of nature and the conservation of the landscape complex; to preserve a gene pool of unique flora and fauna; to protect monuments of nature, culture and history; and to develop tourism. (Ministry Document 1992, 2)

The resolution passed despite heavy opposition from the Minister of Forestry, who feared losing control over Pamir *leskholz* (state forests). The Minister of Agriculture believed that restricting farm rights within the proposed reserve would diminish his ministry budget. The Ministers of Environment, Economics, Culture, and Local Government all supported the measure. The 20 June decision by the Cabinet of Ministers effectually created the Tajik National Park in the Pamir Mountains. The initial legislative boundaries enclosed 1.6 million hectares of some of the wildest mountain, canyon, and plateau landscape in central Asia.

Up to this point, the SEURT proposal affixed Pamir to the park name. As the decree progressed through various ministerial approvals, an important Goskomproda official insisted that foreigners would credit park establishment and jurisdiction to Pamirians—a minority population within Tajikistan but one that comprises a large majority in GBOA of eastern Tajikistan where most of the park lies (Cunha 2007). Although other Goskomproda personnel remained confident that Parliament would ultimately restore the Pamir nomenclature, the recent UNESCO nomination and transcription documents all read "Tajik National Park (Mountains of the Pamir)." Almost all government, scholarly, and lay documents use Tajik National Park. The decree supersedes all prior Soviet-era presidential decrees, parliamentary legislation, and environment ministry proclamations regarding Pamir nature protection efforts. The new park assumed private plots, mineral wealth, surface water, and all natural, cultural, and historical objects.

Although the park is a tangible success for SEURT, transforming a government mandate in Dushanbe into effective management and protection of the Pamir posed significant hurdles. The decree directed Goskomproda to (1) accurately determine appropriate park boundaries, (2) develop a management plan, and (3) devise a strategy for park development. Government officials emphasized the need to generate foreign exchange through tourism. It provided a government account for funding, logistical equipment (office space, field gear, and helicopters), and technical assistance by reassigning select government personnel. The decree allowed park officials to enter what were then restricted zones within Tajikistan—of crucial importance because the southern park boundary lies adjacent to the militarized frontier with Afghanistan, an area off limits since the 1940s Stalin closure.

Immediately after the legislative mandate, Goskomproda tasked Chief Pamir Park Specialist Anvar Buzurukov with updating boundary recommendations, a management plan, and cooperative agreements with five relevant rayon-governing councils. His team completed a refined boundary map in June and the draft management plan soon after (Buzurukov 1992). Increasing ethnic strife, economic upheaval, and the political crisis integral with nascent Tajik independence and civil war, however, left incomplete the majority of cooperative management agreements required by the government mandate. Beginning in late August 1992, political control of Tajikistan changed hands several times. Each pendulum swing led to a purge of government personnel and policies

(Mitchell 2015). A deposed president and a weakened national parliament were in no position to refine and expand the new Tajik National Park or to work on other environmental issues. A second recommendation to incorporate a portion of an updated management plan with new data from Goskomproda also failed in parliament due to ongoing civil unrest (Ministry Document 1993).

At this juncture the dissolving Soviet Empire that began with the failed coup in August 1991 began to affect governing in Tajikistan. Following the Soviet collapse, Tajikistan opted for independence within the Commonwealth of Independent States. One month later, a coalition of ex-Communist leaders led by Rakhman Nabiev consolidated power during national elections in which nine candidates vied for the presidency. The initial euphoria of independence, combined with a flood of new consumer goods into this remote region, quickly gave way to the realities of an independent multicultural state without large subsidies from Moscow. The nationalistic fervor so long suppressed under Communist rule turned violent as tension between government supporters and fractured opposition leaders intensified. Between 1992 and 1997, civil war claimed 50,000 casualties and displaced almost one tenth of the population within and outside of Tajikistan (Kevlihan 2016). The initial violence synchronized with the June 1992 Tajik National Park decree. For the next five years, ongoing conflict derailed almost all government and nongovernmental organization (NGO) efforts to promote, develop, and expand the park (Cunha 1997b).

In 1997 a peace accord between the Tajik government and rebel United Tajik Opposition officially ended the conflict, but independence and war altered the land protection dynamic in the Pamir. Postindependent Tajiks emerged from seventy years of subservient colonialism, where second-class citizenry excluded them from most influential government positions. Although poor, landlocked, and with no oil or nuclear weapons, Tajikistan's geographic nexus as a political and cultural crossroads attracted wartime support from their dissimilar central Asian neighbors—Pakistan, Afghanistan, Iran, Russia, Uzbekistan, Kyrgyzstan, and Turkmenistan. This underscored a neocolonialist aura of a country still reliant on outsiders as peace-building aid from the international community swelled local coffers (Heathershaw 2008, 2009; Nourzhanov and Bleuer 2013). Following the peace accord, government and

SEURT personnel resumed efforts to develop and promote the Tajik National Park. Outmigration of non-Tajiks (mostly Russians) during the five-year conflict, however, included many of their most experienced colleagues. Prior to the war as the reserve took shape, Chief Park Specialist Anvar Buzurukov illustrated how SCNP personnel framed outside cooperation:

> We do not want comments on the management plan. The Tajiks are able to develop the Pamir Park on our own. This is the "roof of the world" and we know this region better than anyone else. The Pamir National Park will become world famous and bring economic development to Tajikistan. We do need financial assistance with equipment like more jeeps, camping equipment, and helicopter fuel. We also want contacts for bringing tourists to the Pamir. (A. Buzurukov, personal communication, 22 July 1992)

Although Tajik nationalism and sovereignty were important, developing tourism remained the primary objective of the new park. From the outset, Buzurukov and other SCNP members explored opportunities to inaugurate cooperative tourist agreements, while guarding their newly established control over this unique resource. Thus emerged an enigma of Tajiks launching a park and tourist industry while lacking meaningful expertise in both.

The Rise of Dual Professionalism

With Tajiks in firm control of their independent state, along with the purge and outmigration of non-Tajiks, a new relationship between public agencies and NGOs emerged. With nearly every SEURT member now holding an influential government post, the organization quickly became an important force in shaping Tajikistan's public policy on protected lands. This dual professionalism descended from the Communist system, where the distinction between public and private effort is more blurred than in Western economies (Cunha 1997a; Ressnick and Wolff 2013). In this case, prominent SEURT members Anvar Buzurukov (Chief Pamir Park Specialist), N. Safarov (Goskomproda Supervising Officer), and O. Gholub (Chief State Wildlife Protection Officer) also fully utilized their government positions to promote the Tajik National Park. They and other SEURT members produced supporting documents (scientific, administrative, public interest, legal, etc.) and lobbied to enlist support of government officials from

Figure 6. Barley harvest in kishlak Barchadev on the Bartang River, Gorno-Badakshan Autonomous Oblast. *Source*: Author photo. (Color figure available online.)

Parliament, the executive branch, and the Academy of Sciences. They petitioned rural *kishlak* (village) leaders during field tours, sponsored research in numerous subfields (scientific, institutional, and tourism), and promoted tourism to directly benefit their private-role aspirations.

Their efforts worked. The legislative mandate ultimately fell primarily to SEURT members within the SCNP branch of Goskomproda. Promoting the park as SEURT representatives while concurrently salaried government employees merged private NGO advocacy with public agency support. The sudden personnel shifts that followed independence further complicated the delineation between public servant and private advocate. Finally, the dramatic turns from Communism to capitalism and dictatorship to democracy exacerbated the legal, economic, and social chaos within Tajikistan (Heathershaw and Herzig 2013). The apparent conflict of interest associated with flourishing dual professionalism triggered violent public uprisings in Russia and neighboring Uzbekistan and Kazakhstan but did not initially impair those advocating for a national park in the Pamir Mountains of Tajikistan. UNESCO, however, did cite the conflict of interest inherent in dual professionalism as one reason to defer Tajikistan's application to the World Heritage List.

The Park Expands

In the five postwar years, a new park leadership team worked to engage with local rayon and kishlak leaders, complete a more detailed inventory of park resources, and update the park management plan. Their revised blueprint established management and use zones within the park. A 2005 Order of the State Directorate of Natural Protected Areas enlarged the park to 2.6 million hectares and approved a plan that reduced management zones from six to four (State Directorate 2005). Thus, thirteen years after creating the Tajik National Park, a more tightly structured and viable reserve emerged with four management zones:

- Core zone (1,685 ha; 65 percent of the park) of wild landscapes and minimal human agency, more than 1,000 active glaciers, glacial landforms, free-flowing rivers, and 300 lakes.
- Limited Economic zone (740,198 ha; 28.3 percent) to buffer the core and allow restricted development. This zone includes five permanent pastoralism and subsistence agriculture settlements on the Bartang River (Figure 6).
- Traditional Use zone (127,665 ha; 5 percent), where approximately 1,000 Kirghiz have established camps near streams and meadows in the eastern Pamir to

236

Figure 7. A demolished home and barn in the western Pamir foothills. *Source*: Author photo. (Color figure available online.)

seasonally graze livestock on the Pamir Plateau inside the eastern boundary of the park.

- Recreation zone (58,400 ha; 2.2 percent) to support tourism, with infrastructure for climbing, trekking, river running, and trophy hunting. This includes climbing camps in the High Pamir.

Reconciling human use with resource protection within these zones presents three issues. The first is traditional grazing and wood collecting in four areas of seasonal settlement (near Lake Karakul on the Pamir Plateau, northern Vanj District, near kishlaks Barchadev and Poimazor, and the Tavildara Valley). To accommodate seasonal grazing of ungulates (sheep and goats everywhere, cattle in Vanj District, yaks on the Pamir Plateau), hay making, and wood collection, 5 percent of the park is a noncontiguous Traditional Use zone. Although well-defined and delimited with input from local residents, monitoring the efficacy of this compromise, particularly in sensitive wetland areas, requires ongoing assessment by qualified scientists (Haslinger et al. 2007).

The second issue involves accommodating the return migration of farmers in the elongated valleys of the Western Pamir foothills—scenic vales akin to those in the U.S. Rocky Mountains. During the 1940s, Communist Party leader Joseph Stalin forcibly relocated these Tajik farmers to communes on floodplains in southwest Tajikistan, where they produced gun cotton for World War II armaments. The partially demolished homes and barns alongside abandoned terraces in the western foothills still

bear witness to this policy (Figure 7). The State Republic Committee for Statistics (1992) estimated that prior to Stalin's relocation program, 1,500 people lived in prospective parkland near Sangvor. The original park proposal allowed return migration of traditional families with prior land and irrigation rights, although determining half-century-old real estate claims remains a difficult and politically sensitive endeavor. Plant and animal life flourished during the fifty years of depopulation. This changed rapidly in the mid-1980s, when perestroika allowed villagers to reestablish their mountain farmsteads. Returning migrants quickly restored the prior cultural landscape but with new machinery and cultural influence from the lowlands superimposed on traditional practices. In addition, exposure to 1980s urban lifestyle and technology significantly altered standard of living expectations. New Tajik government policies that continued the top-down approach from the Soviet period added further ambiguity (Rowe 2010). By the early 1990s as park legislation moved through the Tajik political process, farmers had replanted approximately 25 percent of their former croplands and reconstructed a dozen kishlaks (three within the proposed park) along the Obi Hingou and Obi Mazar rivers near Sangvor. Facing unabated return migration, the SCNP redrew the final park boundaries in late 1992 to exclude the inhabited valleys adjacent to the Pianj River and the low to midelevations of the western foothills.

Accommodating sport trophy hunting is the third issue. In their capacity as dual professionals, park managers solicit clients, market permits, and arrange and sometimes even lead hunting expeditions. Although central Asian wildlife suffers from decades of overgrazing, deforestation, and human predation (Jackson 2012; Berger, Buuveibaatar, and Mishra 2013), the Pamirs are an exception because Stalin's forced migration allowed wildlife fifty years to rebound without human pressure. Post-Soviet sport hunting in the emerging central Asian states is a new and urgent threat to regional biodiversity. In the Pamir, contemporary hunting is either (1) legal, sanctioned and regulated by the government; (2) quasi-legal, authorized and negotiated by government officials though private channels; or (3) illegal, by poachers for meat, body parts, and trophies (Cunha 1997a). The most coveted and therefore expensive mammals in Tajikistan are snow leopard, Marco Polo sheep, and wild hog. Poaching here and elsewhere in central Asia is driven by inflation and the demand for body parts. Illegal sport hunting by foreigners and the inability of park mangers to patrol and enforce regulations in expansive mountain terrain also threaten wildlife. The 2012 Management Plan provides for an initial study to determine the "best site and regulations under which hunting might be permitted, with the purpose to provide income for management and protection of the park" (SANPA 2012, 39). The updated plan also incorporates three recommendations from World Wildlife Fund evaluators: (1) sustained yield harvest plans based on scientific research, (2) legal codification of regulations, and (3) dedicating some permit revenue to monitor the populations of trophy species.

The 2012 Management Plan demonstrates maturation of local control and management capacity. Procuring government funds and user fees remains problematic in Tajikistan, but this concern is shared by park systems the world over (Symes et al. 2016). There are insufficient adequately trained field personnel to enforce sport hunting, wood collecting, and other regulations. The tourist facilities remain critically underdeveloped and inhibited by problems intrinsic to dual professionalism. These unresolved issues pose significant challenges to the environmental and political integrity of Tajik National Park.

UNESCO World Heritage Site

UNESCO World Heritage Sites must offer outstanding universal value and meet at least one of ten selection criteria (see UNESCO World Heritage

Center 2016). Following application to the World Heritage Committee, UNESCO assigns an IUCN technical team to appraise the nominations according the criteria that determine areas of significant world heritage and great human accomplishment or superlative natural phenomena. In 2009 the SCNP nominated the Tajik National Park for inscription to the World Heritage List. The proposed area identified 1,266,500 ha as the core submission, with an additional 1,385,174 ha of parkland to serve as a buffer zone (United Nations Environment Programme World Conservation Monitoring Centre 2009; UNESCO 2010). The application provided evidence to support UNESCO Criteria vii, viii, vi, and viii.

Until this point the Tajik government and Tajik NGOs drove the creation and development of their park, with only limited outside support. Application for supranational designation subjects the proposal to rigorous IUCN evaluation by request of the World Heritage Committee. The evaluations focus on the merits of the proposed property as they relate to UNESCO World Heritage criteria. These include the integrity of legal codification, the site management plan, boundary issues, country management, human use, and potential threats to the site.

IUCN field evaluators did not support Criteria vii and viii as they normally apply to more humid and biologically rich coastal and equatorial zones. Although supporting evidence was presented for Criteria vii and viii, three IUCN recommendations effectively deferred the 2009 application. The first required SCNP to address boundary issues linked to omitting key high plateau, gorges, lakes, and valley zones from the park. They advised "using crests and ridges rather than valley bottoms as natural boundaries" (IUCN 2010, 32). To gain local support, SCNP planners frequently split this biologically rich habitat between park and utilitarian uses but without a convincing management strategy to protect the biota. The second recommendation urged more cooperation with adjacent Kyrgyzstan on a transboundary reserve. As with the national park, Tajik reluctance to collaborate stemmed from their newly independent statehood, nationalism, and the postcolonial hangover mentioned earlier. The final recommendation concerned the risk of seismic-induced failure of landslide-dammed lakes (Alford, Cunha, and Ives 1999).

The park management plan did not meet World Heritage Committee standards because the Tajiks failed to adequately address human threats to the environment. These included trophy and local hunting, tourism impacts, vague plans for transportation corridors and protecting cultural resources, and an insufficient strategy to accommodate the 14,000 people living in five settlement areas, whose needs include wood collecting, livestock grazing, and irrigated farming:

> There is a serious risk that the management plan is largely a collection of recommendations. Following a request for further information, the State Party provided a brief statement confirming that the plan had expired at the end of 2009. It confirms that an order has been made to develop a management plan that "would be based on the principles and categories of the management plan that lost its power." IUCN considers that a fully developed and agreed management plan is required for the property. (IUCN 2010, 32)

These IUCN concerns delayed inscription of the Tajik National Park onto the World Heritage List and countered Tajik aspirations to develop their reserve as a nationalistic project without the technical expertise of outside experts.

During the next three years, SCNP addressed the recommendations in the 2010 technical evaluation. In 2013, their amended application to the World Heritage Committee (Committee for Environment Protection under the Government of the Republic of Tajikistan 2013) and a revised IUCN (2013) evaluation found the park to be of "Outstanding Universal Value based on criteria (vii) and (viii)" (UNESCO 2013, 163). Citing Criteria vii:

> Tajik National Park is one of the largest high mountain protected areas in the Palearctic Realm. The Fedchenko Glacier, the largest valley glacier of the Eurasian Continent and the world's longest outside of the Polar Regions, is unique and a spectacular example at the global level. The visual combination of some of the deepest gorges in the world, surrounded by rugged glaciated peaks, as well as the alpine desert and lakes of the Pamir high plateau adds up to an alpine wilderness of exceptional natural beauty. Lake Sarez and Lake Karakul are superlative natural phenomena. (UNESCO 2013, 163)

Under Criteria viii, the World Heritage Committee cited the importance of the Pamir Mountains as a focus of Eurasian glaciation, the landslide-damned Sarez Lake, and the juxtaposition of mountain and valley terrain with active geomorphic and tectonic processes that "contribute to our fundamental understanding of earth building processes" (UNESCO 2013, 163–64). They also noted that since the

proposed World Heritage Site occupies the entire park, with 78 percent designated as core parkland, and the "remoteness from human settlements, the property is considered to have an outstandingly high level of physical integrity," a designated buffer zone is unnecessary (UNESCO 2013, 164). Moreover, the updated park management plan provides for traditional land use needs while ensuring the conservation of biodiversity.

The World Heritage Committee proclamation directs the SCNP to procure adequate human and financial resources to safeguard the needs of local residents (grazing, firewood, farming) and to develop environmentally and economically sustainable ecotourism (Mislimshoeva et al. 2014). A final recommendation repeated an earlier request to cooperate with neighboring states—with local and outside expertise—to develop opportunities for future transnational reserves (Magin 2005; UNESCO 2013).

Biosphere Reserve: The Next Step

The UNESCO-sponsored Man and the Biosphere (MAB) is a global program of scientific and political cooperation to provide scientific rationale for land use that is ecologically and socioeconomically sustainable. Biosphere Reserves of recognized ecotypes monitor environmental change through conservation, development, and scientific research. Many World Heritage Sites form at least the core of a Biosphere Reserve.

Tajik National Park meets the MAB requirement as a government property with the highest legal protection in Tajikistan. The park and World Heritage Site currently includes two of three MAB zones. The defined core comprises 78 percent of the reserve. This sparsely settled high mountain landscape with deep valleys and extensive glaciers is unlikely to attract additional settlement from central Asia. The existing park buffer satisfies MAB parameters by allowing seasonal grazing and small-scale agriculture. More intense agriculture, pastoralism, small-scale mining, and potential hydropower development exist outside the park buffer and World Heritage Site boundary. Adding MAB oversight and scientific research monitoring will promote sustainable development and regional conservation.

Inclusion into the MAB program will also support the World Heritage Committee inscription recommendations 5, 6, and 7 to secure more human and financial resources for effective long-term protection and management (UNESCO 2013). They further directed the SCNP to cooperate with adjacent states, the World Heritage Center, and Advisory Bodies, "to undertake a regional comparative biodiversity and geodiversity study of Inner Asian high mountains and deserts and to conduct a regional expert workshop with a view to developing opportunities for future transnational potentially serial nominations" (UNESCO 2013, 165). It will also increase the global clientele of tourists (Su and Lin 2014), which remains the primary objective of the Parliamentary Decree and SEURT. Before this second supranational designation, however, SCNP must secure more internal funding and train a qualified staff to enforce comprehensive codified regulations that protect the park's natural and cultural resources. Another key to success is respecting the ethnographically diverse people and lifeways of those who live within the reserve boundaries. These represent the greatest threats to the long-term integrity of the Tajik National Park and ones shared by many reserves in the developing world.

Conclusion

In a span of twenty-four years, a perestroika-inspired coalition of citizen and government advocates created a new protected area in the Pamir Mountains of independent Tajikistan. This unlikely conservation triumph weathered Soviet collapse, civil war, the perils of dual professionalism and intense assessment from IUCN advisory teams. The existing park and World Heritage Site represents a significant milestone in the global movement that began in Yosemite to protect landscapes for the use and enjoyment of all people. Attracting stable funding and developing personnel capacity remain significant threats to the long-term ecological integrity of this new reserve and to achieving modest sustainable development in this isolated region (Breu and Hurni 2003; Kreutzmann and Watanabe 2016).

A successful Tajik National Park is an integral component of a future transfrontier central Asian mountain protected area. This would include Khunjerab National Park in northern Pakistan (1975), Taxkorgan Nature Reserve in Tajik Autonomous County of China (1984), Wakhan National Park in Afghanistan (2014), an undesignated contiguous portion of GBOA in Tajikistan, and a potential reserve in southern Kyrgyzstan. The challenge will be to overcome the military and political rivalries in the Roof of the World (Kreutzmann 2015). If they succeed, SEURT will surely realize its 1980s goal to develop a prominent national park with tourist income parallel to that of Nepal and Pakistan.

Acknowledgments

The author thanks Jack Ives, Deborah Elliott-Fisk, Nigel Allan, Christine Schoenwald-Cox, and Mary Beth Cunha (of the United States), Anvar Buzurukov and Svetlana Blagoveshenskaya (Tajikistan), and Yuri Badenkov (Russia). The comments of anonymous reviewers significantly improved the final article. I also thank Kostea Ahaev, Hariat, several Tajik villagers, and a surgeon in the People's 4th Division Hospital in Dushanbe, who helped me in a time of great need.

Funding

The United Nations University, University of California, Davis, and USAID funded this project.

References

Alford, D., S. Cunha, and J. Ives. 1999. Mountain hazards and development assistance: Lake Sarez, Pamir Mountains, Tajikistan. *Mountain Research & Development* 20 (1): 12–15.

Anonymous. 1982. *Botanical expedition to the Sangvor and Darvaz Pamir—July, August, September 1982.* Dushanbe, Tajikistan: Academy Nauk.

Badenkov, Y. 1991. *Environment, population, and economy of Tajikistan in the XX1st century: A proposal for an international research project for 1990–94.* Moscow: Soviet Academy of Sciences.

Berger, J., B. Buuveibaatar, and C. Mishra. 2013. Globalization of the cashmere market and the decline of large mammals in central Asia. *Conservation Biology* 27 (4): 679–89.

Biodiversity Conservation Center. 2003. Protected natural areas of Tajikistan: Normative and legal basis. *Nature Reserves and National Parks* 42:15–17.

Breu, T., and H. Hurni. 2003. *The Tajik Pamirs: Challenges of sustainable development in an isolated mountain region.* Bern, Switzerland: Centre for Development and Environment, University of Bern, and Geographica Bernensia.

Buzurukov, A. 1990. Temporary statement about the state national park of Tadjik SSR (1st Stage 1991–1993). In *Report of Tadjik SSR State Committee of Nature Protection.* Dushanbe, Tajikistan: Ministry of Environment.

———. 1992. *Pamir National Park: First project of Tajik SSR, collection of scientific and practical papers: Conference on problems of protection of animals and special preservation of natural territories.* Dushanbe, Tajikistan: Ministry of Environment.

————. 1993. Letter to International Mountain Society, S. Cunha, T. Norris, and others, explaining the origin, purpose, function, and projected mission of the Social-Ecological Union of the Republic of Tajikistan. Dushanbe, Tajikistan. (Personal papers on the origin of Pamir [Tajik] National Park, Stephen F. Cunha, Humboldt State University.)

Committee for Environment Protection under the Government of the Republic of Tajikistan. 2013. *Nomination: Tajik National Park (Mountains of the Pamirs)*. Dushanbe, Tajikistan: Committee for Environment Protection under the Government of the Republic of Tajikistan.

Committee for Environment Protection under the Government of the Republic of Tajikistan and National Center for Mountain Regions Development in Kyrgyzstan (CEPT/MRDK). 2011. *Strategy and action plan for sustainable land management in the High Pamir and Pamir-Alai Mountains*. Bishkek, Kyrgyzstan, and Dushanbe, Tajikistan: CEPT/MRDK.

Cunha, S. 1993. An action plan for the proposed mountain protected area in the High Pamirs. In *Parks, peaks, and people*, ed. L. Hamilton, D. Bauer, and H. Takeuchi, 128–31. Honolulu: IUCN-WCPA Mountains Biome.

————. 1997a. The hunting of rare and endangered fauna in the mountains of central Asia. In *Proceedings of the 8th International Snow Leopard Symposium (1995)*, ed. R. Jackson and A. Ahmad, 110–20. Islamabad, Pakistan: Snow Leopard Trust and Worldwide Fund for Nature–Pakistan.

————. 1997b. Summits, snow leopards, farmers, and fighters: Will politics prevent a national park in the High Pamir of Tajikistan? *American Geographical Society Focus* 66 (1): 17–21.

————. 2004. Allah's mountains: Establishing a national park in the central Asian Pamir. In *WorldMinds: Geographical perspectives on 100 problems*, ed. B. Warf, D. Jannelle, and K. Hansen, 25–30. New York: Kluwer.

————. 2007. The Badakshani of the eastern Pamir. In *Disappearing peoples? Indigenous groups and ethnic minorities in South and Central Asia*, ed. B. Brower and B. Johnston, 187–206. Oxford, UK: Berg.

Haslinger, A., T. Breu, H. Hurni, and D. Maselli. 2007. Opportunities and risks in reconciling conservation and development in a post-Soviet setting: The example of the Tajik National Park. *International Journal of Biodiversity Science and Management* 3 (3): 157–69.

Hauser, M. 2011. *The Pamirs* (map). 2nd ed. Zurich, Switzerland: Gecko Maps.

Heathershaw, J. 2008. Seeing like the international community: How peacebuilding failed (and survived) in Tajikistan. *Journal of Intervention and Statebuilding* 2 (3): 329–51.

————. 2009. *Post-conflict Tajikistan: The politics of peacebuilding and the emergence of legitimate order*. London and New York: Routledge.

Heathershaw, J., and E. Herzig, eds. 2013. *The transformation of Tajikistan: The sources of statehood*. London and New York: Routledge.

International Union for the Conservation of Nature (IUCN). 2010. *IUCN technical evaluation: Tajik National Park (Mountains of the Pamirs), Tajikistan*. ID No. 1252. Gland, Switzerland: IUCN.

————. 2013. *IUCN technical evaluation: Tajik National Park (Mountains of the Pamirs), Tajikistan*. ID No. 1252 rev. Gland, Switzerland: IUCN.

Ives, J., and P. Ives, eds. 1987. *The Himalaya-Ganges problem: Proceedings of a Conference*. Tokyo: United Nations University.

Jackson, R. 2012. Fostering community-based stewardship of wildlife in central Asia: Transforming snow leopards from pests into valued assets. In *Rangeland stewardship in central Asia: Balancing improved livelihoods biodiversity conservation and land protection*, ed. V. Squires, 357–80. Dordrecht, The Netherlands: Springer.

Johnson, H. 2008. *The Yosemite grant 1864–1906*. Yosemite, CA: Yosemite Association.

Kevlihan, R. 2016. Insurgency in central Asia: A case study of Tajikistan. *Small Wars & Insurgencies* 27 (3): 417–39.

Kreutzmann, H. 2015. *Pamirian crossroads: Kirghiz and Wakhi of High Asia*. Wiesbaden, Germany: Harrassowitz Verlag.

Kreutzmann, H., and T. Watanabe, eds. 2016. *Mapping transition in the Pamirs: Changing human-environmental landscapes*. Berlin: Springer.

Magin, C. 2005. *World heritage thematic study for central Asia: A regional overview*. Gland, Switzerland: International Union for the Conservation of Nature.

Marton-Lefève, J., T. Badman, and C. Kormos. 2014. World heritage and our protected planet. *World Heritage* 73.

Middleton, R. 2010. *The Pamirs: History, archaeology and culture*. Bishkek, Kyrgyzstan: University of Central Asia.

Ministry Document. 1992. Resolution no. 267 (about the establishment of a Tajik National Park). 20 June, Cabinet of Ministers, Republic of Tajikistan, Dushanbe.

————. 1993. Resolution no. 75 (about the Establishment of Tajikistan National Park). 17 February, Cabinet of Ministers, Republic of Tajikistan, Dushanbe.

Mislimshoeva, B., R. Hable, M. Fezakov, C. Samini, A. Abdulnazarov, and T. Koellner. 2014. Factors influencing households' firewood consumption in the Western Pamirs, Tajikistan. *Mountain Research and Development* 34 (2): 147–56.

Mitchell, J. 2015. Civilian victimization during the Tajik civil war: A typology and strategic assessment. *Central Asian Survey* 34 (3): 357–72.

Nourzhanov, K., and C. Bleuer. 2013. *Tajikistan: A political and social history*. Canberra: Australian National University Press.

Resnick, S., and R. Wolff. 2013. *Class theory and history: Capitalism and communism in the USSR*. New York: Routledge.

Rowe, W. 2010. Agrarian adaptations in Tajikistan: Land reform, water and law. *Central Asian Survey* 29 (2): 189–204.

State Agency for Natural Protection Areas of the Committee for Environment Protection under the Government of the Republic of Tajikistan (SANPA). 2012. *Management plan of Tajik National Park for 2012–2016*. Dushanbe, Tajikistan: SANPA.

State Committee for Nature Protection of Republic of Tajikistan. 1992a. *Scientific basis of preliminary proposals of the Pamir National Park: 1 February 1990–January 1, 1991*. Dushanbe, Tajikistan: Goskomproda.

241

————. 1992b. *Supplement: List of inanimate nature memorials in the territory of the Pamir National Park, including buffer zone (includes notation of some cultural memorials).* Dushanbe, Tajikistan: Goskomproda.

State Directorate. 2005. *Order of state directorate of natural protected areas, no. 147 of 2005.* Dushanbe: Government of Tajikistan.

State Republic Committee for Statistics. 1992. Unpublished census data for Tajikistan (data, maps, explanatory notes). Dushanbe, Tajikistan: State Republic Committee for Statistics.

Stolton, S., P. Shadie, and N. Dudley. 2013. *IUCN WCPA best practice guidance on recognizing protected areas and assigning management categories and governance types.* Gland, Switzerland: International Union for the Conservation of Nature.

Su, Y., and H. Lin. 2014. Analysis of international tourist arrivals worldwide: The role of world heritage sites. *Tourism Management* 40:46–58.

Symes, W., M. Rao, M. Mascia, and L. Carrasco. 2016. Why do we lose protected areas? Factors influencing protected area downgrading, downsizing and degazettement in the tropics and subtropics. *Global Change Biology* 22 (2): 656–65.

Thorsell, J., and L. Hamilton. 2002. *A global overview of mountain protected areas on the World Heritage List.* Gland, Switzerland: International Union for the Conservation of Nature.

UNESCO. 2010. *Report of the decisions adopted by the World Heritage Committee at its 34th Session (Brasilia, Brazil, July–August 2010).* Paris: UNESCO World Heritage Committee.

————. 2013. *Decisions adopted by the World Heritage Committee at its 37th Session (Phnom Penh, Cambodia, June 2013).* Paris: UNESCO World Heritage Committee.

UNESCO World Heritage Center. 2016. The criteria for selection. Paris: UNESCO World Heritage Center. http://whc.unesco.org/en/criteria/ (last accessed 15 June 2016).

United Nations Environment Programme World Conservation Monitoring Centre. 2009. *Tajik National Park (Mountains of the Pamirs) comparative analysis.* Revised 1st draft. Cambridge, UK: UNEP-WCMC.

Harnessing the State: Social Transformation, Infrastructural Development, and the Changing Governance of Water Systems in the Kangra District of the Indian Himalayas

Harry W. Fischer ⓘD

Despite a proliferation of programs and policies aimed at promoting local resource management, we still have limited knowledge of the conditions under which state interventions can be a supportive force in everyday aspects of common pool resource governance. This article explores growing state involvement in community-managed irrigation systems of the Kangra District of Himachal Pradesh, India. Here, agriculture is dependent on water channeled from glacial streams through networks of irrigation canals that have been sustained by local traditions of collective action for centuries. In recent years, however, growing off-farm employment has shifted the center of the agrarian economy and undermined shared norms of collective resource governance, just as state institutions have increasingly identified water systems as an object of development intervention. In this article, I document how irrigation management has been incrementally reinvented through changing institutional arrangements and new infrastructural forms over the past three decades, as existing patterns of collective action have increasingly found expression by leveraging development resources of the state. To the extent that socioeconomic changes associated with broader processes of development are likely to strain commons governance systems in mountain and other regions in the coming years, such collaborative engagements between local collective management and state support systems could become increasingly prevalent. This case suggests the need for new theoretical tools to guide analysis of evolving relationship between communities and state institutions in common pool resource governance.

尽管有为数众多的计画和政策, 旨在提倡地方资源管理, 但关于国家介入共有水资源管理的每日生活面向可能作为支持力量的条件, 我们仍所知有限。本文探讨国家在印度喜马拉雅邦康格拉县的社区管理灌溉系统中逐渐增加的介入。此处的农业, 是依赖透过数百年来由集体行动的地方传统所维系的灌溉网络运送的冰河水。但是近年来, 农场外的就业改变了农业经济的核心, 并削弱了集体资源管理的共享常规, 而国家机构于此同时也逐渐开始指认水资源系统作为发展介入的对象。我于本文中, 记录过去三十年来, 当既有的集体行动模式逐渐展现在操控国家发展的资源之上时, 改变中的制度安排与崭新的基础建设形态如何重塑灌溉管理。有鉴于关乎更为广泛的发展过程的社会经济变迁, 将可能在未来数年间使山区及其他区域的共有治理系统呈现紧张的程度, 此般地方集体管理和国家支持系统间的合作关系将日渐盛行。本案例显示, 我们需要崭新的理论工具, 指引分析社群和国家机构在共有水资源管理中的变迁关系。 关键词: 关键词·共有水资源管理, 去中心化, 喜马拉雅, 灌溉基础建设, 国家。

A pesar de la proliferación de programas y políticas orientadas a promover el manejo de los recursos locales, todavía tenemos limitado conocimiento de las condiciones bajo las cuales las intervenciones del estado pueden ser una fuerza de soporte en los aspectos de la gobernanza de los recursos comunes. Este artículo explora el creciente involucramiento del estado en los sistemas de riego manejados por la comunidad del Distrito Kangra de Himachal Pradesh, India. Aquí la agricultura depende del agua captada desde las corrientes glaciales a través de redes de canales de riego que han sido sostenidos mediante tradiciones locales de acción colectiva, durante siglos. Sin embargo, en años recientes el empleo en aumento que se consigue fuera de las granjas ha variado el centro de la economía agraria y socavado normas compartidas de gobernanza colectiva de los recursos, justo en la medida en que las instituciones del estado crecientemente han identificado los sistemas hidrológicos como un objeto de desarrollo susceptible de intervención. En este artículo, documento la manera como durante las pasadas tres décadas el manejo del riego se reinventa más frecuentemente a través de arreglos institucionales cambiantes y nuevas formas infraestructurales, a medida que los patrones existentes de acción colectiva cada vez más hallan expresión apalancando los recursos de desarrollo del estado. Con el grado con que los cambios socioeconómicos asociados con procesos más amplios de desarrollo puedan manchar los sistemas comunitarios de gobernanza en las montañas y otras regiones en los años venideros, tales compromisos de cooperación entre el manejo colectivo local y los sistemas de ayuda del estado podrían tornarse cada vez más prevalentes. Este caso sugiere la necesidad de nuevas herramientas teóricas que guíen el análisis de relaciones en evolución entre comunidades e instituciones del estado en la gobernanza comunitaria de los recursos.

> The landscape is not ... a fixed record of power but one which is constantly changing and being shaped by new political interests, connections, and constituencies.
>
> —Mosse (2003, 3)

In the Kangra District of India's Himalayan state Himachal Pradesh, intricate systems of earthen irrigation canals carry water from glacial streams to agricultural fields for wet rice cultivation. For centuries, these canals, known locally as *kuhls*, have been managed through local systems of collective action. Yet in the present era, the governance of the kuhls is changing. Growing off-farm employment has shifted the center of the agrarian economy, undermining shared practices of collective action. At the same time, state institutions have increasingly targeted water systems as objects of development intervention. Through the combined effects of these changes, water today is managed by an evolving set of institutional and infrastructural forms that blend local traditions of collective action with various forms of state support. Communities have harnessed the state, which has become an integral force through which access to water is now sustained.

As I argue, the kuhls represent a kind of collaborative engagement between communities and state institutions that might become increasingly common in mountain and other regions in the coming years. Yet to date, we have limited knowledge about how the state might undertake a supportive role in the governance of the commons. Common-pool resource management theory, the dominant analytic for analyzing commons governance, has paid only limited attention to the role of state institutions in structuring local practices of collective action.[1] Scholars working in traditions of critical human–environment scholarship, in turn, have given extensive attention to the ways that external governance regimes often impose new forms of resource exclusion on the poor (Robbins 2012), yet have provided limited evidence of the governance conditions that can help to align state interventions with local needs and aspirations. As others have shown, even attempts to engage communities in resource cogovernance regimes often fail due to the misalignment of formalized procedures with the rich social works that drive collective action in practice (Mosse 2003; Saunders 2014). Communities, of course, are seldom passive; they often mobilize to resist state agendas and to seek recognition for their own management systems (Perreault 2008; Fischer and Chhatre 2013). To date, however, we have limited knowledge about how communities could effectively harness the state—its resources, technical facilities, and institutional forms—to sustain resource management regimes of local control.

Whether we perceive the possibility of a supportive role for state institutions in the governance of the commons hinges on how we understand the nature of the state itself. Developments in the social sciences over the past several decades have increasingly seen the state less as a unified entity exerting power over society from "above" and more as a fractured and variegated collection of actors and institutions with diverse and often contradictory agendas (Sivaramakrishnan 2000). As such, scholars have turned increasing attention to the routinized practices and everyday struggles with which state programs are enacted in everyday life (Painter 2006). The boundaries between state and society are seen as porous: Just as state programs structure myriad aspects of local social interactions, "informal" relations of power permeate many of the routine functions of state administrative bodies (Corbridge et al. 2005; Li 2005). Yet it is not enough to say that state interventions are simply locally negotiated; there is a need for theoretical understanding about the constellation of institutions, social relationships, and material conditions that shape how state power is enacted and at times harnessed at the local level.

Geographers working in traditions of critical human–environment scholarship have made two contributions in recent years that add nuance to how we understand the intersection of state and social power in environmental governance. First, recent work has given increasing attention to the role of nonhuman forces—the material properties of resources, ecologies, and built environments—as "forceful" objects in structuring the exercise and limits of state power (Meehan 2014). Water has been a particular object of focus: As a fluid, hard to control, but essential substance of life (Bakker 2012), even centralized administrative arrangements often depend heavily on localized management practices that lie beyond their direct control (Banister 2014).

Yet local management systems are not always consigned to informal spaces beyond recognition of state institutions. A second body of work has explored how the rescaling of political and administrative functions of the state through decentralization has altered the terms by which local actors encounter a range of state programs (Corbridge et al. 2005; Batterbury and Fernando 2006). In practice, reforms for decentralization have often not resulted in a meaningful redistribution of power to the local level (Ribot, Chhatre, and Lankina 2008). Where local elected bodies have obtained substantive resources and powers of discretion, however, a growing body of research shows that this can provide new opportunities for less powerful individuals to engage in planning processes (Heller, Harilal, and Chaudhuri 2007; Fischer 2016) while also consolidating new channels for local actors to leverage resources from state institutions at higher scales (Chhatre 2008).

In this article, I build on these lines of analysis to explore the changing governance of the kuhls in a locality I refer to as Kohladi.[2] In the following section, I frame kuhl governance as fundamentally a challenge of controlling the flow of water through the mountain landscape, around which local forms of collective action have evolved. Thereafter, I discuss how socioeconomic shifts, emerging forms of decentralized state involvement in water management, and intensifying patterns of citizen engagement with state institutions have transformed kuhl governance over the past three decades. In the conclusion, I discuss how similar trajectories might be reshaping commons governance in many mountain and other societies, with implications for how we understand the mechanics of common-pool resource governance in the present era.

The case material for this article is drawn from in-depth qualitative fieldwork carried out over multiple extended visits to Kohladi over the past five years. I have undertaken detailed interviews with well over 200 area residents, including many individuals directly involved in kuhl governance, both past and present. I have also visited other kuhl systems in Kangra, which has enabled me to see how Kohladi is situated within a broader landscape of changing water systems in the region. Because it is not possible to capture the richness of local stories within the space provided by the special issue, this article seeks primarily to synthesize different accounts into a broad narrative arc that captures the most essential aspects of change within local water governance over the past three decades.[3]

Kuhl Systems and the Unruliness of Water

The Kangra Valley is situated at the base of the high Dhauladhar Range in India's state of Himachal Pradesh. Here, rice must be planted in the midsummer months to avoid the erratic weather of a late fall harvest. Rainfall is scarce in the months preceding the monsoon, yet mountain streams are flush with water from melting snowpack in the high Dhauladhar peaks. During this time, kuhls carry water from glacial streams to agricultural fields for irrigation. Every summer, water is released into fields to moisten the earth for plowing and to provide standing water for rice. Kuhls also supply water to irrigate wheat in the winter, for livestock, to power traditional flourmills, and for washing clothes.

Kuhls are simple irrigation canals, constructed of earthen materials and structured by the landscapes in which they flow. Kuhls intersect with gullies and other minor mountain streams from which they receive additional water; after irrigating agriculture fields, water is recycled into other kuhls below. Kuhls also play a crucial role in draining water following heavy monsoon rains. Kuhls are a central part of a dynamic hydrosocial landscape.

Importantly, kuhls are also fragile. Water erodes earthen walls and deposits silt. Vegetation grows along canal edges, slowing water flow. Torrential monsoon downpours periodically damage weaker sections, especially on steep slopes and around curves. Kuhls require continued vigilance and labor, underscoring the need for coordinated action. It is the ongoing struggle to control water against its natural propensities that stands at the core of kuhl governance, and it is this imperative that has driven the evolving relationship between local social organization and state institutions in the present.

State, Society, and Socioeconomic Change

The kuhls have always been an important link between local production practices and state power. As Baker (2005) noted in his account of kuhls across the Kangra District, the initial construction of many kuhl systems was sponsored by precolonial rulers as a form of patronage. Although the colonial government later sought to codify the distribution of water between localities as formal rights, historical documents also show ongoing legal battles, chronicled in colonial court records, between different communities seeking to renegotiate the terms and distribution of access (Baker 2005). In short, communities were not simply ruled: Governance was the product of ongoing state–society interactions that established the terms by which local management was conducted.

Yet throughout Kangra, everyday management occurred almost entirely through localized systems of collective action. A hereditary water-master, a position known as the *kohli*, would walk around the village twice a year beating a drum; every household would send an adult male to perform maintenance—to rebuild damaged walls, trim vegetation, and scoop out sediment—in exchange for the right to access water. Convention ensured relatively equitable distribution of water among those who participated. Yet, kuhl governance was part and parcel of a broader social and economic system characterized by great asymmetries of power across caste, class, and gender lines. The ultimate control over maintenance and adjudication of disputes tended to lie with the kohli and other powerful actors.

Local patterns of management, however enshrined in cultural practice, have always reflected changes in the broader context in which they have been embedded. Socioeconomic shifts in Kangra over the latter decades of the twentieth century resulted in several broad trajectories of change. To begin with, economic growth, both national and regional, alongside increasing infrastructural integration, has led a growing number of men to seek employment outside of the village—a trend that has been observed in other mountain regions (Azhar-Hewitt 1999). At the same time, the declining importance of agriculture among the upwardly mobile has challenged norms of collective action, even as agriculture remains important for many poorer subsections of society (Baker 2005).

These changes have coincided with increasing everyday engagement with government programs of various forms. Vibrant political mobilization around land reforms in the 1950s and 1970s as well as a high degree of electoral competition in the state legislature since the 1970s helped to establish a culture of robust political engagement over local and regional development agendas (Chhatre and Saberwal 2006). Since the early 1990s, a growing array of government functions have been placed under the control of the village council as part of a broader trend toward decentralization, through which local elected leaders have become key channels of access to various government programs (Fischer 2016). Throughout this period, multiple resource streams have become available from the central government for many different kinds of local infrastructure development—from motorable roads, to electricity, to smaller village paths—including a variety of programs that specifically target water infrastructure (Baker 2005).

In short, at the same time that socioeconomic changes were challenging existing management systems, new kinds of state–society engagements were emerging that, in turn, have given rise to new forms of kuhl governance.

A Brief History of the Kohladi Kuhls

Kohladi is located at the base of the Dhauladhar range. It is a rural area with approximately 1,200 households. It is made up of three local governmental units (*panchayats*), each composed of a handful of hamlets situated on a gently sloping plane between two glacially fed rivers. The Kohladi kuhl system is made up of four main channels—Nandlu, Chitruhl, Chholu, and Bandlu (Figure 1).[4]

In the latter decades of the twentieth century, the Kohladi kuhls were managed by a man named Pramod of a long kohli lineage. By the late 1980s and early 1990s, the government's embrace of small-scale sustainable development efforts led to a growing array of government programs for water development under the command of a variety of administrative authorities. Elders were eager to use these resources to improve existing earthen infrastructure, but they were also concerned that state involvement would erode their autonomy and leave the community at the mercy of unresponsive bureaucracies. In 1991, following the lead of other kuhl systems in Kangra and under provisions set up by the state administration, elders initiated the formation of a kuhl committee built on an institutional template determined by the state, with elected officeholders, a secretary to maintain records, and a treasurer to manage financial affairs. The formalization of management authority was a strategic move to establish a legally recognized body for the transfer of state funds while also retaining formal discretion over future projects.

The management committee did not just formalize an existing management system; it transformed it. To the extent that the kuhl committee's purpose was to

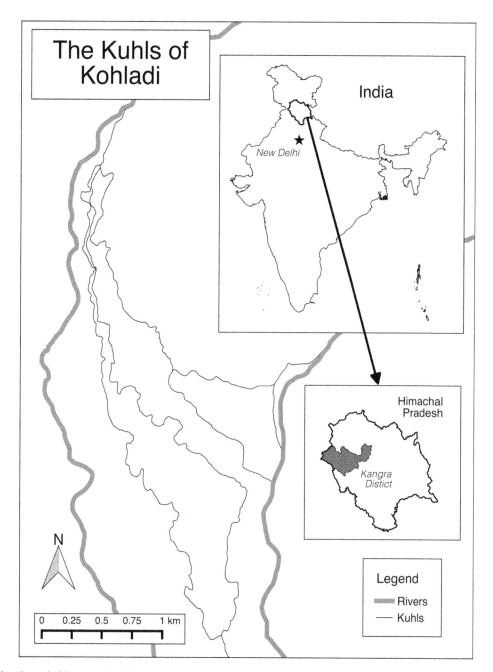

Figure 1. Kohladi is bounded by two glacially fed rivers (left and right). The kuhls meander through the landscape following a downward slope, roughly north to south. Smaller subchannels (not pictured) distribute water to agricultural fields along the way. Kohladi is one of many kuhl systems in the Kangra District of Himachal Pradesh (inset).

facilitate development work, power came to be defined less by landholding or the symbolic authority of the kohli but increasingly through political clout derived from party affiliation. It is no surprise that elections favored established elites with extralocal political connections that provided channels to leverage funds. In the following years, kuhl leaders lobbied a variety of state bodies—the Deputy Commissioner, the Department of Irrigation and Public Health, and the Block Development Office—for resources to undertake works

of different kinds. Various sections of the kuhls were fortified with concrete retaining walls, especially on steep inclines and around curves where earthen walls were prone to rupture during heavy storms. The header of the kuhl system, a particularly vulnerable segment where fast-flowing river water enters the kuhls, was concretized as well, thus reducing the need for regular maintenance.

For officials working in government offices, working through the kuhl committee helped to navigate the local context, thus providing an easy way to achieve administrative targets. Yet the flow of resources was often politicized as well, driven by pressure from political party operatives seeking to consolidate their electoral support by strategically delivering projects to different constituencies. It is no coincidence that community members elected leaders to the kuhl committee of both major parties active in the state—the Congress Party and Bharatiya Janata Party (BJP)— which in turn provided maximum channels to leverage funds. The kohli, for his part, held no official position within the committee (as a hereditary position, he was not subject to election). His symbolic authority was increasingly shared with the kuhl committee chairman.

Although the ability to leverage funds from state bodies brought new party-affiliated elites into the center of governance processes, electoral franchise also provided pathways for a much greater diversity of individuals to gain formal positions within the committee. One such individual was a man named Kamlesh of the low caste *Lohar*—a retired army personnel and aspiring political operative—who, on the basis of his growing connections with the Congress Party, secured election as the committee secretary and subsequently established himself as one of the core leaders of kuhl management. Others from historically marginalized groups came to occupy positions in the committee's secondary tier of hamlet-level representatives. Dinesh, a low caste *Charmar*, vividly recalled how he and other less powerful representatives used their positions to gain increasing leverage over existing elites—"pulling their legs," as he described it—which afforded a measure of both accountability and voice in several areas of governance, including development efforts, maintenance, and the adjudication of disputes.

Existing traditions of management grew increasingly hard to sustain by the late 2000s, however. As has happened in many parts of Kangra, older leaders of the Kohladi kuhls died, including the kohli, and few of the younger generations have emerged to take their place. For younger generations, who have grown up in a society oriented toward off-farm employment as the basis of status and prosperity, the kuhls have far less symbolic importance than in the past. The growing migration of men, especially among upwardly mobile sections of society, has also left agriculture increasingly under the responsibility of women, who have limited knowledge of kuhl maintenance and no precedence of engaging in collective

management. Households that do continue to rely heavily on agriculture have been largely unable to sustain the kuhls by themselves. Although urgent repairs are still often done through decentralized action, regularized maintenance has decreased drastically since the mid-2000s.

Nonetheless, these trends have coincided with another significant change in kuhl governance. Legislation undertaken by the central government has given increasing resources to elected village councils (known as *panchayats*) to undertake water development projects in recent years.[5] Whereas development projects previously implemented by the kuhl committee were at best sporadic, the extensive resources given to panchayats have enabled them to undertake almost continual work, which has helped to compensate for the decline of existing forms of collective action. The Kohladi panchayats have now become central actors in the management of the kuhls.

This, too, has altered patterns of participation and authority within kuhl governance. Unlike the kuhl committee, the panchayat has strong protections for minority groups through affirmative action policies that reserve a proportion of elected village council seats for women on a rotational basis (half of all seats in Himachal since 2010), with similar arrangements for other minority groups such as low castes in proportion to their population. Thus, although the kuhl committee provided pathways for greater engagement of less powerful groups in kuhl governance but largely from the periphery, the growing role of the panchayat has cast a far greater diversity of individuals directly into the center of decision-making processes associated with kuhl governance.

Planning under the panchayat has led to new kinds of work on the kuhls, many of which blend traditional earthen maintenance with concrete development projects. Unlike the older traditions of collective management founded on unpaid communal labor, panchayat work is coordinated as a government project through paid wage labor. Projects must proceed in accordance with various procedural requirements established by the development bureaucracy. Still, panchayat leaders hold chief discretion over what kinds of projects are implemented and where, which in practice has given them significant discretion to formulate projects that target a variety of needs across the broader landscape.

Projects unfold through localized deliberations between various actors involved in kuhl management. Elders previously associated with the kuhls still often seek to influence projects, but now they must do so through elected panchayat leaders. Projects almost invariably evolve through the process of implementation. To give one example, an elder requested Kuldeep, the vice president of the Suri Panchayat, to do maintenance work on the Chholu kuhl. Kuldeep, in turn, undertook extensive earthen renovation but decided also to reserve a portion of the budget to repair a broken retaining wall elsewhere on the kuhl (Figure 2).

The greater diversity of leaders now influencing planning processes has also given greater leverage to poorer subsections of society—including those who continue to rely more heavily on agriculture—to coordinate work within their communities. In another example, Ravi of the low caste *Lohar* worked with Akshay, a representative of a nearby hamlet in the

Chauri Panchayat, to renovate a canal that runs through both of their communities. Akshay undertook a project to widen and deepen the silted kuhl to improve water flow; Ravi continued to monitor progress and provided feedback along the way. At the request of a neighboring household they also built a retaining wall on the kuhl to prevent flooding of common grazing land frequently used by the *Lohar* community (Figure 3).

These projects are only two of a large sample of almost continual projects undertaken by area panchayats since 2008, which together have now touched nearly every part of the Kohladi kuhls. Through piecemeal efforts, Bandlu kuhl has been fully concretized, and the remaining three kuhls have all been the target of significant earthen and concrete work. The kuhl committee still exists, and it helps to coordinate seasonal timings of water access according to convention. Yet, as even its present chairman admits, for all practical purposes the panchayat now stand at the center of authority over the kuhls.

At present, a new chapter in kuhl governance might just be beginning. When I first arrived in Kohladi in 2012, I often heard community members advocating full concretization—that is, laying concrete along the sides and base of the channels—to alleviate the need for seasonal maintenance. In 2014, a new program provided extensive resources for such work.[6] Although Kohladi was not originally identified as a

Figure 2. This retaining wall was originally built in the 1990s, but high-pressure inflows from a seasonal stream to the right (dry as pictured) resulted in significant damage over the years. The lighter section of the wall was repaired in 2013, this time built slightly lower to better enable drainage during monsoon cloudbursts. (Color figure available online.)

Figure 3. This retaining wall protects the kuhl in a vulnerable section where rapid water curves along the edge of a steep downward slope. (Color figure available online.)

project site, an elder named Suresh, whose family had long been involved in kuhl management, heard about the program and lobbied for Kohladi's inclusion. Resources were allocated to concretize the Nandlu kuhl, which flows near his home. As the uppermost kuhl of the system, Nandlu receives the least inflow of water from the broader landscape; the force of its flow diminishes significantly along its path. Concretizing Nandlu, it is hoped, will help to streamline water flow, while also protecting against water loss in leaky and often poorly maintained sections of the channel.

The administrative requirements of the program mandated the creation of a new committee to oversee the management of the kuhls after concretization. A committee for the management of Nandlu was formed, and Suresh was elected chairman due to his role in securing funding for the project. Permission for the work was secured from area panchayats, and at the time of writing, work had just begun. A new phase of the program is expected to start in 2017, which could hold the possibility of concretizing other Kohladi kuhls.

Changing Patterns of Collective Action as a Result of Infrastructural Change

The cumulative effect of these interventions is significant. Although aspects of more traditional forms of collective maintenance persist, many of the most vulnerable segments are now supported by concrete, which has greatly diminished the need for ongoing repairs. Even earthen work is now at times coordinated through paid labor through the panchayat. These changes have helped to compensate for changing patterns of collective action, but they are not without risks. Whereas traditional forms of collective action were ongoing, iterative, and sustained through the kohli's intimate knowledge of the landscape, state-supported projects inject their own bureaucratic imperatives—productivity targets, standardized designs, and administrative safeguards—into everyday management processes. Sometimes flaws in construction require further work, and once a channel is concretized, it cannot be easily undone. Yet to the extent that local actors have retained chief discretion over projects implemented in their community, they have been able to formulate a variety of projects in response to particular water management challenges, each rooted in their knowledge of the local landscape. Unlike bureaucratic actors and their low-level technical staff, local leaders

have a very particular interest in ensuring that work is done according to local needs.

By diminishing the need for regularized maintenance, these infrastructural innovations have fundamentally changed the mechanics by which governance happens. Some community members lament the decline of traditional forms of collective action, but almost all seem to embrace state interventions as a means to protect water access in the present era—particularly as off-farm employment increasingly limits the amount of time households are able to devote to regular maintenance.

Yet, infrastructural changes have not solved the fundamental problems associated with the force of water in the mountain landscape. When concrete breaks, as it periodically does, it still must be repaired. Unlike earthen kuhls, repairs cannot be done by hand, and they are not cheap; they almost always require further mobilizing of government support. Thus, concretizing kuhls does not eliminate the need for collective action but induces new path dependencies in governance. The imperative for water management has brought together state and society through a shifting constellations of formal institutions, state programs, and local patterns of political engagement—not only in the present but also in the future. At the aggregate, these projects have reimagined the relationship between communities and state institutions—and with it, the ways in which social and political power are inscribed in the landscape.

Conclusion

The contested nature of environmental governance is one of the most explored themes in the field of political ecology (Robbins 2012). As a large body of scholarship has shown, communities have often played an active role in engaging with state institutions to protect their own management systems, often seeking formal recognition for them in the process (Perreault 2008; Fischer and Chhatre 2013). Yet, this case reveals a kind of collaborative engagement between communities and state institutions that remains largely untheorized. Through an incremental process of institutional and infrastructural change, communities have harnessed the state, which is now an integral force through which access to water is sustained.

To a large extent, the kuhls of Kangra reflect a broader process of transformation through which the plans and designs of central authorities have come to bear on mountain societies. Decentralization, of course, is not new; state actors have often sought to engage communities in the governance of natural resources, even as far back as the colonial and precolonial eras (Agrawal 2001; Baker 2005). Yet increasing infrastructural integration in conjunction with broader processes of economic change, both regional and national, have brought significant socioeconomic shifts to many mountain communities in recent years (Butz and Cook 2011; Fischer 2016). Although Himalayan societies have long been recognized for their rich traditions of collective resource management (Baker 2005; Agrawal and Chhatre 2006), these transformations are likely to challenge shared interests in commons management in many contexts. To the extent that existing patterns of collective action might no longer be enough, it is increasingly important to understand when and how state support can help to sustain local governance systems.

If external interventions have often been at odds with organic practices of collective action (Mosse 2003; Saunders 2014), this case reveals a gradual process of institutional change, whereby local management and state support systems have coevolved in ways that blend local traditions of management with various forms of state support. In fact, these trends are part and parcel of a broader process of change through which many kinds of government programs—from subsidized food, to state employment schemes, to various forms of infrastructural development—have gained a central presence in everyday life (Fischer 2016). In this case, the formal recognition of local management institutions provided a platform to leverage resources from a variety of state institutions. The protection of local discretion was of particular importance: It gave local actors the authority to formulate projects according to their own knowledge and priorities. It is hard to imagine such a high degree of success, in terms of both continued functionality of the kuhl system and sustained local engagement, were projects to be formulated by bureaucrats disconnected from the local social and hydrological context.

These emerging institutional arrangements have done far more than protect existing traditions or enshrine them with legitimacy. These changes have brought about a new basis of resource management, with new authorities, new technologies, and shifting forms of collective action. These changes have also altered the social basis of management by propelling new actors into the center of governance processes—including individuals from less privileged groups, who continue to rely heavily on agriculture. Today governance is less an outcome of self-organized collective action and increasingly contingent on the character of state–society relationships, as expressed through the ability to leverage resources from a range of state institutions, and bring them to bear on complex and ever-changing mountain geographies.

Acknowledgments

The author would like to thank Ashwini Chhatre, Jesse Ribot, Tom Bassett, Shikha Lakhanpal, Pronoy Rai, and Majed Akhter for their constructive feedback during the development of this article. The article benefitted from discussions at the Illinois–Wisconsin–Minnesota Nature–Society Workshop in October 2013 and the Yale Modern South Asia Workshop in April 2014. Syed Shoaib Ali, Jeetul Bhansingh, and Satya Prasanna all contributed immensely to the development of this work in the field. Most of all, I am grateful to the residents of Kohladi who have generously given their time and support throughout my fieldwork.

Funding

The research was funded by National Science Foundation Grant BCS 11–31073 and a Fulbright-Nehru Research Fellowship.

Notes

1. Although scholarship in this tradition recognizes the role of the state in structuring incentives for management, the locus of analysis continues to be on localized forms of collective action that occur largely independent of direct state intervention. See Baker (2005) for additional discussion.
2. I have adopted pseudonyms for all place names and people to protect the anonymity of those involved.
3. For greater contextual detail about Kangra and Himachal Pradesh more generally, see Fischer (2016, 3–4) and, above all, Baker's (2005) exhaustive account of Kangri kuhls past and present.
4. Communities below Kohladi get a portion of their water from the Kohladi kuhls, but they also draw water from their own sources. Traditionally, they had their own kohlis and their own organized systems of collective action.
5. Chief among the new resources given to local governments has been the National Rural Employment Guarantee Act, which gives local leaders significant discretion to select and implement a variety of small-scale development projects, including many water-related works.

6. The project is supported by funds from the Japanese International Cooperation Agency but is being overseen by the Himachal Pradesh state government.

ORCID

Harry W. Fischer (iD) http://orcid.org/0000-0001-7967-1154

References

Agrawal, A. 2001. State formation in community spaces? Decentralization of control over forests in the Kumaon Himalaya, India. *The Journal of Asian Studies* 60 (1): 9–40.

Agrawal, A., and A. Chhatre. 2006. Explaining success on the commons: Community forest governance in the Indian Himalaya. *World Development* 34 (1): 149–66.

Azhar-Hewitt, F. 1999. Women of the high pastures and the global economy: Reflections on the impacts of modernization in the Hushe Valley of the Karakorum, Northern Pakistan. *Mountain Research and Development* 19 (2): 141–51.

Baker, J. M. 2005. *The kuhls of Kangra: Community-managed irrigation in the Western Himalayas.* Seattle: University of Washington Press.

Bakker, K. 2012. Water: Political, biopolitical, material. *Social Studies of Science* 42 (4): 616–23.

Banister, J. M. 2014. Are you Wittfogel or against him? Geophilosophy, hydro-sociality, and the state. *Geoforum* 57:205–14.

Batterbury, S., and J. L. Fernando. 2006. Rescaling governance and the impacts of political and environmental decentralization: An introduction. *World Development* 34 (11): 1851–63.

Butz, D., and N. Cook. 2011. Accessibility interrupted: The Shimshal road, Gilgit-Baltistan, Pakistan. *The Canadian Geographer* 55 (3): 354–64.

Chhatre, A. 2008. Political articulation and accountability in decentralisation: Theory and evidence from India. *Conservation and Society* 6 (1): 12–23.

Chhatre, A., and V. Saberwal. 2006. *Democratizing nature: Politics, conservation, and development in India.* New Delhi: Oxford University Press.

Corbridge, S., G. Williams, M. Srivastava, and R. Veron. 2005. *Seeing the state: Governance and governmentality in India.* Cambridge, UK: Cambridge University Press.

Fischer, H. W. 2016. Beyond participation and accountability: Theorizing representation in local democracy. *World Development* 86:111–22.

Fischer, H. W., and A. Chhatre. 2013. Environmental citizenship, gender, and the emergence of a new conservation politics. *Geoforum* 50:10–19.

Heller, P., K. N. Harilal, and S. Chaudhuri. 2007. Building local democracy: Evaluating the impact of decentralization in Kerala, India. *World Development* 35 (4): 626–48.

Li, T. M. 2005. Beyond "the state" and failed schemes. *American Anthropologist* 107 (3): 383–94.

Meehan, K. M. 2014. Tool-power: Water infrastructure as wellsprings of state power. *Geoforum* 57:215–24.

Mosse, D. 2003. *The rule of water: Statecraft, ecology, and collective action in South India.* Oxford, UK: Oxford University Press.

Painter, J. 2006. Prosaic geographies of stateness. *Political Geography* 25 (7): 752–74.

Perreault, T. 2008. Custom and contradiction: Rural water governance and the politics of *Usos y Costumbres* in Bolivia's irrigators' movement. *Annals of the Association of American Geographers* 98 (4): 834–54.

Ribot, J., A. Chhatre, and T. Lankina. 2008. Introduction: Institutional choice and recognition in the formation and consolidation of local democracy. *Conservation and Society* 6 (1): 1–11.

Robbins, P. 2012. *Political ecology.* 2nd ed. Malden, MA: Wiley.

Saunders, F. 2014. The promise of common pool resource theory and the reality of commons projects. *International Journal of the Commons* 8 (2): 636–56.

Sivaramakrishnan, K. 2000. Crafting the public sphere in the forests of West Bengal: Democracy, development, and political action. *American Ethnologist* 27 (2):431–61.

Living with Earthquakes and Angry Deities at the Himalayan Borderlands

Mabel Denzin Gergan

The Indian Himalayan Region, a climate change hotspot, is witnessing a massive surge in hydropower development alongside a dramatic rise in natural hazard events. This article explores indigenous people's response to this intersection of concerns around hazards and contentious development beyond more legible instances of social movements or resistance. Through an ethnographic case study located in the Eastern Himalayan state of Sikkim, the site of a 6.9 magnitude earthquake, controversial hydropower projects, and an indigenous antidam protest, I show how people's relationship with a sacred, animate landscape is not easily translatable into the clear goals of environmental politics. Antidam activists and environmentalists link growing ecological precarity in Sikkim to state-led hydropower construction, but for many lay indigenous people, these earthquakes raise deeper cultural anxieties. I demonstrate how these anxieties are grounded in a longer history of the contested relationship between marginalized peoples and hegemonic state and nonstate powers, a relationship that continues in the fraught relationship of the Himalayan margins to the Indian state. I argue that critical engagements with indigenous environmentalism must be in dialogue with diverse interpretations and registers of loss and erasure. In this I follow recent calls to decolonize the Anthropocene that demand that we move beyond a politics of urgency to examine the slow, historical processes of erasure under colonialism and imperialism. I highlight these narratives to argue for a more holistic approach to the uneven impacts of climate change on mountainous environments and their inhabitants.

作为气候变迁热门地点的印度喜马拉雅地区，正见证水力发电建设的大幅兴起，以及自然灾害事件的戏剧性增加。本文探讨原住民族在更为明确的社会运动或反抗场合之外，对于此般围绕着自然灾害和具争议性的发展相互交织的考量之回应。我的田野案例研究包括位于东喜马拉雅的锡金邦这个震度 6.9 级的地震场所、争议性的水力发电计画，以及原住民族的反水坝抗议行动，藉此展现人类与神圣、具有活力的地景之间的关系，并非轻而易举便能转变成环境政治的明确目标。反水坝运动人士和环境保育人士将锡金成长中的生态危险连结至由国家主导的水力发电建设，但对诸多外行的原住民族而言，这些地震引发了更为深层的文化焦虑。我将展现这些焦虑如何植基于被边缘化的人群和霸权国家与非国家权力之间的竞争关系之长远历史，而该关系从喜马拉雅边缘的紧张关系持续至印度国家中。我主张，批判性地涉入原住民族的环境保育主义，必须与多样的诠释和消失与抹除的注记相互对话。于此，我追随晚近对于去殖民人类世的呼吁，该呼吁要求我们超越紧急状态的政治，并检视殖民主义与帝国主义下抹除的缓慢历史过程。我强调这些叙事，主张以更全面的研究方法探讨气候变迁对于山岳环境和其中居民的不均影响。关键词：去殖民, 喜马拉雅地区, 原住民性, 山岳环境, 自然灾害。.

La Región de los Himalaya de la India, un punto caliente del cambio climático, está presenciando la masiva aparición de desarrollo hidroeléctrico simultáneo con un incremento dramático de eventos naturales riesgosos. Este artículo explora la respuesta de la población indígena a este cruce de las preocupaciones sobre riesgos y el desarrollo cuestionado que sobrepasa las instancias más descifrables de los movimientos o la resistencia sociales. Por medio de un estudio etnográfico de caso localizado en el estado himalayo oriental de Sikkim, sitio donde ocurrió un terremoto de magnitud 6.9, de proyectos hidroeléctricos controvertidos y de una protesta indígena contra la represa, muestro cómo la relación de la gente con un paisaje sagrado animado no resulta fácilmente asimilable en las claras metas de la política ambiental. Los activistas contra la represa y los ambientalistas enlazan la creciente precariedad ecológica en Sikkim con la construcción hidroenergética orientada por el estado, aunque para muchos indígenas del común estos terremotos promueven ansiedades culturales más profundas. Yo demuestro la manera como estas ansiedades se anclan en una historia más larga de la cuestionada relación entre los pueblos marginados y las fuerzas hegemónicas del estado u otros actores, relación que se mantiene en la tensa relación existente entre los bordes himalayos y el estado indio. Arguyo que los críticos compromisos con el ambientalismo indígena deben darse en diálogo con las diversas interpretaciones y registros de pérdida y corrección. En este respecto, estoy de acuerdo con recientes llamados a descolonizar el Antropoceno que nos piden movernos más allá de las políticas de apremio para examinar los lentos procesos históricos de corrección bajo el colonialismo y el imperialismo. Subrayo estas narrativas para pedir un enfoque más holístico de los impactos desiguales de cambios climáticos sobre los entornos montañosos y sus pobladores.

Q: What do people say about the earthquakes?
Palzor: People say it's because there is so much lack of peace in the village. Then some people say it's because of *paap* [sin], maybe people have excess paap. Maybe the environment has gotten polluted. Then other people say it's because of the *devi-deorali* [Nepali term for gods and goddesses]. In Lepcha we call them *lingzee*; we believe they live in the mountains and hillside. And slowly we've stopped following these gods and started focusing more on Buddhism, especially this generation, we don't believe [in these gods]. It could be because of that the earthquakes came.

On 18 September 2011, a 6.9 magnitude earthquake shook Sikkim, a small border-land state in the Eastern Himalayas of India. The epicenter of the earthquake was located close to a dam site under construction in the north district of Sikkim. Since 2011 Sikkim has witnessed earthquakes of varying magnitudes every year without fail. Although earthquakes have always been a part of the Himalayan region's geophysical makeup, environmentalists and antidam activists from the indigenous Lepcha tribe argue that the surge in state-led hydropower projects is inducing vulnerabilities[1] in an already unpredictable ecosystem (Kohli 2011; Chopra 2015). As Palzor,[2] a Lepcha youth unaffiliated with the antidam movement suggested, however, beyond the failure of state and institutional mechanisms, for many these earthquakes point to deeper cultural anxieties. Palzor's concerns resonate with those of antidam activists but extend further to incorporate anxieties around abandoning indigenous Lepcha deities in favor of Buddhist ones. I highlight the concerns of lay indigenous Lepchas not to delegitimize the antidam struggle but to argue for a broader understanding of loss and erasure in the Anthropocene (Tolia-Kelly 2016). Theoretically, this article draws on recent calls to decolonize the Anthropocene within geography and related disciplines. Scholars have argued that the "newness" of this epoch and the sense of crisis it evokes can obfuscate the slow ongoing violence unfolding globally since the colonization of the Americas (Danowski, Viveiros de Castro, and Latour 2014; Sundberg 2014; Haraway 2015; Tolia-Kelly 2016). I demonstrate how the concerns of

Palzor and other lay Lepchas reveal deep-seated anxieties around loss of authenticity and self, a collective symptom of the postcolonial condition (Nandy 1989a, 1989b). More specifically, I suggest that critical engagements with indigenous environmentalism must be in dialogue with diverse interpretations and registers of loss and erasure that push our analysis beyond ecological loss (Cruikshank 2007; Tolia-Kelly 2016).

This article builds on my larger research project that examines the shifting relationship between the Indian state and its Himalayan margins and more specifically how these shifts are shaping the political subjectivities of indigenous youth. Like many mountainous environments, the Himalayan region is home to marginalized groups who in the last decade have had to negotiate their place in new transnational contexts of global climate change, environmentalism, and social justice (Cruikshank 2007; Yeh 2015). Hydropower projects driven in part by climate change anxieties around renewable energy (Fletcher 2010) have also fomented social movements that seek to address overlapping concerns around environmental justice and indigenous rights (Chowdhury and Kipgen 2013). The arguments I make in this article are drawn from fifteen months of fieldwork in Sikkim, where I divided my time between Sikkim's capital Gangtok and the Dzongu reserve in North Sikkim. The interviews quoted here were all conducted in Dzongu, which is divided into seven village-level governance units (GPUs). I interviewed antidam activists, local governance officials, and lay Lepchas who were unaffiliated with the antidam movement in each GPU. I am a native speaker of Hindi and Nepali and most interviews were conducted in Nepali. A few interviews with Lepcha elders were conducted in Lepcha and translated with the help of a native speaker.

Living with Indeterminacy at the Margins

Recent scholarship on the Anthropocene evokes a sense of crisis and urgency (Roelvink and Gibson-Graham 2010). The newness of this epoch and narratives of a common humanity as risk and at risk, however, obfuscate how legacies of colonialism and imperialism are inextricably tied to the epoch's genealogy (Danowski, Viveiros de Castro, and Latour 2014; Haraway

2015; Last 2015). More recently even geological proposals for the Anthropocene have had to contend with the devastating consequences of European conquest on indigenous and native populations (Lewis and Maslin 2015). This synergy of concerns around decolonization and the multiple crises this epoch heralds has led to meaningful engagements between postcolonial theory and materiality. Significant in this regard is Spivak's (2003) planetary framework, distinct in its explicitly geophysical language that describes "planet-feeling" as an experience of the uncanny, the *unheimlich*, the undoing of a sense of habitation. For many postcolonial scholars the geophysical dimensions of this epoch such as natural hazards or extreme weather events are emerging as important sites of inquiry. As Last (2015) pointed out, disasters are not always "quasi-apocalyptic manifestations that we see in the media but an extension of things that feel normal ... the apocalypse that is performed daily, routinely, even happily and with permission" (61). The uncertainty and anxiety associated with this epoch also points to an accretion of unaddressed and unexamined historical injustices and erasures (Tolia-Kelly 2016).

Efforts to decolonize the Anthropocene push for recognition of colonial legacies not just in our research sites but also within disciplinary traditions. Within geography, posthumanist scholarship central to discussions on the Anthropocene has been critiqued for its modern secular tendencies, especially in its lack of engagement with indigenous ontological categories (Sundberg 2014; Gergan 2015). In the context of Indigenous research, Blaser (2014) argued for recognition of multiple ontologies, where ontology is understood as an act of storytelling with each story performing the world it narrates. Scientific proposals for the Anthropocene that draw on Enlightenment narratives can then be seen as performing a world with a "linear, authoritarian, and universal narrative of human–environment relations" (Veland and Lynch 2016b, 2). Social histories of marginalized groups, especially those that have historically inhabited volatile environments, can provide important counternarratives in the Anthropocene debates. Writing about drought conditions in Australia, Clark (2008) argued for deeper engagements with the social histories of Aboriginal Australians and that "integrating social history with geological, climatic or evolutionary history has its own potential to destabilize colonial narratives" (739). Similarly, in her research in Alaska and the Yukon Territory, Cruikshank (2007) observed how indigenous oral histories graft together geophysical changes associated with the Little Ice Age along with

accounts of European colonial incursions. Engagements within indigenous ontologies, however, also require an attention to how "indigenous ways of being are not neatly held in some complete alterity, [but] are in capitalism, casinos, mines, cities and law courts" (Sidaway, Woon, and Jacobs 2014, 10). A critical scholarly commitment to indigenous groups must then be cognizant of the multiplicity of narratives and indigenous interlocutors, as well as the limits of liberal projects of recognition (Povinelli 2002; Rappaport 2005; Greene 2009).

In this article I draw attention to the peculiar space of India's mountainous borderlands, home to many indigenous and ethnic groups. The Indian state's interventions in the region have been heavily shaped by the colonial sociology of knowledge and its inherent environmental determinism that viewed the region's geophysicality and by extension its inhabitants as unruly, remote, and exotic (Kennedy 1991; Cohn 1996). Scholars from the region along with Dalit (formerly untouchable) scholars point to the limits of postcolonial scholarship from India, dominated by upper caste scholars and its failure to incorporate their voices (Bhagavan and Feldhaus 2009). Nandy (1989a), commenting on the disorienting nature of colonialism, which leaves the colonized with "someone else's present as their future," demands an attention to "the insurrection of the little cultures" (264) that have been subsumed within the upper caste, elitist historiography (Nandy 1989b, 266). In the context of this article it is also important to note that the spiritual traditions of many "little cultures" have also been erased or subsumed within the "great traditions" of Hinduism and Buddhism with their "perceived purity, historical longevity, and textual authority" (Shneiderman 2010, 309). The fractured nature of the postcolonial experience heightened in the Anthropocene demands an attention to the narratives of little cultures whose histories have undergone multiple iterations of erasure. In what follows, I offer the background context to Sikkim and the Lepcha tribe. I then present two ethnographic sections that draw on interviews with residents of Dzongu. The first section illustrates how concerns raised by lay Lepchas in Dzongu regarding the earthquake signal tensions between Buddhism and shamanic practices. I suggest that this spiritual indeterminacy points to cultural anxieties not addressed by present environmental politics. The second section illustrates how loss and recovery of self are a theme in both Lepcha activist and lay narratives, but the former is tasked with enrolling these claims in a politics of recognition. Finally, I conclude by reflecting on the challenges presented by a politics of urgency that prioritizes a few interlocutors and their specific registers of space and time.

Taming the Sacred Hidden Land

Sikkim, an independent Buddhist monarchy, was annexed to India in 1975, almost three decades after India's own independence from British rule. The swift annexation of Sikkim came as a surprise to the twelfth monarch, Palden Thondup Namgyal, who viewed the Indian state as a powerful yet benevolent neighbor. The story of Sikkim is also the story of several kingdoms and princely states that found their nascent aspirations for sovereignty crushed, replaced instead by a regional identity that could only exist as a margin to an imagined core, the Indian state. Sikkim's early history beginning in the eighth century, though, demonstrates an older history of territorial consolidation when it is believed that Guru Rinpoche (Guru Padmasambhava in some accounts) introduced Buddhism to this *beyul*, a sacred hidden land (Balikci 2008). Accounts of Guru Rinpoche's time in Sikkim reveal that the introduction of Buddhism required the taming of existing supernatural beings, which many interpret as a metaphor for the "taming of the mind, of society, of the environment or even of the country" (Balikci 2008, 88). In the thirteenth century a blood brotherhood treaty was agreed on by three groups, the Tibetan king and the leaders of two indigenous groups, the Lepchas and the Limbus, which assured native populations of the noble and upright intentions of the new Tibetan rulers (Subba 2010). After the establishment of the Namgyal dynasty in the seventeeth century, Buddhism flourished but indigenous groups who were sworn protection under the treaty were ignored and their cultural practices languished. The oral or written histories and mythologies of these groups were completely subsumed under Buddhist Sikkimese nationalism, which centuries later would be subsumed under Indian nationalism. These multiple erasures of histories have led to deep-seated cultural anxieties among several groups, including the Lepchas, that today manifest in a variety of political and cultural demands on the state.

In the early nineteenth century, prized for its strategic geopolitical location, Sikkim was offered protection and minimal external intervention by the British. To balance the pro-Tibetan monarchy, the British started settling Nepalis in the southern and western parts of Sikkim (Hiltz 2003). At present the Nepali community constitute the demographic and political majority. Bhutias, who trace their lineage to the thirteenth-century Tibetan settlers, are also recognized as indigenous to Sikkim and are beneficiaries of affirmative action policies. Most Lepchas in North Sikkim follow a syncretic blend of the Nyingma school of Tibetan Buddhism and Lepcha shamanism with recent years witnessing a steady decline in the latter's influence (Bentley 2007). In 2007, indigenous Lepcha youth from Dzongu spearheaded an antidam campaign that led to the cancellation of three projects planned within the reserve. Although Lepcha youth received tremendous support from transnational activists and scholars within Dzongu, the protests pried on underlying tensions between activist and lay members of the tribe, some of whom were slated to benefit from the projects (McDuie-Ra 2011). My fieldwork in 2014 took place during India's national and Sikkim's state elections, where many of these concerns, including those around growing ecological precarity, took center stage. It is in this charged political climate that the following conversations take place.

Cosmological Indeterminacy

Heightened ecological precarity in mountainous environments is generating a politics of urgency that echoes broader global climate change anxieties. These political framings further political concerns but obscure others that do not carry the same urgency. This section illustrates how concerns raised by lay Lepchas in Dzongu regarding the earthquake signaled tensions between Buddhism and shamanic practices. Loden, my guide and translator,[3] had one of the clearest articulations of this tension. Walking back after an interview with an old *bungthing* [Lepcha shaman] who linked the earthquake to the slow erosion of shamanic practices, Loden expressed concerns around Buddhist ritual practices:

> When there is a death in the family *lamas* [Buddhist monks] come, they do their rituals but none of the family members understand what is being said in the rituals. Once I asked one of them, "What are you saying in your prayers? Maybe you can stay a while after the ritual is over and explain to the family members what was just said." Then some of them got angry at me. They demand a lot of respect and get angry if we don't treat them like royalty.

Loden then wondered out loud how a lama could guide his father's spirit to paradise because he would be performing the rituals in Tibetan but his father only spoke

Figure 1. A Lepcha shaman offering *chi* (millet beer) in a ritual for the safety and protection of a village in Dzongu. (Color figure available online.)

Lepcha. Wouldn't that confuse his father's spirit? In interviews, questions about the earthquake were often explained using the language of sin and pollution. Sin was understood as a spectrum of moral lapses and failures, and most people[4] I interviewed considered hydropower projects sinful for desecrating sacred landscapes. There was less certainty, however, as to whether lack of observance or mistakes made during certain ritual practices could also constitute sin. One of the main sites of this confusion was an irreconcilable difference between Buddhist and shamanic spiritual practices. Shamanism requires animal sacrifice and the ritual spilling of blood, which Buddhism prohibits and considers sinful (Balikci 2008). During fieldwork I was asked to film a shamanic ritual performance that required, among other things, the offering of chickens (Figure 1). Later as we watched the video with chickens being bludgeoned in preparation for the offering, someone remarked in jest, "Gosh, look at all that blood! We must be accumulating a lot of sin. We are such bad Buddhists!" Uncertainties around what constitutes sin in shamanic and Buddhist cosmology were also related to larger anxieties around human culpability in triggering future natural calamities.

In Lepcha cosmology, *Itbu-Rum*, the creator goddess, creates *Matlee Punu*, the earthquake king. She then creates Mt. Kanchendzonga and places him on Matlee Punu's chest to control him (J. Bentley, personal communication, 22 May 2014). Earthquakes and natural disasters in Lepcha cosmology have a stronger correlation to the failure to appease mountain deities like Mt. Kanchendzong through ritual offerings rather than the desecration of sacred landscapes, a more Buddhist cosmology. Ritual offerings to prevent earthquakes were to be performed by *bungthings*, the gatekeeper and mediator between the human and spirit world. The older generation expressed concerns that younger bungthings did not know how to perform rituals of safety and protection. Would performing these rituals incorrectly anger the deities even more? Could the deities be trusted to protect those who performed shamanic rituals or were they capricious, punishing everyone equally? Others, like Kunga, in his early thirties, spoke of bungthings disappearing and their powers diminishing:

> My grandfather was a very strong bungthing. In his *baari* [garden], if people went without asking or picked something, their hands would hurt for days. He was so powerful that if you looked him straight in the eyes, your eyes would also hurt. Even if you looked at the fruit and said, "Oh this is so *daami* [wonderful]!" even then your eyes would start hurting. Nowadays we don't have such powerful bungthings.

Tolia-Kelly (2016) argued that accounts of loss in the Anthropocene seem to privilege ecological loss while less attention is paid to "culturecide . . . the eradication of cultures [which] represent possibilities of alternative ways of living, philosophies, and politics" (2). These narratives of cosmological indeterminacy, which deities to follow, how to correctly perform ritual offerings, disappearing shamans, and diminishing powers point to multiple temporalities and registers of loss and erasure, embedded in geology and social history (Clark 2008). Activist narratives of the earthquake provide an easy resolution to this question: What is causing these earthquakes? Deities angered by dams, of course. Lay narratives, although critical of dams, are less certain of the spiritual cause and effect; instead, their concerns offer a window to examine deeper cultural anxieties brought to the surface by a volatile earth.

On Authenticity and Coping with Loss

> Samten: Maybe the young people will leave the forests because they don't work here anymore. It's not like we can keep shouting at them … it will be bad for them once we die. Slowly people have stopped doing the rituals and there are fewer bungthings. In schools they are teaching Lepcha language but the young ones come and ask me instead of the teachers about the meaning of these words!

Spiritual explanations for the growing intensity and frequency of earthquakes in Sikkim also highlight rapidly shifting relationships between people and the mountainous environments in which they live, work, and worship. Samten, in his eighties, conflated economic, political, and cultural concerns when asked about the earthquake. His response highlights concerns that are prominent among Lepchas around loss of territory, language, and shamanic practices (Bentley 2007). The notion of Lepchas as a vanishing tribe was popularized by colonial authorities, who believed that the tribe was "in great danger of dying out, being driven away from his ancestral glades by the prosaic Nepali and other materialistic Himalayan tribes" (Brown 1917, 4, cited in Kennedy 1991). A Lepcha author, Arthur Foning (1987) from the neighboring Darjeeling district, wrote an influential book, *Lepcha, My Vanishing Tribe*, which further popularized this narrative. The antidam struggle led to a revival of Lepcha cultural practices along with debates over what constitutes authentic indigenous culture. Palden, a twenty-eight-year-old youth, expressed concerns about the erasure of indigenous practices but displayed an openness to the continued practice of both Buddhism and shamanism:

> In our house we practice both Buddhism and Lepcha rituals. If you see in our house, our father has been trained as a lama and he does all the Buddhist rituals but he's never paid much attention to bungthing rituals. Last year he got sick, his one leg was almost paralyzed. He went to the hospital; he went to the *rinpoches* [Buddhist spiritual leaders]. Then we started doing our Lepcha clan ritual and he started getting better. We are nature worshippers, we can't leave that.

Veland and Lynch (2016a) argued that "narratives about who we are, where we live and what we do are epistemic and produce ontologies that to different degrees engage material networks" (7). Loss and recovery of self are a theme in Samten's, Palden's, and Lepcha activist narratives, but it is the activist who is tasked with enrolling these claims in a politics of recognition. Environmental politics in the Himalayan region has thrust together the urgency of climate change politics with a politics of recognition that instead of forging solidarities between groups marginalized by similar processes pits them against each other. In Sikkim, Limbu and Tamang ethnic groups, who were subsumed under the blanket colonial category of Nepal (Subba 2010), were granted tribal status nearly a decade ago. Because they lack recourse to the language of environmental politics, however, their struggle has not received much scholarly attention. How, then, do we speak of loss and erasure given the complex and entangled narratives that are characteristic of the fractured nature of the postcolonial experience? Perhaps a starting point can be found in "narrative recollections and memories about history, tradition and life experience" that offer an alternative to the "crisis-ridden focus of environmental politics" that often "excludes other practices, conceptions, and beliefs of people that don't match this vision" (Cruikshank 2007, 259). As large infrastructural projects and climate change destabilize both ecology and civil society of India's mountainous borderlands (Kohli 2011; Chowdhury and Kipgen 2013; Huber and Joshi 2015), the challenge then is to shift our attention from a politics of urgency and crisis to the *longue durée* of colonial and neocolonial violence and erasure.

The Limits of Politics or What Counts as Politics

In this article I have tried to argue that critical engagements with indigenous environmentalism must be in dialogue with diverse interpretations and registers of loss and erasure beyond those addressed by environmental politics. For marginalized groups inhabiting mountainous environments vulnerable to climate change, along with their livelihoods and homes, their cosmological worlds are also under threat. Tibetan Buddhism and its relationship with Lepcha shamanism has yielded much uncertainty for lay Lepchas. In the wake of the earthquake, lay Lepchas, although critical of hydropower projects, extended their concerns to include this cosmological indeterminacy, which is ultimately tied to anxieties about the tribe's future. The continued practice of Buddhist and shamanic traditions also displays that cosmological worlds have room for flexibility and openness. Bentley (forthcoming), in her research on Lepcha shamanic practices, closely examined an oral account of a magical contest between a Lepcha religious practitioner and a Buddhist protagonist, who some believe is Guru Padmasambhava, who introduced Buddhism to Sikkim. She argued that these accounts cannot be read merely as Buddhism subjugating local beliefs but are to be read as "delineating fields of religious competency and ritual activities." Lepcha shamanic practices have their time and place, as do Buddhist rituals. In the postcolonial context, the politics of marginalized groups is forever enrolled in retrieving former identities that only exist as fragmented knowledge scattered in archives and oral histories. Indigenous environmentalism must respond to the politics of recognition and climate change crisis narrative that requires fixed goals and categories that do not adequately capture nor address the fluidity of place-based attachments.

Rancière (2004) noted that "politics revolves around what is seen and what can be said about it, around who has the ability to see and the talent to speak, around the properties of spaces and the possibilities of time" (13). Climate change anxieties activate a politics of urgency that prioritizes a few interlocutors and their specific registers of space and time. It is not just climate change that is altering the relationship of people to mountainous environments but also powerful narratives employed by scientists, environmentalists, local activists, and academics that influence research and policy initiatives. In the Himalayan context, metanarratives like the Theory of Himalayan Degradation,[5] which blamed environmental crisis in steep-sloped areas on short-sighted local practices like forest clearance and overgrazing, shaped years of environmental research until it was challenged as an oversimplification of ground realities (Ives 2004). Writing about the discursive context of the theory's crisis narrative, Blaikie and Muldavin (2004) argued that the Theory of Himalayan Degradation "drew upon notions of backwardness, technological incompetence, and neo-Malthusianism" (521) that had roots in colonial forestry practices in India and other postcolonial

contexts. In a similar move, Rangan (2000), writing about the iconic "tree-hugging" Chipko movement in the Western Himalayas, analyzed local discontent with environmentalists after the success of the movement, which resulted in new antilogging laws and loss of economic opportunities. She wrote, "Every narrative is an exercise in establishing a particular morality; and narrators often succeed … when their narratives exercise a limited and limiting morality which renders most social and material practices, save their chosen few, as irrelevant, inauthentic, or illegitimate" (41). If ontologies are indeed stories that narrate particular worlds, Blaser (2014) advised caution as we proceed, as "the stories being told cannot be fully grasped without reference to their world-making effects … the corollary of all this is that, indeed, some 'ethnographic subjects' can be wrong … they 'world' worlds we do not want to live in or with" (54). The challenge then is to build solidarities within and among groups being marginalized by neocolonial and neoliberal projects while supporting ontologies that are narrating life-giving, communal worlds.

Acknowledgments

I am grateful to all who offered helpful suggestions, from the abstract to the final draft of this article: Ahsan Kamal, Sara Smith, Pavithra Vasudevan, Benjamin Rubin, and Conor Harrison. I am especially thankful to Mark Fonstad, Jennifer Cassidento, and two anonymous reviewers for such a constructive review experience. The nuanced feedback and suggestions from the two anonymous reviewers pushed me to carefully consider the theoretical and methodological gaps in this article. It would be impossible to name all of the people who helped me during my fieldwork in Sikkim. I forged some wonderful relationships during my fieldwork that nourished me with food for thought and laughter when I most needed it. Special thanks are due to Jenny Bentley and her intimate knowledge of Lepcha religious traditions, as it was our many conversations that pushed me to think through these questions. I am indebted to many people in Gangtok, Mangan, and Dzongu who opened their homes to me and answered my questions with patience, grace, and good humor.

Funding

This article draws on fieldwork research that was made possible from the generous grants I received over the course of my graduate career from the National

Science Foundation, Center for Global Initiatives, the Carolina Asia Center, and the Department of Geography at the University of North Carolina at Chapel Hill.

Notes

1. Although regional geologists and scientists have rejected the possibility of dam-induced seismicity, the location of earthquake epicenters near dam sites has led to a strong association between earthquakes and dams in the Sikkimese public imaginary.
2. Everyone I spoke to has been given pseudonyms and all identifying details have been removed.
3. Loden, in his early twenties, was the local guide for an ecotourism homestay in one of the seven GPUs where I conducted interviews. As I could converse in Nepali but not Lepcha, Loden would help me with Lepcha speakers, mostly among the older generation. His ability to speak Lepcha was quite significant because many young people can no longer speak Lepcha fluently. Of all of the quotations included here, only Samten's interview required translation; I translated all the others from Nepali to English. This footnote was included after an anonymous reviewer rightly pointed out the importance of "paying nuanced attention to what people say and how they say it" to any decolonizing endeavor.
4. I say "most people" here because I did not ask people directly whether hydropower projects were considered sinful given the sensitive nature of the issue.
5. Personal communication with Jenny Bentley, 2014.
6. I would like to attribute this insight on the danger of metanarratives to Barbara Brower's presentation, "What's the Harm in a Name? Anthropocene et al. and the History of Himalayan Scholarship" in the paper session I co-organized with Mitul Baruah, "Interrogating the Anthropocene in the Himalayan Region: Hazards, Infrastructure, and Environmental Justice" at the annual meeting of the American Association of Geographers, March 29–April 2, 2016.

References

Balikci, A. 2008. *Lamas, shamans and ancestors: Village religion in Sikkim.* Boston: Brill's Tibetan Studies Library.

Bentley, J. 2007. Change and cultural revival in a mountain community of Sikkim. *Bulletin of Tibetology* 43 (1–2): 59–79.

———. Forthcoming. Narrations of contest/contested narrations.

Bhagavan, M. B., and A. Feldhaus, eds. 2009. *Claiming power from below: Dalits and the subaltern question of India.* New York: Oxford University Press.

Blaikie, P. M., and J. S. Muldavin. 2004. Upstream, downstream, China, India: The politics of environment in the Himalayan region. *Annals of the Association of American Geographers* 94 (3): 520–48.

Blaser, M. 2014. Ontology and indigeneity: On the political ontology of heterogeneous assemblages. *Cultural Geographies* 21 (1): 49–58.

Brown, P. 1917. *Tours in Sikhim and the Darjeeling District.* Calcutta: W. Newman and Co.

Chopra, R. 2015. Assessment of environmental degradation and impact of hydroelectric projects during the June 2013 disaster in Uttarakhand. Special Committee Report for the Supreme Court of India.

Chowdhury, A. R., and N. Kipgen. 2013. Deluge amidst conflict: Hydropower development and displacement in the north-east region of India. *Progress in Development Studies* 13 (3): 195–208.

Clark, N. 2008. Aboriginal cosmopolitanism. *International Journal of Urban and Regional Research* 32 (3): 737–44.

Cohn, B. S. 1996. *Colonialism and its forms of knowledge: The British in India.* Princeton, NJ: Princeton University Press.

Cruikshank, J. 2007. *Do glaciers listen?: Local knowledge, colonial encounters and social imagination.* Vancouver, BC: University of British Columbia Press.

Danowski, D., E. Viveiros de Castro, and B. Latour. 2014. The thousand names of Gaia: From the Anthropocene to the Age of the Earth. Paper presented at Rio de Janeiro, September15–19. https://thethousandnamesofgaia.files.word press.com/2014/07/position-paper-ingl-para-site.pdf (last accessed 1 June 2016).

Fletcher, R. 2010. When environmental issues collide: Climate change and the shifting political ecology of hydroelectric power. *Peace & Conflict Review* 5 (1): 14–30.

Foning, A. R. 1987. *Lepcha, my vanishing tribe.* New Delhi, India: Sterling.

Gergan, M. D. 2015. Animating the sacred, sentient and spiritual in post-humanist and material geographies. *Geography Compass* 9 (5): 262–75.

Greene, S. 2009. *Customizing indigeneity: Paths to a visionary politics in Peru.* Palo Alto, CA: Stanford University Press.

Haraway, D. 2015. Anthropocene, capitalocene, plantatiocene, chthulucene: Making kin. *Environmental Humanities* 6 (1): 159–65.

Hiltz, J. 2003. Constructing Sikkimese national identity in the 1960s and 1970s. *Bulletin of Tibetology* 32 (2): 69–86.

Huber, A., and D. Joshi. 2015. Hydropower, anti-politics, and the opening of new political spaces in the eastern Himalayas. *World Development* 76:13–25.

Ives, J. 2004. *Himalayan perceptions: Environmental change and the well-being of mountain peoples.* London and New York: Routledge.

Kennedy, D. 1991. Guardians of Edenic sanctuaries: Paharis, Lepchas, and Todas in the British mind. *South Asia: Journal of South Asian Studies* 14 (2): 57–77.

Kohli, K. 2011. Inducing vulnerabilities in a fragile landscape. *Economic and Political Weekly* 46 (51): 19–22.

Last, A. 2015. Fruit of the cyclone: Undoing geopolitics through geopoetics. *Geoforum* 64:56–64.

Lewis, S. L., and M. A. Maslin. 2015. Defining the anthropocene. *Nature* 519 (7542): 171–80.

McDuie-Ra, D. 2011. The dilemmas of pro-development actors: Viewing state-ethnic minority relations and intra-ethnic dynamics through contentious development projects. *Asian Ethnicity* 12 (1): 77–100.

Nandy, A. 1989a. *Intimate enemy.* Oxford, UK: Oxford University Press.

————. 1989b. Shamans, savages and the wilderness: On the audibility of dissent and the future of civilizations. *Alternatives: Global, Local, Political* 16:263–77.

Povinelli, E. A. 2002. *The cunning of recognition: Indigenous alterities and the making of Australian multiculturalism.* Durham, NC: Duke University Press.

Rancière, J. 2004. *The politics of aesthetics: The distribution of the sensible,* trans. G. Rockhill. London: Continuum.

Rangan, H. 2000. *Of myths and movements: Rewriting Chipko into Himalayan history.* London: Verso.

Rappaport, J. 2005. *Intercultural utopias: Public intellectuals, cultural experimentation, and ethnic pluralism in Colombia.* Durham, NC: Duke University Press.

Roelvink, G., and J. K. Gibson-Graham. 2010. An economic ethics for the Anthropocene. *Antipode* 41 (1): 320–46.

Shneiderman, S. 2010. Are the central Himalayas in Zomia? Some scholarly and political considerations across time and space. *Journal of Global History* 5 (2): 289–312.

Sidaway, J. D., C. Y. Woon, and J. M. Jacobs. 2014. Planetary postcolonialism. *Singapore Journal of Tropical Geography* 35 (1): 4–21.

Spivak, G. C. 2003. *Death of a discipline.* New York: Columbia University Press.

Subba, T. B. 2010. Indigenizing the Limbus: Trajectory of a nation divided into two nation states. In *Indigeneity in India,* ed. B. G. Karlsson and T. B. Subba, 143–58. London: Kegan Paul.

Sundberg, J. 2014. Decolonizing posthumanist geographies. *Cultural Geographies* 21 (1): 33–47.

Tolia-Kelly, D. P. 2016. Anthropocenic culturecide: An epitaph. *Social and Cultural Geography* 17 (6): 786–92.

Veland, S., and A. H. Lynch. 2016a. Arctic ice edge narratives: Scale, discourse and ontological security. *Area* Advance online publication. doi:10.1111/area.12270.

————. 2016b. Scaling the Anthropocene: How the stories we tell matter. *Geoforum* 72:1–5.

Yeh, E. T. 2015. How can experience of local residents be "knowledge"? Challenges in interdisciplinary climate change research. *Area* 48 (1): 34–40.

The Sacred Mountain Shiveet Khairkhan (Bayan Ölgiy aimag, Mongolia) and the Centering of Cultural Indicators in the Age of Nomadic Pastoralism

Esther Jacobson-Tepfer and James E. Meacham

Located in the upper valley of Tsagaan Gol, in northwestern Mongolia's Altai Mountains, the sacred mountain Shiveet Khairkhan is surrounded by archaeological monuments extending in time from the Bronze Age (early third millennium BCE) through the Turkic Period (sixth to ninth centuries CE). The character of the high valley it centers and the extended physical context including rivers and glaciated mountains call to mind a sacred diagram involving a mountainous landscape, directionality, and color symbolism. Such general associations with Buddhist concepts would not be the reason Shiveet Khairkhan is considered sacred, however. The wealth of archaeology around the mountain's base and lining the Tsagaan Gol river valley indicates that this status might go back for several thousand years, to a period much earlier than Buddhism. The material presented here derives from two decades of original archaeological survey and documentation and draws on the approaches of several different disciplines. By considering this topic in terms of integrated approaches, it is possible to suggest the complexity of Shiveet Khairkhan within its larger cultural and geographical context and to explore the ways in which this mountain might have become designated as sacred.

Shiveet Khairkhan 圣山座落于蒙古阿尔泰山西北部查干河的上游河谷，并被从青铜时代（西元前三千年初期）延续至突厥时期（西元第六至第九世纪）的考古历史遗迹所包围。其所集中的高河谷特徵和延伸的包含河流和冰原山岳的物理脉络，使人联想起包含山岳地景、方向性与颜色象徵主义的神圣图示。但此般与佛教概念相关的广泛连结，却不是 Shiveet Khairkhan 被认为神圣的原因。围绕着山脚和沿着查干河谷的丰沛考古学，显示此般地位或许可回溯至甚早于佛教的数千年前时期。本文所展现的材料，来自二十年的原始考古调查与纪录，并运用若干不同领域的方法。透过以整合性的方法考量此一主题，便可能主张 Shiveet Khairkhan 在其更为广泛的文化及地理脉络中的复杂性，并探讨这座山或能够成为被指认的圣地之方式。关键词：阿尔泰山，青铜时代，圣山，屹立之石，突厥。

Localizada en el valle superior del Tsagaan Gol, en los Montes Altai del noroeste de Mongolia, la montaña sagrada de Shiveet Khairkhan se halla rodeada de monumentos arqueológicos que se prolongan en el tiempo desde la Edad de Bronce (a principios del tercer milenio de la AEC), a través del Período Turco (siglos VI al IX de la EC). Aquella montaña destaca el carácter del alto valle, y el contexto físico extendido, incluyendo ríos y montañas glaciadas, trae a la mente un diagrama sagrado que involucra un paisaje montañoso, direccionalidad y simbolismo de color. Tales asociaciones generales con los conceptos budistas no serían, sin embargo, razón para que las Shiveet Khairkhan sean consideradas como sagradas. La riqueza arqueológica que abunda alrededor de la base de la montaña y a lo largo del valle del río Tsagaan Gol indica que aquel estatus podría remontarse por varios miles de años, a un período muy anterior al del budismo. El material que se presenta aquí proviene de dos décadas de estudios arqueológicos originales y de documentación, y se apoya en los enfoques de varias disciplinas diferentes. Al considerar este tópico en términos de enfoques integrados, es posible sugerir la complejidad de la Shiveet Khairkhan dentro de su más amplio contexto cultural y geográfico, y explorar los modos como esta montaña podría haber llegado a designarse como sagrada.

The sacralization of mountains occurs in many cultures, and for very different reasons. From an ancient period, Chinese sacred mountains were associated with the cult of the emperor; the Icelandic Vikings understood the sacred mountain to be the abode of the dead, and in Navajo traditions mountains represent the edge of the significant world (Bernbaum 1990). Within ancient Egypt, the sacred mountain seems to have been the place of creation, and within Hurrian and Hittite traditions the mountain was the abode of the gods (Clifford 1972). Among the ancient Hebrews, Mount Zion was the locus of the Temple and of the four rivers of life (Clifford 1972). The understanding that mountains offer access to a higher realm, that they represent a perfected world, or that they are the dwelling places of the gods is most vividly rooted in religious traditions that emerged from India: Vedism and its successors, Hinduism and Buddhism. Buddhism first entered Mongolia in the thirteenth century, but in the sixteenth century the Tibetan form was adopted as the state religion by Altan Khan. In its doctrinal and artistic expression, this tantric tradition carried the fused concepts of the sacred mountain and of the mandala as a diagram of the perfected universe.[1] Most significantly, from the beginning of Buddhism's history within Mongolia, sacred mountains were associated with the preservation of natural habitat—places for the preservation of pure water, uncut forests, properly used pastures, and the protection of wild animals. With time, the mountains designated as sacred by the state came to be also understood as the abodes of the gods or even as the body of the god (Wallace 2015).

It is certain, however, that the veneration of mountains in Mongolia had long been rooted in more ancient shamanic traditions wherein the mountain refers to the World Tree, the Cosmic Mountain, and the vertical axis of a shaman's ascent and descent into other realms.[2] Indeed, there are several structural forms found across Mongolia that indicate the great antiquity of an understanding that the significant universe is centered by a vertical axis and organized by reference to the four cardinal directions. These forms—all of far greater antiquity than Buddhism—include the stone mounds known as *ovoo* found ubiquitously throughout mountainous Mongolia and associated with mountain worship; the Bronze Age memorial structures known as *khirigsuur* in the form of centered and symmetrical stone diagrams; and the standing stone as a cosmic axis within the four quarters (Jacobson-Tepfer 2015).

Mongolia's most famous sacred mountains are those associated with state worship and rituals (Wallace 2015). Although there are no such designated peaks within northwest Bayan Ölgiy aimag (Figure 1), there are several mountains that have over time been understood as sacred but for unclear reasons. Given the forced disruption in traditional beliefs during Mongolia's period of socialism and the complete loss of ancient texts for this region, we cannot know either when or why these mountains became sacred. In some cases the basis for such status can be inferred from its physical location or aspect. For example, Öndör Khairkhan Uul (Lofty Sacred Mountain; 3,914 m) is one of the tallest and most impressive mountains on the western border with China. Tsengel' Khairkhan Uul (Sacred Mountain of Great Joy; 3,943 m) is a massive mountain, its glaciated and rounded summit visible from many directions. Towering over the valley of Sogoo Gol is Khuren Khairkhan Uul (Dark Brown Sacred Mountain; 3,154 m). This peak is distinguished by its paradoxical beauty: On the one hand, it seems to be utterly bare of any significant vegetation; on the other hand, depending on the time of day and season, the mountain changes hue from brown to rose-orange to green and purple. On the far west where the glaciated Tavan Bogd ridge divides Mongolia from Russia and China rises Tsookhor Khairkhan Uul (3,786 m), and a little to the southeast can be found Tsagaan Khairkhan Uul (3,828 m). Why among all of the high peaks in this area these two mountains should have been given a sacred designation is not known, nor is it known when that designation came about. That uncertainty attends most of Mongolia's sacred peaks, however, most recognizable by the designation *khairkhan*. One thing is certain: Although probably predating the advent of Buddhism in Mongolia, the term itself is redolent of Buddhist values—of something or some being that is merciful, gracious, and holy.

Shiveet Khairkhan Uul (Grassy Sacred Mountain; 3,349 m; Figure 2) rises at the high western end of the valley of the white river, Tsagaan Gol. On the west the mountain is framed by the glaciated peaks of Tavan Bogd and its adjacent ridge (Figure 3), and to the east the Tsagaan Gol valley stretches in a deep, wide, and stony trench to the river's confluence with Khovd Gol (Figure 4). From a satellite image, Shiveet Khairkhan has the appearance of a large pyramid set at the end of a major valley and framed by two streams that converge at the lowest end of the mountain's eastern flank (Figure 5). The spatial relationship of the mountain to its extended landscape is certainly

Sacred Mountains: Bayan Ölgiy aimag

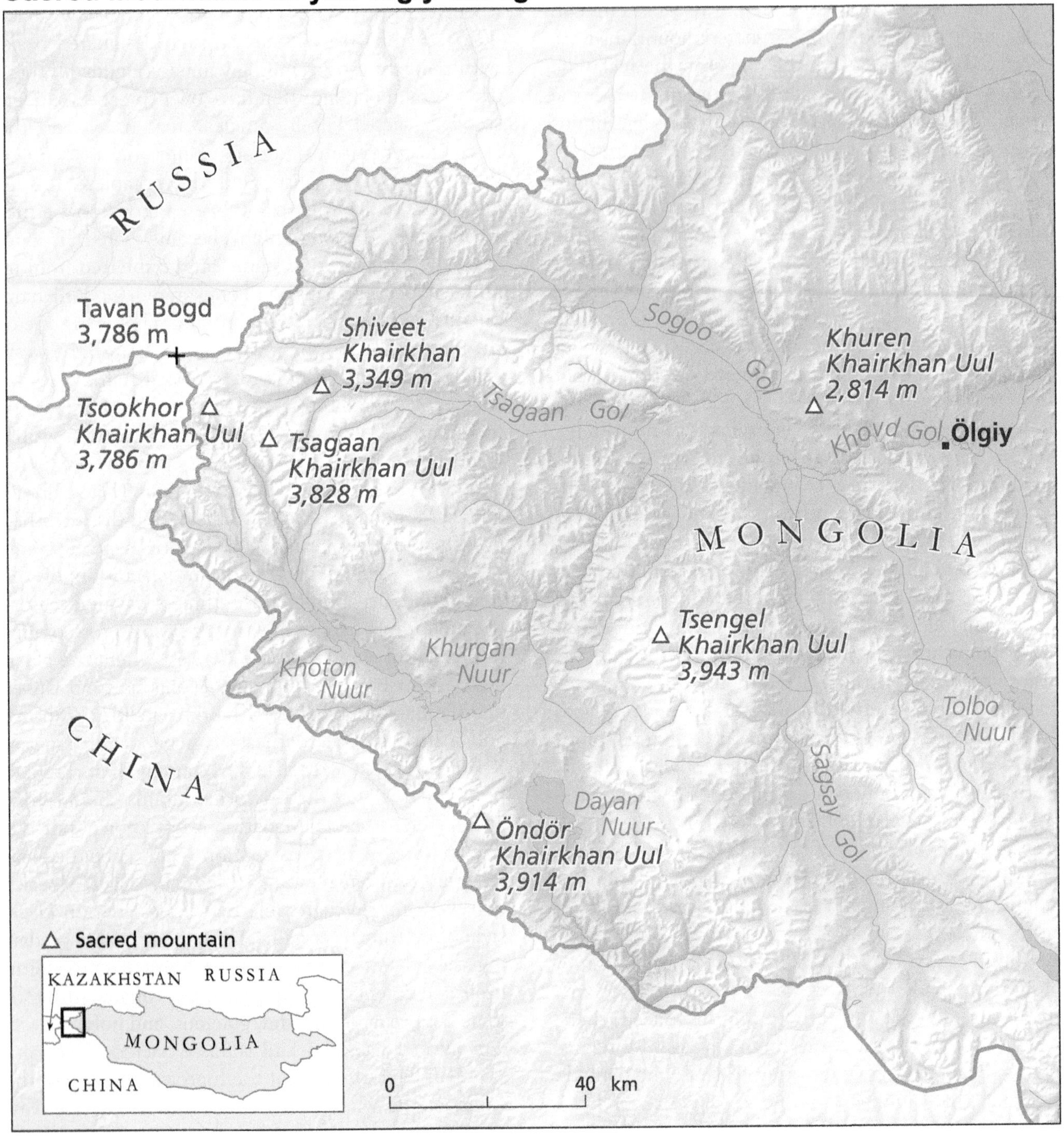

Figure 1. Bayan Ölgiy Aimag: Study region with location of select sacred mountains. (Color figure available online.)

significant, balanced between Tavan Bogd on the west and Tsagaan Gol on the east. Shiveet Khairkhan Uul is neither the highest nor the most impressive of Bayan Ölgiy's sacred peaks, but by virtue of its dark and stony character, its topography, and extended landscape, the mountain's physical appearance is arresting. Local herders today associate certain values with the mountain that might point back several millennia in time. The mountain's summit is highly valued for winter pasture for yaks and horses: despite the intense cold of that elevation, the winds coming down from Tavan Bogd sweep the mountain clean of snow, revealing the thick grass for which the mountain is named. The springs found around the mountain's base

264

Figure 2. Shiveet Khairkhan Uul seen from the east. *Source:* Photo by G. Tepfer. (Color figure available online.)

Figure 4. The flow of Tsagaan Gol eastwards, seen from a terrace on Shiveet Khairkhan's east slope. Moraines in the center of the photograph hide the confluence of the two streams, Tsagaan Salaa and Khar Salaa. *Source:* Photo by G. Tepfer.

are considered to have medicinal value; whether or not that is true, the water is indeed unusually clear and sweet.

Seen from the east, Shiveet Khairkhan has the shape of a large triangular boat filling the valley with its black mass. The mountain's sides are formidable: On the south, east, and north, the slopes are too steep and unstable for human access. Indeed, the rumble of rockslides is regularly heard from the mountain's steep slopes. On the south and east sides, these slides end against large boulders, whereas on the north they stabilize around islands of small larch. The only feasible approach to the top is from the west, but local herders insist that climbing the mountain is prohibited because of its sacred nature, and hunting the large herd of ibex on the summit is also not allowed. The summit itself takes the form of a long, grassy incline to the east, but in many places the broad crown is broken

by a startling array of jagged rock shards poking crazily into the sky. From this promontory opens a 360° view across the mountains of Russian Altai, northern China, and Bayan Ölgiy.

The spatial order at the center of which Shiveet Khairkhan rises is affirmed by the two rivers flowing around it. On the north the mountain is flanked by the Tsagaan Salaa (White or Milky Stream), so named because of the deep milky coloration derived from its source in the glaciers of Tavan Bogd. On the south the mountain is bracketed by the Khar Salaa (the Black or Clear Stream). These two streams join together directly in front of Shiveet Khairkhan's eastern face, there forming the Tsagaan Gol. That river, in turn, flows east to join Khovd Gol, the largest and most important river in western Mongolia. Thus, Shiveet Khairkhan could be conceived as a pivot centering high peaks on the west (Tavan Bogd) and a major river on the east; a white river on the north and a black river on the south. The resulting physical order recalls the cosmic order embedded in Buddhist mandala and suggests a primary basis for the mountain's sacred status.[3]

A large ovoo set on the mountain's east side reaffirms the transformation of the mountain's physical context into a diagram of sacred space (Figure 6). Around the central pile of stones festooned with branches and the cloth offerings of pilgrims extend neat rows of stones in the four cardinal directions. The whole becomes, in effect, a diagram echoing and distilling the order of the mountain in its landscape. If the ovoo appears to reflect the Buddhist concept of sacred space embodied, for example, in the mountainous form of a *stupa* it nonetheless has roots in a deeper

Figure 3. Upper Tsagaan Salaa (tributary to the Tsagaan Gol) and Tavan Bogd, looking west from the west flank of Shiveet Khairkhan Uul. *Source:* Photo by G. Tepfer.

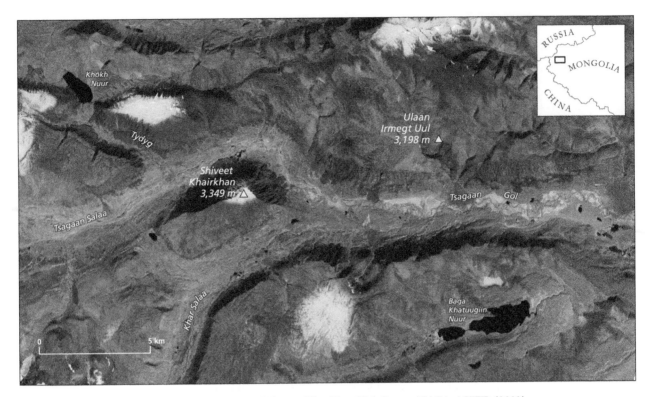

Figure 5. Satellite image of the Upper Tsagaan Gol and Shiveet Khairkhan Uul. *Source:* NASA ASTER (2003).

impulse: at the center of the central pile can be glimpsed a large stone of the kind erected in the late Bronze Age (second–early first millennia BCE).

There are no written records of the existence of Shiveet Khairkhan earlier than the twentieth century, and because the population of this high mountain area has changed constantly over the last few thousand years, there is no cultural memory except as it is carried through the stone monuments. Their sheer number and spatial organization suggest, however, that the

perception of a cosmic order reflected in the mountain and its topography, the notion of what is sacred, and the impulse to sacralize that place might have roots going back to the Bronze Age—well before Buddhism was adopted in Mongolia in the sixteenth century.

It is certain that during the Bronze Age the Tsagaan Gol valley and its adjoining slopes were greener and even supported some forest (Jacobson-Tepfer, Meacham, and Tepfer 2010). Rock art in the upper valley indicates the earlier existence here of elk, moose, bear, and even reindeer—animals that cannot be supported by the present environment. It seems that by the end of the Bronze Age about 3,000 years ago, the lower valley and all the surrounding uplands had developed their dry and harsh character. In the present—and probably for the last two millennia—the lower third of the Tsagaan Gol valley is a rocky plain covered with thin grass through which the white water of the river meanders. Adjoining slopes are high and uninviting and the valley itself is choked by long tongues of glacial debris and disorderly lateral moraines (Figure 7). About two thirds of the way up the valley, at the *bag* (administrative center) known as Tsagast Nuur, there are more frequent signs of good pasture, particularly where streams come down from adjacent slopes. This point is marked by an increasing number of mounds and altars; this is also the point at which Shiveet Khairkhan heaves clearly into full view.

Figure 6. Shiveet Khairkhan and the large oboo at the base of its eastern flank. *Source:* Photo by G. Tepfer.

Figure 7. Looking south over moraines within the Tsagaan Gol valley. On the left bank several khirigsuur can be seen and on the right bank the entrance of Khatugiin Gol into the Tsagaan Gol valley can be glimpsed. *Source:* Photo by G. Tepfer.

From that point on and higher, where the two tributaries of the Tsagaan Gol diverge, begin to appear rich pastures with abundant water. The valley of the Tsagaan Salaa leading west to Tavan Bogd is narrow, but in its upper stretch the adjoining slopes give onto high winter pasture. The valley of the Khar Salaa is a mosaic of streams, marshy areas, and low rocky promontories extending up to the stream's source at the base of Tsagaan Khairkhan Uul and Rashaany Ikh Uul (3,668 m; Figure 8). As rich as the upper valley appears now, what we see is certainly far drier than in the deep past. At this point, only shrubs, willows, and stunted larch are scattered on north-facing slopes; the few Siberian pine that we saw along the Khar Salaa in the early years of our work have entirely disappeared. Nonetheless, it is possible to imagine that at least

Figure 8. Valley of the Khar Salaa looking southwest to Tsagaan Khairkhan Uul. *Source:* Photo by G. Tepfer.

some of the sacred aspect of Shiveet Khairkhan relates to the abundance of fine pasture in these uplands—within a region now generally harsh and stony.

There are, of course, many other valleys that drew ancient herders to the high mountains in search of good pasture. There is no other valley, however, so extensively marked by ancient archaeology. Monuments appear at the lowest end of the valley, on terraces on the north side of the river (Figure 9A). These overlook the confluence of Tsagaan Gol and the Khovd and the distant summit of Tsengel' Khairkhan Uul. Thereafter mounds and altars appear sporadically on the north side of the valley until a point opposite the confluence of Khatugiin Gol and Tsagaan Gol where mounds, khirigsuur, and altars of many kinds become much more densely arrayed. This concentration coincides with the point at which the last tongue of a massive moraine settles into the stony flat of the valley (Figure 7); this is also where one begins to see Shiveet Khairkhan in the distance. From here on to the west the tempo at which mounds appear along the sides of the valley markedly increases until at the settlement of Tsagast Nuur their varied array becomes quite striking: large khirigsuur, four-cornered mounds, mounds edged by vertical stones, virtual dwellings, and a great variety of altars. From this point on to the west the pyramidal shape of Shiveet Khairkhan dominates the view to the west.

The upper valley's dense array of monuments includes surface structures and standing stones dating from the Bronze Age through the Turkic Period. The most numerous ancient monuments, however, are the petroglyphic images (rock art) pecked and engraved into the surfaces of bedrock and boulders. In effect the surface monuments and rock art of the upper Tsagaan Gol valley form a vast museum of ancient culture extending over a period of several thousand years. The density of these monuments suggests that although pasture must have been the primary reason ancient herders journeyed up to this valley, the mountain's sacred character might have also been a significant reason.

The surface structures clustered around Shiveet Khairkhan and lining the edges of Tsagaan Gol are not unique to this valley. They are all of types known elsewhere across Mongolia and the Altai-Sayan uplift (Jacobson-Tepfer, Meacham, and Tepfer 2010). It is the quantity of monuments here in the upper Tsagaan Gol that is unusual and the opportunity they offer to understand the relationship of one temporal and cultural layer to another.[4] The layers of monuments offer insight into the way in which ancient cultures thought

A. Monument Concentrations: Tsagaan Gol Valley

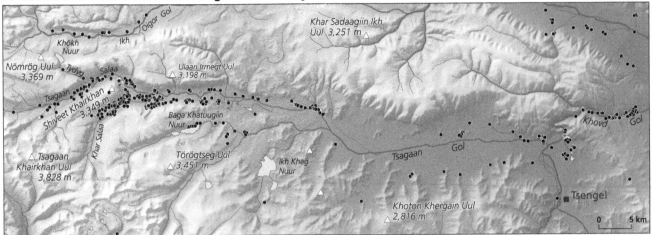

B. Surface Structures

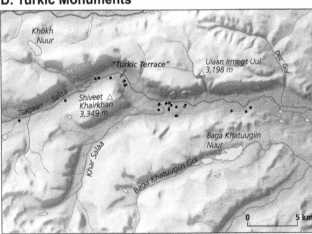

C. Standing Stones

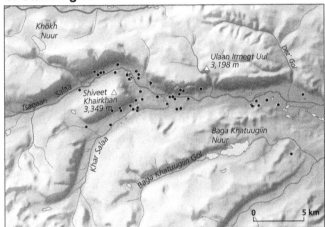

D. Turkic Monuments

Context Map

Figure 9. (A) Distribution of monument concentrations in the Tsagaan Gol valley. (B) Bronze–Early Iron Age surface structures, Upper Tsagaan Gol. (C) Bronze Age–Early Iron Age standing stones, Upper Tsagaan Gol. (D) Turkic stone figures, enclosures, and altars, Upper Tsagaan Gol.

historically, seeing the traces of the past as legitimizing or sacralizing their own structures and sense of place. At the same time, there are general principles in the distribution of monument types suggesting that in each period location reflected varied understandings of the significance of directionality and orientation (Jacobson-Tepfer, Meacham, and Tepfer 2010; Jacobson-Tepfer 2015).

268

Bronze–Early Iron Age Surface Structures

As is the case throughout much of the Altai region, the oldest monuments appear to be simple mounds arranged on high terraces overlooking a river. It is probable that these mounds covered bodies laid directly on the surface of the ground.[5] They are often accompanied by circular structures that might have been solely of ritual significance. Another structure known as *khirigsuur* takes the form of a central mound and an outer wall, round or square, joined to the center by stone rays. In some cases the central mound takes the form of one or more boulders (e.g., Mongolian Altai Image Collection [MAIC] 2013, BLKH_00003_TG). Khirigsuur are usually (but not always) located in a broad area rather than on a narrow terrace. They are almost certainly burials dating to the Bronze Age, but no khirigsuur excavated in this part of Mongolia has included any furnishings or even bodies by which the structure could be precisely dated.

Within the array of burial monuments are mounds built over burial chambers that can be dated to the early Iron Age (MAIC 2013, BRMD_00003_TG, BRMD_00005_TG). Even though no such structures have been excavated in this valley, the monument type has been amply investigated in other parts of the Mongolian Altai and extensively in the Russian Altai (Kubarev 1987; Polos'mak 2001).

Figure 9B indicates that the concentration of Bronze Age surface structures is on terraces on the north and south sides of Shiveet Khairkhan and on the long moraine stretching to the east of the mountain. Within that array, most would appear to be oriented to either Shiveet Khairkhan, the river, or both. The only exceptions in this group of monuments are Iron Age burials; these are always arranged according to a north–south axis, regardless of where they appear in the valley.

Bronze Age–Early Iron Age Standing Stones

In this region of the Altai, standing stones take one of several forms. Stones of massive size, often of human height or more, are prominent markers within the landscape. As the map here indicates, many are found on high terraces where they might be invisible from the valley floor. In only one area, the northeast edge of Shiveet Khairkhan, are they found on a broad terrace abutting the river. Stones can be erected singly or in small groups arranged from north to south. The stones are usually accompanied by the remnants of square altars within which they were originally erected and by small circular altars on their east. Thus, the stones create a spatial diagram fixing the axes north to south and west to east around the central vertical axis. Here in the upper valley as in many other places in the Altai Mountains, massive stones are often found in high passes where they seem oriented to the flow of the river eastward (Figure 10). While affirming that primary orientation, others appear to frame Shiveet Khairkhan (MAIC 2013, TG0455).

Deer stones are a subclass of massive standing stones. There are only three identified to date in this valley, and with only one exception (MAIC 2013, DRST_00001_TG) they are all of a fairly simple type. On the basis of studies of deer stones across Mongolia, we know that those here can be dated to the late Bronze Age or, more probably, to the early Iron Age (Savinov 1994; Jacobson-Tepfer 2001).

Turkic Monuments: Stone Figures, Enclosures, and Altars

Standing stones carved with a distinctly human aspect were occasionally fashioned in the late Bronze Age, but only in the Turkic Period did anthropomorphic image stones become common. There are several in the upper Tsagaan Gol valley, and at least two are of fine quality. These image stones are always located on terraces in conjunction with a particular kind of enclosure, square or rectangular in form. As the map here indicates (Figure 9D), there is a concentration of these monuments at the base of the northeast flank of Shiveet Khairkhan, on a terrace we have dubbed the Turkic Terrace. Other Turkic monuments seem to

Figure 10. Standing stone overlooking Tsagaan Salaa. The north flank of Shiveet Khairkhan is visible on the middle right. *Source:* Photo by G. Tepfer.

follow the downstream flow of Tsagaan Gol, just after the confluence of the river's two tributaries. Yet other enclosures and altars can be found along the length of the Tsagaan Gol on its left bank. Despite the relationship of the Turkic Terrace to Shiveet Khairkhan, the easterly direction was the determining factor in the location and orientation of these and all Turkic monuments (Figure 11). Nonetheless, the unusual number of Turkic monuments around the mountain indicates that Turkic herders regularly returned to this high valley. They left not only monuments commemorating important individuals in a lineage but also many self-images in the valley's rock art.

This article does not allow us to examine fully the logic of the distribution of surface monuments around Shiveet Khairkhan or to elaborate on such issues as proximity and view shed. A few comments regarding specific sites might here suggest the wealth of issues involved in an examination of the whole complex.

The site identified in our survey, TS 5, is an uneven terrace at the confluence of the Tygyd and the Tsagaan Salaa. Distributed over the terrace are several mounds and khirigsuur, all muted as if periodically worn by floods over several millennia. The location certainly derives significance from its proximity to a confluence of streams and probably from the view shed to the base of Shiveet Khairkhan.

The site we have identified in our field notes as TS 10 is one of three terraces on the north side of Shiveet Khairkhan. All three terraces are dominated by Turkic monuments and include the Turkic Terrace referred to earlier, but TS 10 is the most densely used and includes not simply altars and image stones from the Turkic Period but also a number of Bronze Age standing stones and surface structures. One might argue

that there is no special reason for the intensity with which this terrace was used other than the fact that it is flat and expansive. Its location directly at the foot of Shiveet Khairkhan, however, in immediate proximity to one of the best springs emerging from its base and with a direct view down the river to the east, would indicate that the sacred mountain, the White River, and the direction east endowed this terrace with particular significance beginning in the Bronze Age.

Another site identified as TG 1 is located on the high moraine just east of the lower Khar Salaa and its confluence with Tsagaan Salaa. This plain is unusually dense with surface structures including large khirigsuur, altars, and a number of Turkic enclosures, and on the south edge is found a major concentration of rock art. The highest point on this moraine is marked by a curious circular structure—more likely an altar than a burial mound (MAIC 2013, RNKH_00007_TG). White stones mark the south side of the circle, black stones mark the north side, and the circular center is a combination of stones. Certainly this peculiar organization must be directly related to the larger topographic association of color and directionality: The Black River (Khar Salaa) is on the south and the White River (Tsagaan Salaa) is on the north. In other words, the circular stone setting directly reverses, or balances, the color associations of the rivers on either side of Shiveet Khairkhan. At the same time, the bicolored disk lies directly east of the mountain and also in the view shed of the large ovoo on the mountain's east flank (Figure 6). One can only assume that these associations of color and directionality were intentional and were related to Shiveet Khairkhan.

Rock Art

The distribution of rock art in the upper Tsagaan Gol valley is indicated in a series of maps shown in Figure 12. A comparison of these maps indicates that petroglyphic materials from the Bronze Age are particularly numerous, followed by those from the Early Iron Age. Least numerous of all are those referred to as pre-Bronze Age. Although it is not altogether clear why the petroglyphs of each period are distributed as they are, certain patterns suggest a direct relationship between petroglyphic concentrations, period, and possible seasonal settlement areas. Whatever the logic here, it is clear that the ancient herders who journeyed to this upper valley stayed long enough within its relatively confined borders to produce a phenomenal amount of imagery. If we take into consideration that

Figure 11. Four Turkic period image stones on a terrace on the northeast flank of Shiveet Khairkhan. Here seen from the west, these stones face east down the Tsagaan Gol valley. Raised banks in the midground hide the river. *Source:* Photo by G. Tepfer.

A. Rock Art Concentrations

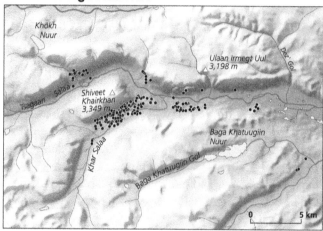

B. Pre-Bronze Age Rock Art

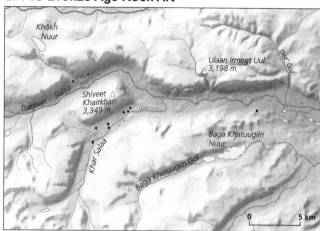

C. Bronze Age Rock Art

D. Early Iron Age Rock Art

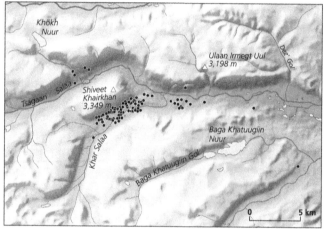

E. Turkic Period Rock Art

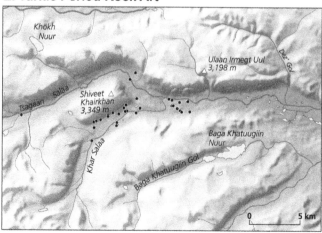

Context Map

Figure 12. (A) Rock art concentrations, Upper Tsagaan Gol. (B) Distribution of rock art in the Pre-Bronze Age. (C) Distribution of rock art in the Bronze Age. (D) Distribution of rock art in the Early Iron Age. (E) Distribution of rock art in the Turkic Period.

within any year the auspicious months (climatologically speaking) for producing petroglyphs were very few—at best June through September—then the number of decorated surfaces is even more impressive.

Conclusions

Without written texts to check our assumptions it is futile to insist on why a certain place or a particular mountain has long been considered sacred. This is certainly the case with Shiveet Khairkhan Uul, for which there is no mention in any known historical sources. Then, of course, there are no relevant texts or maps for this region except those that might relate to administrative concerns of the Chinese Manchu Dynasty when it controlled much of northern Asia (Haltod 1966). One is forced to use contingent data to determine that any particular place was regarded in the past with the sacred status it appears to hold in the present.

All evidence indicates that Shiveet Khairkhan Uul was held in considerable respect since herders first began to go into the upper Tsagaan Gol valley in the Bronze Age. There is no other way of accounting for the major number of surface monuments in that valley. Their numbers indicate the periodic presence of relatively large groups of people and the organization of burial mounds and ritual structures from the Bronze and Iron Ages and from the Turkic Period suggest that the same lineages might have returned to the upper valley over several generations.[6] One can surmise that fine pasture drew herders and their flocks to this valley as did the presence of excellent streams and springs. The valley must also have been considered a good hunting ground: the petroglyphic record indicates that in an earlier period fragments of middle Holocene forests still supported wild game such as elk and bear, and on the rocky ridges around the valley could be found abundant ibex and argali. There is little to support the idea that this upper valley served as a defensive position: rock art representations of combat are few and far between, and the topography of the valley offers no easy way out—no escape route.

A few other high valleys in Bayan Ölgiy offered excellent pasture, water, and the presence of game, and several of these have a significant number of monuments dating back to the Bronze Age, even if the numbers are not comparable to what we have in the Tsagaan Gol valley. Clearly there was some other aspect that singled out this one valley, in particular. We would argue that the distinguishing factor was the topographic character of Shiveet Khairkhan: the way in which it rises in the center of coordinated directional axes and the way in which the rivers—Khar Salaa and Tsagaan Salaa, the Black and White streams—rise in the high peaks to the west and join to form the Tsagaan Gol's flow to the east. The pyramidal shape of the mountain adds to the manner in which it and its surrounding topography become a metaphor for a sacred diagram. We are familiar with these formulations from the mandala of later Buddhism, but we are also aware that even the ancient Buddhist diagrams were rooted in much older understandings regarding cosmic order, the significance of the central vertical axis and the four cardinal directions. These principles are well attested by Bronze Age khirigsuur and standing stones of the Altai and Sayan regions. It is clear that long before Buddhism came to Mongolia, within much of Asia there was a greater understanding of ordered space as a sacred diagram. The mountain Shiveet Khairkhan Uul and its surrounding topography must have been understood from early in the Bronze Age as the center of a sacred realm and as the reiteration on earth of cosmic order.

Acknowledgments

The content of this article emerged from many years of fieldwork in the Tsagaan Gol valley in conjunction with Russian, Mongolian, Kazakh, and Uriankhai colleagues. Among these individuals, we are particularly indebted to our late friend, Mantai. The visualization of this high valley would not be possible without the work of Gary Tepfer, project photographer. We also want to express appreciation to Alethea Steingisser, Cartographic Production Manager, and Dylan Molnar, Student Cartographer, both in the InfoGraphics Lab, Department of Geography, University of Oregon, for their work in the preparation of maps used in this article.

Notes

1. See, for example, the Tangka entitled "The Kingdom of Shambhala," nineteenth century, with the mandala-like organization of mountains centered by Kalachakra (Rhie and Thurman 1991, 157–58). A mandala "refers to a protected environment, a realm where the mind is free to manifest its highest imagination" (Rhie and Thurman 1991, 33).
2. See the discussion of the shaman's journey and sources for that material in Jacobson-Tepfer (2015, 341–48).
3. In 2010 this valley was entered into UNESCO's listing of World Heritage sites as part of the serial nomination, "Petroglyphic Complexes of the Mongolian Altai."
4. Throughout this discussion we do not speak of precise dates for surface monuments here ascribed to the Bronze Age because there are none yet ascertained. In general, the amount of scientific excavation in this region of Mongolia has been limited and little has been published. We are on firmer ground with mounds and burial mounds associated with the Late Bronze Age and the Early Iron Age. Extensive research on Turkic Period monuments also allows us to be fairly certain of stone figures, enclosures, and altars dating to that period.

5. Because these monuments have rarely been excavated, we have no firm understanding of their date. On the basis of a number of contingent factors, however, it is certain that they are among the earliest surface monuments in this part of the Altai. See, for example, MAIC (2013, UNMD_00015_KV).
6. Burial mounds from these periods are arranged in groups, as along a terrace (Bronze Age), or in lines running north to south (Iron Age, Turkic Period). It is generally understood that these discreet groups probably refer to families or, more likely, lineages.

References

Bernbaum, E. 1990. *Sacred mountains of the world.* San Francisco: Sierra Club Books.

Clifford, R. J. 1972. *The cosmic mountain in Canaan and the Old Testament.* Cambridge, MA: Harvard University Press.

Haltod, M. 1966. *Mongolischen Ortsnamen.* Wiesbaden, Germany: Franz Steiner Verlag.

Jacobson-Tepfer, E. 2001. Cultural riddles: Stylized deer and deer stones of the Mongolian Altai. *Bulletin of the Asia Institute* 15:31–56.

———. 2015. *The hunter, the stag, and the mother of animals.* Oxford, UK: Oxford University Press.

Jacobson-Tepfer, E., J. Meacham, and G. Tepfer. 2010. *Archaeology and landscape in the Mongolian Altai: An atlas.* Redlands, CA: ESRI.

Kubarev, V. D. 1987. *Kurgany Ulandryka* [The kurgans of Ulandryk]. Novosibirsk, Russia: Nauka.

Mongolian Altai Image Collection (MAIC). 2013. Eugene: University of Oregon. http://oregondigital.org/sets/maic

Polos'mak, N. V. 2001. *Vsadniki Ukoka* [The horsemen of the Ukok Plateau]. Novosibirsk, Russia: INFOLIO.

Rhie, M. M., and R. A. F. Thurman. 1991. *Wisdom and compassion: The sacred art of Tibet.* New York: Harry M. Abrams.

Savinov, D. G. 1994. *Olennyi kamni v kul'ture kochevnikov Evrazii* [Deer stones in the culture of the Eurasian nomads]. St. Petersburg, Russia: St. Petersburg University Press.

Wallace, V. A. 2015. Buddhist sacred mountains, auspicious landscapes, and their agency. In *Buddhism in Mongolian history, culture, and society,* ed. V. A. Wallace, 221–40. Oxford, UK: Oxford University Press.

Mountain Agriculture for Global Markets: The Case of Greenhouse Floriculture in Ecuador

Gregory Knapp

Mountain agriculture has been conceptualized in terms of altitudinal zones, verticality, and agroecosystems, but an alternative framework is that of adaptive dynamics, conceptualizing farming in terms of choice between options based on optimizing returns in different frameworks of rational decision making in different production zones. In this framework, production zones are not defined solely in terms of altitude but also in terms of soil, slope, and access to irrigation. A recent option in the irrigated production zone has been greenhouse floriculture, which has become one of the most globally competitive agricultural exports in equatorial mountains. In Ecuador, greenhouse floriculture expanded in the 1990s partly in response to favorable trade agreements but also due to diffusion of technologies from multiple sources and local entrepreneurship. Interviews with various actors and fieldwork provide details on greenhouse adaptive strategies and suggest that this agroindustrial activity has proven unusually resilient to changes in global trade patterns and changes in climate. It has provided an option for employment that has stemmed outmigration and encouraged some immigration of labor. At the same time, there are concerns regarding impacts on water resources and regarding pesticide impacts. Excessively static or ecosystemicist conceptions of mountain environments and agricultural strategies fail to anticipate the full range of possibilities for development in the diverse production zones of high-altitude regions. These possibilities also help to contest assertions about the inevitable decline of mountain agriculture in the face of modernization and globalization.

高山农业以海拔植被区、垂直度与农业生态系统进行概念化，但另类的架构则是调适的动态，以根据在不同的生产区中进行不同理性决策的架构中最优化报酬的选项之间的选择，对农耕进行概念化。在此一架构中，生产区并不仅以海拔进行定义，而是同时以土壤、斜率和取得灌溉的管道定义之。在灌溉生产区的一个晚近选项便是温室园艺，并已成为赤道山区最具全球竞争力的农作出口。在厄瓜多尔，温室园艺于 1990 年代开始扩张，部分是对有利的贸易协议之回应，同时也是由多重来源的技术扩散和在地创业精神所导致。与各种行动者进行的访谈和田野工作，提供了温室调适策略的细节，并主张此一农用工业活动，已被証实对全球贸易模式改变和气候变迁具有异常的回复力。它提供了抵抗对外移民的就业选择，并鼓励了部分的对内劳动移入。于此同时，亦有有关其对水资源的冲击和杀虫剂的影响之关注。对山地环境与农业策略进行过度静态或生态系统主义式的概念化，将无法预期在高海拔区域的多样生产区中的全部发展可能性。这些可能性同时有助于对抗有关高山农业在面临现代化与全球化时必将衰落的主张。 关键词：关键词：调适动态，安地斯山脉，厄瓜多尔，园艺，山岳。

La agricultura de montaña ha sido conceptualizada en términos de zonas altitudinales, verticalidad y agroecosistemas, aunque un marco alternativo es el de la dinámica adaptativa, en el que la agricultura se conceptualiza en términos de selección entre opciones basadas en la optimización de rendimientos en marcos diferentes para la toma de decisiones racionales, en diferentes zonas de producción. En este marco, las zonas de producción no se definen únicamente en términos de altitud sino también en términos de suelo, inclinación y acceso a la irrigación. Una opción reciente en la zona de producción basada en irrigación ha sido la floricultura de invernadero, la cual se ha convertido en uno de los renglones agrícolas de exportación globalmente más competitivos de las montañas ecuatoriales. En Ecuador, la floricultura de invernadero se expandió en la década de los 1990 parcialmente en respuesta a acuerdos comerciales favorables, aunque también debido a la difusión de tecnologías desde múltiples fuentes, y al emprendimiento local. Entrevistas con algunos actores y trabajo de campo proporcionan detalles sobre las estrategias adaptativas de invernadero y sugieren que esta actividad agroindustrial ha resultado atípicamente resiliente frente a cambios en los patrones comerciales globales y a los cambios climáticos. Ha generado una opción de empleo contra la migración hacia el exterior y estimulado cierta inmigración de empleo. Pero al mismo tiempo, hay preocupación por los impactos que puedan sobrevenir a los recursos hídricos y preocupación por el impacto de pesticidas. Las concepciones excesivamente estáticas o ecosistémicas de los entornos montañosos, y las estrategias agrícolas, fallan en anticipar el amplio espectro de posibilidades de desarrollo en las diversas zonas de producción de las regiones de mayor altitud. Estas posibilidades también ayudan a contrarrestar afirmaciones acerca del inevitable declive de la agricultura de montaña frente a la modernización y la globalización.

Mountains have commonly been viewed as regions of limitation, where human activities and agriculture have to cope with restraints distinct from those of lowlands. This article discusses the most globally successful high montane commercial agricultural industry, greenhouse floriculture (Borsdorf and Stadel 2015), using the results of field interviews to help understand the emergence, success, and potential resilience of this agribusiness (Figure 1).

Some have asserted that mountains are unsuitable for modern (capital-intensive, globalized) agriculture (Foley et al. 2011). Mountain farmers could be seen to be constrained to adapt to hypoxia, cold stress, low plant productivity, aridity, thin soils, and steep slopes (Moran 2008). Some of these limitations affect humans, hypoxia in particular. Others affect agriculture. The regular decline of mean temperature with elevation has been associated with a variety of schemes of altitudinal zonation of plants and crops since Humboldt (Zimmerer 2011). High mountain environments have also been studied as agroecosystems in terms of energy flows (Baker and Little 1976) and as coupled human and environment systems (Brondizio, Ostrom, and Young 2009; Collins et al. 2011).

In addition to altitudinal zonation and ecosystemicist approaches, some have chosen to study mountain agriculture in terms of adaptive dynamics, the active, fluid choice of strategies and tactics by groups of actors with given objectives in particular environments (Bennett 1969; Denevan 1983; Knapp 1991). The objectives pursued by actors are determined in part by their positionality in socioeconomic formations, as smallholders (Netting 1993) or commercially oriented entrepreneurs, for example. Multiple sites and scales of power can be framed with actor network theory (Latour 2005), postdevelopment theory (Escobar 2010), agent ecology (Vayda and Walters 1999), policy-oriented development geography (Liverman and Vilas 2006; Bebbington, Abramovay, and Chiriboga 2008), or political ecology, including its historical and feminist variants (Robbins 2012). Most of these perspectives grant relative autonomy to differing actors and institutions at various scales, albeit constrained by larger scale political economic trajectories.

Production Zones and Greenhouse Floriculture

Mayer (1985) provided the helpful concept of *production zone*, going beyond the vertical layer cake to define specific mountain areas suitable in terms of

Figure 1. Inside an organic rose plantation greenhouse, Biogarden Chimborazo, 27 June 2007. (Color figure available online.)

soil, slope, aspect, and water for particular crop combinations. The term can refer to a specific area of land or be generalized over larger scales. This concept is related to the concept of *farm spaces* used by Zimmerer (1996) in his study of high mountain agriculture and agrobiodiversity. Production zones in particular places are intimately related to social and cultural patterns, including land tenure regimes, commons, rules regulating access to resources, patterns of temporary and permanent migration and remittances, and practices relating to sale of produce, craft items, and the like. Even in small areas, there are often wide variations in control over resources, with actors of different sociopolitical status, insertion into national and transnational markets, and access to higher scales of action. Zones should be seen as not static but subject to change as a product of endogenous decision making, cooperation, or conflict (Young 2008). The state, changing market realities, and nongovernmental organizations (NGOs) might be involved in deliberate trajectories of modernization. Although sustainable development has become a fashionable term in the Global North, South Americans sometimes prefer the concept of *territorial development*, explicitly linking government with economic change in particular areas (Bebbington, Abramovay, and Chiriboga 2008).

Zimmerer and Bell (2015) highlighted the distinction between valley and upland zones as governing landscape perceptions and policies in the Andes. The humid flats and wetlands were of key agricultural importance prior to the Spanish conquest (Knapp 1981). Pre-Columbian indigenous peoples built canals to create additional high-value irrigated production zones in highland Ecuador (Knapp 1991, 1992), Peru, Bolivia, Chile, and Argentina (Palomino Meneses 1986; Gelles 2010). After the Spanish conquest, the wetlands were typically used as pasture, and the irrigation systems were used for food crops and (at lower elevations) for sugarcane and other tropical crops (Knapp 1991). Starting in the 1960s, greenhouse floriculture emerged as an export activity in the high-altitude wetland and irrigated production zones of Colombia and, later, of Ecuador. Studying the emergence and characteristics of this agricultural activity sheds light on adaptive dynamics in mountains and the prospects for sustainability of this new activity. Although floriculture is now important in both Africa and South America (Rikken 2011), this article focuses on the equatorial Andes of South America and, in particular, Ecuador.

Emergence and Resilience of Floriculture in Ecuador

Multiple methodologies might be appropriate for teasing out patterns, processes, and relationships in mountain agricultural dynamics, including historical and archival study, field measurements and mapping, and open-ended interviews with actors in a variety of roles and scales. My original focus on Ecuadorian agriculture was on the long-term practices and trajectories of smallholder farmers (Knapp 1991). Beginning in 2005, I began interviewing actors in floriculture, including owners, workers, officials in floriculture organizations, employees of supplier companies, academics, NGOs, government agencies, activists, and others. I collected maps of plantations and consulted archival information in municipalities. The results of these studies, as well as other published research, document a remarkable trajectory of agricultural growth in a high mountain environment.

Andean commercial floriculture emerged in Colombia in the mid- to late 1960s (Ziegler 2007); carnation plantations were also established in Ecuador around this time but failed commercially (Gasselin 2000). Entrepreneur Mauricio Dávalos Guevara established one of the first rose plantations in Ecuador in 1983 (Morillo 2000). The annual value of cut flower exports from Ecuador has grown from $14 million in 1990 to $766 million in 2012 (Expoflores 2013).

This expansion has involved a relatively small number of producers and a relatively small area of land. Data have been collected with varying criteria by the Ecuadorian census, by growers' associations, and by independent marketing organizations (Knapp 2015). As of 2014 the agricultural census found that about 6,867 hectares were planted to flowers, with about 4,796 of these in roses, 602 in Gypsophila (baby's breath), and 141 in carnations (Instituto Nacional de Estadística y Censos 2014). The grower association Expoflores, however, believes that these figures are inflated, including some land that is not actually harvested. Recently there have been perhaps 700 to 1,000 farms (*fincas*); the higher numbers include very small farms for local sale. Some enterprises (*empresas*) own multiple *fincas*, with a small number of operations responsible for much of international sales (Gasselin 2000; S. Lopez, Expoflores, personal communication, 4 August 2009; Expoflores 2013; S. Saa and S. Lopez, Expoflores, personal communication, 3 July 2013; J. Pozo, Expoflores, personal communication, 18 July 2014).

In the broader context, Ecuador suffered a series of crises including war with Peru in 1995, a major El Niño in 1998, and governance issues leading to dollarization in 2000. Ecuador experimented with neoliberal policies including privatization, deregulation, and trade liberalization during the presidential term of Sixto Durán Ballén (1992–1996). The Andean Trade Preference Act in 1991 rewarded Ecuador for cooperating with the drug war; it was expanded by the Andean Trade Promotion and Drug Eradication Act (ATPDEA) in 2002. The result was to remove tariffs on flower (and other) exports to the United States. These were significant not just in terms of making Ecuadorian flowers more competitive in the U.S. market but in removing a time-consuming barrier in the commodity chain; flowers could be loaded onto trucks at the point of arrival without costly delays in processing paperwork and paying fees.

The mere existence of ATPDEA does not explain the growth of Ecuadorian floriculture, however. Part of the explanation is environmental (Sawers 2005). Roses benefit from bright light and low variations in daily and seasonal temperatures (Ziegler 2007). Being located on the Equator, days and nights are of roughly equal length year-round. At high elevations, flower growth is slower and bloom size larger, and large amounts of sugar accumulate in stems, helping assure long shelf life. In Ecuador, flower plantations have been established at up to 3,450 m elevation, although most greenhouses are at between 2,400 and 3,050 m, and open air flower plantations ("summer flowers") are between 1,800 and 2,400 m (Gasselin 2000; S. Saa, Expoflores, personal communication, 4 August 2009). Widespread areas of gently sloping terrain and fertile soils are also helpful for flower production; however, plantations often improve unpromising soils (see later).

Other important preconditions were also present in Ecuador. Water was available via wetlands and irrigation systems that were deployed since pre-Columbian times. Land could be had through preexisting estate agriculture (remnants of old haciendas after land reform) devoted to pastures and dairy; ownership often passed from traditional elite families to those interested in profit maximization. Labor could be obtained from local farm families on minifundia with limited commercial options other than national and transnational temporary or permanent migration (Kyle 2001).

Capital was an issue from the beginning; many banks were reluctant to lend to this untried industry, especially after the 1999 crisis, although over time some learned to provide help (S. Lopez, Expoflores, personal communication, 4 August 2009). Some already wealthy estate owners turned to floriculture. Others pursued partnerships to raise funds, including funds from international investors. Transnational or national corporate frameworks of management have not proven to be very successful in Ecuador; Dole, for example, attempted to assemble a larger scale set of plantations but eventually withdrew, partly because of labor conflicts (H. Chiriboga, Biogarden Chimborazo, personal communication, 28 June 2007).

Technology has involved the greenhouses, irrigation systems, agrochemicals, and suitable rose cuttings for planting. Greenhouses serve functions in controlling the temperature between 22°C and 30°C and ameliorating frost risk in flats (Figure 1). Greenhouses also help in maintaining humidity at around 70 percent (thus controlling fungus and mildew), in maximizing benefit from aerosol applications of pesticides, and providing some protection against dust, including light volcanic ash falls. Greenhouse plastic is produced in Ecuador and can cost up to $15,000 per hectare; it needs to be replaced every two to three years (H. Chiriboga, Biogarden Chimborazo, personal communication, 28 June 2007).

Israriego built on initial Israeli efforts in strawberry cultivation and became a large-scale producer of efficient irrigation equipment starting in 1986; by 2006 it was supplying about 70 percent of the flower plantation irrigation market. Irrigation systems can cost up to $12,000 per hectare (R. Zapata, Israriego, personal communication, 3 August 2006). Plantations can be inserted into preexisting irrigation networks, paying fees under the table for water rights; water rights in Ecuador are not legally subject to sale but have routinely been subject to purchase for many years. Groundwater could also be pumped. Water demand of greenhouse flowers is normally around 0.5 L per second per hectare, peaking at 1 L per second per hectare during the summer dry season (Gasselin 2000; R. Zapata, Israriego, personal communication, 14 August 2006; H. Chiriboga, Biogarden Chimborazo, personal communication, 28 June 2007). Water is typically stored in open-air tanks to protect against variation in supply.

Entrepreneurship and managerial skills have been crucial to the development of the industry. In general, it did not count on much help from government, universities, or NGOs. The bulk of the owners are Ecuadorian, although there are Colombian, Russian, U.S., and other operators. Because the operators are in competition with each other, it has been difficult for the industry to speak with a single voice or even to collect reliable data on the national level. The multiplicity and independence of operators, however, has allowed for experimentation, flexibility, and rapid adaptation to changing conditions, including changing markets (H. Chiriboga, Biogarden Chimborazo, personal communication, 28 June 2007; S. Lopez, Expoflores, personal communication, 4 August 2009; J. Pozo, Expoflores, personal communication, 18 July 2014).

Greenhouse floriculture, for roses, involves planting 55,000 to 75,000 roses per hectare, with density increasing in recent years. Many growers buy high-quality genetic stock for roses from multinational firms that charge $1 per cutting. Soils are prepared prior to planting with a variety of amendments including cacao husks, cut-up rose stems, and even mined páramo soils. The drip irrigation system includes precisely measured, computerized injection of liquid fertilizers. Operation of the plantations is labor intensive, involving 12.5 employees per hectare for roses; workers earn at least minimum wage, about $340 a month in 2014 (H. Chiriboga, Biogarden Chimborazo, personal communication, 28 June 2007; S. Saa and S. Lopez, Expoflores, personal communication, 3 July 2013; J. Pozo, Expoflores, personal communication, 18 July 2014). A 2011 survey indicated that workers averaged $400 a month, supervisors $900 a month, and managers $1,800 to $5,000 a month, comparable to rates in Colombia (S. Saa and S. Lopez, Expoflores, personal communication, 3 July 2013).

The growing crop is sprayed with cocktails of pesticides. About fifty different pesticides are deployed in Ecuadorian floriculture (Archivo Municipio de Cayambe, Dirección de Ambiente [AMCDA] 2005), including insecticides, fungicides, and antimicrobials. Ideally, the pesticides are applied by male workers who might only perform the job for a certain period of time, are provided with protective clothing, and receive periodic health checks; they are also paid more. Harvested flowers are taken to the postharvest facility, sometimes by recycled banana plantation tramway systems. In the postharvest area, flowers are dunked in a cleansing bath and then cut into stems that are packed in cartons of a dozen roses each. In a cold room boxes are packed with twelve cartons and are quickly loaded onto refrigerated trucks that take the flowers to the Quito airport. From the airport, the flowers are shipped internationally to airports capable of handling cut flowers. For example, in the United States, Miami is the most important port of entry, with Los Angeles also having facilities. Various

middlemen then take over the wholesale marketing of flowers to retail outlets such as supermarkets and florists. Retail vendors seldom inform consumers of the origin of the flowers, and most customers are still unaware of the Andean origin of their blooms.

A typical rose plant produces three blooms every eighty days or so, depending on altitude (Gasselin 2000; H. Chiriboga, Biogarden Chimborazo, personal communication, 28 June 2007). A hectare with its 55,000 to 75,000 rose bushes might produce $200,000 worth of marketable roses per year, averaging about US$0.30 each at the airport, although some roses are sold at a higher price (H. Chiriboga, Biogarden Chimborazo, personal communication, 28 June 2007; S. Saa and S. Lopez, Expoflores, personal communication, 3 July 2013; J. Pozo, Expoflores, personal communication, 18 July 2014). Plantations have at certain periods been highly profitable, such as the 20 to 30 percent profits reported for the late 1990s (Gasselin 2000); at that time the system was the agricultural system with the highest labor productivity in Ecuador, yielding $7,200 to $10,800 in revenue per worker (Gasselin 2000). Flower production is now the most profitable agricultural land use and is widespread in the high montane irrigated production zones with paved road access to the Quito airport.

Impacts and Resilience

The industry has had positive impacts. It has been estimated that 55,000 persons are employed directly in the industry (J. Pozo, Expoflores, personal communication, 18 July 2014). Although receiving minimum wage, individuals working on plantations are participating in the formal economy and have access to plantation and government programs. Workers, especially female workers, reinvest income in their farms (e.g., dairy cattle), health care, and education. Gender relations in smallholder households have become more equal as a result (Korovkin 2003). There is evidence that the availability of off-farm work to smallholders increases agrobiodiversity (Skarbø 2014). Greenhouse technology is now being used by family farmers to grow such crops as tomatoes and babaco for local markets and in some cases to produce flowers for export (Mena-Vásconez, Boelens, and Vos 2016).

Concerns about negative environmental impacts have included the effects of pesticides and possible impacts on water (Tenenbaum 2002; Harari 2004; Sawers 2005–2006; Bergman 2008). Until very recently, the industry was minimally regulated at the national level. In Cayambe, however, plantations have been required to submit environmental impact statements documenting pesticide use and water quality (AMCDA 2005). There have been concerns about unionization and treatment of labor (Lyall 2014). Local organizations are concerned about solving these and other problems, at the same time maintaining the presence of an industry vital to local employment (S. Cabezas, President, Federation of Popular Organizations of Ayora, personal communication, 18 July 2006). Growers have been experimenting with organic flower production, which has also proven profitable. Sales have been limited, however, by poor customer awareness of the importance of organic production for worker health (F. Falconi, personal communication, 29 June 2014).

Climate change is another concern. The equatorial Andes have experienced increases in mean temperature in recent decades (Rabatel 2013); farmers have successfully planted maize and potatoes at higher elevations than previously (Skarbø and VanderMolen 2016). Increasing temperatures should be associated with increased potential evapotranspiration. Some Andean studies suggest that precipitation has or will increase to compensate for increased evapotranspiration (Buytaert and De Bièvre 2012). Local smallholder farmers have claimed that precipitation has decreased or that its patterns have changed (VanderMolen 2011), but this is difficult to verify with data, in part because of the poor quality of historical data coupled with the deactivation of many meteorological stations due to neoliberal reforms. The rise in temperatures does not appear to be a concern for floriculture. Growers report more rain in the summer and less during the winter *veranillo* or secondary dry season; these have not affected production and, indeed, lower winter humidity might be helpful for flower production. Winds are a concern for greenhouse plastic; winds have reportedly become less severe in the summer but have gone up a bit in winter, slightly affecting Valentine's Day production (S. Saa and S. Lopez, Expoflores, personal communication, 3 July 2013; J. Pozo, Expoflores, personal communication, 18 July 2014).

Flower plantations depend on irrigation. Although Ecuadorian glaciers have been shrinking (Rabatel et al. 2013), most irrigation systems in Ecuador do not derive from glaciers. Most Andean canals are earthen, and water supply is dramatically improved with cement lining and improved diversion structures (VanderMolen 2011). Under conditions of traditional agriculture these improvements are uneconomic and have some adverse consequences for sustainability and users who rely on subsurface flow. Under current conditions of expanding commercial agriculture, however, such improvements are easily paid for, and advanced drip irrigation technology provides further help through efficient use of water. Due to technological improvements, water supplied by many local canals has been increasing (F. Trujillo, water board, La Victoria Canal, personal communication, 14 June 2006; J. Pozo, Exploflores, personal communication, 18 July 2014).

Volcanoes are a potentially serious environmental factor. Ash falls can destroy greenhouse plastic, disrupt transportation (including air transportation to international markets), and damage flowers in open fields. More serious activity could spur a lahar, which could displace hundreds of thousands of people and affect far more than the flower industry. Flower plantation owners are, however, more concerned about social and political factors. Locally, labor costs have been rising with the recently improving Ecuadorian economy and expansion of the state under Correa. Although most workers are not in unions, workers increasingly have options. In some cases workers have migrated from poorer parts of Ecuador (Chimborazo) and even Peru, but these labor sources are limited. Plantation managers have told me that they are concerned with competition from Africa with a similar climate but much lower labor costs (e.g., Latin Flor plantation, personal communication, 30 June 2015). The dollarized economy, with a strong dollar, further increases costs and makes it difficult to compete with countries that are not dollarized, especially Ethiopia and Kenya. Although many other mountain regions produce commercial flowers, no other equatorial country has Ecuador's geographic advantage in producing large, high montane rose blooms (Rikken 2011). Ecuador is also more favorably situated for speedy access to U.S. airports.

Ecuador's trade policy is also a concern. In 2013, the regime renounced ATPDEA, and subsequent trade legislation in the United States has benefited minor flower crops such as baby's breath, but not roses, which as of 2015 have to pay a duty of 6.8 percent. This duty has the potential to reduce demand from major U.S. chains (D. E. Marko, trade negotiator, personal communication, 3 July 2013). This has affected exports to the United States, but Ecuador has pursued markets in Europe (with which it now has a trade agreement), Russia, and Asia (Rikken 2011).

Discussion and Conclusion

Along with quinoa (so far on a much smaller scale; see Bazile, Bertero, and Nieto 2015), flower plantations can be seen as the first internationally successful export agriculture in high mountain zones. Excessively static or ecosystemicist conceptions of mountain environments and agricultural strategies fail to anticipate the full range of possibilities for development in the diverse production zones of high-altitude regions. Even in the context of "traditional" smallholder agriculture, flexibility is high and options exist to cope with social and environmental change (Knapp and Cañadas 1988; VanderMolen 2011; Mena-Vásconez, Boelens, and Vos 2016). In the context of commercial agriculture, flowers have proven to be profitable and production might be resilient, at least in the medium term. Greenhouses and modern drip irrigation systems provide buffers against weather and climate fluctuations, and the industry has proven adept in finding alternative markets. The flower industry has absorbed local, regional, and transnational labor, magnifying its impacts. Although this study has focused on Ecuador, flowers have proven to be a competitive commercial crop in mountains worldwide. As quinoa has shown, commercial possibilities can emerge relatively suddenly in other production zones as well (Bazile, Bertero, and Nieto 2015).

These possibilities also help to contest assertions about the inevitable decline of mountain agriculture in the face of modernization and environmental change. Concerns about the future of high Andean floriculture still exist, of course; the industry might founder on issues of trade, exchange rates, or health or labor concerns, for example. The industry still has the agroecological drawbacks of (relatively) large-scale operations, labor forces, and agrochemical uses. Organic alternatives, however, already are being deployed, and the greenhouse technology has diffused to smaller scale commercial farms for such crops as babaco and tomatoes as well as flowers. Although tradition still matters (Borsdorf and Stadel 2015), policymakers, politicians, activists, and scholars should also keep in mind adaptive dynamics as a productive approach to the study of high mountain livelihoods and productive options as involving both stasis and change (Young 2008).

Acknowledgments

The anonymous reviewers and the special issue editor, Mark Fonstad, provided valuable comments. Space does not permit listing the numerous individuals in Ecuador who generously helped with this project, but I appreciate them all.

Funding

This research was supported by a Fulbright-Hays Research Abroad Grant and by the Teresa Lozano Long Institute of Latin American Studies at the University of Texas at Austin from funds granted to the Institute by the Andrew W. Mellon Foundation.

References

Archivo Municipio de Cayambe, Dirección de Ambiente (AMCDA). 2005. *Estudios de Impacto Ambiental* [Environmental impact studies]. Unpublished study, accessed via the municipal archive of Cayambe, Ecuador.

Baker, P., and M. Little, eds. 1976. *Man in the Andes: A multidisciplinary study of high-altitude Quechua.* Stroudsburg, PA: Dowden, Hutchison and Ross.

Bazile, D., D. Bertero, and C. Nieto, eds. 2015. *State of the art report of quinoa in the world in 2013.* Rome: FAO and CIRAD.

Bebbington, A., R. Abramovay, and M. Chiriboga. 2008. Social movements and the dynamics of rural territorial development in Latin America. *World Development* 36 (12): 2874–87.

Bennett, J. 1969. *Northern plainsmen.* Chicago: Aldine.

Bergman, C. 2008. A rose is not a rose. *Audubon* 110 (1): 46–53.

Borsdorf, A., and C. Stadel. 2015. *The Andes: A geographical portrait.* New York: Springer.

Brondizio, E. S., E. Ostrom, and O. Young. 2009. Connectivity and the governance of multilevel socio-ecological systems: The role of social capital. *Annual Review of Environment and Resources* 34:253–78.

Buytaert, W., and B. De Bièvre. 2012. Water for cities: The impact of climate change and demographic growth in the tropical Andes. *Water Resource Research* 48:W08503.

Collins, S. L., S. R. Carpenter, S. M. Swinton, D. E. Orenstein, D. L. Childers, T. L. Gragson, N. B. Grimm, et al. 2011. An integrated conceptual framework for long-term social-ecological research. *Frontiers in Ecology and the Environment* 9:351–57.

Denevan, W. 1983. Adaptation, variation, and cultural geography. *The Professional Geographer* 35:399–407.

Escobar, A. 2010. Latin America at a crossroads. *Cultural Studies* 24 (1): 1–65.

Expoflores. 2013. *Ecuador, el sector floricultor, análisis de la situación actual* [Analysis of the real situation of the floriculture sector in Ecuador]. Quito: Ecuador: Expoflores.

Foley, J., N. Ramankutty, K. A. Brauman, E. S. Cassidy, J. S. Gerber, M. Johnston, N. D. Mueller, et al. 2011. Solutions for a cultivated planet. *Nature* 478:337–42.

Gasselin, P. 2000. Le temps des roses: La floriculture et les dynamiques agraires de la region agripolitaine de Quito (Equateur) [The time of the rose: Floriculture and agrarian dynamics in the vicinity of Quito (Ecuador)]. Unpublished PhD thesis, Institut National Agronomique (INA) Paris.

Gelles, P. H. 2010. Cultural identity and indigenous water rights in the Andean highlands. In *Out of the mainstream: Water rights, politics and identity*, ed. R. Boelens, D. Getches, and A. Guevara-Gil, 119–44. London: Earthscan.

Harari, R. 2004. *Seguridad, salud y ambiente en la floricultura* [Security, health and environment in floriculture]. Quito, Ecuador: IFA-PROMSA.

Instituto Nacional de Estadística y Censos. 2014. Encuesta de superficie y producción agropecuaria continua [Survey of the area and production of permanent agriculture]. http://www.ecuadorencifras.gob. ec/documentos/web-inec/Estadisticas_agropecuarias/ espac/espac_2014/ (last accessed 30 November 2015).

Knapp, G. 1981. El nicho ecológico llanura húmeda en la economía prehistórica de los Andes de Altura [The ecological niche of humid flatlands in the prehistoric economy of the High Andes]. *Sarance* 9:83–96.

———. 1991. *Andean ecology: Adaptive dynamics in Ecuador.* Boulder, CO: Westview.

———. 1992. *Riego precolonial y tradicional en la Sierra Norte del Ecuador* [Traditional and prehistoric irrigation in the Northern Andes of Ecuador]. Quito, Ecuador: Ediciones Abya Yala.

———. 2015. Mapping flower plantations in the equatorial high Andes. *Journal of Latin American Geography* 14 (3): 229–44.

Knapp, G., and L. Cañadas. 1988. Conclusions and implications for policies of rural development. In *The impact of climatic variations on agriculture: Vol. 2. Assessments in semi-arid regions*, ed. M. L. Parry, T. R. Carter, and N. T. Konijn, 485–88. Dordrecht, The Netherlands: Kluwer Academic.

Korovkin, T. 2003. Cut-flower exports, female labor, and community participation in highland Ecuador. *Latin American Perspectives* 30 (4): 18–42.

Kyle, D. 2001. *Transnational peasants: Migration, networks and ethnicity in Andean Ecuador.* Baltimore: Johns Hopkins University Press.

Latour, B. 2005. *Reassembling the social: An introduction to actor-network theory.* Oxford, UK: Oxford University Press.

Liverman, D., and S. Vilas. 2006. Neoliberalism and the environment in Latin America. *Annual Review of Environmental Resources* 31:327–63.

Lyall, A. 2014. Assessing the impacts of fairtrade on worker-defined forms of empowerment on Ecuadorian flower plantations. Final report commissioned by Fairtrade International and Max Havelaar-Foundation. http:// www.fairtrade.net/fileadmin/user_upload/content/2009/ resources/140212-Worker-Empowerment-Ecuador-Flower-Plantations-final.pdf (last accessed 20 November 2015).

Mayer, E. 1985. Production zones. In *Andean ecology and civilization*, ed. S. Masuda, I. Shimada, and C. Morris, 45–84. Tokyo: University of Tokyo Press.

Mena-Vásconez, P., R. Boelens, and J. Vos. 2016. Food or flowers? Contested transformations of community food security and water use priorities under new legal and market regimes in Ecuador's highlands. *Journal of Rural Studies* 44:227–38.

Moran, E. 2008. *Human adaptability: An introduction to ecological anthropology.* 3rd ed. Boulder, CO: Westview.

Morillo, W. 2000. Los pioneros cuentan sus experiencias [The pioneers relate their experiences]. *Marketing Flowers Internacional* 17:17–25.

Netting, R. M. 1993. *Smallholders, householders: Farm families and the ecology of intensive, sustainable agriculture.* Stanford, CA: Stanford University Press.

Palomino Meneses, A. 1986. Antiguedad y actualidad del riego en los Andes [Antiquity and status of irrigation in the Andes]. *Allpanchis* 18:27–28.

Rabatel, A., B. Francou, A. Soruco, J. Gomez, B. Caceres, J. L. Ceballos, R. Basantes, et al. 2013. Current state of glaciers in the tropical Andes: A multi-century perspective on glacier evolution and climate change. *The Cryosphere* 7:81–102.

Rikken, M. 2011. The global competitiveness of the Kenyan flower industry. Presentation at the Fifth Video Conference on the Global Competitiveness of the Flower Industry in Eastern Africa, World Bank Group, Kenya Flower Council, and ProVerde. http:// www.kenyaflowercouncil.org/pdf/VC5%20Global%20 Competitiveness%20Kenyan%20Flower%20Industry% 20-%20ProVerde.pdf (last accessed 25 November 2015).

Robbins, P. 2012. *Political ecology: A critical introduction.* 2nd ed. Malden, MA: Wiley-Blackwell.

Sawers, L. 2005. Nontraditional or new traditional exports: Ecuador's flower boom. *Latin American Research Review* 40 (3): 40–67.

———. 2005–2006. Sustainability and Ecuador's flower export boom. *International Journal of Environmental, Cultural, Economic and Social Sustainability* 1 (2): 17–21.

Skarbø, K. 2014. The cooked is the kept: Factors shaping the maintenance of agro-biodiversity in the Andes. *Human Ecology* 42:711–26.

Skarbø, K., and K. VanderMolen. 2016. Maize migration: Key crop expands to higher altitudes under climate change in the Andes. *Climate and Development* 8 (3): 245–55.

Stewart, A. 2007. *Flower confidential: The good, the bad, and the beautiful in the business of flowers.* Chapel Hill, NC: Algonquin Books of Chapel Hill.

Tenenbaum, D. 2002. Would a rose not smell as sweet? *Environmental Health Perspectives* 110 (5): 240–47.

VanderMolen, K. 2011. Percepciones de cambio climático y estrategias de adaptación en las comunidades agrícolas de Cotacachi (Debate Agrario-Rural) [Perceptions of climate change and adaptive strategies in the agricultural communities of Cotacachi (agrarian and rural debates)]. In *Ecuador debate: Problemas y perspectivas del extravismo*, 145–57. Quito, Ecuador: Centro Andino de Acción Popular.

Vayda, A. P., and B. B. Walters. 1999. Against political ecology. *Human Ecology* 27 (1): 167–79.

Young, K. R. 2008. Stasis and flux in long-inhabited locales: Change in rural Andean landscapes. In *Land-change science in the tropics: Changing agricultural landscapes,* ed. A. Millington, and W. Jepson, 11–32. New York: Springer.

Ziegler, C. 2007. *Favored flowers: Culture and economy in a global system*. Durham, NC: Duke University Press.

Zimmerer, K. S. 1996. *Changing fortunes: Biodiversity and peasant livelihood in the Peruvian Andes*. Berkeley: University of California Press.

———. 2011. Mapping mountains. In *Mapping Latin America: A cartographic reader*, ed. J. Dym, and K. Offen, 125–30. Chicago: University of Chicago Press.

Zimmerer, K. S., and M. G. Bell. 2015. Time for change: The legacy of a Euro-Andean model of landscape versus the need for landscape connectivity. *Landscape and Urban Planning* 13:104–16.

Mountainous Terrain and Civil Wars: Geospatial Analysis of Conflict Dynamics in the Post-Soviet Caucasus

Andrew M. Linke, Frank D. W. Witmer, Edward C. Holland, and John O'Loughlin

Existing research on the relationship between mountainous terrain and conflict has generally been implemented using crude metrics capturing the actions and motivations of armed groups, both insurgent and government. We provide a more geographically nuanced investigation of two specific propositions relating mountainous terrain to violent conflict activity. Our study covers five wars in the Caucasus region: the second North Caucasus war in Chechnya and neighboring republics (1999–2012); Islamist and Russian government conflict in the same area (2002–2012); fighting between Armenians and Azerbaijanis in Nagorno-Karabakh (1990–2012); and battles between Georgia and separatists in South Ossetia (1991–2012) and Abkhazia (1992–2012). Our analysis of insurgent and government violence reciprocity illustrates some expected patterns of what we call the operational costs of context. By varying the dimensions for our units of analysis—the context within which violent interactions take place—however, we arrive at differing conclusions. Our research represents a meaningful and transparent engagement with the influences of the well-known and understudied modifiable areal unit problem (MAUP) in geographically sensitive analysis.

山区和冲突间的关系之研究, 一般透过运用粗糙的指标, 同时捕捉作为反抗和政府组织的武装团体的行动及意图。我们为连结山地与暴力冲突活动的两个主张, 提供地理上更为细缴的探讨。我们的研究涵盖高加索地区的五场战役: 在车臣及周围的共和国发生的第二次北高加索战役 (1999 年至 2012 年); 在同一地区中的伊斯兰与俄罗斯政府间的冲突 (2002 年至 2012 年), 亚美尼亚人与阿塞拜疆人在纳戈尔诺-卡拉巴赫发生的战争 (1991 年至 2012 年), 以及格鲁吉亚和分离主义者在南奥赛梯的战役 (1992 年至 2012 年)。我们对于反抗和政府暴力互动的分析, 描绘出我们称之为脉络操作成本的若干预期模式。但透过多样化分析单元的各个面向——暴力互动所发生的脉络——我们却得到了不同的结论。我们的研究, 呈现对具有地理敏感度的分析中为人所熟知且未被充分研究的可调整地区单元问题 (MAUP) 进行有意义且透明的涉入。关键词: 高加索, 冲突, 可调整地区单元问题, 政治地理学, 空间分析。

La investigación existente sobre las relaciones entre terreno montañoso y conflicto ha sido implementada, en general, con el uso de métricas crudas para captar las acciones y motivaciones de los grupos armados, tanto de insurgentes como de los gobiernos. Lo que nosotros entregamos es una investigación de matices más geográficos sobre dos proposiciones específicas que relacionan el terreno montañoso con las actividades del conflicto violento. Nuestro estudio cubre cinco guerras en la región del Cáucaso: la segunda guerra del Norte del Cáucaso en Chechenia y las repúblicas vecinas (1999–2012); el conflicto islamista con el gobierno ruso en la misma área (2002–2012); la lucha entre armenios y azerbaiyanos en Nagorno-Karabakh (1990–2012); y las batallas entre Georgia y los separatistas en Osetia del Sur (1991–2012) y Abkhazia (1992–2012). Nuestro análisis de la reciprocidad en violencia de insurgencia y gobierno ilustra algunos de los patrones esperados de lo que nosotros denominamos costos operacionales del contexto. Sin embargo, variando las dimensiones de nuestras unidades de análisis—el contexto dentro del cual tienen lugar las interacciones violentas—llegamos a diferentes conclusiones. Nuestra investigación representa un compromiso significativo y transparente con las influencias del bien conocido como poco estudiado problema de la unidad areal modificable (MAUP, acrónimo en inglés) en análisis geográficamente sensible.

Scholars in the field of conflict studies have increasingly adopted geographical statistical analyses for their research. Unfortunately, many still engage with geography superficially. We believe that the study of mountainous terrain and civil war violence especially suffers from conceptual–empirical incompatibility. We highlight the important difference between studying civil war in the aggregate and analyzing violence dynamics that take place within civil wars. Our study improves on existing research in the literature with better data and statistical methodologies designed to capture the geographical contexts within which violence emerges and develops over time. Our analysis of violence in the North and South Caucasus calls into question any simplistic narrative about how rugged terrain relates to conflict dynamics.

Conflict analysts incorporate geography into their research in several ways. The first is through a concern for spatial and temporal disaggregation of research questions and statistical methods. Investigations of riots in London (Baudains, Johnson, and Braithwaite 2013), government-opposition attacks in Iraq (Linke, Witmer, and O'Loughlin 2012), Bosnian civil war events (Weidmann and Ward 2010), or Islamist insurgent activity in the North Caucasus (Zhukov 2012) all demonstrate the merits and utility of localizing violence research.

A second area of attention for geographical conflict research centers on diffusion or contagion effects. Studies in this vein adopt epidemiological language describing conflict as a force spreading across regions according to underlying political or economic processes (Houweling and Siccama 1985; Buhaug and Gleditsch 2008; Schutte and Weidmann 2011; Linke, Schutte, and Buhaug 2015). Geographers emphasize that diffusion also takes place within and across social network structures in addition to territorial connections (Medina and Hepner 2011; Radil, Flint, and Chi 2013).

Third, the compositional quality of a location (whether a town or region) might influence conflict patterns, and these characteristics can be inherently geographical. The geographic distribution of ethnic communities (Toft 2003; Weidmann and Saleyhan 2012), for example, has important implications for representation in a country's political institutions and can therefore translate into intergroup disputes. Contentious politics might or might not become violent depending on the distribution of territorial homelands, political accommodations, or as a function of social interactions that take place within demographically diverse versus homogenous areas.

A survey of the violent conflict literature reveals a prevailing reliance on this third understanding of geography. Some exceptions to the simplified idealization of geography exist, including Daly (2012), who, in her study of Colombia's civil conflict, embraced "a reorientation away from physical geography and back to the human and social geography that determines if rebellion is organizationally feasible" (473; see also Buhaug and Gates 2002). More specifically, conflict researchers often confine geography to physical geographic considerations instead of also including human geography. In particular, mountainous terrain and forest cover are commonly identified as correlates of violence in the classical civil war literature (Guevara 1961; McColl 1969; Grundy 1971; Fearon and Laitin 2003; Do and Iyer 2010; Nemeth, Mauslein, and Stapley 2014). The difference between social and physical understandings of geography across disciplines is linked with an understanding that "place" (the human geography emphasis) is more than only "space" (which tends to dominate in political science; O'Loughlin 2000). Our current goal is to advance the study of mountainous terrain influences on violence between government and nonstate actors. In doing so, we study group interaction dynamics of conflicts (the endogenous elements) against the background of particular elevation profile contexts (the exogenous elements).

In their study of civil war violence in sub-Saharan Africa, Tollefsen and Buhaug (2015) tested the effects of opposing actors' accessibility for intrastate armed conflict. They included structural variables like road networks, distances to capital cities, and mountainous terrain in addition to "sociocultural inaccessibility," which is related to demography and institutional exclusion of ethnic communities. The expected relationship to mountainous terrain is that armed opposition to the state thrives where there is sanctuary for organizational activities of insurgents. Sanctuary can, of course, be political if it is related to international borders (Saleyhan 2009), but it can also be social if it is related to identity politics and information sharing (e.g., denouncing militant activities to a counterinsurgent campaign, as in Lyall 2010).

Terrain and conflict research can be improved by focusing on scales of analysis and geographical context. There is currently limited evidence of a correlation between mountains and conflict (Buhaug and Rød 2006; Hegre and Sambanis 2006; Rustad et al. 2008) despite anecdotal accounts and selective narratives of such a link. The majority of research, however, is carried out with crude measurements and with a single unit of analysis (whether subnational or at a country level). Exceptions include O'Loughlin, Witmer, and Linke (2010), who aggregated insurgent and government force violent events in Afghanistan and compared trends along flat and hilly terrain profiles. Also using subnational analysis, Tollefsen and Buhaug (2015) found positive statistical associations between inaccessibility due to poor transport and the risk of violent intrastate conflict. We follow these two approaches and make several improvements by: (1) focusing more closely on reciprocal engagements between government and insurgent forces; (2) investigating a comparatively limited range of cases, which reduces the potential for unobserved influences; and (3) using a more geographically precise event data analysis.

The five North and South Caucasus wars that we study have their origins in the shared legacy of the Soviet state and its federal system. For each, the rubric of third-tier polities in the Soviet federal hierarchy (autonomous Soviet socialist republics (SSRs) and autonomous oblasts) opposing second-tier units (then union republics, which are now independent states) explains the origins of political tension. These conflicts "came from the peculiar existence of nations within nations, a phenomenon which may be referred to as 'matrioshka nationalism'" (Bremmer 1997, 11–12). The Soviet national-territorial arrangement promoted regional autonomy movements, which at the end of the USSR turned to conflict as a mechanism to achieve their political aims (Cornell 2002).

In Nagorno-Karabakh (see Figure 1), a majority Armenian region that was part of the Azerbaijani SSR, local parliamentarians issued a resolution on the transfer of the region to the Armenian republic in February 1988. Moscow's reaction was tentative and local interests responded decisively. According to de Waal (2003, 15), "The slow descent into armed conflict began" the day the resolution passed; the war in Nagorno-Karabakh continued with periods of intense fighting until 1994 and left approximately 25,000 dead. All Azerbaijanis were displaced from the region. Border skirmishes between the two sides continue to this day. A similar politics of territorial designation emerged in South Ossetia and Abkhazia (Georgia), and Chechnya (Russia). In each case, local nationalist leaders put forward a movement toward independence from the Soviet republics. In South Ossetia, fighting between paramilitaries ran from January 1991 through June 1992 and claimed roughly 1,000 lives. The war in Abkhazia, during 1992 and 1993, "was a failed attempt

to subordinate and incorporate this previously autonomous region into a newly unified and centralized Georgian state" (O'Loughlin, Kolossov, and Toal 2011, 4). The outcomes of these wars in terms of fixed delimited borders and extensive displacement of ethnic Georgians were formalized following the August 2008 conflict between Georgia and the two de facto states, aided by their Russian patron. The first war in Chechnya began in December 1994 and ended in a tentative peace agreement in August 1996. When Chechen militants—most notably Shamil Basayev—sought a more decisive resolution to the first war by territorial expansion and invaded neighboring Dagestan, the Russian government responded with substantial force. The conflict later developed into a regional-scale insurgency that adopted Islam as its motivating ideology. In the early years of the second Chechen war, fighting was particularly intense in and near the republic's capital of Grozny; the insurgency subsequently diffused to the neighboring republics of Ingushetia and Dagestan (O'Loughlin, Holland, and Witmer 2011).

Identifying Conflict Processes

The central theme in terrain-related conflict research is that mountainous regions favor insurgency as an organizational mode of conflict (Fearon and Laitin 2003). *Insurgencies* are defined as "a technology of military conflict characterized by small, lightly armed bands practicing guerilla warfare from rural base areas" (Fearon and Laitin 2003, 75). Because these groups are small and lightly armed, they move easily to camps and exploit clandestine networks to hide while conducting operations. Heavily armed, slow-moving governmental forces, in contrast, typically experience difficulty in their efforts to project power into isolated regions; government forces are paradoxically burdened by equipment that should ensure their military dominance.

Boulding's (1962) loss-of-strength gradient is the key conceptual link between social and physical geography in this line of research. Distance from population centers and peripheral locations for rebel activities play an important role in determining the potential for rebel organization, recruitment, and training (Buhaug and Gates 2002; Cunningham, Gleditsch, and Saleyhan 2009). State militaries, police, and other forces are more likely to be weak in rural and geographically marginalized areas of a country (Grundy 1971; Hegre 2008; Pickering 2012). Our cases exemplify these political circumstances (e.g.,

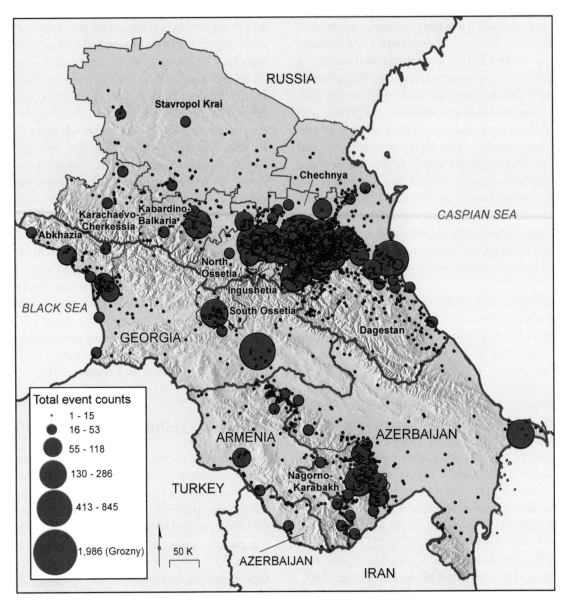

Figure 1. Graduated circles represent the number of violent events aggregated to the nearest kilometer (Universal Transverse Mercator Zone 38N), which avoids excessive overlap in the graphic. The terrain hillshade is generated from Shuttle Radar Topography Mission elevation data. (Color figure available online.)

Georgia's wars in both South Ossetia and Abkhazia), where contested sovereignty arises (Kolstø 2006).

Grundy (1971) argued that in the context of guerilla war, "a few square miles of mountainous jungle may be as strategically invulnerable as, let us say, a hundred square miles of prairie or, perhaps, a thousand square miles of flat plain crisscrossed by roads and telephone wires and dotted with airstrips and radio transmitters" (45). Fearon and Laitin (2003, 85) found that, controlling for political and economic factors, a country that is "half mountainous" (90th percentile of their sample) has a 13.2 percent risk of experiencing civil war; they found that a similar country that is not mountainous

should expect a 6.5 percent risk of major armed conflict. Similar recent research finds that mountainous terrain is associated with proportionally more terrorist attacks (Nemeth, Mauslein, and Stapley 2014). Activity by the government side, however, is completely absent from this analysis, which is a serious limitation of the study. It is not clear from the current literature how exactly collective violence on the ground relates to terrain.

Our specific propositions are based on an understanding of constraints that shape both government and nonstate actor capabilities. As the stated goal of each is to confront the other, our interest is in clarifying the structural conditions that either party uses to its advantage.

We rely on the notion of reciprocity; a plausible scenario is that where one actor conducts a strike against an opponent, the other party reacts by conducting operations nearby in location and time (see Linke, Witmer, and O'Loughlin [2012] for a more comprehensive exposition of reciprocity dynamics). We do not rely on a strict definition for the location of possible reactions and instead examine the effects across dozens of models using variable definitions of the range of response from 10 to 50 km^2 and including *rayon* (county) administrative units.[1] Our objective is not explaining the onset of the Caucasus wars at the most general level but focusing on their dynamics across terrains and sociodemographic contexts.

We anticipate that each conflict actor will experience operational costs in particular contexts or settings. Such costs for insurgents might include ease of accessing military equipment. Governments, in turn, suffer the burden of operational costs where terrain is impassable and guerilla fighters can hide, equip, and muster support for their cause undeterred. These conditions represent what we call the "operational costs of context," which vary substantially by mountainous terrain profiles. Observable implications of the theory can be tested in two specific propositions:

1. In high neighboring terrain regions, insurgent reciprocity (violent action) for a government-initiated event will be stronger than government reciprocity. Insurgents have a "sanctuary" advantage in such areas.
2. In low neighboring terrain regions, government reciprocity (violent action) for an insurgent-initiated event will be stronger than insurgent reciprocity. Government forces have an occupier control advantage in such areas.

Our expectations are illustrated graphically in Figure 2 in a straightforward schematic. For each type of region (I for higher neighboring terrain and II for lower), we test how well insurgent violence predicts government violence (result a) and vice versa (result b). The sign of the expected relationship is shown in parentheses.

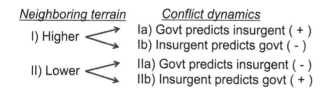

Figure 2. Expected action–reaction dynamics for government and insurgent violence in low and high neighboring terrain contexts.

To take into account modifiable areal unit problem (MAUP) concerns (Openshaw 1983), we use four spatial units of analysis. Figure 3 shows the definition of each areal unit across our study region. MAUP issues are not only a technical dilemma but also represent theoretical questions about the dimensions of geographical context and the ranges of social interaction (for a demonstration of MAUP's importance in conflict analysis, see Linke and O'Loughlin 2015).

Conflict Events, Elevation, and Social Control Variable Data

Descriptive statistics for all of our data at 25 km^2 are presented in Table 1 (see our Supplemental Material for the statistics at other spatial resolutions). Original conflict events data for our research were gathered from media sources and coded following the format of the Armed Conflict and Location Event Data project (Raleigh et al. 2010). Research assistants searched Lexis-Nexis archives for reporting of events that involved violence (including terms such as *attack*, *strike*, and *bomb*, among many others). All reports are stored and checked against duplicate stories to verify the information's accuracy. Each conflict incident is then entered in a data set with the date, perpetrating actor, location, type of event (e.g., suicide bombing), and notes providing any additional information. Unknown actors are included in our data if the violent incident was reported with location, time, and event type explained in detail. The North Caucasus data have been used in other related studies (e.g., O'Loughlin, Holland, and Witmer 2011; O'Loughlin and Witmer 2011); South Caucasus data were coded for this extended analysis. Figures 4 and 5 show the distribution of violent events over time.

Because we focus on interactive dynamics of insurgent and government forces, we aggregate actors into broader classifications. Police, secret police, border patrols, and military forces of any internationally recognized state are classified as government actors. We allow for this broad definition of government actors because the police and border guards often perpetrate violent seizures, and patrol activities can also lead directly to confrontations. Our designation for insurgent actors includes all known Islamist groups, ethno-nationalist political parties and movements, and unknown but nonstate perpetrators of violence.

We do not require that insurgent and government forces interact directly in a single incident. In other words, a militant might detonate a suicide bomb that

a) 10km Grids
b) 25km Grids
c) 50km Grids
d) Rayons (Counties)

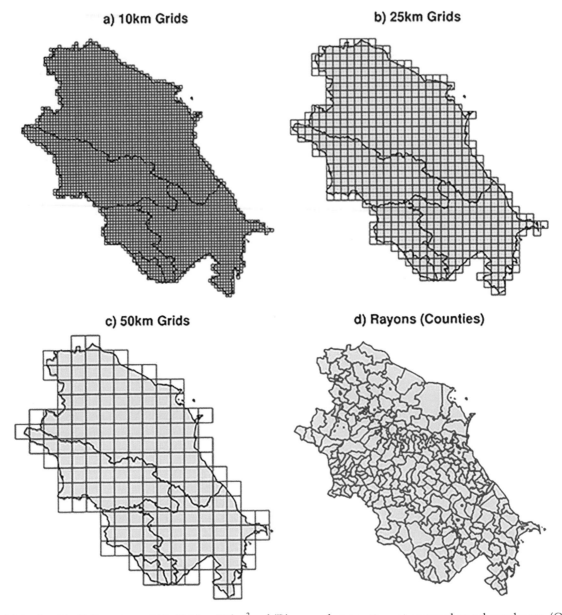

Figure 3. Multiple grid cell dimensions of (A–C) 10 to 50 km² and (D) *rayon* administrative units mapped over the study area. (Color figure available online.)

kills civilians in a marketplace in Vladikavkaz, North Ossetia (as occurred in September 2010). Alternatively, a police sweep through an area that results in unarmed civilians being killed will appear in our data even if the police sweep never encountered members of an Islamist *jamaat* (group or congregation). The political violence literature explains that in cases of contested territorial control, violence is often used against civilians with strategic purpose (Kalyvas 2006). A terrorist attack that does not directly target the president could still clearly undermine the legitimacy of a regime or sway the opinion of residents to encourage defection and denunciation.

We apply a geographic projection of the conflict event locations to Universal Transverse Mercator (UTM) Zone 38N and merge all outcome, predictor, and control variable data in grid cells and rayons that are defined by a monthly temporal resolution. Mean elevation is calculated from the 30 m Shuttle Radar Topography Mission (SRTM) data for each unit of analysis (Farr et al. 2007). We use first-order neighbor contiguity to measure variables for neighboring units of analysis. We are particularly interested in the conflict dynamics that operate within terrain zones and not only in "controlling away" the effects of elevation or estimating the direct influence of elevation on

Table 1. Summary statistics for all variables used in our models at the 25 km² resolution

Variables	Min	Median	M	Maximum	SD
All events	0.000	0.000	0.123	101.000	1.324
Government events	0.000	0.000	0.072	72.000	0.890
Government events spatial lag	0.000	0.000	0.073	17.375	0.470
Rebel events	0.000	0.000	0.047	36.000	0.529
Rebel events spatial lag	0.000	0.000	0.047	7.500	0.266
Titular percent	0.000	84.296	65.244	100.000	37.800
Employed percent	2.589	14.587	14.941	38.447	5.116
Urban percent	0.000	25.159	27.047	100.000	20.773
Population size (1,000s)	2.424	6.552	6.493	10.078	1.049
Forest cover (percent)	0.000	5.572	16.093	84.556	20.267
Border distance (km)	1.058	3.812	3.669	5.802	1.217
Distance to road (km)	0.263	3.004	3.274	14.621	1.454

conflict absent any conditional effects. Therefore, we create two subsets of the data set based on the difference between the elevation of a given unit and the average of the neighboring units (a threshold of 50 m or greater was used because most locations in the study area are surrounded by higher terrain). We refer to these as higher neighboring elevation and lower neighboring elevation. A map of the designation for 25 km² grid cells is shown in Figure 6.

We strive to control for possible alternative explanations of conflict, including poverty (Buhaug et al. 2011), excluded ethnicity status (Wimmer 2002),

population size (Raleigh and Hegre 2009), and infrastructure such as roads (Zhukov 2012). Percentage titular measures the proportion of the nominal ethnic group (e.g., Georgians) in each of the four countries and their de facto territories. We include this variable because we expect areas with low levels of titular nationals to be more likely to engage in violent struggles for autonomy. We collect these data from the most recent publicly available census information. South Ossetia was a notable exception; percentage titular is 20 percent based on estimates made by the International Crisis Group (2010) after the August 2008 war there. In each of the three de facto states, the titular nationality of the parent states—Georgian for Abkhazia and South Ossetia and Azerbaijani for Nagorno-Karabakh—is used for consistency across cases and in partial acknowledgment of the undetermined status of these polities.

Where possible, census data are also used to calculate percentage urban population within subnational units to control for the relationship between city location and observed violence. In Georgia, Azerbaijan, and Armenia, percentage urban is reported in the most recent censuses. The Russian state statistical agency, Goskomstat, maintains a database of economic indicators for municipalities. In the de facto states, measures of percentage urban are reported inconsistently across data sources. Nagorno-Karabakh Republic includes this information in the regional-level results of the 2005 census. For Abkhazia and

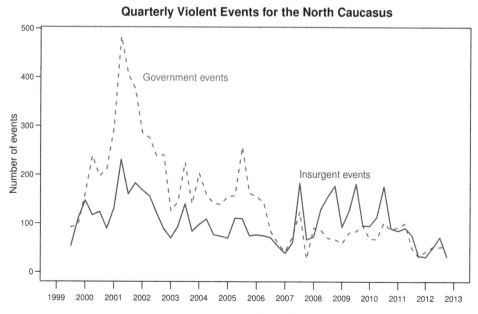

Figure 4. North Caucasus insurgent (solid) and, separately, government (dashed) violence by three-month periods between 1999 and 2012. (Color figure available online.)

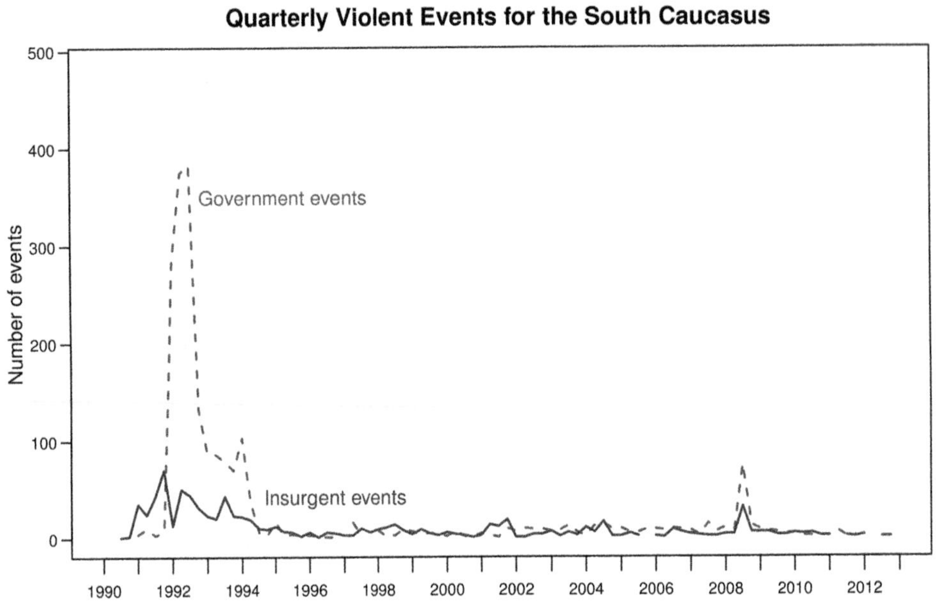

Quarterly Violent Events for the South Caucasus

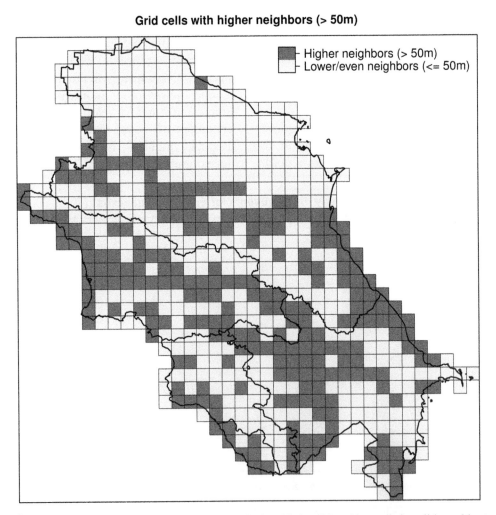

Figure 5. South Caucasus insurgent (solid) and, separately, government (dashed) violence by three-month periods between 1990 and 2013. (Color figure available online.)

Grid cells with higher neighbors (> 50m)

Figure 6. By 25 km² grid cell units of observation, the designation of higher (dark cells) and lower (light cells) neighboring terrain variables across the study area. Of 641 grid cells, 242 are designated as higher neighboring terrain and 399 as lower. (Color figure available online.)

South Ossetia, we generate estimates on the percentage urban from the results of two separate surveys conducted in the regions in March and November 2010, respectively (O'Loughlin, Kolossov, and Toal 2011; Toal and O'Loughlin 2013).

Percentage employed is intended as a proxy for the level of economic development or wealth in the subnational units as well as a measure of state capacity. To more fully evaluate the latter condition, we collect data on the percentage of people employed in either government or private business. Constructing the variable in this fashion leaves out respondents who indicate that they are self-employed. Where these data were not available in the most recent census, we use the most current available data from other governmental sources. For the de facto states of Abkhazia and South Ossetia, this metric is based on the results of our two 2010 surveys.

Our population data are from the Center for International Earth Science Information Network (CIESIN 2004) for the year 2000 and are static throughout the time series. We aggregate the population raster image within our grid cells using an area-weighted zonal statistic. We log transform these data for our analysis because they are highly skewed.

We have three structural and physical geographical controls. Percentage forest cover is calculated as the mean tree cover in each unit of analysis for the year 2000. This metric is derived from Landsat imagery with each unit assigned the percentage of closed vegetation canopy taller than 5 m in height (Hansen et al. 2013). Distance to an international border is calculated by creating a 2 km^2 raster layer where each pixel represents the distance to the nearest international border. From this raster image, the mean distance to a border is measured for each unit. This variable is log-transformed in our analysis. Distance to a road is similarly calculated using a finer resolution raster layer and taking the mean value within every unit. The road data are from version one of the Global Roads Open Access Data Set (gROADS; CIESIN 2013).

Methods

Our estimation captures the effect of violence perpetrated by either the government or insurgents at time $t-1$ (an action) on the opposing party's behavior during time t (the reaction). Each proposition for the corresponding terrain type calls for a comparison of

coefficients from two regression models:

$$Y_{Git} = \beta_0 + \beta_1 Z_{Git-1} + \beta_2 Z_{Iit-1} + \beta_3 X_{it} \\ + f_1(UTM\_E, \ UTM\_N) + f_2(M) + R_{it} + e_{it}$$

(1)

and

$$Y_{Iit} = \beta_0 + \beta_1 Z_{Git-1} + \beta_2 Z_{Iit-1} + \beta_3 X_{it} \\ + f_1(UTM\_E, \ UTM\_N) + f_2(M) + R_{it} + e_{it},$$

(2)

where Y_{Git} is the outcome measurement of violent events by government (G) forces in spatial unit i for month t. In Equation 1, coefficient β_1 captures the influence of the control variable for prior government events, Z_G, in the neighboring area at $t-1$. β_2 is the quantity of interest, measuring how strongly previous insurgent violence (Z_{Iit-1}) predicts government violent events. Vector β_3 captures the influence of the matrix of controls X. β_0 is the model intercept and e_{it} measures unexplained error. A thin plate spline smoothing function, f_1, is applied to the easting and northing UTM location coordinates and f_2 is a similar spline for the month identifier, M. Each model estimate includes fixed effects (R) for the republic (e.g., Dagestan) or country (e.g., Georgia). Equation 2 differs from Equation 1 in two ways. First, the outcome Y_{Iit} is the count of insurgent events (I) per unit month (it) instead of government incidents. Second, the quantity of interest is β_1, measuring the influence of prior nearby government events (in Equation 1 this was a control variable for prior activity).

Most observational data are characterized by spatial dependencies (Anselin 1988) and we address this in our estimation. First, each model includes a space–time-lagged measurement of the actor-specific conflict event outcome. Second, we use a generalized additive model (GAM) with a spatial smoothing term for the location coordinates of each unit of observation (see Wood [2004, 2006], and an application to conflict analysis in Zhukov [2012]). Similar to the implementation in Wood (2004), our GAM method controls for the effect of location on the outcome of interest.

Our overdispersed event counts outcome variable calls for a negative binomial functional form; this distribution requires a theta (θ) dispersion parameter that we estimate from an identically specified generalized linear model (these initial GLM model results are not reported). We cluster standard errors at the unit of

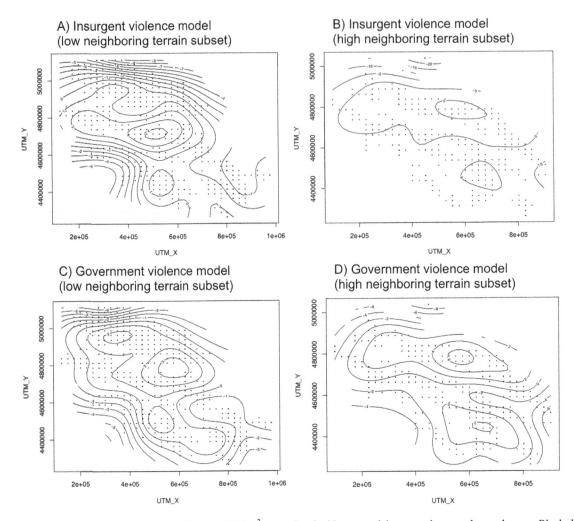

Figure 7. Spatial smoothing terms (or "splines") of our 25 km² generalized additive models mapped across the study area. Black dots represent the centroid of each grid cell. The models presented in A through D correspond to our main propositions.

analysis. In Figure 7 we map the coefficients of the spatial smoothing term for the 25 km² grid cell resolution as an example of the methodology. The influences of locational context are clearly visible across models (Figures 7A–7D). See the Supplemental Material for the temporal smoothing term.

Results

Our spatial analysis results are presented in Table 2. The main estimate to interpret for each actor is the term measuring the spatial lag of the opposing side's activity at time $t - 1$ (shown in bold). Results from logistic and Poisson regression analyses are presented in our Supplemental Material as robustness tests.

To evaluate the propositions, we compare model results across terrain subsets. Model numbers in Table 2 correspond directly to Figure 2. The estimate

for government reactions to insurgent events in the higher neighboring terrain subset (Model Ib) is compared with the insurgent reactions to government events also in the higher neighboring terrain subset (Model Ia). Using a 25 km² resolution we find that every government-initiated conflict event is associated with 9.5 percent more insurgent violence in the following time period (0.091 increase in log event count). In contrast, government reactions to insurgent events in this context are not statistically significant, lending support to our first proposition. Our analysis thus suggests that nonstate armed actors enjoy a strategic advantage in areas with higher nearby terrain.

The related second proposition, which posits that governments would enjoy advantages that result in stronger reciprocity in lower neighboring terrain regions, has no support. The estimates of both insurgent reactions (Model IIa) and government reactions (Model IIb) in areas with lower neighboring elevation

Table 2. Spatial generalized additive model results including all indicators of interest, control variables, and model diagnostics for each spatial resolution

| | Higher neighboring terrain (I) | | | | | | Lower neighboring terrain (II) | | | | | |
| | Insurgent event predictors (Model Ia) | | | Government event predictors (Model Ib) | | | Insurgent event predictors (Model IIa) | | | Government event predictors (Model IIb) | | |
	Est.	SE	p	Est.	SE	p	Est.	SE	p	Est.	SE	p
						25 km² spatial resolution						
(Intercept)	−12.923	2.015	0.000	−12.332	1.891	0.000	−15.096	2.454	0.000	−10.390	2.137	0.000
Spatial lag insurgent events	0.403	0.058	0.000	**0.012**	**0.059**	0.834	0.572	0.104	0.000	**−0.089**	**0.136**	**0.512**
Spatial lag government events	**0.091**	**0.033**	**0.006**	0.304	0.031	0.000	**−0.048**	**0.068**	**0.481**	0.372	0.077	0.000
Titular percentage	−0.002	0.007	0.823	0.007	0.006	0.243	−0.036	0.011	0.001	−0.017	0.008	0.048
Employed percentage	0.099	0.047	0.036	0.087	0.034	0.011	−0.049	0.059	0.409	−0.042	0.042	0.315
Urban percentage	0.012	0.010	0.202	0.006	0.008	0.454	0.021	0.013	0.108	0.015	0.010	0.124
Population (ln)	0.797	0.152	0.000	0.784	0.161	0.000	1.063	0.178	0.000	0.879	0.162	0.000
Forest cover (ln)	−0.005	0.006	0.404	−0.011	0.006	0.074	0.003	0.013	0.815	0.011	0.011	0.333
Border distance (ln)	−0.601	0.197	0.002	−0.548	0.197	0.005	0.697	0.514	0.175	−0.106	0.272	0.695
Road distance (ln)	−0.210	0.153	0.169	−0.228	0.153	0.136	−0.067	0.140	0.634	−0.097	0.135	0.473
θ	0.574			0.262			0.125			0.091		
AIC	14884.4			18648.7			7728.1			10345.9		
AUC	0.945			0.942			0.938			0.928		
N	55,598			55,598			81,550			81,550		
DV events	4,953			7,812			1,466			2,059		
						10 km² spatial resolution						
(Intercept)	−12.523	1.582	0.000	−11.338	1.608	0.000	−13.474	1.810	0.000	−12.866	1.842	0.000
Spatial lag insurgent events	0.969	0.307	0.002	**0.363**	**0.265**	**0.171**	0.895	0.177	0.000	**0.276**	**0.201**	**0.170**
Spatial lag government events	**0.240**	**0.132**	**0.069**	0.970	0.107	0.000	**0.003**	**0.085**	**0.970**	0.409	0.097	0.000
Titular percentage	−0.006	0.007	0.373	0.013	0.007	0.045	−0.005	0.009	0.571	−0.005	0.011	0.640
Employed percentage	0.096	0.026	0.000	0.106	0.026	0.000	0.031	0.034	0.365	0.052	0.034	0.124
Urban percentage	0.029	0.006	0.000	0.019	0.006	0.001	0.010	0.010	0.277	0.006	0.009	0.482
Population (ln)	0.379	0.103	0.000	0.406	0.121	0.001	0.676	0.175	0.000	0.547	0.218	0.012
Forest cover (ln)	−0.002	0.005	0.616	−0.009	0.005	0.057	0.008	0.006	0.169	0.006	0.005	0.167
Border distance (ln)	0.520	0.193	0.007	0.063	0.202	0.754	0.386	0.204	0.059	0.126	0.169	0.457
Road distance (ln)	−0.237	0.085	0.005	−0.220	0.075	0.003	−0.356	0.102	0.001	−0.273	0.072	0.000
θ	0.100			0.100			0.100			0.100		
AIC	18,006.2			23,711.3			18,893.6			25,076.6		
AUC	0.954			0.956			0.951			0.946		
N	252,701			252,701			561,833			561,833		
DV events	2,999			4,797			3,442			5,082		

(Continued on next page)

Table 2. Spatial generalized additive model results including all indicators of interest, control variables, and model diagnostics for each spatial resolution (*Continued*)

| | Higher neighboring terrain (I) | | | | | | Lower neighboring terrain (II) | | | | | |
| | Insurgent event predictors (Model Ia) | | | Government event predictors (Model Ib) | | | Insurgent event predictors (Model IIa) | | | Government event predictors (Model IIb) | | |
	Est.	SE	p	Est.	SE	p	Est.	SE	p	Est.	SE	p
50 km² spatial resolution												
(Intercept)	−2.853	2.603	0.273	−9.052	2.243	0.000	−12.145	2.756	0.000	−7.895	2.462	0.001
Spatial lag insurgent events	0.273	0.076	0.000	0.039	0.064	0.540	0.292	0.096	0.002	−0.071	0.067	0.291
Spatial lag government events	0.050	0.034	0.142	0.218	0.027	0.000	0.032	0.046	0.495	0.266	0.040	0.000
Titular percentage	−0.050	0.014	0.001	−0.036	0.009	0.000	−0.033	0.015	0.026	0.015	0.010	0.136
Employed percentage	−0.072	0.082	0.377	−0.008	0.061	0.899	0.050	0.081	0.537	0.069	0.061	0.260
Urban percentage	0.042	0.010	0.000	0.021	0.009	0.020	−0.017	0.023	0.447	−0.018	0.017	0.314
Population (ln)	0.978	0.157	0.000	0.473	0.160	0.003	1.847	0.328	0.000	1.422	0.305	0.000
Forest cover (ln)	−0.026	0.012	0.023	0.003	0.012	0.810	0.037	0.014	0.008	0.029	0.015	0.053
Border distance (ln)	−0.734	0.496	0.139	0.611	0.366	0.095	−1.251	0.373	0.001	−1.156	0.370	0.002
Road distance (ln)	−0.118	0.348	0.733	0.197	0.218	0.367	−0.345	0.213	0.106	−0.570	0.170	0.001
θ	0.608			0.300			0.274			0.144		
AIC	10,387.0			12,428.8			3,897.8			5,548.6		
AUC	0.937			0.932			0.925			0.922		
N	18,083			18,083			18,654			18,654		
DV events	5,600			8,161			759			1,700		
Rayons resolution												
(Intercept)	−11.763	2.121	0.000	−11.529	2.244	0.000	−15.587	2.555	0.000	−11.083	2.132	0.000
Spatial lag insurgent events	0.240	0.027	0.000	0.099	0.035	0.005	0.831	0.098	0.000	0.530	0.066	0.000
Spatial lag government events	0.157	0.030	0.000	0.270	0.023	0.000	0.107	0.049	0.029	0.346	0.038	0.000
Titular percentage	0.000	0.008	0.991	0.003	0.007	0.713	−0.013	0.011	0.261	−0.011	0.008	0.181
Employed percentage	0.019	0.019	0.299	0.015	0.018	0.399	−0.050	0.024	0.037	−0.043	0.022	0.056
Urban percentage	0.004	0.004	0.390	0.003	0.004	0.460	0.001	0.005	0.871	−0.001	0.004	0.812
Population (ln)	0.577	0.117	0.000	0.496	0.101	0.000	0.631	0.189	0.001	0.459	0.155	0.003
Forest cover (ln)	0.011	0.011	0.339	0.000	0.009	0.967	0.005	0.010	0.648	0.009	0.009	0.345
Border distance (ln)	−0.230	0.368	0.532	−0.154	0.409	0.707	0.948	0.492	0.054	0.595	0.432	0.168
Road distance (ln)	0.200	0.137	0.146	0.395	0.122	0.001	0.340	0.124	0.006	0.165	0.118	0.161
θ	0.407			0.328			0.125			0.247		
AIC	14,399.8			17,003.8			8,305.51			10,889.7		
AUC	0.921			0.917			0.925			0.927		
N	33,497			33,497			31,478			31,478		
DV events	4,427			6,641			1,983			3,231		

Note: AIC = Akaike information criterion; AUC = area under the receiver–operator characteristics curve; DV = dependent variable violent event count.

profiles are not statistically significant at conventional levels. One reason for a government's inability to reciprocate after insurgent violence is that in low terrain areas, insurgents retreat beyond the range that is captured in our units of analysis. In other words, government reciprocity might not be expected in the immediate vicinity of an insurgent incident.

The possibility that movements of either actor across the operational territory could influence our results, as in the preceding hypothetical scenario, is sound justification for examining our conclusions across alternative spatial resolutions of analysis. Our results in Table 2 show that at 10 km^2 and at 50 km^2 support for the first proposition disappears; insurgent reactions to government violence are no longer statistically significant in high neighboring terrain areas (at $p \leq 0.1$ the 10 km^2 result is noteworthy). Although this exercise calls into question the robustness of a 25 km^2 test of the first proposition, our rejection of the second proposition is consistent across the different spatial resolution aggregations. Because government forces still have statistically insignificant reactions to insurgent violence in low neighboring terrain regions at 10 km^2 and at 50 km^2, we are reassured that our conclusion for this proposition is not heavily biased by the MAUP.

In the bottom panel of Table 2 we present our estimates for rayons. At this spatial resolution each proposition estimate is statistically significant. This difference in model results could be due to the strategic importance of administrative unit borders for both actors. Model IIb shows that for every insurgent event in lower neighboring terrain regions, there is 69 percent more government violence (0.530 increase in log event count). In lower neighboring terrain regions, government events lead to a comparatively small increase in insurgent-led conflict of 11.2 percent (Model IIa; 0.107 increase in log event count). These results strongly support the second proposition, in contrast to our earlier models. Governments appear to have an advantage in regions that do not have high-elevation surrounding rayons. The first proposition, which has tenuous support across grid cell analyses earlier (only the 25 km^2 resolution), is supported for the rayon-scale dynamics of violence. Comparing Models Ia and Ib, insurgent reactive violence is substantially stronger in areas with high neighboring terrain than government responses; a government event is associated with 16.9 percent more insurgent violence (0.157 increase in log event count), whereas an insurgent event correlates with a 10.4 percent increase in government-led conflict events (0.099 increase in log

event count). Although each of the estimates is statistically significant, the greater magnitude of insurgent reciprocity in areas with high-elevation terrain nearby supports the first proposition.

Conclusions

Our study complicates any overly simplistic narratives of conflict actor behavior relative to mountainous terrain and also questions several assumptions that researchers make about the contexts within which civil conflict takes place. All of Russia has often been coded as experiencing civil war at a country level due to fighting in the North Caucasus region, a small area relative to the entire country. Our ability to study intrastate armed conflict within the area where the conflict is taking place isolates specific trends that are unobservable at comparatively coarse spatial resolutions but are expected and intuitive. Our finding that operational costs of context might result in mountainous terrain suiting armed insurgents against government forces is conditional on the definition of geographical context; the distances to which either party is expected to travel within units of observation bring us to different conclusions about conflict behavior. Although mountainous terrain might allow nonstate armed forces to organize, train, and supply, this does not necessarily translate into the blow-by-blow advantage for insurgents in areas near and within mountainous regions, as our results for certain spatial resolutions (10 km^2 and 50 km^2) have shown.

Examining whether our conclusions hold for other regions of the world is a promising path for future research. There is strong evidence, though, that alternative definitions of context, which here is operationalized as the spatial resolution of analysis units, will reveal variable effect estimates for conflicts in any region of the world and for most indicators of interest. As a result, researchers should either provide strong justifications for the reasoning that leads them to adopt their preferred units of analysis or provide transparent results for alternative boundaries. In our ongoing work, we also plan to investigate whether reciprocity between government and nonstate armed actors is characterized by varying temporal dimensions of influence.

The inductive style of analysis we carry out—one that probes effect estimates across a range of geographical scales—is the most transparent approach to studying violent conflict using geographical data at spatially disaggregated scales. Although the physical geographical setting for violence between actors in a civil war setting defines operational limitations, we also stress the influence of MAUP issues for the quantitative study of violence. Presenting research audiences with a single universal effect estimate for any dynamic of violence might ignore messy social realities that shape conflict processes place by place and region by region across the world.

Acknowledgments

We acknowledge the work of many undergraduate research assistants who coded and georeferenced the events data. We thank two reviewers for comments that forced us to improve this article. Nancy Thorwardson prepared Figure 1 for publication. Supporting information documents and replication materials are available online through the corresponding author's Web site.

Funding

The authors thank the National Science Foundation's Human and Social Dynamics program (grant numbers 0433927 and 0827016 to Principal Investigator John O'Loughlin) for the financial support that made possible both field work in the North and South Caucasus between 2005 and 2012 and the violent events data collection.

Supplemental Material

Supplemental analysis for this article can be accessed on the publisher's Web site at http://dx.doi.org/10.1080/24694452.2016.1243038. Table S1 presents descriptive statistics for our data sets at alternative spatial resolutions (10 km^2 and 50 km^2). Figure S1 shows the temporal smoothing term for month ID that corresponds with the spatial smoothing term of our main model presented in Figure 7. Table S2 shows estimates for a Poisson functional form of our model of conflict event count outcomes. In Table S3 we present the results for a binary version of the outcome variable in a logistic regression estimate.

Note

1. We adopt the term *rayon* to refer to the county-scale units in the North Caucasus as well as the subnational units in the South Caucasus, which are variously termed in these countries (Georgia = municipality; Armenia = *marz*; and Azerbaijan = rayon).

References

Anselin, L. 1988. *Spatial econometrics: Methods and models.* Dordrecht, The Netherlands: Kluwer.

Baudains, P., S. Johnson, and A. Braithwaite. 2013. Geographic patterns of diffusion in the 2011 London riots. *Applied Geography* 45:211–19.

Boulding, K. 1962. *Conflict and defense: A general theory.* New York: Harper and Brothers.

Bremmer, I. 1997. Post-Soviet nationalities theory: Past, present, and future. In *New states, new politics: Building the post-Soviet nations,* ed. I. Bremmer and R. Taras, 3–26. New York: Cambridge University Press.

Buhaug, H., and S. Gates. 2002. The geography of civil war. *Journal of Peace Research* 39 (4): 417–33.

Buhaug, H., and K. S. Gleditsch. 2008. Contagion or confusion? Why conflicts cluster in space. *International Studies Quarterly* 52 (2): 215–33.

Buhaug, H., K. S. Gleditsch, H. Holtermann, G. Østby, and A. F. Tollefsen. 2011. It's the local economy, stupid! Geographic wealth dispersion and conflict outbreak location. *Journal of Conflict Resolution* 55 (5): 814–40.

Buhaug, H., and J. K. Rød. 2006. Local determinants of African civil wars, 1970–2001. *Political Geography* 25 (3): 315–35.

Center for International Earth Science Information Network (CIESIN). 2004. Center for International Earth Science Information Network (2004) gridded population of the world. http://sedac.ciesin.columbia.edu/gpw/index.jsp (last accessed 1 December 2015).

———. 2013. *Global Roads Open Access data set.* Version 1 (gROADSv1). Palisades, NY: NASA Socioeconomic Data and Applications Center.

Cornell, S. 2002. Autonomy as a source of conflict: Caucasian conflicts in theoretical perspective. *World Politics* 54 (2): 345–76.

Cunningham, D. E, K. S. Gleditsch, and I. Saleyhan. 2009. It takes two: A dyadic analysis of civil war duration and outcome. *Journal of Conflict Resolution* 53 (4): 570–97.

Daly, S. Z. 2012. Organizational legacies of violence: Conditions favoring insurgency onset in Colombia, 1964–1984. *Journal of Peace Research* 49 (3): 473–91.

de Waal, T. 2003. *Black garden: Armenia and Azerbaijan through peace and war.* New York: New York University Press.

Do, Q.-T., and L. Iyer. 2010. Geography, poverty, and conflict in Nepal. *Journal of Peace Research* 47 (6): 735–48.

Farr, T. G., P. A. Rosen, E. Caro, R. Crippen, R. Duren, S. Hensley, M. Kobrick, et al. 2007. The Shuttle Radar Topography Mission. *Review of Geophysics* 45 (2): RG2004.

Fearon, J., and D. Laitin. 2003. Ethnicity, insurgency, and civil war. *American Political Science Review* 97 (1): 75–90.

Grundy, K. W. 1971. *Guerrilla struggle in Africa: An analysis and preview.* New York: Grossman.

Guevara, C. 1961. *Guerrilla warfare.* New York: Monthly Review Press.

Hansen, M. C., P. V. Potapov, R. Moore, M. Hancher, S. A. Turubanova, A. Tyukavina, D. Thau, et al. 2013. High-resolution global maps of 21st-century forest cover change. *Science* 342:850–53.

Hegre, H. 2008. Gravitating toward war: Preponderance may pacify but power kills. *Journal of Conflict Resolution* 52 (4): 566–89.

Hegre, H., and N. Sambanis. 2006. Sensitivity analysis of empirical results on civil war onset. *Journal of Conflict Resolution* 50 (4): 508–35.

Houweling, H. W., and J. G. Siccama. 1985. The epidemiology of war, 1816–1980. *Journal of Conflict Resolution* 29 (4): 641–63.

International Crisis Group. 2010. South Ossetia: The burden of recognition. *Europe Report No. 205*, 7 June 2010. http://www.crisisgroup.org/~/media/Files/europe/205%20South%20Ossetia%20-%20The%20Burden%20of%20Recognition.ashx (last accessed 29 November 2015).

Kalyvas, S. 2006. *The logic of violence in civil war*. New York: Cambridge University Press.

Kolstø, P. 2006. The sustainability and future of unrecognized quasi-states. *Journal of Peace Research* 43 (6): 723–40.

Linke, A. M., and J. O'Loughlin. 2015. Reconceptualizing, measuring, and evaluating distance and context in the study of conflicts: Using survey data from the North Caucasus of Russia. *International Studies Review* 17 (1): 1–19.

Linke, A. M., S. Schutte, and H. Buhaug. 2015. Population attitudes and the diffusion of political violence in sub-Saharan Africa. *International Studies Review* 17 (1): 107–25.

Linke, A. M., F. D. W. Witmer, and J. O'Loughlin. 2012. Space–time Granger analysis of the war in Iraq: A study of coalition and insurgent action–reaction. *International Interactions* 38 (4): 402–25.

Lyall, J. 2010. Are co-ethnics more effective counterinsurgents? Evidence from the second Chechen war. *American Political Science Review* 104 (1): 1–20.

McColl, R. 1969. The insurgent state: Territorial bases of revolution. *Annals of the Association of American Geographers* 59 (4): 613–31.

Medina, R., and G. Hepner. 2011. Advancing the understanding of sociospatial dependencies in terrorist networks. *Transactions in GIS* 15 (5): 38–72.

Nemeth, S. C., J. A. Mauslein, and C. Stapley. 2014. The primacy of the local: Identifying terrorist hot spots using geographic information systems. *Journal of Politics* 76 (2): 304–17.

O'Loughlin, J. 2000. Geography as space and geography as place: The divide between political science and political geography continues. *Geopolitics* 5 (3): 126–37.

O'Loughlin, J., E. C. Holland, and F. D. W. Witmer. 2011. The changing geography of violence in the North Caucasus of Russia, 1999–2011: Regional trends and local dynamics in Dagestan, Ingushetia and Kabardino-Balkaria. *Eurasian Geography and Economics* 52 (5): 596–630.

O'Loughlin, J., V. Kolossov, and G. Toal. 2011. Inside Abkhazia: A survey of attitudes in a de facto state. *Post-Soviet Affairs* 27 (1): 1–36.

O'Loughlin, J., and F. D. W. Witmer. 2011. The localized geographies of violence in the North Caucasus of Russia, 1999–2007. *Annals of the Association of American Geographers* 101 (1): 178–201.

O'Loughlin, J., F. D. W. Witmer, and A. M. Linke. 2010. The Afghanistan–Pakistan wars, 2008–2009: Microgeographies, conflict diffusion, and clusters of violence. *Eurasian Geography and Economics* 51 (4): 437–71.

Openshaw, S. (1983). *The modifiable areal unit problem*. Norfolk, UK: Geo Books.

Pickering, S. 2012. Proximity, maps and conflict: New measures, new maps and new findings. *Conflict Management and Peace Science* 29 (4): 425–43.

Radil, S. M., C. Flint, and S.-H. Chi. 2013. A relational geography of war: Actor–context interaction and the spread of World War I. *Annals of the Association of American Geographers* 103 (6): 1468–84.

Raleigh, C., and H. Hegre. 2009. Population size, concentration, and civil war: A geographically disaggregated analysis. *Political Geography* 28 (4): 224–38.

Raleigh, C., A. M. Linke, H. Hegre, and J. Karleson. 2010. Introducing ACLED: An armed conflict location and event dataset. *Journal of Peace Research* 47 (5): 651–60.

Rustad, S. C. A., J. K. Rød, W. Larsen, and N. P. Gleditsch. 2008. Foliage and fighting: Forest resources and the onset, duration, and location of civil war. *Political Geography* 27 (7): 761–82.

Saleyhan, I. 2009. *Rebels without borders: Transnational insurgencies in world politics*. Ithaca, NY: Cornell University Press.

Schutte, S., and N. Weidmann. 2011. Diffusion patterns of violence in civil war. *Political Geography* 30 (3): 143–52.

Toal, G., and J. O'Loughlin. 2013. Inside South Ossetia: A survey of attitudes in a de facto state. *Post-Soviet Affairs* 29 (2): 136–72.

Toft, M. D. 2003. *The geography of ethnic violence: Identity, interests, and the indivisibility of territory*. Princeton, NJ: Princeton University Press.

Tollefsen, A. F., and H. Buhaug. 2015. Insurgency and inaccessibility. *International Studies Review* 17 (1): 6–25.

Weidmann, N. B., and I. Saleyhan. 2012. Violence and ethnic segregation: A computational model applied to Baghdad. *International Studies Quarterly* 57 (1): 52–64.

Weidmann, N. B., and M. D. Ward. 2010. Predicting conflict in space and time. *Journal of Conflict Resolution* 54 (6): 883–901.

Wimmer, A. 2002. *Nationalist exclusion and ethnic conflict: Shadows of modernity*. New York: Cambridge University Press.

Wood, S. 2004. Stable and efficient multiple smoothing parameter estimation for generalized additive models. *Journal of the American Statistical Association* 99 (467): 673–86.

———. 2006. *Generalized additive models: An introduction with R*. Boca Raton, FL: Chapman and Hall.

Zhukov, Y. 2012. Roads and the diffusion of insurgent violence: The logistics of conflict in Russia's North Caucasus. *Political Geography* 31 (3): 144–56.

Making Mountain Places into State Spaces: Infrastructure, Consumption, and Territorial Practice in a Himalayan Borderland

Galen Murton ⓘ

This article looks at a trans-Himalayan borderland to see how new road development projects affect social and sovereign relationships across mountain landscapes between Chinese Tibet and Mustang, Nepal. Research asked about local experiences with new forms of motorized transport and popular consumption of Chinese-manufactured commodities to understand what factors led the Nepali state to undertake new bureaucratic projects in a historically peripheral space. Employing a dialectic framework of mobility and containment, a materialist-territorial analysis reveals how transborder infrastructure development affects trade relations and consumption practices in the Nepal–China borderlands and, in turn, how these dynamics condition state-making processes at social and geopolitical levels. Following the cross-scalar trajectory of one rural road project from local grassroots initiative to national development program to international transportation network, I argue that the economic interests of a place-based project with regional cultural connections set in motion an expanding presence of Nepali state apparatuses in a trans-Himalayan borderland space.

本文聚焦横跨喜马拉雅的边境地带, 探讨崭新的道路建设计画, 如何影响中国西藏和尼泊尔木斯塘山岳地景间的社会和主权关系。本研究质问崭新的汽车运输和对中国製造商品的大众消费的在地经验, 藉此理解是什麽样的因素导致尼泊尔政府在历史上的边陲空间中采用崭新的官僚计画。运用流动性和围堵的辩证架构, 唯物论的领土分析揭露了跨越国界的基础建设如何影响尼泊尔—中国边境的贸易关系和消费实践, 以及这些动态如何回头决定社会和地理层级的国家打造过程之条件。我透过追寻一个乡村道路建设计画的在地草根倡议到国家发展计画以至国际运输网络的跨尺度轨迹, 主张以地方为基础、且有着区域文化连结的计画之经济利益, 驱动了尼泊尔国家机器在跨喜马拉雅边境空间中的扩张。 关键词: 关键词·中国, 消费, 喜马拉雅, 流动, 尼泊尔, 领域性。

Este artículo explora una zona fronteriza transhimalaya para ver cómo los nuevos proyectos de desarrollo de vías afectan las relaciones sociales y la soberanía a través de los paisajes montañosos situados entre el Tíbet chino y Mustang, Nepal. En la investigación se preguntó acerca de las experiencias locales con nuevas formas de transporte motorizado y el consumo popular de mercaderías de manufactura china para entender qué factores llevaron al estado nepalí a emprender nuevos proyectos burocráticos en un espacio históricamente periférico. Empleando un marco dialéctico de movilidad y contención, un análisis materialista-territorial revela cómo el desarrollo de infraestructura transfronteriza afecta las relaciones comerciales y las prácticas de consumo en la zona fronteriza Nepal–China y, a la vez, cómo esta dinámica condiciona los procesos de construcción de estado a los niveles sociales y geopolíticos. Siguiendo la trayectoria de un proyecto de carreteras rurales a partir de una iniciativa eminentemente local y básica hasta el programa nacional de desarrollo y a la red internacional de transporte, sostengo que los intereses económicos de un proyecto de asiento local con las conexiones culturales regionales desatan una presencia en expansión del andamiaje estatal nepalí en un espacio fronterizo transhimalayo.

A Tibetan cultural region and administrative district of northern Nepal, the former Kingdom of Mustang has long bridged the ecological, cultural, and economic worlds of both Nepal and China. Resting north of the Great Himalayan Divide but penetrated by rivers that flow to the Bay of Bengal, Mustang's high mountain landscape supports trade corridors that link the Tibetan Plateau with the plains of India (van Spengen 2000). Merchants, pilgrims, and armies have traversed these routes for centuries, giving rise to a unique social landscape that largely transcends modern demarcations of a bordered world (Fürer-Haimendorf 1988; Scott 2009). Working against this mobile history, however, geopolitics of the mid-twentieth century repositioned Mustang as a sensitive borderland space and introduced new terms of isolation and containment for the district. More recently, border policies have changed again, as new road systems and market dynamics reposition local populations as consumer subjects in cash-based political economies. As a consequence of these trends, both Chinese and Nepali authorities have introduced new bureaucratic institutions to regulate and solidify transborder trade in Mustang, such that new forms of governance increasingly advance into spaces where significant state presence has been historically absent.

This article looks at a trans-Himalayan borderland to examine how new road development projects affect social and sovereign relationships between Chinese Tibet and Mustang, Nepal. Research asked about local experiences with new forms of motorized transport and what factors led the Nepali state to undertake new bureaucratic projects in Mustang. On the basis of a materialist-territoriality analysis (Giddens 1986; Agnew and Corbridge 2002; Hetherington 2007), I argue that the development of road infrastructure in Mustang has produced three interdependent outcomes: (1) new consumer subjectivities predicated on transborder trade; (2) new mobilities that reinscribe long-standing social hierarchies; and (3) new political economies that make space for the Nepali state to take shape.

Employing a dialectic framework of mobility and containment, this study shows how transborder infrastructure development and trade relations condition territorial processes in the Nepal–China borderlands at social and geopolitical levels. *Mobility* implies a centrifugal force (Kristof 1959) and refers to the ability of both goods and people to move across the Nepal–China border as well as a historical practice formative to place-based identities (Sheller and Urry 2006; Cresswell and Merriman 2012; Harris 2013; Cresswell 2016). Aspects of mobility include Mustang's place along the historical trans-Himalayan Salt Trade; the provenance, cultivation, and legacy of Tibetan Buddhist monastic institutions across the district; Mustang's key position in Nepal's strategic road network development program; the current proliferation of private and motorized vehicle systems in Mustang; and the dynamics of transborder trade and commerce that connect Mustang's villages with Chinese manufacturing centers as well as international remittance economies. Conversely, *containment* is the processes of bordering and delimitations of exclusion that made Mustang into an exceptional borderland space; containment is essentially centripetal and refers to the ways in which mobilities are foreclosed. Local experiences with the state in Mustang

have long been characterized by containment: suzerainty over the Kingdom of Mustang (Lo) by the region and state's successive regimes of rule, Nepal–China bordering policies of the 1960s, prohibition of foreigners in Mustang until 1992, Chinese fencing of the border in 1999 and 2000, and complex Sino–Nepali regulations over trade between Mustang and Tibet.

In dialectical terms, roads and border infrastructures are synthetic forms produced by a convergence of mobility (thesis) with containment (antithesis). Everyday experiences in Mustang take place at and are strongly conditioned by the dialectical intersections of mobility and containment. These intersectional places are the roads, the border posts, the trade fairs, and even the domestic spaces that mediate social relations between populations in Mustang and communities in Chinese Tibet as well as central Nepal. Most important for this article, it is at these points of convergence that the Nepali state increasingly and conspicuously takes shape—especially in the borderlands—in the form of institutional bodies that function to expand commerce, tax trade, regulate transport, and cultivate tourism.

Data for this study were generated through multisite, participatory fieldwork in twenty villages in Upper Mustang in spring and summer 2015, both on and off the main road through Mustang–Nepal Road F042. Methods were actively mobile and a key approach to participatory observation included travel by motorcycle, Jeep, truck, horse, and foot throughout Upper and Lower Mustang; this enabled grounded experiences with and observations of new vehicle road systems on the material and social life of multiple Mustang communities. Building on these mobile insights, qualitative and ethnographic data were produced from semistructured interviews ($n = 100$) and survey questionnaires ($n = 60$) and research informants were identified through both random and snowball sampling.

Geopolitics, Roads, and Histories of Mobility and Containment in Mustang

For centuries, Mustang has been seen and treated as a peripheral space by the Nepali state, such that little infrastructure or social services reached what was viewed from Kathmandu as a marginal borderland. An ancient Tibetan Buddhist kingdom with strong ties to the powerful, ruling Sakya monastic centers of southern Tibet, Mustang was incorporated into the nascent Nepali state in 1789 through an alliance with Gorkhali unifiers King Prithvi Narayan Shah and then his son, Regent Bahadur Shah (Dhungel 2002). Despite this eighteenth-century integration with the Kathmandu-based Hindu monarchy of Nepal, Mustang largely maintained its status as a semiautonomous kingdom from central Nepal rule through numerous cycles of governance, including Shah kings, Rana plutocrats, Panchayat rulers, and early democratic transitions. From the eighteenth century up until recent decades, Mustang's citizenry was neglected by the Nepali state despite the incorporation of the Mustang Kingdom into the Nepal polity. This neglect was especially evident in the lack of central investment and development in Mustang, particularly in the sectors of education, public health, and transportation.

Beginning in the 1950s, Mustang became a zone of concern for political leadership in both Kathmandu and Beijing. Following the Chinese People's Liberation Army (PLA) occupation of Tibet in 1951, Tibetan refugees took flight over the China–Nepal borders, and although thousands continued to refugee communities in India, many others remained in Nepal. Due to its close proximity to Tibetan territory, strong kinship and monastic relations with Tibetan populations and social–physical landscape congruent to the Tibetan Plateau, Mustang quickly became a major settlement area for Tibetan refugees. By 1960, this space of refuge also became the primary base for the CIA-funded Tibetan guerrilla resistance movement, Chushi Gangdruk (Four Rivers, Six Ranges), also known as the Khampa rebellion (McGranahan 2010).[1]

Controversy generated by the Tibetan resistance movement ultimately led to an effective closure of Mustang and reinforced a spatial and political marginalization for local communities over the course of two decades. As a result of the Tibetan resistance and subsequent pressure from the Chinese Embassy in Kathmandu following the surrender, arrest, and assassination of Chushi Gangdruk leaders, Mustang became even more geopolitically sensitive and an officially restricted state space. Due to a translation of political conflict across the Mustang–Tibet border into a formal Nepali state policy of isolation, Mustang remained closed to the outside world until 1992. This closure was maintained in Nepali state law, such that foreign travel was forbidden to the upper, northern reaches of Mustang and state services were barely delivered. Although modest numbers of urban and lowland Nepalis paid visits to the district between the 1960s and 1990s, livelihoods based on transborder pastoralism and trade were severely disrupted, resulting in the suspension of historical exchange patterns and curtailment of long-standing kinship practices between Mustang and Tibet.[2]

Figure 1. An ancient stupa on the approach to the Korala marks the route of the trans-Himalayan Salt Trade that today serves as the Mustang–Tibet road. (Color figure available online.)

Today, new road developments in Mustang are helping to rewrite trans-Himalayan border histories. Marked by ancient stupas that served as both meeting sites and border points (see Figure 1), the Mustang–Tibet road over the Korala pass ascends one of the oldest formal border crossings in the Tibet–Himalaya region and follows what was for centuries a central artery of the trans-Himalayan Salt Trade. Despite Mustang's central location on one of the most important trade routes between Tibet and Nepal, and between China and India more broadly, open roads (especially those capable of handling motorized transport) are extremely new to Mustang.

Modernizing road projects in Mustang bridge local, national, and regional scales, and four dimensions of the transborder road construction process bear close attention: location, production, connection, and circulation. Beginning in the mid-twentieth century, residents and leaders from Mustang appealed to the Nepali government to build the district a road, to better connect their communities with the rest of the country, and to facilitate improved access to commerce, education, medicine, and urban opportunities; the road was long promised by Kathmandu but never delivered (Thapa 2008). Eventually, Mustang communities took matters into their own hands, constructing a road between the capital, Lo Monthang, and the Nepal–China border at the Korala in the early 2000s. According to one of Mustang's highest ranking political leaders, over the course of several years between 2000 and 2004, each Village Development Committee from the upper villages of Mustang contributed annual dues of between 1 and 5 lakh Nepali rupees (~$1,000–$5,000 per annum) toward the construction of the road. Labor for the project was sourced both within the district and from outside. Despite conflicting reports about the financier and the labor involved for construction, it is important to note that the first road to and from Mustang connected the district not with somewhere else in central Nepal but instead to China via southern Tibet. As a result of this project and the expedience of motorized traffic, systems and scales of trade between Mustang and Tibet quickly began to change. New speeds and courses of travel have generated new currents of commodity flows into Mustang, and administrative bodies have been assigned to manage the increasing movement of trade volumes across the borderlands.

Betraying a twentieth-century history of social and political containment, new transport infrastructure in Mustang—predicated on Nepal Road F042—has produced new practices of mobility that are shifting social and economic relations in Mustang with both China and Nepal. Despite great expectations that roads can and will bring unprecedented levels of prosperity to rural Nepal (World Bank 2006), the physical viability of the Mustang roads remains marginal and they suffer from disconnection and routine breakage. For example, river crossings at several chokepoints require the offloading and transfer of cargo between vehicles or an otherwise precarious crossing of the Kali Gandaki River by Jeeps and wagons but only during low-water periods. This obstacle is a key reason why the Mustang road is most heavily traveled during the late winter and early spring months, when the Kali Gandaki gorge serves as a "winter road" and vehicles are able to ford the river more safely. Compounding this tenuous accessibility, however, are the perennial challenges of road maintenance exacerbated by Mustang's rugged mountain landscape and monsoonal climate patterns. Landslides are common in the spring and summer months, particularly in the lower reaches of the Mustang roads between Jomson and Beni/Pokhara. Even in the higher reaches, though, where Mustang's landscapes are situated in the rain shadow of the Annapurna and Dhaulagiri Himalaya, the earth remains unstable and the roads particularly prone to closure.

State Imaginaries of Infrastructure and Rural Realities of Roads

Road infrastructure development is a fundamental project of national state-making, and the impacts of roads are both heightened and complicated in borderland regions. One of only three operational linkages across the Nepal Himalayas, the Mustang–Tibet road, now integrated with Nepal Road F042, is a key component of the Government of Nepal's Strategic Road Network (Figure 2).[3] Specifically, the Strategic Road Network is vigorously promoted as one of the greatest new routes to economic prosperity and human development for the landlocked country (Post Report 2016). This is outlined in the Government of Nepal's latest Five-Year Plan for Transportation Infrastructure that privileges major road (and even highly ambitious and remote rail) construction as a cornerstone of twenty-first-century development and modernization (Ministry of Transportation 2016). Despite chronic problems with maintenance and the near impossibility of keeping the road open year-round in the near term, the Mustang roads remain key components of this national agenda. State thinking of this kind resonates with Harvey and Knox's (2015) insights on infrastructural imaginaries in the Peruvian highlands, where road development is seen as a "convergence of imaginative and material practices that bring into being new social and material relations" (6).

When looking at roads, Larkin (2013) revealed the ways in which technologies converge into systems and how these systems shape fundamental dimensions of everyday life. That is, even as infrastructures enable the movement of people and goods, they also control those movements and thus help to rewrite relations between such things (Larkin 2013). Especially in modern spaces of state-led development, where technologies are liberally promoted for their ability to "improve" life (Li 2007), the infrastructures that these technologies build become reified objects that obscure the actual social relations on which they are based. An "amalgam of technical, administrative, and financial techniques," road infrastructures "emerge out of and store within them forms of desire and fantasy and can take on fetish-like aspects that sometimes can be wholly autonomous from their technical function" (Larkin 2013, 328–29). Larkin's insights politicize infrastructural systems and thereby problematize the popular discourse behind road development and *bikas*—or development—in Nepal (Pigg 2009).

Recent geography literature has called for a more critical engagement between transportation literature and mobilities studies (Cidell and Lechtenberg 2016; Kwan and Schwanen 2016). Whereas research on transportation geography predominately examines

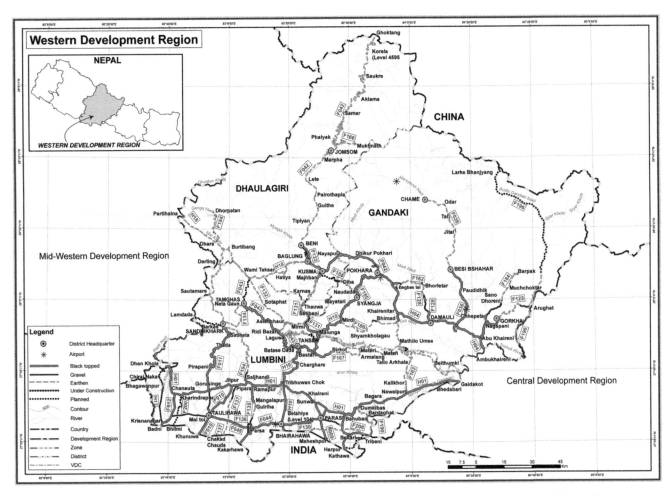

Figure 2. Map of national road development projects in Mustang District and Western Development Region of Nepal (Department of Roads, Ministry of Physical Infrastructure and Transportation, Government of Nepal). (Color figure available online.)

interactions at macrosocietal levels and through technological lenses, a mobilities approach—with particular emphasis on the social interaction of people, movement, and place (Cresswell 2010)—brings politics to bear on the key themes of transportation studies, including systems, networks, and technologies of transport. Taking the analysis further, Cidell and Lechtenberg (2016) argued that transportation is a political project and that transportation studies must be engaged as such: "Although transportation collapses space and time by altering the positionality of places relative to each other, it does so unevenly, shaping those places in the process" (259).

Today in Mustang, as in so many other places around the world, road developments are at once promoted as great equalizers even as they, in fact, reinforce longstanding social and economic stratifications. Jigme, a research informant in Mustang, succinctly described this reality to me during one of our conversations in the village of Kagbeni: "If you are poor, the road makes you

[feel] poorer." As a man without the means to own a private vehicle, and without the cash liquidity to charter a Jeep to carry goods to his store from bazaars in southern Tibet or central Nepal, he has become increasingly dependent on public transportation networks that do not keep reliable schedules. In addition to creating new costs that he formerly avoided by walking with his goods loaded onto mules, Jigme found that new roads have made life harder in Mustang. Although cargo costs are ultimately lower for truck transport than for mule caravans, the unpredictability of road travel adds physical and psychological hardship to what was previously a more dependable journey: "Before [when we walked], you knew how long a trip to Beni or Pokhara would take, and you planned for that [ten to twelve days]. Today, it might take you two days, or it might take you two weeks. We never know."

At Jigme's end of the socioeconomic spectrum, other locals begrudge the road more personally. One seasonal tea shop owner told me, "Before, it did not cost money to travel [you either walked or rode a horse]. But today, if you don't have money, you cannot travel like the others." The transition from a barter-based to a cash-based economy has shifted the terms of engagement by which Mustang's residents experience and struggle for access to the roads and the vehicles themselves. Another shop owner, Dorje, reminisced more fondly about the days when the only choice of transport was mule or yak caravan. In those days, he could predictably plan for a trip from Kagbeni to Pokhara, knowing that it would be long and arduous, but that ten to twelve days was a reliable and dependable timeline for the trip. Today, in contrast, although Dorje can quickly (and expensively) reach Pokhara by Jeep or bus, he does not control his own movement. If the Jeep does not come on time—or, for that matter, does not come that day or the next—he is stuck, with a load that he can not carry alone and subject to expenses that he must absorb or else go hungry and cold.

Jigme's and Dorje's comments point to the ways in which transport systems condition one's agency and how new roads trace old tracks of social division. When a truck breaks down and you have no other means of travel, you have little recourse than to pay for lodging until another vehicle comes by—and that might be in three hours or three days. If you are poor, road travel makes everything more difficult and thereby makes you (feel) poorer. Conversely, if you own your own vehicle or have the means to charter a private Jeep or at least sufficient disposable income to pay for lodging when necessary, motorized transport is perceived and lived very differently. For those with money, roads are expedient, and travel by vehicle comes easily; but for those without cash, it is quite the opposite experience. Ultimately, this demonstrates how roads are indeed vectors of both mobility and containment for individual and state bodies alike.

Commodity Circulations, Consumption Practices, and Making State Territory

The Chinese and Tibetan goods that one finds in Mustang today include the traditional and the modern, and the paths they follow are both old and new. In addition to mass-produced beverages and prepackaged foods, over 90 percent of informants' domestic spaces were characterized by Chinese-made goods such as carpets, ceramic mugs, dishware, vacuum thermoses, furniture, ready-made clothing, solar panels, and motorcycles. Although Chinese products are now

commonplace in Mustang, the imports come in just three major deliveries per year and via two main routes. Some goods are imported directly from the Tibetan Plateau along the newly developed track of the trans-Himalayan Salt Route—or the northernmost end of Nepal Road F042. This includes packaged food, canned drinks, kitchen tables, and metal stoves, as well as the staples of a highland Himalayan diet: salt, tea, meat, and butter (or *tsa*, *cha*, *sha*, and *mar*). In general, these products enter Mustang directly from Tibet only during brief periods in the spring and fall, when the Nepal–China border is opened for two to three weeks for the semiannual *tsongra* trade fair. Driven over the Korala pass from Likse to Lo Monthang, the import of these goods has jumped scale with the advent of the motor road between Tibet and Mustang. It is particularly in response to these imports that new bureaucracies were introduced in 2015 for the collection of customs and excise taxes on the Nepali side of the border.

The other vector for Chinese exports and imports to Mustang follows a more circuitous path and yet overall volume is ultimately higher. This route links Mustang with the markets of southern Tibet via central Nepal through road linkages between the new Beni-Jomsom road (southern extension of Nepal Road F042), the Kathmandu-Pokhara Highway (Nepal Highway H04), and the Arniko-Friendship Highway or Kyirong-Rasuwa Road (both of which run from southern Tibet to Kathmandu). Because the Beni-Jomsom section is heavily prone to monsoonal landslides and the Jomsom-Lo Monthang section is also broken in several places, this route is only used for significant cargo transfer in the winter season, when low water levels along the partially frozen Kali Gandaki gorge effectively create a "winter road" in uppermost Mustang. Due to these tenuous conditions, the vast majority of both Chinese and Nepali goods carried to Mustang traverse this route just once per year, after Tibetan *Losar* New Year in the late winter and early spring. These shipments travel via overladen wagons rumbling north up the Kali Gandaki valley, carrying prodigious loads of Chinese manufactured goods, Nepali agricultural products, and Indian imports alike.

Since the advent of new road systems into Mustang, there have been major reductions in the cost of goods across the district. This situation can be attributed to three interdependent factors: (1) the ubiquity of mass-produced Chinese products throughout Mustang (and the country), (2) recent reductions in overland shipping costs to Mustang from both southern Tibet and central Nepal, and (3) increasing levels of purchasing power due to Mustang's remittance economy. First, the economies of scale behind Chinese manufacturing and shipping have significantly reduced the real prices (and quality) of many Tibet-sourced goods in Mustang, particularly carpets, kitchenware, and furniture. Despite the long distances traveled, informants in Mustang consistently reported that Chinese and Tibetan products purchased in central Nepal were often less expensive than goods purchased directly from Tibet (in the next breath, however, informants also complained of the inferior quality of modern Chinese manufacturing in comparison to the Tibetan craftsmanship that Mustang is accustomed to). Second, the time, labor, and cost of transporting goods from both Tibet and central Nepal to Mustang has plummeted in recent years as a result of improved road conditions and larger vehicle fleets. Ten years ago, it cost at least 100 Nepali rupees (US$1) to ship four kilograms of cargo from Pokhara to Lo Monthang; today, that same cargo can be shipped for less than 60 Nepali rupees (US$0.60). Reductions in the friction of distance for shipping to Mustang have not only lowered prices but also more broadly altered consumer habits and livelihood practices across the region. Finally, a majority of families in Mustang now have relatives residing in New York City. Remittances from U.S.-based relatives, augmented by other relatives based in Kathmandu or other global cities, have transformed political economies in Mustang by injecting unprecedented levels of cash into local economic relationships. Unsurprisingly, however, Mustang's higher caste and higher class families are generally the ones with (more) relatives abroad. Thus, remittances tend to follow already established social hierarchies as cash enables new consumer and transport mobility for some and a lack of cash reinscribes historical terms of marginalization and containment for others.

In *Capitalism's Eye*, Hetherington (2007) took a "consumer view of the subject" to analyze how uneven social relations made within a capitalist system function to transform and territorialize space. In contrast to a classical Marxist producer view, this perspective of and from the consumer lays bare the cultural foundations of capitalist society as everyday practices of consumption. In aggregate and at the market, consumer practices thus constitute a political process at the nexus of culture, power, and place. A social space where capital circulates, controlling this nexus is a paramount interest of the state.

Territoriality is a practice of mastering space conducted by the state through the control of capital flows. Looking at the spatial operations of power, Sack (1983) identified territoriality as a spatial strategy to "affect, influence, or control actions, interactions, or access by asserting and attempting to enforce control over a specific geographic area" (55). This power is predicated on what Elden (2013) called the power of territory: a bundle of political techniques employed to measure land and control terrain. Agnew and Corbridge (2002) conceptualized the practice of territoriality as that of "mastering space." Building on Giddens's (1986) emphasis on the role of capital in the everyday processes of state and institutional formation, spaces are produced through a convergence of capital circulations and social practices—and with predictable bureaucratic responses of control.

Political economies are represented and maintained by practices of consumption and the control of political economies is a chief revenue stream, and thus concern, for the state. When and where political economy operates is thus a space to be mastered, especially under conditions of exponential market growth generated by new transport networks. Therefore, the circulation of a certain critical mass of capital—embodied in various forms of goods, property, and cash—leads to the formation of state institutions responsible for the control of territory where new political economies operate. As can be seen in Mustang, state territoriality is increasingly exercised through economic regulation and tax collection.

As a result of new trade practices, commodity flows, and border policies, new state institutions are rapidly taking shape in Mustang. These include the first full-time residency in Lo Monthang of the Government of Nepal's Border Commissioner to Mustang and increasing numbers of regional tax officers as well as the training and assignment of dozens of troops from Nepal's paramilitary Armed Police Force to the district. Although transborder trade between Mustang and Tibet has long been tax-exempt, formal duties on Chinese imports were first implemented in 2015 on the basis of 5 lakh Nepali rupees (~US$5,000) per truckload; this rate is expected to increase in 2016. To handle greater trade volumes, plans have been drawn to expand Mustang's small and dilapidated Government of Nepal Customs and Quarantine Houses into the first and only commercial dry port for overland import–export trade in Nepal's Western Development Region. Located in a place once and long considered peripheral to the Kathmandu state center, the circulation and consumption of Chinese commodities in Mustang has (re)produced a borderland space where state making is an expansive territorial reaction to economic transformation.

Conclusion

According to Harvey and Knox (2015), roads are productive for analysis because of "what they can tell us about how infrastructural relations simultaneously make national territories, international corridors, regional circuits, and specific localities" (25). Road projects in Mustang demonstrate this interconnection of scales and the interdependence of local, regional, and international relations. First, it was a local grassroots initiative rather than a national development scheme that, after decades of demands, finally produced the first road between Lo Monthang and regional markets. Second, this initial road connected Mustang not with Nepal but with China. Third, the road rapidly accelerated the import of Chinese goods from Tibet into Mustang. Fourth, the road and associated consumption patterns consequently set the groundwork for significant additional infrastructural and bureaucratic projects in Mustang, financed and undertaken by Nepali state actors as well as other international donors (including both China and India).

Following the road, what began as a locally constructed route to improve transborder mobility and evolved into a regional effort to grow business enterprises has led further to national and international-level master planning to develop the state and its constituents through neoliberal modes of tourism and taxation in a trans-Himalayan borderland space. These social–infrastructural projects are touted by Kathmandu as the latest and best way for Mustang to manage its ongoing development strategy through "sustainable livelihoods," bringing *bikas* or development to the nation through new vectors of mobility. As constituents in Mustang evaluate the prospect of making Mustang a new pilgrimage and trade corridor between Nepal and Tibet, however, critics not only challenge the national discourse of development but also point to the social disruption and environmental risks of more trade and tourist traffic on Mustang's fragile alpine landscape.

This article examined material dynamics between Chinese Tibet and Mustang, Nepal, to better understand how territoriality functions in concert with consumer practice. Situated in Himalayan mountain spaces at the intersection of border politics, infrastructure development, and material culture, the study reveals how political economy operates in the production and territorialization of state space and the ways that everyday practices of consumption function to reinforce new spatial productions and power relations. As massive amounts of Chinese commodities become increasingly available in Mustang—and elsewhere in Nepal—it is apparent that the mobility of everyday material things undergirds the territorial power of the Nepali state. Yet these mobilities remain inseparable from countertensions of containment, as access to roads—or the lack thereof—reinforces long-standing social hierarchies and experiences of marginalization from the high Himalaya to broader mountain landscapes.

Acknowledgments

I thank the editors of this special issue and two anonymous reviewers for constructive and insightful comments on an earlier draft. I also received helpful feedback from fellow participants at the 2016 Dissertation Workshop at New York University.

Funding

Research for this article was made possible by grants from the Social Science Research (International Dissertation Research Fellowship), the U.S. Department of Education Fulbright Hays Program (Doctoral Dissertation Research Abroad Grant# P022A1400), and both the Graduate School and the Department of Geography at the University of Colorado Boulder.

ORCID

Galen Murton ⓘ http://orcid.org/0000-0003-1706-891X

Notes

1. Tellingly, Mustang was designated as the base for Chushi Gangdruk not simply for its relative proximity to Tibet but also due to its periphery to the Nepali state and the weak rule of the Nepali crown across the district (McGranahan 2010). In fact, it was largely under pressure exerted by the United States that King Mahendra acquiesced to allow Tibetan military operations to proceed in Mustang. Although the Nepali crown did not directly support this resistance movement, it was not until 1972—following Washington's détente with

Beijing and cessation of support for Chushi Gangdruk—that King Mahendra leveraged the voice of the Dalai Lama to intervene and cease the Khampa operations once and for all (Shakya 1999).

2. Although the Mustang–Tibet border was historically characterized by a free-range fluidity, beginning in the 1960s it also became a prime site of regional geopolitical intrigue. Although there were no major Nepali or Chinese state structures at the border, by 1960 the area did feature one of fifteen Indian checkposts that spanned the entire Sino-Nepal boundary zone (Rose 1971). Following the Chinese Communist Party's occupation of Tibet in 1949, Indian checkposts were established in the 1950s to monitor the activities of the PLA in southern Tibet (Rose 1971; Garver 2001). Observations on PLA activities were collected at these checkposts and intelligence was then relayed via coded messages to the Indian Embassy in Kathmandu and onward to Indian security offices in Delhi (Cowan 2015). By the end of the 1960s, however, Nepali institutions largely replaced these Indian checkposts.

3. This infrastructure development scheme is an ambitious project that endeavors to construct a total of six transborder roads between northern Nepal and Tibetan China within the next five to ten years (Department of Roads 2014a, 2014b). The larger interstate project behind the Strategic Road Network proposes that the Mustang–Tibet road will soon augment the current Sino-Nepal trade corridors of the Arniko-Friendship Highway and the Kyirong-Rasuwa corridor. The Mustang road's contribution to the broader Strategic Road Network depends on its connection with central Nepal via a newer division of roads that link Lo Monthang with the district capital of Jomsom as well as Jomsom with the large central cities of Beni and Pokhara. Although this road network between Lo Monthang and Pokhara remains punctuated by broken sections due to monsoonal landslides, missing bridges, and general obstacles to maintenance, it is now possible for both people and goods to make in just two days what twenty years ago was a two-week journey.

References

Agnew, J., and S. Corbridge. 2002. *Mastering space*. London and New York: Routledge.

Cidell, J., and D. Lechtenberg. 2016. Developing a framework for the spaces and spatialities of transportation and mobilities. *Annals of the American Association of Geographers* 106 (2): 257–65.

Cowan, S. 2015. The Indian checkposts, Lipu Lekh, and Kalapani. *The Record Nepal* 15 December 2015. http://recordnepal.com/wire/indian-checkposts-lipu-lekh-and-kalapani (last accessed 28 June 2016).

Cresswell, T. 2010. Towards a politics of mobility. *Environment and Planning D: Society and Space* 28 (1): 17–31.

———. 2016. *Geographies of mobilities: Practices, spaces, subjects*. London and New York: Routledge.

Cresswell, T., and P. Merriman. 2012. *Geographies of mobilities*. Surrey, UK: Ashgate.

Department of Roads. 2014a. *Strategic Road Network 2013_14 map*. Kathmandu: Government of Nepal Ministry of Physical Infrastructure and Transport. http://www.dor.gov.np/SSRN_2013_14.php (last accessed 28 June 2016).

———. 2014b. *Western development region*. Kathmandu: Government of Nepal Ministry of Physical Infrastructure and Transport. http://www.dor.gov.np/SSRN_2013_14.php (last accessed 28 June 2016).

Dhungel, R. 2002. *The kingdom of Lo (Mustang)*. Kathmandu, Nepal: Tashi Gephel Foundation.

Elden, S. 2013. *The birth of territory*. Chicago: University of Chicago Press.

Fürer-Haimendorf von, C. 1988. *Himalayan traders*. Delhi, India: Time Books International.

Garver, J. W. 2001. *Protracted contest*. Seattle: University of Washington Press.

Giddens, A. 1986. *The constitution of society*. Berkeley: University of California Press.

Harris, T. 2013. *Geographical diversions*. Athens: University of Georgia Press.

Harvey, P., and H. Knox. 2015. *Roads*. Ithaca, NY: Cornell University Press.

Hetherington, K. 2007. *Capitalism's eye*. London and New York: Routledge.

Kristof, L. 1959. The nature of frontiers and boundaries. *Annals of the Association of American Geographers* 49 (3): 269–82.

Kwan, M. P., and T. Schwanen. 2016. Geographies of mobility. *Annals of the American Association of Geographers* 106 (2): 243–56.

Larkin, B. 2013. The politics and poetics of infrastructure. *Annual Review of Anthropology* 42 (1): 327–43.

Li, T. M. 2007. *The will to improve*. Durham, NC: Duke University Press.

McGranahan, C. 2010. *Arrested histories*. Durham, NC: Duke University Press.

Ministry of Transportation. 2016. *5 year plan for transportation infrastructure development*. Kathmandu: Government of Nepal Ministry of Physical Infrastructure and Transport.

Pigg, S. L. 2009. Investing social categories through place: Social representations and development in Nepal. *Comparative Studies in Society and History* 34 (3): 491.

Post Report. 2016. Roads to China border top budget priority. *The Kathmandu Post* 4 May 2016. http://kathmandupost.ekantipur.com/news/2016-05-04/roads-to-china-border-top-budget-priority.html (last accessed 28 June 2016).

Rose, L. E. 1971. *Nepal: Strategy for survival*. Berkeley: University of California Press.

Sack, R. D. 1983. Human territoriality: A theory. *Annals of the Association of American Geographers* 73 (1): 55–74.

Scott, J. C. 2009. *The art of not being governed: An anarchist history of upland Southeast Asia*. New Haven, CT: Yale University Press.

Shakya, T. 1999. *The dragon in the land of snows: A history of modern Tibet since 1947*. New York: Columbia University Press.

Sheller, M., and J. Urry. 2006. The new mobilities paradigm. *Environment and Planning A* 38 (2): 207–26.

Thapa, M. 2008. *Mustang bhot in fragments*. Kathmandu, Nepal: Himal Books.

van Spengen, W. 2000. *Tibetan border worlds*. London and New York: Routledge.

World Bank. 2006. *Infrastructure at the crossroads: Lessons from 20 years of World Bank experience*. Washington, DC: The International Bank for Reconstruction and Development/The World Bank.

GALEN MURTON is a PhD Candidate in the Department of Geography at the University of Colorado Boulder, Boulder, CO 80309. E-mail: galen.murton@colorado.edu. His research interests include the social and political implications of road developments across the borderlands of High Asia and the geopolitics of infrastructure investment in postdisaster environments.

Khumbi yullha and the *Beyul*: Sacred Space and the Cultural Politics of Religion in Khumbu, Nepal

Lindsay A. Skog

Many parts of the Himalaya are at once indigenous people's homelands, national parks or conservation areas, world-renowned trekking and mountaineering destinations, and the sites of ongoing ecological and socioeconomic development interventions. In addition, for many residents, protective territory deities reside in nearby peaks, and valleys between provide sacred places of refuge. Like in mountain regions elsewhere, these meanings represent overlapping and entwined claims of authority and territory from the state, indigenous communities, development agencies, and religious institutions. In this article I consider the ways in which resident Sherpas in Khumbu, Nepal, negotiate the overlapping spaces, authorities, and territories associated with understandings of the region as *Khumbi yullha's*—a local deity—territory and the Nyingma Buddhist institutional claim to the region as a *beyul*—a sacred, hidden valley refuge, which development actors, both inside and outside the Khumbu Sherpa community, have attempted to mobilize as a sacred landscape supporting environmental conservation initiatives. Based on eighteen months of fieldwork in 2009 to 2010 and 2013, I focus on the spatiality of the cultural politics of religion in Khumbu in competing claims of territory from the Buddhist monastic institution and localized practices and the ways in which such constructions shape the outcomes of intervention programs.

喜马拉雅的诸多部分, 同时是原住民族的家园, 国家公园或保育地区, 世界知名的徒步旅行和登山地点, 以及持续不断的生态和社会经济发展介入的场域。此外, 对许多居民而言, 提供保护的领土神灵便驻守在邻近的山顶上, 而其间的谷地, 则提供了避难的神圣地点。诸如其他地方的山岳地区, 这些意义, 呈现出国家、原住民社群、发展组织和宗教机构之间重叠且相互交缠的主权及领域主张。我于本文中, 考量尼泊尔坤布的夏尔巴人协商关于将区域理解为在地神灵坤布尤拉 (Khumbi yullha) 的领域, 以及宁玛 (Nyingma) 佛教机构宣称该区域作为神圣且隐匿的山谷避难处之 "秘境" 的重叠空间、主权和领土, 其中来自于坤布夏尔巴社群内部与外部的发展组织, 同时尝试动员上述对圣地地景的理解作为支持环境保育的动机。我根据在 2009 年至 2010 年与 2013 年间的十八个月田野工作, 聚焦坤布在佛教修道院机构与在地实践之间相互竞争的领土宣称的宗教文化政治空间性, 以及此般构造形塑介入计画之结果的方式。
关键词: 山岳, 尼泊尔, 宗教信仰, 宗教地景, 领土。

Muchas partes de los Himalayas son a la vez patrias de población indígena, parques nacionales o áreas de conservación, destinos de excursionismo y montañismo de renombre mundial y escenarios de intervencionismo de desarrollos ecológicos y socioeconómicos actuales. Además, para muchos residentes, en los montes circundantes residen deidades protectoras del territorio, y los valles que quedan en medio proveen lugares sagrados de refugio. Como ocurre en las regiones montañosas de otras partes, estas significaciones encarnan reclamos de autoridad y territorio que se traslapan y entrelazan, tanto del estado como de las comunidades indígenas, agencias de desarrollo e instituciones religiosas. En este artículo registro las maneras como los residentes sherpas de Khumbu, Nepal, negocian los espacios superpuestos, las autoridades y territorios asociados con el entendimiento de la región como el territorio de Khumbi yullha—una deidad local—y el reclamo institucional de Nyingma budista sobre la región como un *beyul*—un oculto refugio sagrado del valle, al que actores del desarrollo, tanto dentro como fuera de la comunidad sherpa de Khumbu, han intentado movilizar como un paisaje sagrado que apoya iniciativas de conservación ambiental. Con base en dieciocho meses de trabajo de campo entre 2009 y 2010 y en 2013, me concentro en la espacialidad de las políticas culturales sobre religión en Khumbu, en reclamos en competencia por territorio de la institución monástica budista y las prácticas y maneras locales dentro de las cuales tales construcciones configuran los resultados de los programas intervencionistas.

Nyingma Buddhists believe that *Guru Rinpoche* (Sh./Tib.; Sk. *Padmasambhava*), a revered Indian Buddhist teacher, hid 108 *beyul*, or sacred valleys, among the high peaks of the Himalaya. In these valley refuges, Nyingmapa practitioners and Nyingma traditions, the oldest among the

Tibetan Buddhist sects, would be protected in times of turmoil and need.[1] Although many beyul remain unrevealed, monastic scholars and academics have identified twenty to twenty-two opened, or revealed, beyul, including Khumbu, Nepal (Reinhard 1978; Diemberger 1993, 1996, 1997; Childs 1999; L. N. Sherpa 2008). Yet, intertwined meanings, histories, and beliefs comprise understandings of Himalayan sacred and animate landscapes, where localized spirits reside in streams, rocks, and trees alongside mountain-dwelling Buddhist deities. Thus, understandings of some Himalayan mountain landscapes as beyul are only one layer of more complex sacred spaces imbued with multiple meanings.

Khumbu, a 1,500 km[2] area nestled in the heart of the Himalaya and in the shadow of Mount Everest (Sh. *Jomolungma*; Nep. *Sagarmatha*), rises from approximately 2,800 m at its southern boundary to 8,848 m on the summit of Mount Everest. Snow-capped peaks and ridges surround the region, which indeed give the impression of a hidden valley. The region was declared Sagarmatha (Mt. Everest) National Park (SNP) in 1976 and recognized as a United Nations Educational, Scientific and Cultural Organization (UNESCO) World Heritage site in 1979. A resident population of more than 2,500 ethnic Sherpas reside in six main villages (Khumjung, Khunde, Nauche, Thame Og, Pangboche, and Phortse) alongside an ever-growing population of non-Sherpa national park employees, military and police officials, schoolteachers, health care workers, and domestic workers.[2] In addition, non-Sherpa Nepali and Tibetan high-altitude workers, porters, and leasing lodge managers are drawn to Khumbu's lucrative trekking and mountaineering industry. As a result, Khumbu is at once a Sherpa homeland, a mountaineer's playground, a foreign trekker's Shangri-la, a national park, a UNESCO World Heritage site, a beyul, and more. Like mountain landscapes elsewhere, the complexity of meanings and multiple spatialities in Khumbu produce overlapping claims of authority and territory from both near and far. This article specifically considers the ways in which multiple sacred spatialities in Khumbu are the result of a cultural politics of religion, which shapes residents' perceptions of and relations to space in the region. I conclude by reflecting on how such spatialities affect the outcomes of environmental conservation and indigenous political interventions.

In 2008, amid concerns over both cultural and environmental sustainability in Khumbu, The Mountain Institute (TMI)—a North American nonprofit institute operating in Nepal with the mission to promote sustainable livelihood development, environmental conservation, and cultural preservation—produced a short documentary exploring the ways in which understandings of Khumbu as a beyul could support environmental conservation efforts in the region. Interspersed among images of Khumbu's snow-capped peaks and ridges, wildlife, and sacred sites, the documentary featured interviews with TMI personnel, the *Rinpoche* (Sh. head monk) at Tengboche Monastery—who has long been concerned about the growing impacts of tourism on Sherpa culture and the Khumbu environment—and several Khumbu Sherpas, representing a cross section of social standings in Khumbu. The film described the beyul as an umbrella including all of Khumbu's numerous small-scale sacred sites, such as mountain deity abodes and sacred springs. As the film continued, the interviews turned to concerns over the loss of beyul knowledge among Khumbu Sherpas, with several interviewees explaining that they knew little about the beyul concept or that they had never heard of it (T. R. Sherpa 2007).

For three years, TMI's program aimed to "spread the benefits of tourism more equitably among the local people in the Everest region of Nepal, while preserving indigenous culture and environment through education and awareness-building" (TMI 2010). In addition to encouraging tourism and trekking in the less frequently visited areas of Khumbu, the program worked to buttress and promote Sherpa culture and identity by educating Khumbu Sherpas and visitors to the region about the beyul concept. Overall, the program worked to link Buddhist and environmental ethics by calling on Khumbu Sherpas to observe and enforce a set of environmentally friendly behavioral taboos-cum-conservation practices associated with the beyul concept (L. N. Sherpa 2003, 2005; T. R. Sherpa 2007; TMI 2008).

Concomitant to, but separate from, TMI's program, Khumbu community leaders working with a foreign long-term Khumbu researcher advanced a declaration supporting Khumbu's designation as an Indigenous and Community Conserved Area (ICCA; Stevens 2008). The International Union for the Conservation of Nature (IUCN) endorses such a designation to recognize effective local governance and conservation of natural resources by indigenous peoples and as a corrective to state-designated protected areas, which have often violated the rights of resident indigenous peoples. The declaration mobilized understandings of Khumbu as a beyul, as well as community relations with *Khumbi yullha*, a local mountain deity, in support of the designation. The controversial declaration was ultimately deemed illegal by the government of Nepal, as it threatened the authority of the national park and the state, in Khumbu (Stevens 2008).

Although TMI's efforts are distinct from those to recognize Khumbu as an ICCA, their parallel timing follows growing movements in both the local and global history of comanagement in protected areas. Calls among activists to move away from Yellowstone-model fortress conservation that excluded indigenous peoples and local residents from both policy and planning and local resource use in the 1970s recognized the role of indigenous and local peoples in environmental conservation and protection and ushered in efforts toward greater incorporation and comanagement strategies between local residents and distanced planners (Stevens 1997; Neumann 1998; Brockington 2002; Dowie 2011). The IUCN and World Wildlife Fund (WWF) spearheaded many of these efforts, which ultimately formed the basis of attempts to establish new types of national parks and protected areas that incorporate indigenous and local voices.

In this context, SNP was established and the park was lauded as an example of international and local collaboration. Yet, local input and participation was limited and, over time, several points of contention arose among state planners, park authorities, and local residents, especially over forest management (Brower 1991; Stevens 1997). Despite such conflicts, traditional resource management practices persisted in some parts of Khumbu, including some practices associated with sacred sites and forests (Brower 1991; Stevens 1997; L. N. Sherpa 2003, 2005).

The perceived potential of the beyul concept to support environmental conservation, development, and indigenous rights led one of my Sherpa key informants to refer to the beyul as the "perfect package for Khumbu Sherpas." Yet, despite the seemingly fertile ground for articulations between the concept of Khumbu as a sacred landscape and global discourses of environmental conservation and development, many Khumbu Sherpa informants' understandings of Khumbu as a sacred place and as a Sherpa territory are more closely tied to everyday practices and rituals worshipping and appeasing Khumbi yullha, rather than the beyul. This article aims to make sense of this seeming contradiction by examining the spatiality of the cultural politics of religion in Khumbu.

This research is based on eighty-eight semistructured interviews conducted during eighteen months of fieldwork in 2009 to 2010 and 2013. Of these, I interviewed six nongovernmental organization program managers, directors, or associates involved in environmental conservation and indigenous politics in Nepal. I conducted eighty-two interviews with Khumbu Sherpas in both Khumbu and Kathmandu.

Mobilizing Sacred Landscapes

Investigations exploring the numerous ways in which religious sites and landscapes act as mechanisms in natural resource management and conservation demonstrate how taboos and practices associated with both small-scale religious sites and larger scale religious landscapes, indigenous knowledges and beliefs, and religious traditions conserve biodiversity in sacred groves and surrounding sacred lakes (Gold and Gujar 1989; Sharma, Rikhari, and Palni 1999; Allison 2004; Ramakrishnan 2005) and support forest conservation practices in sacred forests and groves (Stevens 1993; Ingles 1995; L. N. Sherpa 2003, 2005; Arora 2006; Spoon and Sherpa 2008; see also Berkes 1999; Dove, Sajise, and Doolittle 2011). By the 2000s, scholars and activists took an interest in the relationships among religion, spirituality, environmental conservation, and development as an interdisciplinary subfield (Sponsel 2001, 2012; Tucker and Grim 2001; Taylor 2010). In the global arena, programs sponsored by WWF, Conservation International, and the International Centre for Integrated Mountain Development, often in collaboration with state governments, demonstrate that environmental conservationists, indigenous rights activists, and scholars remain particularly interested in the ways in which religious landscapes can be mobilized to support environmental conservation efforts and indigenous rights claims (Dove, Sajise, and Doolittle 2011; Sponsel 2012; WWF 2016).

By the early 2000s, efforts to mobilize the beyul concept to support environmental conservation and indigenous rights in Khumbu emerged from the intersection of calls for increased local participation in protected area management and recognition of the importance of sacred values to environmental conservation (UNESCO 2003, 2006). The beyul documentary visionary, himself a Khumbu Sherpa, captured this convergence by stating,

> We have to capitalize on people's culture, people's belief systems, and all those human side of things in order to get stronger support from the local people. Because without the support of the people who live inside the

national park and around the national park the enforcement of rules and regulations alone is not sufficient to protect all the species and valuable endangered species that we have. (T. R. Sherpa 2007)

By incorporating sociocultural factors, including religion, ethnicity, and gender, into considerations of environmental conservation agendas, as well as calling for collaboration between conservation managers and local communities, including indigenous peoples, much of this work offers correctives to disenfranchising and exclusionary approaches to environmental conservation (Agrawal and Gibson 1999; Yeh and Coggins 2014). Beyond such correctives, the case study of Khumbu's religious and animate landscapes demonstrates the ways in which narratives of sacred landscapes today allow environmental conservation and indigenous rights leaders to stake claims more closely aligned with community interests. Yet, as this case study also demonstrates, mobilizations of sacred landscapes in efforts to support environmental conservation and indigenous rights risk simplifying complex landscapes and belying the cultural politics from which they are forged.

The Cultural Politics of Religion

The historical negotiations among localized, shamanistic, and animistic practices in the Himalaya and the rise of monastic Buddhist practices and institutions is well-trodden territory among Himalayan scholars (Mumford 1989; Childs 1999; Ramble 2007). Ortner (1978, 1989) examined the cultural politics of this interface through the rise of the Nyingma monastic community and the remaking of rituals in Khumbu. I build on her work to demonstrate the ways in which struggles between Khumbu lay and monastic communities over the authority to direct the symbolic representations and material practices constitutive of Sherpa life worlds play out spatially in Khumbu.

I conceptualize Sherpa religion as the complex of Sherpa practices aimed at maintaining relations with place-based deities and spirits, as well as monastic practices and rituals associated with Nyingma Tibetan Buddhist traditions. I recognize that religion has been critiqued as an etic category of classification that, at times, encourages problematic comparisons and generalizations across diverse epistemologies (Asad 1993; Masuzawa 2005; Mandair 2009). I use it here, however, both in the sense of institutionalized world religions, in this case monastic Buddhism, and in the sense evoked by McGuire's (2008) lived religion, in which religion

is conceptualized "at the individual level, as an ever-changing, multifaceted, often messy—even contradictory—amalgam of beliefs and practices that are not necessarily those religious institutions consider important" (4). Holding both of these uses of the term *religion* simultaneously foreshadows the cultural political struggles of negotiating lived beliefs and practices with institutionalized monastic authority.

Khumbu's Animate Landscape

A heterogeneous set of water and land spirits, as well as numerous mountain deities, animates the Khumbu landscape (i.e., Sh. *lu, tsen*). Such spirits and deities are associated to varying degrees with mundane and everyday concerns such as health, prosperity, success, wealth, and social harmony. Household rituals, seasonal and annual community-wide rituals, and numerous everyday practices and ritual performances, including taboos against polluting sites where lu and tsen are thought to reside, maintain beneficial relations with Khumbu's various spirits and deities, as well as work to produce and reproduce Khumbu as a Sherpa territory. For instance, it is common for Sherpas to perform a *puja* (Sh. ceremony) to appease the land spirits and request their protection prior to disturbing land to build a new structure. Similarly, Sherpas perform an annual community invocation of the local spirits and mountain deities for the protection and success of the vulnerable crops throughout the summer growing season.

Among Khumbu's myriad mountain deities, Sherpas identify Khumbi yullha as the most powerful of the tsen—a *yullha*, or territory deity (Karmay 1996; L. N. Sherpa 2008). Khumbi yullha resides atop a mountain also named Khumbi yullha (both are often shortened to Khumbila), with the villages of Khunde, Khumjung, and Nauche at its base. Khumbi yullha is most often depicted with white hair, wearing a white scarf, and riding a red or brown horse. Postcard-size images of Khumbi yullha adorn nearly every household shrine in Khumbu and can be found in village *gondes* (Sh. temples) and monasteries. Khumbu Sherpas understand Khumbi yullha as a personal benefactor who protects individual and household health and safety, fulfills desires, and brings wealth. For Khumbu Sherpas, Khumbi yullha also acts as a protector deity for each village and Khumbu as a whole. As such, he protects villages from natural disasters while also bestowing general success and wealth on the villages (cf. Makley 2014). A local lama explained, "Khumbi yullha helps this place—no landslides, no thieves, no killing each other, no bad disease." On the other hand, Sherpas also fear Khumbi yullha, who could withdraw protection or even cause harm if he is angered.

Although Khumbi yullha is generally described as Khumbu's protective territory deity, in general, residents from Nauche, Khunde, and Khumjung demonstrate different relationships with Khumbi yullha than residents of Khumbu's other villages. Tenzing Sonam,[3] a middle-aged Sherpa man living in Nauche, exemplifies the relationships expressed by Nauche, Khumjung, and Khunde residents:

Khumbila [Khumbi yullha] is the protector of Khumbu. We consider him as god also, as protector of Khumbu. And the sustaining of this economy in Khumbu, we believe it is blessed by Khumbi yullha. Khumbila, you have seen at that place [A rock above Nauche where Nauche men go to perform pujas for Khumbi yullha]—last time we went there carrying this flag. It's to worship Khumbila. We went there to worship Khumbila yearly in order to purify this land, in order to purify this Khumbu, and everything that has been, you know. We have been suffering from the problems, from the negative things, from the natural disasters. So we believe that after the blessing of Khumbila, we will be relieved from this, all the disaster and things, you know. So, Khumbila is a god for us.

In his comments, Tenzing Sonam detailed both a specific relationship that he maintains with Khumbi yullha and the broader relations the village maintains as a whole through village rituals. In contrast, informants in Thame perform pujas for Khumbi yullha as a familial and individual protector deity, not as a village protector. Unlike Tenzing Sonam, other informants explained they know very little about Khumbi yullha; however, most also describe him as "Khumbu's god" or "Khumbu's protector." Thus, although some Khumbu residents appear more knowledgeable about Khumbi yullha than others and demonstrate individual, as well as village, relations with the deity, informants throughout Khumbu acknowledge his authority over maintaining both the well-being and economic success of the region.

The set of everyday practices and ritual performances associated with Khumbi yullha, and to a lesser extent with Khumbu's other land and water spirits, governs Khumbu Sherpas' social and human–environment relations and demonstrate the authority Khumbi yullha holds in shaping Khumbu Sherpa life worlds. Sherpas' relations with Khumbi yullha, as well as other place-based deities and spirits, are productive of space—an animate landscape and a Sherpa territory where Khumbu Sherpas recognize the authority of Khumbi yullha in ensuring well-being and success. Yet this space overlaps with the space produced by the Buddhist monastic community in the form of the beyul.

The Buddhist Beyul

In the past, many gods and spirits roamed the world without homes and responsibilities. The Tibetan King decided to build a monastery, Samyé Monastery; however, the gods and spirits were opposed to this and dismantled each day's work during the night. Knowing that Guru Rinpoche, a powerful Buddhist teacher, was more powerful than the troublesome local gods, the lama asked that Guru Rinpoche to come from India, where he had been giving blessings and receiving education. Guru Rinpoche traveled from India to Tibet through the Himalaya. Along his journey he gave each local god a home and responsibilities, including Khumbi yullha, whom he made responsible for Khumbu's protection. On arriving in Tibet, Guru Rinpoche instructed the lama to continue building the monastery, assuring him that the gods, now committed to protecting Buddhist teachings and peoples, would allow the construction to continue, even ordering the formerly pesky gods to assist in the process. Khumbi yullha became quite lazy, tired, and angry while building the monastery, however. As punishment, Guru Rinpoche made the dirt in Khumbi yullha's land different from the dirt in other Himalayan landscapes and no longer waterproof.

The establishment of the first Buddhist monastery in Tibet, Samyé Monastery, with patronage from the Buddhist King Trisong Detsen, in roughly 779 AD, marked a moment of violent reconfiguration of the relations between Buddhism and Bön in Tibet (Dowman 1997). The oft-told story of Khumbi yullha's conversion from a localized, pesky, and, as some describe, wrathful deity to a protector of Buddhism and Buddhist peoples is not uncommon. Further, in similar yullha conversion stories, Guru Rinpoche is sometimes said to have converted the area into a beyul and charged the yullha to protect the new beyul (Pommaret 1996). Yet, there is more to this story than the seemingly simple origin of a territory deity; the story of Khumbi yullha's conversion is also a territorial claim by the Buddhist monastic community.

The first Nyingma Buddhist monastery in Khumbu, Tengboche Monastery, was established in 1912 (Ortner 1989), but Khumbu Sherpas have been maintaining relations with the spirits and deities residing in Khumbu's rivers, trees, rocks, and mountains for much longer. The everyday practices and rituals in many Sherpa households demonstrate a closer affinity to Bön, the pre-Buddhist religious tradition of the current Tibetan ethnic region, and reveal a complicated religious history informed in part, by Bön and Nyingma Buddhist traditions. Bön is best characterized as an animistic and shamanistic tradition thought to have declined in Tibet in the eighth century as the Buddhist tradition emerged. At that time, Bön practitioners were persecuted and exiled; however, Bön beliefs were so entrenched in Tibetan popular religion that they could not be completely eradicated and were surreptitiously incorporated into Buddhist practices (Karmay 1998). Bön practice reemerged in Tibet in approximately the eleventh century with a closer affiliation to Buddhist practice and within the Buddhist framework, especially among the Nyingma sect.

The founding of Tengboche Monastery and, later, Thame Monastery marked the beginning of an ongoing power struggle, a cultural politics of religion, between lay Sherpas' concerns with protection rituals and exorcisms focused on local deities and demons and the concerns of the monastic institutions oriented toward "higher" Buddhist practices and the universal Buddhist pantheon (Ortner 1989). Ortner (1989, 1999) demonstrates that with the foundings of the celibate monastic institutions, social and monastic forces in Khumbu attempted to eliminate popular, lower ritual practices.[4] At the time, Sherpa popular religion led by married lamas and shamans was accorded a low status by the celibate monastic institutions practicing "high" Buddhism.

Tshering, an elder in Nauche, captured the boundaries some Khumbu Sherpas construct between lay practices oriented toward localized deities and spirits, at times involving shamans, and the monastic community:

> The monastery and monks those only thinking "what have the Buddhas told"? They follow by the Buddha's rule, but local area we must follow everybody, any kind [of deity and spirit], we have to follow for the Buddhas, we have to follow by the shamans, we have to do everything. ... The monks, they say no gods, but there are some gods, but no demons, no ghosts. But inside the book [Buddhist texts] they have [gods, demons, ghosts]. When the offering time comes they have even one small tree, they also have one god [a local spirit]. Only small farm, they also have water god [the monastery makes an offering to the local water spirit living on their agricultural land]. Where the small stone, they also have own god. When the offering time comes, they will make offering.

Even as this captures the boundary between Khumbu's lay and monastic communities, Tshering's description of the localized water and tree spirits at the monastery illustrates the ways in which lay practices are remaking monastic practices, and his reference to following the Buddha demonstrates the reverse. It is important here to see the ways in which Sherpas actively negotiate the multiple religious authorities in Khumbu.

Khumbu's Buddhist monastic institution claims territorial authority in two ways. Through claims of protection and security in the beyul, the Nyingma Buddhist monastic community claims authority over Khumbu's territory, thus rendering it a Nyingma Buddhist territory. Second, the claim of Khumbu being a beyul was not only an assertion of authority over territory in Khumbu; it is *through* the claim of the beyul that the Buddhist monastic authority asserted its authority. By taming Khumbi yullha and granting him authority over Khumbu, the Buddhist monastic institution is claiming the ability to determine authority in Khumbu—itself a claim of authority.

The construction of Beyul Khumbu by the Nyingma monastic institution is not the only way in which to understand the beyul concept as a territorial claim; mobilizations of the beyul concept to support environmental conservation and indigenous politics are themselves efforts to construct territory and in doing so are claims of authority. In near opposition to constructions of Khumbu as Khumbi yullha's territory, constructions of Khumbu as a beyul, as an idealized space produced from and symbolic of a Nyingma Buddhist ideology, might best be understood as a space produced from the authority of development actors—often elites from both within and outside of Khumbu—and even a claim to territory (cf. Coggins 2014). By mobilizing Khumbu as a sacred landscape, development actors and programs are claiming the authority to determine how that space best fits their agendas and, in doing so, forging, in movement, new understandings of Khumbu as a sacred place.

Conclusion

Understanding the ways in which a cultural politics of religion in Khumbu shapes constructions of space is vital to understanding the risks and outcomes of conservation mechanisms and efforts to support indigenous rights mobilizing sacred landscapes. This article is not an assessment or evaluation of the impact of TMI's efforts or those to declare Khumbu an ICCA or similar projects; rather, this work is intended to illuminate the dynamics shaping the outcomes of such projects and thus contributes to broader academic and policy-oriented conversations about the politics and practices of territory, natural resource management, and rights-based development.

It might be expedient to dismiss simplified mobilizations of sacred landscapes as strategic deployments of essentialized understandings of human–environment relations and religion. For Spivak (1988), groups might find such "strategic essentialism" advantageous, despite persistent internal differences and struggles, in presenting themselves in a way that achieves specific goals. On the other hand, as Heatherington (2010) pointed out, cultural politics and internal difference might undermine such strategies and, thus, shape the outcome of such mobilizations. Similarly, Brosius (1997) argued that such generalizations, although useful in making indigenous knowledge accessible to outsiders, homogenize diverse practices and create new cultural and material forms. Thus, as demonstrated in Khumbu, representations of sacred landscapes found useful for political gain or positioning might create their own problematic essentialism by belying the histories and social political struggles at the foundation of those claims.

I suggest that attention to the micropolitics of religious and sacred landscapes might, in some cases, avoid problematically essentializing narratives and constructions of space and potentially reveal even greater opportunities toward meeting community goals. Beyond the concerns just raised, this research compels us to consider the ways in which certain processes have the power to essentialize. That is, at the intersection of sacred and religious landscapes, environmental conservation, and indigenous politics, who is granted authority through their ability to determine how understandings of the sacred are mobilized and to what ends?

Acknowledgments

I am grateful to the two anonymous reviewers for their insights and suggestions to improve this article.

Funding

This research was generously supported by the National Science Foundation (Doctoral Dissertation Research Improvement Grant BCS-1303147 and CAREER Grant BCS-0847722), the Society of Women Geographers (Pruitt National Dissertation Fellowship), the IIE Fulbright program and the United States Education Foundation–Nepal, the University of Colorado at Boulder, the Tokyo Foundation, and the Association of American Geographers.

Notes

1. I indicate Sherpa words as Sh., Tib. for Tibetan, Nep. for Nepali, and Sk. for Sanskrit. For ease of reading and pronunciation, on first usage I present foreign language terms in a simplified or common rendering.
2. The 2011 census recorded 2,572 Sherpas in Nauche and Khumjung Village Development Districts (Central Bureau of Statistics, National Planning Commission Secretariat, Government of Nepal 2014); however, it is likely that many Khumbu Sherpas are registered in Kathmandu, which would bring the total population closer to 5,400 as reported by Stevens (2008).
3. I use pseudonyms for all research participants quoted here.
4. Makley (2014) described a similar tension, which has always existed between monastic and lay traditions in the Tibetan ethnic region of Amdo, in which monastic leaders deride lay mediums, shamans, for propagating low religion and "baser desires."

References

Agrawal, A., and C. C. Gibson. 1999. Enchantment and disenchantment: The role of community in natural resource conservation. *World Development* 27 (4): 629–49.

Allison, E. 2004. Spiritually motivated natural resource protection in Eastern Bhutan. In *The spider and the piglet*, ed. K. Ura and S. Kinga, 528–61. Thimphu, Bhutan: The Centre for Bhutan Studies.

Arora, V. 2006. The forest of symbols embodied in the Tholung sacred landscape of North Sikkim, India. *Conservation and Society* 4 (1): 55–83.

Asad, T. 1993. *Genealogies of religion: Discipline and reasons of power in Christianity and Islam*. Baltimore: Johns Hopkins University Press.

Berkes, F. 1999. *Sacred ecology*. London and New York: Taylor & Francis.

Brockington, D. 2002. *Fortress conservation: The preservation of the Mkomazi Game Reserve, Tanzania*. Bloomington: Indiana University Press.

Brosius, J. P. 1997. Endangered forest, endangered people: Environmentalist representations of indigenous knowledge. *Human Ecology* 25 (1): 47–69.

Brower, B. 1991. *Sherpa of Khumbu: People, livestock, and landscape*. Delhi: Oxford University Press.

Central Bureau of Statistics. National Planning Commission Secretariat. Government of Nepal. 2014. *National Population and Housing Census, SoluKhumbu, 2011*.

Childs, G. 1999. Refuge and revitalization: Hidden Himalayan sanctuaries (*sbas-yul*) and the preservation of Tibet's imperial lineage. *Acta Orientalia* 60:126–58.

Coggins, C. 2014. Animate landscapes: Nature conservation and the production of agropastoral sacred space in Shangrila. In *Mapping Shangrila: Contested landscapes in the Sino-Tibetan borderlands*, ed. E. Yeh and C. Coggins, 205–28. Seattle: University of Washington Press.

Diemberger, H. 1993. Gangla Tshechu, Beyul Khenbalung: Pilgrimage to hidden valleys, sacred mountains and springs of life water in southern Tibet and eastern Nepal. In *Anthropology of Tibet and the Himalaya*, ed. C. Ramble and M. Brauen, 60–72. Zurich: Ethnological Museum of the University of Zurich.

———. 1996. Political and religious aspects of mountain cults in the hidden valley of Khenbalung: Tradition, decline and revitalization. In *Reflections of the mountain*, ed. A. M. Blondeau and E. Steinkellner, 219–31. Wien, Germany: Verlag der Osterreichischen Akademie der Wissenschaften.

———. 1997. Beyul Khenbalung, the Hidden Valley of the Artemisia: On Himalayan communities and their sacred landscape. In *Mandala and landscape*, ed. A. W. Macdonald, 287–334. New Delhi: D. K. Printworld.

Dove, M., P. E. Sajise, and A. A. Doolittle, eds. 2011. *Beyond the sacred forest: Complicating conservation in Southeast Asia*. Durham, NC: Duke University Press.

Dowie, M. 2011. *Conservation refugees: The hundred-year conflict between global conservation and native peoples*. Cambridge, MA: MIT Press.

Dowman, K. 1997. *The sacred life of Tibet*. New Delhi: HarperCollins India.

Gold, A., and B. Gujar. 1989. Of gods, trees, and boundaries: Divine conservation in Rajasthan. *Asian Folklore Studies* 48 (2): 211–29.

Heatherington, T. 2010. *Wild Sardinia: Indigeneity and the global dreamtimes of environmentalism*. Seattle: University of Washington Press.

Ingles, A. W. 1995. Religious beliefs and rituals in Nepal. In *Conserving biodiversity outside protected areas: The role of traditional agro-ecosystems*, ed. P. Halladay and D. Gilmour, 205–24. Gland, Switzerland: IUCN.

Karmay, S. 1996. The Tibetan cult of mountain deities and its political significance. In *Reflections of the mountain*, ed. A. M. Blondeau and E. Steinkellner, 59–75. Wien, Germany: Verlag der Osterreichischen Akademie der Wissenschaften.

———. 1998. *The arrow and the spindle: Studies in history, myths, rituals and beliefs in Tibet*. Kathmandu, Nepal: Mandala Book Point.

Makley, C. 2014. The amoral other: State-led development and mountain deity cults among Tibetans in Amdo Rebgong. In *Mapping Shangrila: Contested landscapes in the Sino-Tibetan borderlands*, ed. E. Yeh and C. Coggins, 229–54. Seattle: University of Washington Press.

Mandair, A. S. 2009. *Religion and the specter of the West: Sikhism, India, postcoloniality, and the politics of translation*. New York: Columbia University Press.

Masuzawa, T. 2005. *The invention of world religions: Or, how European universalism was preserved in the language of pluralism*. Chicago: University of Chicago Press.

McGuire, M. B. 2008. *Lived religion: Faith and practice in everyday life*. New York: Oxford University Press.

The Mountain Institute. 2008. *Exploring roles and linkages: Sacred sites, conservation and livelihoods*. Proceedings of the Workshop on Conservation, Sacred Natural Sites and Conservation Linkages held in Kathmandu from 4–5 December 2008. Kathmandu, Nepal: Department of National Park and Wildlife Conservation, The Mountain Institute, Eco-Himal, and Mountain Spirit.

———. 2010. Building livelihoods along Beyul trails. http://www.mountain.org/map/building-livelihoods (last accessed 5 July 2016).

Mumford, S. 1989. *Himalayan dialogue: Tibetan lamas and Gurung shamans in Nepal*. Madison: University of Wisconsin Press.

Neumann, R. P. 1998. *Imposing wilderness: Struggles over livelihood and nature preservation in Africa*. Vol. 4. Berkeley: University of California Press.

Ortner, S. 1978. *Sherpas through their rituals*. New Delhi: Vikas Publishing.

———. 1989. *High religion: A cultural and political history of Sherpa Buddhism*. Princeton, NJ: Princeton University Press.

———. 1999. *Life and death on Mt. Everest: Sherpas and Himalayan mountaineering*. New Delhi: Oxford University Press.

Pommaret, F. 1996. On local and mountain deities in Bhutan. In *Reflections of the mountain*, ed. A. M. Blondeau and E. Steinkellner, 39–56. Wien, Germany: Verlag der Osterreichischen Akademie der Wissenschaften.

Ramakrishnan, P. S. 2005. Mountain biodiversity, land use dynamics and traditional ecological knowledge. In *Global change and mountain regions: An overview of current knowledge*, ed. U. M. Huber, H. K. M. Bugmann, and M. A. Reasoner, 551–61. Dordrecht, The Netherlands: Springer.

Ramble, C. 2007. *The navel of the demoness: Tibetan Buddhism and civil religion in highland Nepal*. New Delhi: Oxford University Press.

Reinhard, J. 1978. Khembalung: The hidden valley. *Kailash* 6 (1): 5–35.

Sharma, S., H. C. Rikhari, and L. M. S. Palni. 1999. Conservation of natural resources through religion: A case study of the central Himalaya. *Society & Natural Resources* 12:599–622.

Sherpa, L. N. 2003. Sacred Beyuls and biological diversity conservation in the Himalayas. In *International workshop on the importance of sacred natural sites for biodiversity conservation*, ed. C. Lee and T. Schaaf, 93–97. Paris: UNESCO.

———. 2005. Sacred hidden valleys and ecosystems conservation in the Himalaya. In *Conserving cultural and biological diversity: The role of sacred natural sites and cultural landscapes*, ed. T. Schaaf and C. Lee, 68–72. Paris: UNESCO.

———. 2008. *Through a Sherpa window*. Kathmandu, Nepal: Vajra.

Sherpa, T. R. 2007. *Beyul: The sacred hidden valleys*. [Film] Kathmandu, Nepal: Mila Productions.

Spivak, G. C. 1988. Subaltern studies: Deconstructing historiography. In *Selected subaltern studies*, ed. R. Guha and G. C. Spivak, 3–32. New York: Oxford University Press.

Sponsel, L. 2001. Is indigenous spiritual ecology a new fad? Reflections from the historical and spiritual ecology of Hawai'i. In *Indigenous traditions and ecology: The interbeing of cosmology and community*, ed. J. Grim, 159–74. Cambridge, MA: Harvard University Press.

———. 2012. *Spiritual ecology: A quiet revolution*. Santa Barbara, CA: ABC-CLIO.

Spoon, J., and L. N. Sherpa. 2008. Beyul Khumbu: The Sherpas and Sagarmatha (Mount Everest) National Park and Buffer Zone, Nepal. In *Protected landscapes and cultural and spiritual values*, ed. J.-M. Mallarach, 68–79. Heidelberg, Germany: Kasparek Verlag.

Stevens, S. 1993. *Claiming the high ground*. Berkeley: University of California Press.

———. 1997. *Conservation through cultural survival: Indigenous peoples and protected areas*. Washington, DC: Island.

———. 2008. *The Mount Everest region as an ICCA: Sherpa conservation stewardship of the Khumbu Sacred Valley, Sagarmatha (Chomolungma/Mt. Everest) National Park and Buffer Zone*. http://www.iccaconsortium.org/wp-content/uploads/images/media/grd/mount_everest_nepal_report_icca_grassroots_discussions.pdf (last accessed 5 July 2016).

Taylor, B. R. 2010. *Dark green religion: Nature spirituality and the planetary future*. Berkeley: University of California Press.

Tucker, M. E., and J. A. Grim. 2001. Introduction: The emerging alliance of world religions and ecology. *Daedalus* 130 (4): 1–22.

United Nations Educational, Scientific and Cultural Organization (UNESCO). 2003. *International workshop on the importance of sacred natural sites for biodiversity conservation held in Kunming and Xishuangbanna Biosphere Reserve, People's Republic of China, 17–20 February 2003*. Paris: UNESCO.

———. 2006. *Conserving cultural and biological diversity: The role of sacred natural sites and cultural landscapes, Tokyo, Japan, 30 May–2 June 2005*. Paris: UNESCO.

World Wildlife Fund (WWF). 2016. Eastern Himalayas. http://www.worldwildlife.org/places/eastern-himalayas (last accessed 5 July 2016).

Yeh, E., and C. Coggins, eds. 2014. *Mapping Shangrila: Contested landscapes in the Sino-Tibetan borderlands*. Seattle: University of Washington Press.

Index

Note: Page numbers in *italics* refer to figures
     Page numbers in **bold** refer to tables

Printed and bound by CPI Group (UK) Ltd, Croydon, CR0 4YY

01/11/2024

01782603-0020